Springer Series in Computational Mathematics

5

Editorial Board

Vivette Girault Pierre-Arnaud Raviart

Finite Element Methods for Navier-Stokes Equations

Theory and Algorithms

With 21 Figures

Springer-Verlag
Berlin Heidelberg New York Tokyo

Professor Vivette Girault
Professor Pierre-Arnaud Raviart
Université Pierre et Marie Curie, Paris VI
Laboratoire associé 189, Analyse Numérique
Tour 55–65, 5ème étage
4, place Jussieu
F-75230 Paris Cedex 05
France

Mathematics Subject Classification (1980):
35A15, 35G20, 35Q10, 46E35, 65N15, 65N30, 76D05, 76D07

ISBN-13: 978-3-642-64888-5 **e-ISBN-13: 978-3-642-61623-5**
DOI: 10.1007/978-3-642-61623-5

Library of Congress Cataloging-in-Publication Data
Girault, Vivette, 1943–
Finite element methods for Navier-Stokes equations.
(Springer series in computational mathematics; 5)
Extended version of: Finite element approximation of the Navier-Stokes equations. 1979.
Bibliography: p.
Includes index.
1. Viscous flow. 2. Navier-Stokes equations – Numerical solutions. 3. Finite element method. I. Raviart, Pierre-Arnaud, 1939–. II. Girault, Vivette, 1943–. Finite element approximation of the Navier-Stokes equations. III. Title. IV. Series.
QA929.G77 1986 531′.0533 85-30387

Softcover reprint of the hardcover 1st edition 1986

Typesetting: Asco Trade Typesetting Ltd., Hong Kong
Printing and bookbinding: Konrad Triltsch, Würzburg
2141/3140-543210

Preface

The material covered by this book has been taught by one of the authors in a post-graduate course on Numerical Analysis at the University Pierre et Marie Curie of Paris. It is an extended version of a previous text (cf. Girault & Raviart [32]) published in 1979 by Springer-Verlag in its series: Lecture Notes in Mathematics.

In the last decade, many engineers and mathematicians have concentrated their efforts on the finite element solution of the Navier-Stokes equations for incompressible flows. The purpose of this book is to provide a fairly comprehensive treatment of the most recent developments in that field. To stay within reasonable bounds, we have restricted ourselves to the case of stationary problems although the time-dependent problems are of fundamental importance. This topic is currently evolving rapidly and we feel that it deserves to be covered by another specialized monograph. We have tried, to the best of our ability, to present a fairly exhaustive treatment of the finite element methods for inner flows. On the other hand however, we have entirely left out the subject of exterior problems which involve radically different techniques, both from a theoretical and from a practical point of view. Also, we have neither discussed the implementation of the finite element methods presented by this book, nor given any explicit numerical result. This field is extensively covered by Peyret & Taylor [64] and Thomasset [82]. Finally, we have tried as much as possible to make this text self-contained and therefore we have either proved or recalled all the theoretical results required.

This book is divided into four chapters and a technical appendix. The first chapter is devoted to the theoretical aspects of the Stokes equations for an incompressible fluid flow. It includes a thoroughly complete, detailed and mostly original study of the function spaces $H(\text{div}; \Omega)$ and $H(\mathbf{curl}; \Omega)$ closely associated with the Stokes problem. In particular, the reader will find here a fundamental decomposition of vector fields in two and three dimensions. The existence and uniqueness of the solution of the Stokes problem are fully proved. Finally, a number of algorithms to dissociate the velocity from the pressure are introduced.

Chapter II deals with the finite element approximation of the Stokes problem in the primitive variables (velocity and pressure). It can serve as a good introduction to the subject of mixed finite element methods, which plays an important part in a wide range of applications. It describes most of the finite element methods available in this context and introduces some new three-dimensional elements. An original feature of this chapter is that it provides a unified treatment of the so-called B.B. compatibility condition between the velocity and pressure spaces.

Although the finite element methods of Chapter II are the most popular, they do not satisfy exactly the incompressibility condition. On the contrary, Chapter III is devoted to the study of exactly incompressible finite element methods. It solves the Stokes problem using other variables such as the stream function and the vorticity or the stream function and the gradient of velocity tensor in two dimensions or even the vector potential and the vorticity in three dimensions. This chapter provides a number of useful and (seldom known) techniques for analyzing accurately these finite element schemes. Such techniques are not restricted to the Stokes problem but may be adapted to other mechanical situations like the bending of plates in elasticity.

Chapter IV is devoted to the theory and approximation of the full Navier-Stokes problem. The existence and uniqueness theorems are entirely standard but the approximation is presented in a fairly new light. Its originality consists in extending systematically the results of the previous two chapters to this nonlinear situation. The basic result is a general theorem concerning the approximation of branches of nonsingular solutions of nonlinear problems. When it is applied to the Navier-Stokes equations, it enables one to recover optimal rates of convergence. We end this chapter by describing a number of useful algorithms for handling the Navier-Stokes nonlinearity.

Finally, the appendix presents an up-to-date summary of the finite element theory which is constantly used throughout this book.

We wish to thank our colleagues C. Bernardi, M. Crouzeix, O. Pironneau, G. Raugel and L. Tartar for many fruitful and exciting discussions. We are particularly grateful to R. Verfürth for reading the manuscript and providing very helpful suggestions. For the material preparation of this work, we are above all gratefully indebted to our colleagues of the Computing Science Department who provided the microcomputer with which the manuscript was typed. We thank also Mme. Ruprechts for typing a part of the manuscript.

Paris, March 1986

Vivette Girault
Pierre-Arnaud Raviart

Table of Contents

Chapter I. Mathematical Foundation of the Stokes Problem

§ 1. Generalities on Some Elliptic Boundary Value Problems

This paragraph contains a short survey on the Dirichlet's and Neumann's problems for the harmonic and biharmonic operators.

1.1. Basic Concepts on Sobolev Spaces

Our purpose here is to recall the main notions and results, concerning the classical Sobolev spaces, which we shall use later on. Although they are stated without proof, these results are complete, rigorous and fairly general. Some of them, like the trace theorems, will only play a small part as theoretical tools in subsequent proofs and readers who are not familiar with such specialized mathematics need not dwell on them. But others, like the Sobolev Imbedding Theorem will be of constant use. The reader will find more details in references like Nečas [58] or Adams [1].

To simplify the discussion, we shall work from now on with real-valued functions, but of course every result stated here will carry on to complex-valued functions.

Let Ω denote an open subset of $\mathbb{R}^N$ with boundary Γ. We define $\mathscr{D}(\Omega)$ to be the linear space of infinitely differentiable functions, with compact support on Ω. Then, we set

$$\mathscr{D}(\bar{\Omega}) = \{\phi|_\Omega;\ \phi \in \mathscr{D}(\mathbb{R}^N)\}$$

or equivalently, if $\mathcal{O}$ denotes any open subset of $\mathbb{R}^N$ such that $\bar{\Omega} \subset \mathcal{O}$,

$$\mathscr{D}(\bar{\Omega}) = \{\phi|_\Omega;\ \phi \in \mathscr{D}(\mathcal{O})\}.$$

Now, let $\mathscr{D}'(\Omega)$ denote the dual space of $\mathscr{D}(\Omega)$, often called the space of distributions on Ω. We denote by $\langle .,. \rangle$ the duality pairing between $\mathscr{D}'(\Omega)$ and $\mathscr{D}(\Omega)$ and we remark that when f is a locally integrable function, then f can be identified with a distribution by

$$\langle f, \phi \rangle = \int_\Omega f(x)\phi(x)\,dx \qquad \forall \phi \in \mathscr{D}(\Omega).$$

In other words, $\langle .,. \rangle$ is an extension of the scalar product of $L^2(\Omega)$. Now, we can define the derivatives of distributions. Let $\alpha = (\alpha_1, \ldots, \alpha_N) \in \mathbb{N}^N$ and set

$$|\alpha| = \sum_{i=1}^{N} \alpha_i.$$

For u in $\mathscr{D}'(\Omega)$, we define $\partial^\alpha u$ in $\mathscr{D}'(\Omega)$ by:

$$\langle \partial^\alpha u, \phi \rangle = (-1)^{|\alpha|} \langle u, \partial^\alpha \phi \rangle \qquad \forall \phi \in \mathscr{D}(\Omega);$$

when u is α times differentiable, $\partial^\alpha u$ coincides with the usual notion of derivative:

$$\partial^\alpha u = \frac{\partial^{|\alpha|} u}{\partial x_1^{\alpha_1} \ldots \partial x_N^{\alpha_N}}.$$

For each integer $m \geqslant 0$ and real p with $1 \leqslant p \leqslant \infty$, we define the Sobolev space:

$$W^{m,p}(\Omega) = \{v \in L^p(\Omega);\ \partial^\alpha v \in L^p(\Omega) \quad \forall |\alpha| \leqslant m\},$$

which is a Banach space for the norm:

$$(1.1) \qquad \|u\|_{m,p,\Omega} = \left(\sum_{|\alpha| \leqslant m} \int_\Omega |\partial^\alpha u(x)|^p \, dx \right)^{1/p} \qquad p < \infty$$

or

$$\|u\|_{m,\infty,\Omega} = \max_{|\alpha| \leqslant m} \left(\operatorname*{ess\,sup}_{x \in \Omega} |\partial^\alpha u(x)| \right), \qquad p = \infty.$$

The space $W^{m,p}(\Omega)$ is separable for $1 \leqslant p < \infty$ and reflexive for $1 < p < \infty$. We also provide $W^{m,p}(\Omega)$ with the following seminorm

$$(1.2) \qquad |u|_{m,p,\Omega} = \left(\sum_{|\alpha| = m} \int_\Omega |\partial^\alpha u(x)|^p \, dx \right)^{1/p}, \quad \text{for } p < \infty,$$

and we make the above modification when $p = \infty$. If u belongs to $W^{m,p}(\mathcal{O})$ for every measurable, compact proper subset $\mathcal{O}$ of Ω we say that u is locally in $W^{m,p}(\Omega)$ and we write

$$u \in W^{m,p}_{\text{loc}}(\Omega).$$

When $p = 2$, $W^{m,2}(\Omega)$ is usually denoted by $H^m(\Omega)$, and if there is no ambiguity, we drop the subscript $p = 2$ when refering to its norm and seminorm. $H^m(\Omega)$ is a Hilbert space for the scalar product:

$$(1.3) \qquad (u, v)_{m,\Omega} = \sum_{|\alpha| \leqslant m} \int_\Omega \partial^\alpha u(x) \partial^\alpha v(x) \, dx.$$

In particular, we write the scalar product of $L^2(\Omega)$ with no subscript at all.

Parallel to the Sobolev spaces, we recall the familiar definition of $\mathscr{C}^m$-functions:

$\mathscr{C}^0(\Omega)$ denotes the space of continuous functions defined in Ω and

$$\mathscr{C}^m(\Omega) = \{u \in \mathscr{C}^0(\Omega);\ \partial^\alpha u \in \mathscr{C}^0(\Omega) \quad \forall |\alpha| \leqslant m\}.$$

As the $\mathscr{C}^m$-functions are not necessarily bounded we also introduce the space

$$\mathscr{C}^m(\bar{\Omega}) = \{u \in \mathscr{C}^m(\Omega);\ \partial^\alpha u \text{ are bounded and uniformly continuous on } \Omega \quad \forall 0 \leqslant |\alpha| \leqslant m\}.$$

Likewise, we define the space $\mathscr{C}^{m,1}(\bar{\Omega})$:

$$\mathscr{C}^{m,1}(\bar{\Omega}) = \{u \in \mathscr{C}^m(\bar{\Omega});\ \partial^\alpha u \text{ are Lipschitz-continuous in } \bar{\Omega} \quad \forall 0 \leqslant |\alpha| \leqslant m\}.$$

For $m \geqslant 0$, $\mathscr{C}^m(\bar{\Omega})$ and $\mathscr{C}^{m,1}(\bar{\Omega})$ are Banach spaces for the respective norms:

$$\|u\|_{\mathscr{C}^m(\bar{\Omega})} = \max_{0 \leqslant |\alpha| \leqslant m} \sup_{x \in \Omega} |\partial^\alpha u(x)|,$$

$$\|u\|_{\mathscr{C}^{m,1}(\bar{\Omega})} = \|u\|_{\mathscr{C}^m(\bar{\Omega})} + \max_{0 \leqslant |\alpha| \leqslant m} \sup_{\substack{x, y \in \Omega \\ x \neq y}} \frac{|\partial^\alpha u(x) - \partial^\alpha u(y)|}{\|x - y\|},$$

where $\|x\|$ denotes the Euclidean norm of $\mathbb{R}^N$.

As $\mathscr{D}(\Omega) \subset W^{m,p}(\Omega)$, we define

$$W_0^{m,p}(\Omega) = \overline{\mathscr{D}(\Omega)}^{W^{m,p}(\Omega)},$$

i.e. $W_0^{m,p}(\Omega)$ is the closure of $\mathscr{D}(\Omega)$ for the norm $\|\,.\,\|_{m,p,\Omega}$. When $m \geqslant 1$ and Ω is a proper subset of $\mathbb{R}^N$ then $W_0^{m,p}(\Omega)$ is generally a proper subspace of $W^{m,p}(\Omega)$ and we shall characterize its functions further on. On the other hand, when $m = 0$ we have the following result.

Lemma 1.1. *The space $\mathscr{D}(\Omega)$ is dense in $L^p(\Omega)$ for $1 \leqslant p < \infty$.*

The next theorem, called the Poincaré-Friedrichs inequality, asserts that the mapping $v \to |v|_{m,\Omega}$ is a norm on $H_0^m(\Omega)$, equivalent to $\|\,.\,\|_{m,\Omega}$.

Theorem 1.1. *If Ω is connected and bounded at least in one direction, then for each integer $m \geqslant 0$, there exists a constant $K = K(m, \Omega) > 0$ such that*

$$\|v\|_{m,\Omega} \leqslant K |v|_{m,\Omega} \quad \forall v \in H_0^m(\Omega). \tag{1.4}$$

For $1 \leqslant p < \infty$, we denote by $W^{-m,p'}(\Omega)$ the dual space of $W_0^{m,p}(\Omega)$ normed by:

$$\|f\|_{-m,p',\Omega} = \sup_{\substack{v \in W_0^{m,p}(\Omega) \\ v \neq 0}} \frac{\langle f, v\rangle}{\|v\|_{m,p,\Omega}}, \tag{1.5}$$

where p' satisfies

$$1/p + 1/p' = 1. \tag{1.6}$$

The following lemma characterizes the functionals of $W^{-m,p'}(\Omega)$.

Lemma 1.2. *Let p and p' satisfy* (1.6) *with* $1 \leqslant p < \infty$. *A distribution f belongs to* $W^{-m,p'}(\Omega)$ *if and only if there exist functions* $f_\alpha \in L^{p'}(\Omega)$, *for* $|\alpha| \leqslant m$, *such that*

$$f = \sum_{|\alpha| \leqslant m} \partial^\alpha f_\alpha.$$

Nearly all properties of Sobolev spaces on a domain Ω require some regularity of the boundary Γ. It is important to define this concept of regularity with precision. The following definition is taken from Grisvard [42].

Definition 1.1. Let Ω be an open subset of $\mathbb{R}^N$. We say that its boundary Γ is continuous (resp. Lipschitz-continuous, of class $\mathscr{C}^m$, of class $\mathscr{C}^{m,1}$ for some integer $m > 0$) if for every $x \in \Gamma$ there exists a neighborhood $\mathcal{O}$ of x in $\mathbb{R}^N$ and new orthogonal coordinates $y = (y', y_N)$ where $y' = (y_1, \ldots, y_{N-1})$, such that:

i) $\mathcal{O}$ is a hypercube in the new coordinates:

$$\mathcal{O} = \{y;\ -a_j < y_j < a_j,\ 1 \leqslant j \leqslant N\}.$$

ii) There exists a continuous (resp. Lipschitz-continuous, $\mathscr{C}^m$, $\mathscr{C}^{m,1}$) function ϕ defined in

$$\mathcal{O}' = \{y';\ -a_j < y_j < a_j,\ 1 \leqslant j \leqslant N-1\}$$

that satisfies:

$$|\phi(y')| \leqslant a_N/2 \qquad \forall y' \in \mathcal{O}',$$

$$\Omega \cap \mathcal{O} = \{y;\ y_N < \phi(y')\}, \qquad \Gamma \cap \mathcal{O} = \{y;\ y_N = \phi(y')\}.$$

Essentially, this definition means that locally Ω is below the graph of some function ϕ, Γ is represented by the graph of ϕ and the regularity of Γ is determined by that of ϕ. It is important to point out that, with this definition, a domain with a continuous boundary is never on both sides of Γ at any point of Γ. In particular, domains with cuts or cusps are forbidden, but boundaries with corners are allowed. The most straightforward example of domain with a Lipschitz-continuous boundary is a bounded polyhedron in $\mathbb{R}^3$ or a bounded polygon in $\mathbb{R}^2$.

To shorten the text, we shall say that Ω is Lipschitz-continuous when it has a Lipschitz-continuous boundary.

Note that a Lipschitz-continuous boundary has almost everywhere a unit normal vector $\mathbf{n}$. Furthermore, for $m \geqslant 1$ a $\mathscr{C}^{m,1}$ boundary Γ has a normal vector that belongs to $\mathscr{C}^{m-1,1}(\Gamma)^N$; if Ω is bounded, this normal vector can be extended to a vector field that belongs to $\mathscr{C}^{m-1,1}(\bar{\Omega})^N$. Likewise, if Ω is a bounded domain with a Lipschitz-continuous boundary Γ, the distance function:

$$d(x, \Gamma) = \inf_{y \in \Gamma} \|x - y\|$$

belongs to $W^{1,\infty}(\Omega)$.

The next theorem shows that smooth functions are dense in $W^{m,p}(\Omega)$.

Theorem 1.2. *Let Ω be an open Lipschitz-continuous subset of $\mathbb{R}^N$.*

1°) *The space $\mathscr{D}(\bar{\Omega})$ is dense in $W^{m,p}(\Omega)$ for all integers $m \geqslant 0$ and real p with $1 \leqslant p < \infty$.*

2°) *Let $u \in W^{m,p}(\Omega)$ and let $\tilde{u}$ denote its extension by zero outside Ω. If $\tilde{u} \in W^{m,p}(\mathbb{R}^N)$ then $u \in W_0^{m,p}(\Omega)$.*

3°) *If in addition Γ is bounded and $m \geqslant 1$, there exists a continuous linear extension operator P from $W^{m,p}(\Omega)$ into $W^{m,p}(\mathbb{R}^N)$:*

$$Pu|_{\Omega} = u \qquad \forall u \in W^{m,p}(\Omega).$$

We now come to the fundamental Sobolev Imbedding Theorem which essentially relates different Sobolev spaces and spaces of smooth functions.

Theorem 1.3. *Let Ω be an open Lipschitz-continuous subset of $\mathbb{R}^N$ and let $p \in \mathbb{R}$ with $1 \leqslant p < \infty$ and m and $n \in \mathbb{N}$ with $n \leqslant m$. The following imbeddings hold algebraically and topologically:*

$$(1.7) \qquad W^{m,p}(\Omega) \subset \begin{cases} W^{n,q}(\Omega) & \text{if } 1/q = 1/p - (m-n)/N > 0, \\ W^{n,q}_{\text{loc}}(\Omega) & \forall q \in [1, \infty) \text{ if } 1/p = (m-n)/N, \\ \mathscr{C}^n(\Omega) & \text{provided } 1/p < (m-n)/N. \end{cases}$$

Moreover, if Ω is bounded, the last inclusion holds in $\mathscr{C}^n(\bar{\Omega})$ and the imbedding of $W^{m,p}(\Omega)$ into $W^{n,q'}(\Omega)$ is compact for all real q' that satisfy:

$$(1.8) \quad \text{or} \quad \begin{cases} 1 \leqslant q' < Np/(N-(m-n)p) & \text{whenever } N > (m-n)p, \\ 1 \leqslant q' < \infty & \text{when } N = (m-n)p. \end{cases}$$

In addition, these compact imbeddings are also valid for negative n or m.

This theorem will be used constantly in the sequel. For instance, in the next paragraph we shall use the fact that $L^2(\Omega)$ is compactly imbedded in $H^{-1}(\Omega)$. As an immediate application, we have the following corollary about multiplication in $W^{m,p}(\Omega)$.

Corollary 1.1. *Assume that Ω is a bounded Lipschitz-continuous open subset of $\mathbb{R}^N$. Let m_1, m_2 and m be three non negative integers and p_1, p_2 and p be three real numbers in $[1, \infty)$ such that $m_1 \geqslant m$, $m_2 \geqslant m$ and either*

$$m_1 + m_2 - m \geqslant N(1/p_1 + 1/p_2 - 1/p) \geqslant 0, \quad m_i - m > N(1/p_i - 1/p) \quad i = 1, 2$$

or

$$m_1 + m_2 - m > N(1/p_1 + 1/p_2 - 1/p) \geqslant 0, \quad m_i - m \geqslant N(1/p_i - 1/p) \quad i = 1, 2.$$

Then the mapping $u, v \to u \cdot v$ is a continuous bilinear map from $W^{m_1,p_1}(\Omega) \times W^{m_2,p_2}(\Omega)$ into $W^{m,p}(\Omega)$.

Before studying the trace of functions of $W^{m,p}(\Omega)$, it is convenient to extend the notion of Sobolev spaces to nonintegral values of m. There are several definitions of fractional Sobolev spaces which unfortunately are not equivalent. Here we shall use mostly the following one.

Definition 1.2. Let Ω be an open subset of $\mathbb{R}^N$, $m \geqslant 0$ be an integer and s and p be two real numbers with $1 \leqslant p < \infty$ and $s = m + \sigma$ where $\sigma \in \mathbb{R}$ with $0 < \sigma < 1$. We denote by $W^{s,p}(\Omega)$ the space of all distributions u defined in Ω such that

$$u \in W^{m,p}(\Omega)$$

and

$$\int_\Omega \int_\Omega \frac{|\partial^\alpha u(x) - \partial^\alpha u(y)|^p}{\|x - y\|^{N+\sigma p}} dx\, dy < +\infty \qquad \forall |\alpha| = m.$$

Likewise, we denote by $W^{s,\infty}(\Omega)$ the subspace of functions u in $W^{m,\infty}(\Omega)$ such that

$$\max_{|\alpha|=m} \operatorname*{ess\,sup}_{\substack{x,y \in \Omega \\ x \neq y}} \frac{|\partial^\alpha u(x) - \partial^\alpha u(y)|}{\|x - y\|^\sigma} \leqslant +\infty.$$

It can be shown that $W^{s,p}(\Omega)$ is a Banach space for the norm:

$$\|u\|_{s,p,\Omega} = \left\{ \|u\|^p_{m,p,\Omega} + \sum_{|\alpha|=m} \int_\Omega \int_\Omega \frac{|\partial^\alpha u(x) - \partial^\alpha u(y)|^p}{\|x - y\|^{N+\sigma p}} dx\, dy \right\}^{1/p} \tag{1.9}$$

with the obvious modification when $p = \infty$. Like in the integral case, we define for $s > 0$:

$$W_0^{s,p}(\Omega) = \overline{\mathscr{D}(\Omega)}^{W^{s,p}(\Omega)}$$

and we denote by $W^{-s,p'}(\Omega)$ the dual space of $W_0^{s,p}(\Omega)$ with p and p' related by (1.6). It turns out that several previous results carry over to fractional Sobolev spaces with few modifications. More precisely, the statements of Sobolev's Imbedding Theorem 1.3 and that of its Corollary 1.1 are valid for fractional-order Sobolev spaces. The density part of Theorem 1.2 carries over without modification to $s > 0$, while the extension part is valid for $s > 0$ when Ω is bounded. In addition, we have the following outstanding Interpolation Theorem of Lions & Peetre [54] which we shall use over and over.

Theorem 1.4. *Let Ω be a bounded Lipschitz-continuous open subset of $\mathbb{R}^N$. Let $\theta \in [0, 1]$ and let s_i and t_i be two pairs of real numbers with $0 \leqslant t_i \leqslant s_i$ for $i = 1, 2$ and let $\mathscr{L}_i$ and $\mathscr{L}_\theta$ denote respectively $\mathscr{L}(W^{s_i,p}(\Omega); W^{t_i,p}(\Omega))$ and $\mathscr{L}(W^{(1-\theta)s_1+\theta s_2,p}(\Omega); W^{(1-\theta)t_1+\theta t_2,p}(\Omega))$ for some real p with $1 < p < \infty$. Let π be an operator in $\mathscr{L}_1 \cap \mathscr{L}_2$; then π also belongs to $\mathscr{L}_\theta$ and there exists a constant C such that*

$$\|\pi\|_{\mathscr{L}_\theta} \leqslant C \|\pi\|^{1-\theta}_{\mathscr{L}_1} \|\pi\|^{\theta}_{\mathscr{L}_2}. \tag{1.10}$$

Also, when $p = 2$ there is an alternate definition of $W^{s,2}(\Omega)$ using Fourier transforms which yields algebraically and topologically the same space as Definition 1.2.

Definition 1.3. 1°) For real $s > 0$ we set:

$$H^s(\mathbb{R}^N) = \{v \in L^2(\mathbb{R}^N); (1 + \|\sigma\|^2)^{s/2}\hat{v}(\sigma) \in L^2(\mathbb{R}^N_\sigma)\} \tag{1.11}$$

with the norm:

$$\|v\|_{s,\mathbb{R}^N} = \{\|v\|^2_{0,\mathbb{R}^N} + \|(1 + \|\sigma\|^2)^{s/2}\hat{v}(\sigma)\|^2_{0,\mathbb{R}^N_\sigma}\}^{1/2}, \tag{1.12}$$

where $\hat{v}$ denotes the Fourier transform of v.

2°) When Ω is an open subset of $\mathbb{R}^N$, we define

$$H^s(\Omega) = \{v \in L^2(\Omega); \exists \tilde{v} \in H^s(\mathbb{R}^N) \text{ with } \tilde{v}|_\Omega = v\} \tag{1.13}$$

with the norm:

$$\|v\|_{s,\Omega} = \inf_{\tilde{v} \in H^s(\mathbb{R}^N),\, \tilde{v}|_\Omega = v} \|\tilde{v}\|_{s,\mathbb{R}^N}. \tag{1.14}$$

As mentioned above, we have the following relation between $H^s(\Omega)$ and $W^{s,2}(\Omega)$.

Lemma 1.3. *Let Ω be a bounded Lipschitz-continuous open subset of $\mathbb{R}^N$. Then algebraically and topologically, we have*

$$W^{s,2}(\Omega) = H^s(\Omega) \qquad \forall s > 0.$$

In addition, when s is an integer (1.11) *defines the classical Sobolev space $H^m(\Omega)$ with an equivalent norm.*

Now, we are in a position to examine the boundary values of functions in $W^{s,p}(\Omega)$. We assume that Ω is a bounded open subset of $\mathbb{R}^N$ with a boundary Γ that is at least Lipschitz-continuous. Let us first define what we mean by the space $W^{s,p}(\Gamma)$. According to Definition 1.1, we can view Γ locally as an $N - 1$ dimensional submanifold of $\mathbb{R}^N$ by means of the mapping

$$\Phi(y') = (y', \phi(y'))$$

from $\mathcal{O}'$ onto $\Gamma \cap \mathcal{O}$. Then we set the following definition.

Definition 1.4. Let Ω be a bounded open subset of $\mathbb{R}^N$ with a boundary Γ of class $\mathscr{C}^{k,1}$ for some integer $k \geqslant 0$. A distribution u on Γ belongs to $W^{s,p}(\Gamma)$ for $s \leqslant k + 1$ if $u \circ \Phi$ belongs to $W^{s,p}(\mathcal{O}' \cap \Phi^{-1}(\Gamma \cap \mathcal{O}))$ for all possible $\mathcal{O}$ and ϕ fulfilling the assumptions of Definition 1.1.

Let $(\mathcal{O}_j, \Phi_j)_{1 \leqslant j \leqslant J}$ be any atlas of Γ such that each pair $(\mathcal{O}_j, \Phi_j)$ satisfies the hypotheses of the above definition. Then one possible Banach norm for $W^{s,p}(\Gamma)$

is the functional:

(1.15)
$$u \to \left\{ \sum_{j=1}^{J} \| u \circ \Phi \|^p_{s,p,\mathcal{O}'_j \cap \Phi_j^{-1}(\Gamma \cap \mathcal{O}_j)} \right\}^{1/p}.$$

However, this norm is clumsy and we shall replace it by more convenient equivalent norms. For example, when $s = 0$ we shall use the more familiar L^p-norm:

$$\| u \|_{0,p,\Gamma} = \left\{ \int_\Gamma |u(x)|^p \, ds(x) \right\}^{1/p}$$

where ds denotes the surface measure of Γ. At this stage, it is worthwhile to point out that the density result of Theorem 1.2 as well as the Sobolev's Imbedding Theorem 1.3 (with dimension $N - 1$) are also valid on Γ.

Now, let u be a function of $\mathscr{D}(\bar{\Omega})$ and let us denote its boundary values by $\gamma_0 u$. The following trace theorem extends the operator γ_0 to functions in $W^{s,p}(\Omega)$.

Theorem 1.5. *Let Ω be like in Definition* 1.4 *and let $p \geqslant 1$ and $s \geqslant 0$ be two real numbers such that $s \leqslant k + 1$, $s - 1/p = l + \sigma$ where $l \geqslant 0$ is an integer and $0 < \sigma < 1$. Then the mapping $u \to \gamma_0 u$ defined on $\mathscr{D}(\bar{\Omega})$ has a unique linear continuous extension as an operator from*

$$W^{s,p}(\Omega) \quad \textit{onto } W^{s-1/p,p}(\Gamma).$$

Moreover, in $W^{1,p}(\Omega)$ we have:

$$\mathrm{Ker}(\gamma_0) = W_0^{1,p}(\Omega).$$

For such s and p, this theorem suggests the following norm on $W^{s-1/p,p}(\Gamma)$ which can be proved to be equivalent to (1.15):

(1.16)
$$\| f \|_{s-1/p,p,\Gamma} = \inf_{\substack{v \in W^{s,p}(\Omega) \\ \gamma_0 v = f}} \| v \|_{s,p,\Omega}.$$

When $p = 2$, (1.16) is a Hilbert norm. In this text, we shall mainly use the spaces $H^{1/2}(\Gamma)$ and $H^{3/2}(\Gamma)$ corresponding to $p = 2$ and respectively $s = 1$ or 2. We shall also be interested in $H^{-1/2}(\Gamma)$, the dual space of $H^{1/2}(\Gamma)$ equipped with the obvious dual norm:

(1.17)
$$\| f^* \|_{-1/2,\Gamma} = \sup_{\substack{f \in H^{1/2}(\Gamma) \\ f \neq 0}} \frac{\langle f^*, f \rangle}{\| f \|_{1/2,\Gamma}},$$

where again $\langle .,. \rangle$ denotes the duality pairing between $H^{-1/2}(\Gamma)$ and $H^{1/2}(\Gamma)$. Here again, we observe that $\langle .,. \rangle$ is an extension of the scalar product of $L^2(\Gamma)$ in the sense that when $f^* \in L^2(\Gamma)$, we can identify $\langle f^*, f \rangle$ with

$$\int_\Gamma f^*(x) f(x) \, ds(x).$$

Apart from the boundary value operator γ_0, we shall also require the trace of the normal derivative $\gamma_1 u$ defined for u in $\mathscr{D}(\bar{\Omega})$ by

$$\gamma_1 u = \partial u/\partial n = \sum_{i=1}^{N} \gamma_0(\partial u/\partial x_i)n_i, \tag{1.18}$$

where $\mathbf{n} = (n_1, \ldots, n_N)$ denotes the *unit outward normal* to Γ. Then we can complete as follows the statement of Theorem 1.5.

Theorem 1.6. *We keep the assumptions of Theorem* 1.5 *with* $l \geqslant 1$. *The mapping*:

$$u \to \{\gamma_0 u, \gamma_1 u\}$$

defined on $\mathscr{D}(\bar{\Omega})$ *has a unique linear continuous extension as an operator from*

$$W^{s,p}(\Omega) \quad \textit{onto } W^{s-1/p,p}(\Gamma) \times W^{s-1-1/p,p}(\Gamma).$$

Moreover, in $W^{2,p}(\Omega)$ *we have the following characterization*:

$$\operatorname{Ker} \gamma_0 \cap \operatorname{Ker} \gamma_1 = W_0^{2,p}(\Omega).$$

Remark 1.1. If the boundary of Ω has corners, its normal vector has jumps and it is obvious that $\partial u/\partial n$ is rough no matter how smooth u can be. Nevertheless, it is possible to extend slightly the statement of Theorem 1.6. Assume that Ω is a bounded two-dimensional polygon; let Γ_j denote the sides of Γ and $\mathbf{n}_j$ the corresponding exterior unit normal, $1 \leqslant j \leqslant J$. Then the mapping:

$$u \to (\partial u/\partial n_j;\ 1 \leqslant j \leqslant J)$$

is linear, continuous and *surjective* from $W^{k+2,p}(\Omega)$ onto

$$\prod_{j=1}^{J} W^{k+1-1/p,p}(\Gamma_j),$$

for each integer $k \geqslant 0$ and real $p \in (1, \infty)$. Note that there is no matching condition for $\partial u/\partial n$ at the vertices of Γ.

The situation of the boundary values of u is a bit more complicated because there is usually a matching condition at the vertices of Γ. For the sake of simplicity, we shall just take u in $W^{2,p}(\Omega)$. Denote the vertices of Γ by S_j with the convention that $S_{J+1} = S_1$. The mapping:

$$u \to (h_j = u|_{\Gamma_j};\ 1 \leqslant j \leqslant J)$$

is linear, continuous and *surjective* from $W^{2,p}(\Omega)$ onto the subspace of

$$\prod_{j=1}^{J} W^{2-1/p,p}(\Gamma_j)$$

defined by the compatibility conditions

$$h_j(S_{j+1}) = h_{j+1}(S_{j+1}), \qquad 1 \leqslant j \leqslant J.$$

We close this section with two useful applications of Green's formula.

Lemma 1.4. *Let Ω be a bounded open subset of $\mathbb{R}^N$ with a Lipschitz-continuous boundary Γ.*

1°) *For u and v in $H^1(\Omega)$ and for $1 \leqslant i \leqslant N$, we have:*

$$\int_\Omega u(\partial v/\partial x_i)\,dx = -\int_\Omega (\partial u/\partial x_i)v\,dx + \int_\Gamma \gamma_0(uv)n_i\,ds. \tag{1.19}$$

2°) *If in addition $u \in H^2(\Omega)$ we have:*

$$\sum_{i=1}^N \int_\Omega \frac{\partial u}{\partial x_i}\frac{\partial v}{\partial x_i}\,dx = -\sum_{i=1}^N \int_\Omega \frac{\partial^2 u}{\partial x_i^2} v\,dx + \sum_{i=1}^N \int_\Gamma \gamma_0\left(\frac{\partial u}{\partial x_i} v\right) n_i\,ds. \tag{1.20}$$

Adopting the usual notations:

$$\Delta u = \sum_{i=1}^N \frac{\partial^2 u}{\partial x_i^2}, \qquad \mathbf{grad}\, u = (\partial u/\partial x_1, \ldots, \partial u/\partial x_N),$$

(1.20) becomes:

$$(\mathbf{grad}\, u, \mathbf{grad}\, v) = -(\Delta u, v) + \int_\Gamma (\partial u/\partial n)\gamma_0 v\,ds. \tag{1.21}$$

1.2. Abstract Elliptic Theory

This section gives a brief account of a fundamental tool used in studying linear partial differential equations of elliptic type.

Let V be a real Hilbert space with norm denoted by $\|\,.\,\|_V$; let V' be its dual space and let $\langle .,. \rangle$ denote the duality pairing between V' and V. Let $(u, v) \to a(u, v)$ be a real bilinear form on $V \times V$, l an element of V' and consider the following problem:

Find $u \in V$ such that

$$a(u, v) = \langle l, v\rangle \qquad \forall v \in V. \tag{P}$$

The following theorem is due to Lax & Milgram [49].

Theorem 1.7. *We assume that a is continuous and elliptic on V, i.e. there exist two constants M and $\alpha > 0$ such that*

$$|a(u, v)| \leqslant M \|u\|_V \|v\|_V \qquad \forall u, v \in V \tag{1.22}$$

and

$$a(v, v) \geqslant \alpha \|v\|_V^2 \qquad \forall v \in V. \tag{1.23}$$

Then Problem (P) *has one and only one solution u in V. Moreover, the mapping $l \to u$ is an isomorphism from V' onto V.*

Corollary 1.2. *When a is symmetric—i.e.* $a(u, v) = a(v, u)\ \forall u, v \in V$—*then the solution u of* (P) *is also the only element of V that minimizes the following quadratic functional* (*also called energy functional*) *on V*:

$$J(v) = (1/2)a(v, v) - \langle l, v\rangle. \tag{1.24}$$

1.3. Example 1: Dirichlet's Problem for the Laplace Operator

In all the examples, we assume that Ω is bounded and Γ Lipschitz-continuous. Consider the following non-homogeneous Dirichlet's problem:

Given f in $H^{-1}(\Omega)$ *and g in* $H^{1/2}(\Gamma)$, *find a function u such that*:

$$(\mathrm{D})\quad \begin{cases} (1.25) & -\Delta u = f \quad \text{in } \Omega, \\ (1.26) & u = g \quad \text{on } \Gamma. \end{cases}$$

Let us formulate this problem in terms of Problem (P). We set $V = H_0^1(\Omega)$ and

$$a(u, v) = (\mathbf{grad}\, u, \mathbf{grad}\, v).$$

It is clear that a is continuous in $H_0^1(\Omega)^2$, and owing to Theorem 1.1,

$$a(v, v) = \|\mathbf{grad}\, v\|_{0,\Omega}^2 = |v|_{1,\Omega}^2 \geqslant C\|v\|_{1,\Omega}^2.$$

Besides that, since $H^{1/2}(\Gamma)$ is the range space of γ_0 in $H^1(\Omega)$, let u_0 in $H^1(\Omega)$ satisfy $u_0 = g$ on Γ, and examine the following problem:

Find u in $H^1(\Omega)$ *such that*

$$(\mathrm{D}')\quad \begin{cases} (1.27) & u - u_0 \in H_0^1(\Omega), \\ (1.28) & a(u - u_0, v) = \langle f, v\rangle - a(u_0, v) \quad \forall v \in H_0^1(\Omega). \end{cases}$$

Since a is continuous, the mapping $v \to \langle f, v\rangle - a(u_0, v)$ belongs to $H^{-1}(\Omega)$. Therefore, thanks to the Lax & Milgram's Theorem, Problem (D′) has one and only one solution u in $H^1(\Omega)$.

It remains only to prove that u may be characterized as the unique solution of Problem (D). Taking $v \in \mathscr{D}(\Omega)$ in (1.28) gives:

$$a(u, v) = -\langle \Delta u, v\rangle = \langle f, v\rangle \quad \forall v \in \mathscr{D}(\Omega).$$

Hence u satisfies

$$(\mathrm{D}_1)\quad \begin{cases} (1.27) & u - u_0 \in H_0^1(\Omega), \\ (1.25) & -\Delta u = f \quad \text{in } H^{-1}(\Omega). \end{cases}$$

Conversely, every solution of (D_1) is a solution of (D′) by the density of $\mathscr{D}(\Omega)$ in $H_0^1(\Omega)$. But

$$u - u_0 \in H_0^1(\Omega) \quad \text{iff } u = g \quad \text{on } \Gamma,$$

therefore Problems (D_1) and (D) are the same.

As far as the regularity of u is concerned, we know from the Lax & Milgram's Theorem that the mapping $l \to u - u_0$ is an isomorphism from $H^{-1}(\Omega)$ onto $H_0^1(\Omega)$. Therefore,

$$\|u - u_0\|_{1,\Omega} \leqslant C_2 \|l\|_{-1,\Omega}.$$

Clearly,

$$\|l\|_{-1,\Omega} \leqslant \|f\|_{-1,\Omega} + \|u_0\|_{1,\Omega}.$$

Hence

$$\|u\|_{1,\Omega} \leqslant C_3\{\|f\|_{-1,\Omega} + \|u_0\|_{1,\Omega}\} \quad \forall u_0 \in H^1(\Omega) \quad \text{such that} \quad u_0 = g \quad \text{on } \Gamma.$$

From definition (1.16) this implies that

$$\|u\|_{1,\Omega} \leqslant C_3\{\|f\|_{-1,\Omega} + \|g\|_{1/2,\Gamma}\}.$$

Thus, we have proved the following proposition:

Proposition 1.1. *Problem* (D) *has one and only one solution u in $H^1(\Omega)$ and there exists a constant $C = C(\Omega)$ such that*

$$\|u\|_{1,\Omega} \leqslant C\{\|f\|_{-1,\Omega} + \|g\|_{1/2,\Gamma}\}, \tag{1.29}$$

i.e. u depends continuously upon the data of (D).

When f and g are more regular, it is natural to expect that the solution u of Problem (D) is also smoother. The next theorem states the precise regularity of u. Its proof, which is far outside the scope of this book, can be found for example in Grisvard [42].

Theorem 1.8. 1°) *Let Ω be a bounded open subset of $\mathbb{R}^N$ with a $\mathscr{C}^{k+1,1}$ boundary Γ for some integer $k \geqslant 0$. Suppose that the data f and g of Problem* (1.25) (1.26) *satisfy*

$$f \in W^{k,p}(\Omega), \qquad g \in W^{k+2-1/p,p}(\Gamma)$$

for some real p with $1 < p < \infty$. Then $u \in W^{k+2,p}(\Omega)$ and there exists a constant $C = C(k, p, \Omega)$ such that

$$\|u\|_{k+2,p,\Omega} \leqslant C\{\|f\|_{k,p,\Omega} + \|g\|_{k+2-1/p,p,\Gamma}\}. \tag{1.30}$$

2°) *When Ω is a two-dimensional bounded polygon with no reentrant corner, there exists a real $p_\Omega > 2$ depending on the greatest inner angle of Γ such that*

$$u \in W^{2,p}(\Omega), \qquad 1 < p < p_\Omega,$$

whenever $f \in L^p(\Omega)$ and $(g|_{\Gamma_j};\ 1 \leqslant j \leqslant J) \in \prod_{j=1}^J W^{2-1/p,p}(\Gamma_j)$ satisfies the matching conditions of Remark 1.1.

3°) *If Ω is a bounded, convex polyhedron in three dimensions, the conclusion of* 2°) *is still valid for the homogeneous Dirichlet problem* ($g = 0$).

Remark 1.2. As an immediate application of this theorem with $g = 0$, we see that when Ω is a bounded convex polygon in $\mathbb{R}^2$ (or polyhedron in $\mathbb{R}^3$) then the mapping $u \to \Delta u$ is an isomorphism from $W^{2,p}(\Omega) \cap W_0^{1,p}(\Omega)$ onto $L^p(\Omega)$ for all $p \in (1, 2 + \varepsilon]$ for some $\varepsilon > 0$. When the boundary of Ω is $\mathscr{C}^{1,1}$, this isomorphism holds for all $p \in (1, \infty)$.

1.4. Example 2: Neumann's Problem for the Laplace Operator

Here, we assume in addition that Ω is connected and we deal with the non-homogeneous Neumann's problem:

Find u such that:

(N)
$$\begin{cases} (1.31) \quad -\Delta u = f \quad \text{in } \Omega, \\ (1.32) \quad \partial u/\partial n = g \quad \text{on } \Gamma, \\ \text{where } f \in L^2(\Omega) \text{ and } g \in H^{1/2}(\Gamma) \text{ satisfy the relation:} \\ (1.33) \quad \int_\Omega f\,dx + \langle g, 1 \rangle_\Gamma = 0. \end{cases}$$

Since Problem (N) only involves the derivatives of u, it is clear that its solution is never unique. We turn the difficulty by seeking u in the quotient space $H^1(\Omega)/\mathbb{R}$ equipped with the quotient norm

$$\|\dot{v}\|_{H^1(\Omega)/\mathbb{R}} = \inf_{v \in \dot{v}} \|v\|_{1,\Omega}. \tag{1.34}$$

The theorem below states an important property of this space; its proof can be found in Nečas [58].

Theorem 1.9. *Let Ω be a bounded, connected and Lipschitz-continuous open subset of $\mathbb{R}^N$. The space $H^1(\Omega)/\mathbb{R}$ is a Hilbert space for the quotient norm* (1.34). *Moreover, on this space the functional $\dot{v} \to |v|_{1,\Omega}$ is a norm equivalent to* (1.34).

With this space, we can put Problem (N) into the abstract setting of Problem (P). Let $V = H^1(\Omega)/\mathbb{R}$,

$$a(\dot{u}, \dot{v}) = (\mathbf{grad}\, u, \mathbf{grad}\, v) \quad \forall u \in \dot{u}, \quad v \in \dot{v}$$

and

$$l: \dot{v} \to (f, v) + \langle g, v \rangle_\Gamma \quad \forall v \in \dot{v}. \tag{1.35}$$

Note that the right-hand side of (1.35) is independent of the particular $v \in \dot{v}$ thanks to the compatibility condition (1.34). Furthermore, $l \in V'$ because, owing to (1.16), we have:

$$|(f, v) + \langle g, v \rangle_\Gamma| \leqslant (\|f\|_{0,\Omega} + \|g\|_{-1/2,\Gamma}) \inf_{v \in \dot{v}} \|v\|_{1,\Omega}.$$

Thus

(1.36) $$\|l\|_{V'} \leqslant \|f\|_{0,\Omega} + \|g\|_{-1/2,\Gamma}.$$

Obviously, $a(\dot{u}, \dot{v})$ is continuous on $V \times V$, and by virtue of Theorem 1.9

$$a(\dot{v}, \dot{v}) = |v|_{1,\Omega}^2 \geqslant C_1 \|\dot{v}\|_{H^1(\Omega)/\mathbb{R}}^2.$$

Hence, by the Lax & Milgram's Theorem, the following problem:

(N′) $\begin{cases} \textit{Find } u \textit{ in } H^1(\Omega)/\mathbb{R} \textit{ satisfying} \\ (1.37) \qquad a(\dot{u}, \dot{v}) = \langle l, \dot{v} \rangle \quad \forall \dot{v} \in H^1(\Omega)/\mathbb{R}, \end{cases}$

has a unique solution $\dot{u} \in H^1(\Omega)/\mathbb{R}$.

Let us interpret Problem (N′). When v is restricted to $\mathscr{D}(\Omega)$, (1.37) yields:

(1.31) $$-\Delta u = f \quad \text{in } L^2(\Omega) \quad \forall u \in \dot{u}.$$

Next, by taking the scalar product of (1.31) with v and comparing with (1.37), we find:

(1.38) $$(\mathbf{grad}\, u, \mathbf{grad}\, v) = -(\Delta u, v) + \langle g, v \rangle_\Gamma \quad \forall v \in H^1(\Omega).$$

Therefore, Problem (N′) is equivalent to:

Find u in $H^1(\Omega)$ satisfying (1.31) *and* (1.38).

It remains to interpret (1.38) as a boundary condition. At the present stage this cannot be done without assuming that $u \in H^2(\Omega)$. Then Green's formula (1.21) yields:

$$\int_\Gamma (\partial u/\partial n) v \, ds = \langle g, v \rangle_\Gamma \quad \forall v \in H^1(\Omega),$$

i.e.

$$\partial u/\partial n = g \quad \text{on } \Gamma.$$

Therefore, Problems (N) and (N′) are equivalent. Of course, this is not entirely satisfactory inasmuch as the existence of a solution of Problem (N) is subjected to the regularity of the solution of (N′). Although this regularity does generally hold, the more powerful tools of the next paragraph will eliminate this extra smoothness assumption.

Now, let us examine the dependence of the solution $\dot{u}$ of Problem (N′). According to the Lax & Milgram's Theorem 1.7, (1.36) and the equivalence Theorem 1.9, we obtain:

$$|u|_{1,\Omega} \leqslant C_2(\|f\|_{0,\Omega} + \|g\|_{-1/2,\Gamma}).$$

We have thus proved the following result.

Proposition 1.2. *Problem* (N′) *has a unique solution $\dot{u}$ in $H^1(\Omega)/\mathbb{R}$ and this solution is continuous with respect to the data:*

(1.39) $$|u|_{1,\Omega} \leq C(\|f\|_{0,\Omega} + \|g\|_{-1/2,\Gamma}) \quad \forall u \in \dot{u}.$$

Moreover, when $\dot{u} \in H^2(\Omega)/\mathbb{R}$ *then it is also the only solution of Problem* (N).

As for the Dirichlet's problem, the solution of Problem (N′) is more regular when its data has extra smoothness. The precise result, which is also given by Grisvard [42], closely resembles Theorem 1.8.

Theorem 1.10. 1°) *Let* Ω *be like in Theorem* 1.8 *and assume that the data* f *and* g *of Problem* (1.37) *satisfy*:

$$f \in W^{k,p}(\Omega), \qquad g \in W^{k+1-1/p,p}(\Gamma) \qquad 1 < p < \infty.$$

Then $\dot{u} \in W^{k+2,p}(\Omega)/\mathbb{R}$ *and there exists a constant* $C = C(k, p, \Omega)$ *such that*

(1.40) $$\|\dot{u}\|_{W^{k+2,p}(\Omega)/\mathbb{R}} \leq C\{\|f\|_{k,p,\Omega} + \|g\|_{k+1-1/p,p,\Gamma}\}.$$

2°) *When* Ω *is a two-dimensional bounded polygon with no reentrant corner, there exists a real* $p_\Omega > 2$ *depending on the maximum inner angle of* Γ *such that*

$$\dot{u} \in W^{2,p}(\Omega)/\mathbb{R}, \qquad 1 < p < p_\Omega,$$

provided $f \in L^p(\Omega)$ *and* $(g|_{\Gamma_j};\ 1 \leq j \leq J) \in \prod_{j=1}^{J} W^{1-1/p,p}(\Gamma_j)$.

3°) *If* Ω *is a bounded, convex polyhedron in* $\mathbb{R}^3$ *the conclusion of* 2°) *is valid for the homogeneous Neumann problem* ($g = 0$).

1.5. Example 3: Dirichlet's Problem for the Biharmonic Operator

Consider the non-homogeneous problem:

For f *given in* $H^{-2}(\Omega)$, g_1 *given in* $H^{3/2}(\Gamma)$ *and* g_2 *in* $H^{1/2}(\Gamma)$, *find* u *such that*:

(B)
$$\begin{cases} (1.41) & \Delta^2 u = f \quad \text{in } \Omega, \\ (1.42) & u = g_1 \quad \text{on } \Gamma, \\ (1.43) & \partial u/\partial n = g_2 \quad \text{on } \Gamma. \end{cases}$$

The function space naturally attached to this problem is $H_0^2(\Omega)$ and the bilinear form is:

$$a(u, v) = (\Delta u, \Delta v).$$

This form is elliptic on $H_0^2(\Omega)$ because the mapping $v \to \|\Delta v\|_{0,\Omega}$ is a norm on $H_0^2(\Omega)$ equivalent to the norm $\|\,.\,\|_{2,\Omega}$. Indeed, for v in $\mathcal{D}(\Omega)$, we can easily show by integrating by parts and interchanging derivatives that

(1.44) $$\|\Delta v\|_{0,\Omega}^2 = |v|_{2,\Omega}^2.$$

By density, the same result holds for the functions of $H_0^2(\Omega)$. The equivalence follows from Poincaré's Theorem 1.1.

According to Theorem 1.6, if Γ is $\mathscr{C}^{1,1}$, there exists a function u_0 in $H^2(\Omega)$ such that

(1.45) $$u_0 = g_1 \quad \text{on } \Gamma, \qquad \partial u_0/\partial n = g_2 \quad \text{on } \Gamma.$$

Thus we turn to the following problem:

(B′) $$\begin{cases} \textit{Find } u \textit{ in } H^2(\Omega) \textit{ such that} \\ (1.46) \qquad u - u_0 \in H_0^2(\Omega), \\ (1.47) \qquad a(u - u_0, v) = \langle f, v \rangle - a(u_0, v) \quad \forall v \in H_0^2(\Omega). \end{cases}$$

By the Lax & Milgram's Theorem 1.7, Problem (B′) has exactly one solution u in $H^2(\Omega)$. Owing to (1.45) and (1.46), u satisfies the boundary conditions:

$$u = g_1 \quad \text{on } \Gamma, \qquad \partial u/\partial n = g_2 \quad \text{on } \Gamma.$$

Besides that, by restricting the test functions of (1.47) to $\mathscr{D}(\Omega)$, we find

$$\Delta^2 u = f \quad \text{in } H^{-2}(\Omega).$$

Therefore, u is a solution of (B).

Conversely, as in the case of the Laplace operator, we can show that Problem (B) has at most one solution in $H^2(\Omega)$.

From (1.47) and the equivalence of norms, we derive the bound

$$\|u\|_{2,\Omega} \leqslant C_1(\|f\|_{-2,\Omega} + \|u_0\|_{2,\Omega}) \qquad \forall u_0 \text{ satisfying (1.45)},$$

i.e.

$$\|u\|_{2,\Omega} \leqslant C_2(\|f\|_{-2,\Omega} + \|g_1\|_{3/2,\Gamma} + \|g_2\|_{1/2,\Gamma}).$$

These results are summed up in the proposition below:

Proposition 1.3. *If Γ is $\mathscr{C}^{1,1}$, Problem* (B) *has exactly one solution u in $H^2(\Omega)$, bounded as follows:*

(1.48) $$\|u\|_{2,\Omega} \leqslant C(\|f\|_{-2,\Omega} + \|g_1\|_{3/2,\Gamma} + \|g_2\|_{1/2,\Gamma}).$$

The above analysis does not allow for corners since it assumes that Γ is $\mathscr{C}^{1,1}$. This hypothesis plays a crucial part in the lifting operator: $(g_1, g_2) \to u_0$. Of course, if these non-homogeneous data are given directly in the form of a function u_0 in $H^2(\Omega)$ such that

$$u = u_0 \quad \text{on } \Gamma \quad \text{and} \quad \partial u/\partial n = \partial u_0/\partial n \quad \text{on } \Gamma$$

then Proposition 1.3 applies to a Lipschitz-continuous domain with u_0 instead of g_1 and g_2. Otherwise, we must alter a little the statement of Problem (B). Suppose that Ω is a two-dimensional bounded polygon. In view of Remark 1.1, let Γ_j and S_j for $1 \leqslant j \leqslant J$ denote respectively the sides and vertices of Γ. Let us take J functions:

$$h_j \in H^{3/2}(\Gamma_j) \qquad 1 \leqslant j \leqslant J$$

satisfying the matching conditions $h_j(S_{j+1}) = h_{j+1}(S_{j+1})$ and J functions:

$$g_j \in H^{1/2}(\Gamma_j) \qquad 1 \leqslant j \leqslant J$$

and consider the problem:

Find u such that

$$(\mathrm{B}'') \quad \begin{cases} (1.41) & \Delta^2 u = f \quad \text{in } H^{-2}(\Omega), \\ (1.49) & u = h_j \quad \text{on } \Gamma_j \\ (1.50) & \partial u/\partial n = g_j \quad \text{on } \Gamma_j \end{cases} \quad 1 \leqslant j \leqslant J.$$

By virtue of Remark 1.1, we know that there exists a function u_0 in $H^2(\Omega)$ such that

$$u_0 = h_j, \qquad \partial u_0/\partial n = g_j \quad \text{on } \Gamma_j, \qquad 1 \leqslant j \leqslant J$$

and

$$\| u_0 \|_{2,\Omega} \leqslant C_3 \left\{ \sum_{j=1}^{J} (\| h_j \|^2_{3/2,\Gamma_j} + \| g_j \|^2_{1/2,\Gamma_j}) \right\}^{1/2}.$$

Hence the conclusion of Proposition 1.3 (with the functions h_j and g_j instead of g_1 and g_2) applies also to Problem (B″) in this case.

When the boundary of Ω is sufficiently smooth, it is possible to derive more information about the regularity of u.

Theorem 1.11. *Let Ω be a bounded open subset of $\mathbb{R}^N$ with a boundary Γ of class $\mathscr{C}^{k+3,1}$ for an integer $k \geqslant -2$ and assume that the data f, g_1 and g_2 of the biharmonic Problem* (B) *satisfy*:

$$f \in H^k(\Omega), \qquad g_1 \in H^{k+7/2}(\Gamma), \qquad g_2 \in H^{k+5/2}(\Gamma).$$

Then $u \in H^{k+4}(\Omega)$ and there exists a constant $C = C(k, \Omega)$ such that

$$\| u \|_{k+4,\Omega} \leqslant C\{ \| f \|_{k,\Omega} + \| g_1 \|_{k+7/2,\Gamma} + \| g_2 \|_{k+5/2,\Gamma}\}. \tag{1.51}$$

But even when Γ has corners, the conclusion of Proposition 1.3 can be refined for the biharmonic problem with homogeneous boundary conditions.

Theorem 1.12. *Assume that Ω is a two-dimensional bounded polygon with no re-entrant corner. Then the mapping $u \to \Delta^2 u$ is an isomorphism from $H^3(\Omega) \cap H_0^2(\Omega)$ onto $H^{-1}(\Omega)$.*

This last result is fundamental to establish the regularity of the solution of the Stokes problem in a plane, convex polygon (cf. Grisvard [43]).

§ 2. Function Spaces for the Stokes Problem

A rigorous analysis of the Stokes problem requires special function spaces involving the divergence and curl of vector fields. In almost all publications, the crucial properties of these spaces stem from a powerful and difficult theorem proved by De Rham [68] which says essentially:

$$\text{(2.0)} \quad \begin{cases} \textit{if a distribution vector field } u \textit{ satisfies } \langle u, \phi \rangle = 0 \textit{ for all divergence-free} \\ \textit{functions of } \mathscr{D} \textit{ then } u = \mathbf{grad}\, S \textit{ for some distribution } S. \end{cases}$$

However, this theorem is far from necessary and we present in this paragraph a different approach inspired from Tartar [78].

2.1. Preliminary Results

The following theorem, due to Peetre [63] and Tartar [78], will be used several times in this text.

Theorem 2.1. *Let E_1, E_2, E_3 be three Banach spaces, $A \in \mathscr{L}(E_1; E_2)$ and B a compact operator in $\mathscr{L}(E_1; E_3)$ such that*

$$\text{(2.1)} \qquad \|u\|_{E_1} \cong \|Au\|_{E_2} + \|Bu\|_{E_3} \qquad \forall u \in E_1.$$

Then the following properties hold.

1°) *The dimension of* Ker(A) *is finite; the mapping A is an isomorphism from $E_1/$*Ker(A) *onto $\mathscr{R}(A)$; $\mathscr{R}(A)$ is a closed subspace of E_2.*

2°) *There exists a constant C_0 such that, if F is a Banach space and $L \in \mathscr{L}(E_1; F)$ vanishes on* Ker(A), *then*

$$\text{(2.2)} \qquad \|Lu\|_F \leqslant C_0 \|L\|_{\mathscr{L}(E_1;F)} \|Au\|_{E_2} \qquad \forall u \in E_1.$$

3°) *If G is a Banach space and $M \in \mathscr{L}(E_1; G)$ satisfies*

$$\text{(2.3)} \qquad Mu \neq 0 \qquad \forall u \in \operatorname{Ker}(A) - \{0\},$$

then

$$\text{(2.4)} \qquad \|u\|_{E_1} \cong \|Au\|_{E_2} + \|Mu\|_G.$$

Proof. 1°) Here we are going to use a well known result (cf. Taylor [80]) which says that if the unit sphere is compact in a normed linear space V then V is finite-dimensional. Let us apply this result to Ker(A). First observe that by virtue of (2.1), $\|u\|_{E_1}$ and $\|Bu\|_{E_3}$ are two equivalent norms on Ker(A). Now, let (u_n) be a bounded sequence of Ker(A). Since B is compact, we can extract a subsequence (u_μ) such that Bu_μ converges in E_3. Therefore (u_μ) is a Cauchy sequence in E_1, and hence a convergent sequence in Ker(A). Thus the unit sphere is compact in Ker(A) and hence the dimension of Ker(A) is finite.

Now that we know that Ker(A) is a finite-dimensional subspace of E_1, we can introduce the quotient space:

$$X = E_1/\mathrm{Ker}(A)$$

which is a Banach space for the familiar quotient norm:

$$\|\dot{u}\|_X = \inf_{u \in \dot{u}} \|u\|_{E_1}.$$

In addition, because Ker(A) is finite-dimensional, the above infimum is attained, i.e. each class $\dot{u}$ has a representative $\tilde{u}$ such that

$$\|\dot{u}\|_X = \|\tilde{u}\|_{E_1} \qquad \forall \dot{u} \in X.$$

As A is a linear, continuous and one-to-one mapping from X onto $\mathscr{R}(A)$, to establish the announced isomorphism it suffices to prove that A has a continuous inverse, i.e. there exists a constant $C > 0$ such that:

$$\|\dot{u}\|_X \leqslant C\|A\dot{u}\|_{E_2} \qquad \forall \dot{u} \in X. \tag{2.5}$$

This is achieved by contradiction: assume that there exists a sequence $(\dot{u}_n)$ in X such that

$$\|\dot{u}_n\|_X = 1 \quad \text{and} \quad \lim_{n\to\infty} A\dot{u}_n = 0 \quad \text{in } E_2.$$

Hence

$$\|\tilde{u}_n\|_{E_1} = 1 \quad \text{and} \quad \lim_{n\to\infty} \|A\tilde{u}_n\|_{E_2} = 0.$$

Therefore, we can extract a subsequence $(\tilde{u}_\mu)$ such that $(B\tilde{u}_\mu)$ converges in E_3. Then (2.1) implies that $(\tilde{u}_\mu)$ is a Cauchy sequence in E_1. Thus, there exists u in Ker(A) such that $\lim \tilde{u}_\mu = u$ in E_1. This implies that $\dot{u}_\mu \to \dot{0}$ which contradicts our hypothesis.

Finally, since A is an isomorphism from X onto $\mathscr{R}(A)$ and since X is a Banach space, it follows immediately that $\mathscr{R}(A)$ is also a Banach space. Therefore $\mathscr{R}(A)$ is a closed subspace of E_2.

2°) Since L vanishes on Ker(A), we can write

$$Lu = L\dot{u} = LA^{-1}Au \qquad \forall u \in E_1.$$

Therefore

$$\|Lu\|_F \leqslant \|L\|_{\mathscr{L}(E_1;F)}\|A^{-1}\|_{\mathscr{L}(\mathscr{R}(A);X)}\|Au\|_{E_2} \qquad \forall u \in E_1,$$

thus yielding (2.2) with $C_0 = \|A^{-1}\|_{\mathscr{L}(\mathscr{R}(A);X)}$.

3°) Finally, let us prove that there exists a constant $C > 0$ such that

$$\|Au\|_{E_2} + \|Mu\|_G \geqslant C\|u\|_{E_1} \qquad \forall u \in E_1.$$

Again, we proceed by contradiction. Let (u_n) be a sequence in E_1 such that

$$\lim_{n\to\infty} (\|Au_n\|_{E_2} + \|Mu_n\|_G) = 0 \quad \text{and} \quad \|u_n\|_{E_1} = 1.$$

Then, there is a subsequence (u_μ) such that (Bu_μ) converges in E_3. Hence (u_μ) is a Cauchy sequence in E_1 and therefore

$$\lim_{\mu\to\infty} u_\mu = u$$

with

$$Au = 0, \qquad Mu = 0 \quad \text{and} \quad \|u\|_{E_1} = 1.$$

But then, (2.3) implies that $u = 0$; this leads to a contradiction. □

Now, let Ω be a bounded subset of $\mathbb{R}^N$ with a Lipschitz-continuous boundary Γ. From now on, we shall often deal with vector-valued functions. We shall distinguish vectors by means of bold-face characters and extend naturally all the previous norms to vectors as follows: if $\mathbf{v} = (v_1, \ldots, v_N)$ then

$$\|\mathbf{v}\|_{m,p,\Omega} = \left(\sum_{i=1}^{N} \|v_i\|_{m,p,\Omega}^p\right)^{1/p}.$$

The next theorem is part of a general result of functional analysis due to Nečas [57]. Its proof is long and delicate because it only assumes the Lipschitz-continuity of the boundary. When the boundary is smooth there is an alternate, easier proof that can be found in Duvaut & Lions [26]. Here we omit either proof for they are both outside the scope of this book.

Theorem 2.2. *Let Ω be a bounded, Lipschitz-continuous open set. There exists a constant $C > 0$, depending only on Ω, such that*

$$\|p\|_{0,\Omega} \leqslant C\{\|p\|_{-1,\Omega} + \|\mathbf{grad}\, p\|_{-1,\Omega}\} \quad \forall p \in L^2(\Omega). \tag{2.6}$$

As an immediate consequence, we have:

Corollary 2.1. 1°) *Under the assumptions of Theorem 2.2, the range of the gradient operator:* $\mathbf{grad} \in \mathscr{L}(L^2(\Omega); H^{-1}(\Omega)^N)$ *is a closed subspace of* $H^{-1}(\Omega)^N$.

2°) *If in addition Ω is connected, there exists a constant $C > 0$, depending only on Ω, such that*

$$\|\dot{p}\|_{L^2(\Omega)/\mathbb{R}} \leqslant C\|\mathbf{grad}\, \dot{p}\|_{-1,\Omega} \quad \forall \dot{p} \in L^2(\Omega)/\mathbb{R}. \tag{2.7}$$

3°) *Let ω be an open subset of Ω with positive measure. There exists a constant $C_\omega > 0$ depending only on Ω and ω such that*

$$\|p\|_{0,\Omega} \leqslant C_\omega\{\|p\|_{0,\omega} + \|\mathbf{grad}\, p\|_{-1,\Omega}\} \quad \forall p \in L^2(\Omega). \tag{2.8}$$

Proof. The idea is to apply Theorem 2.1 with $E_1 = L^2(\Omega)$, $E_2 = H^{-1}(\Omega)^N$, $E_3 = H^{-1}(\Omega)$, $A = \mathbf{grad}$ and $B =$ the canonical imbedding of $L^2(\Omega)$ into $H^{-1}(\Omega)$ which is compact according to Theorem 1.3. Clearly,

$$\|p\|_{-1,\Omega} + \|\mathbf{grad}\, p\|_{-1,\Omega} \leqslant C_1\|p\|_{0,\Omega} \quad \forall p \in L^2(\Omega).$$

Therefore, (2.1) follows immediately from (2.6), thus proving the first conclusion.

Likewise, since Ω is connected, $\mathbb{R} = \mathrm{Ker}(\mathbf{grad})$ and (2.5) proves (2.7).

Finally, let $G = L^2(\omega)$ and $M: L^2(\Omega) \to L^2(\omega)$ be the identity mapping. Because ω has positive measure, $Mc \neq 0$ for all constants $c \neq 0$, and (2.4) implies (2.8). □

The second Corollary gives an important result of regularity.

Corollary 2.2. *Let Ω be connected and satisfy the assumptions of Theorem 2.2. If*

$$p \in L^2_{\mathrm{loc}}(\Omega) \quad \text{and} \quad \mathbf{grad}\, p \in H^{-1}(\Omega)^N,$$

then $p \in L^2(\Omega)$.

Proof. We set

$$X = \{p \in L^2_{\mathrm{loc}}(\Omega); \mathbf{grad}\, p \in H^{-1}(\Omega)^N\}$$

and

$$[p]_\omega = \|p\|_{0,\omega} + \|\mathbf{grad}\, p\|_{-1,\Omega},$$

where $\omega \subset\subset \Omega$ has positive measure. Then $[\cdot]_\omega$ is a norm on X and we infer from (2.8) that $[\cdot]_\omega$ and $\|\cdot\|_{0,\Omega}$ are two equivalent norms on $L^2(\Omega)$. Thus $L^2(\Omega)$ is a Banach space for the norm $[\cdot]_\omega$ and hence it suffices to prove that $L^2(\Omega)$ is dense in X for $[\cdot]_\omega$.

1°) Assume for the moment that Ω is strictly star-shaped with respect to one of its points, say y. This amounts to say that, by taking y as origin,

$$\theta\bar{\Omega} \subset \Omega \quad \forall\theta \in [0,1) \quad \text{and} \quad \bar{\Omega} \subset \theta\Omega \quad \forall\theta > 1.$$

Here, we take $\theta > 1$ and we set $\Omega_\theta = \theta\Omega$. For a continuous function ϕ on Ω we make the change of variable $\phi \to \phi_\theta$ defined on Ω_θ by:

$$\phi_\theta(x) = \phi(x/\theta) \quad \forall x \in \Omega_\theta$$

which we extend to distributions, $u \in \mathscr{D}'(\Omega) \to u_\theta \in \mathscr{D}'(\Omega_\theta)$ by:

$$\langle u_\theta, \phi\rangle = \theta^N \langle u, \phi_{1/\theta}\rangle \quad \forall\phi \in \mathscr{D}(\Omega_\theta).$$

Then it is easy to check that

$$\mathbf{grad}(u_\theta) = (1/\theta)(\mathbf{grad}\, u)_\theta \quad \forall u \in \mathscr{D}'(\Omega),$$

$$\lim_{\theta\to 1} \|u_\theta - u\|_{0,\Omega} = 0 \quad \forall u \in L^2(\Omega),$$

$$\lim_{\theta\to 1} \|u_\theta - u\|_{-1,\Omega} = 0 \quad \forall u \in H^{-1}(\Omega).$$

Hence, if $p \in X$ then $p_\theta \in L^2(\Omega)$ for all $\theta > 1$ and in view of the above remarks, we readily derive that

$$\lim_{\theta\to 1}\,[p_\theta - p]_\omega = 0.$$

Therefore $p \in L^2(\Omega)$.

2°) In the general case, we use the following property (cf. for example Bernardi [8]):

A bounded, Lipschitz-continuous open set is the union of a finite number of star-shaped, Lipschitz-continuous open sets.

Clearly, it suffices to apply the above argument to each of these sets to derive the desired result on the entire domain. □

2.2. Some Properties of Spaces Related to the Divergence Operator

Unless otherwise specified, we assume in this section that Ω is a *bounded* subset of $\mathbb{R}^N$ with a *Lipschitz-continuous* boundary Γ.

For $\mathbf{v} = (v_1, \ldots, v_N)$, we define the divergence operator by:

$$\operatorname{div} \mathbf{v} = \sum_{i=1}^{N} (\partial v_i/\partial x_i).$$

Note the identity:

$$\operatorname{div}(\mathbf{grad}\, v) = \Delta v.$$

Let us introduce the following spaces of divergence-free functions:

$$\mathscr{V} = \{\boldsymbol{\phi} \in \mathscr{D}(\Omega)^N;\ \operatorname{div} \boldsymbol{\phi} = 0\}, \qquad V = \{\mathbf{v} \in H_0^1(\Omega)^N;\ \operatorname{div} \mathbf{v} = 0\}.$$

Here we equip $H_0^1(\Omega)^N$ with the norm $|.|_{1,\Omega}$, equivalent to $\|.\|_{1,\Omega}$ by virtue of Poincaré's Theorem 1.1. Since V is a closed subspace of $H_0^1(\Omega)^N$, we have the decomposition:

$$H_0^1(\Omega)^N = V \oplus V^\perp,$$

where $V^\perp$ denotes the orthogonal of V in $H_0^1(\Omega)^N$ for the scalar product $(\mathbf{grad}\, u, \mathbf{grad}\, v)$ associated with $|.|_{1,\Omega}$.

The following lemma establishes a first coarse version of De Rham's Theorem (2.0).

Lemma 2.1. *If* $\mathbf{f} \in H^{-1}(\Omega)^N$ *satisfies*

(2.9) $$\langle \mathbf{f}, \mathbf{v}\rangle = 0 \quad \forall \mathbf{v} \in V$$

then there exists $p \in L^2(\Omega)$ *such that*

$$\mathbf{f} = \mathbf{grad}\, p.$$

When Ω *is connected,* p *is unique up to an additive constant.*

Proof. The proof is based on the Closed Range Theorem of Banach. First, we note that $-\mathbf{grad} \in \mathscr{L}(L^2(\Omega); H^{-1}(\Omega)^N)$ is the dual operator of $\operatorname{div} \in \mathscr{L}(H_0^1(\Omega)^N; L^2(\Omega))$. Then, since $\mathscr{R}(\mathbf{grad})$ is a closed subspace of $H^{-1}(\Omega)^N$, we can apply the Closed Range Theorem of Banach (cf. for instance Yosida [84]):

$$\mathscr{R}(\mathbf{grad}) = (\operatorname{Ker}(\operatorname{div}))^0 = V^0,$$

where V^0 denotes the polar set of V:

$$V^0 = \{\mathbf{y} \in H^{-1}(\Omega)^N;\ \langle \mathbf{y}, \boldsymbol{\phi} \rangle = 0 \quad \forall \boldsymbol{\phi} \in V\}. \tag{2.10}$$

This is precisely the statement of our lemma. □

The next corollary gives a characterization of $V^\perp$. Previously, we require a definition.

Definition 2.1. Let $(-\Delta)^{-1} \in \mathscr{L}(H^{-1}(\Omega)^N; H_0^1(\Omega)^N)$ denote Green's operator related to Dirichlet's homogeneous problem for $-\Delta$ in $\mathbb{R}^N$, i.e. $\mathbf{u} = (-\Delta)^{-1}\mathbf{f}$ iff $\mathbf{u}$ is the solution of:

$$-\Delta \mathbf{u} = \mathbf{f} \quad \text{in } \Omega, \qquad \mathbf{u} = \mathbf{0} \quad \text{on } \Gamma.$$

Corollary 2.3. *We have*

$$V^\perp = \{(-\Delta)^{-1}\, \mathbf{grad}\, q;\ q \in L^2(\Omega)\}.$$

Proof. First, it is easy to check that $(-\Delta^{-1})\,\mathbf{grad}\, q \in V^\perp$ for all q in $L^2(\Omega)$. Conversely, let $\mathbf{u} \in V^\perp$ and consider the linear functional l defined on $H_0^1(\Omega)^N$ by $\langle l, \mathbf{v} \rangle = (\mathbf{grad}\,\mathbf{u}, \mathbf{grad}\,\mathbf{v})$. It belongs to $H^{-1}(\Omega)^N$ and vanishes on V. Therefore, in view of Lemma 2.1, there exists $p \in L^2(\Omega)$ such that:

$$\langle l, \mathbf{v} \rangle = (\mathbf{grad}\,\mathbf{u}, \mathbf{grad}\,\mathbf{v}) = \langle \mathbf{grad}\, p, \mathbf{v} \rangle \qquad \forall \mathbf{v} \in H_0^1(\Omega)^N.$$

Hence $\mathbf{u}$ has indeed the expression: $\mathbf{u} = (-\Delta^{-1})\,\mathbf{grad}\, p$. □

Now, we observe that Green's formula (1.19) yields:

$$\int_\Omega \operatorname{div} \mathbf{v}\, dx = 0 \qquad \forall \mathbf{v} \in H_0^1(\Omega)^N.$$

Thus the range space of div is contained in a proper, closed subspace of $L^2(\Omega)$ which for the sake of convenience we denote by $L_0^2(\Omega)$:

$$L_0^2(\Omega) = \left\{p \in L^2(\Omega);\ \int_\Omega p\, dx = 0\right\}.$$

The usefulness of $L_0^2(\Omega)$ arises in particular from the fact that

$$\|q\|_{0,\Omega} = \inf_{c \in \mathbb{R}} \|q + c\|_{0,\Omega} = \|\dot{q}\|_{L^2(\Omega)/\mathbb{R}} \qquad \forall q \in L_0^2(\Omega). \tag{2.11}$$

Thus $L_0^2(\Omega)$ can be identified isometrically with $L^2(\Omega)/\mathbb{R}$. The next corollary states further that the range space of div is exactly $L_0^2(\Omega)$.

Corollary 2.4. *Let Ω be connected. Then*
1°) *the operator* **grad** *is an isomorphism of $L_0^2(\Omega)$ onto V^0;*
2°) *the operator* div *is an isomorphism of $V^\perp$ onto $L_0^2(\Omega)$.*

Proof. 1°) We know that $\mathbf{grad} \in \mathscr{L}(L_0^2(\Omega); V^0)$; furthermore Lemma 2.1 asserts that this mapping is a bijection. As V^0 and $L_0^2(\Omega)$ are both Banach spaces, it follows that **grad** is an isomorphism.

2°) By virtue of 1°), and since div is the dual operator of $-\mathbf{grad}$, we have that div is an isomorphism from $(V^0)'$ onto $(L_0^2(\Omega))'$. Now, it suffices to prove that V^0 can be identified with $(V^\perp)'$. Let $\mathbf{g}$ be any element of $(V^\perp)'$ and let us extend $\mathbf{g}$ to $H_0^1(\Omega)^N$ by setting:

$$\langle \tilde{\mathbf{g}}, \mathbf{v} \rangle = \langle \mathbf{g}, \mathbf{v}^\perp \rangle \qquad \forall \mathbf{v} \in H_0^1(\Omega)^N,$$

where $\mathbf{v}^\perp$ denotes the orthogonal projection of $\mathbf{v}$ on $V^\perp$. Clearly, $\tilde{\mathbf{g}} \in V^0$ and the linear mapping $\mathbf{g} \to \tilde{\mathbf{g}}$ maps isometrically $(V^\perp)'$ onto V^0. This permits to identify $(V^\perp)'$ and V^0. □

Remark 2.1. The proof of Corollary 2.4 2°) is valid in abstract situations (cf. Lemma 4.1).

Again, let $\mathbf{n}$ denote the unit exterior normal to Γ. The following lemma gives a lifting operator of boundary values by means of divergence-free functions.

Lemma 2.2. *Let Ω be connected. For each $\mathbf{g} \in H^{1/2}(\Gamma)^N$ satisfying $\int_\Gamma \mathbf{g} \cdot \mathbf{n}\, ds = 0$ there exists a function $\mathbf{u} \in H^1(\Omega)^N$, unique up to an additive function of V, such that*

$$\operatorname{div} \mathbf{u} = 0 \quad \textit{in } \Omega, \qquad \mathbf{u} = \mathbf{g} \quad \textit{on } \Gamma.$$

Moreover, we have

$$\inf_{\mathbf{v} \in V} \|\mathbf{u} + \mathbf{v}\|_{1,\Omega} \leqslant C \|\mathbf{g}\|_{1/2,\Gamma}, \tag{2.12}$$

where the constant $C > 0$ is independent of $\mathbf{u}$ *and* $\mathbf{g}$.

Proof. Obviously such a function $\mathbf{u}$ must be unique in $[H^1(\Omega)^N]/V$. Let us check its existence. Let $\mathbf{w}$ be any function of $H^1(\Omega)^N$ that satisfies $\mathbf{w} = \mathbf{g}$ on Γ. Then Green's formula (1.19) gives:

$$\int_\Omega \operatorname{div} \mathbf{w}\, dx = \int_\Gamma \mathbf{g} \cdot \mathbf{n}\, ds = 0.$$

Therefore $\operatorname{div} \mathbf{w} \in L_0^2(\Omega)$ and owing to Corollary 2.4 2°) there exists a unique $\mathbf{v}$ in $V^\perp$ such that

$$\operatorname{div} \mathbf{v} = \operatorname{div} \mathbf{w},$$

and

$$|\mathbf{v}|_{1,\Omega} \leqslant C_1 \|\operatorname{div} \mathbf{w}\|_{0,\Omega}.$$

Clearly,

$$\mathbf{u} = \mathbf{w} - \mathbf{v}$$

is the required function. It remains to check the bound (2.12). First, observe that

$$\|\mathbf{u}\|_{1,\Omega} \leqslant C_2 \|\mathbf{w}\|_{1,\Omega}$$

with a constant $C_2 > 0$ independent of **u** and **w**. Then we derive (2.12) with the same constant C_2 by taking the infimum on both sides of this inequality and using the norm (1.16) on $H^{1/2}(\Gamma)$. □

The following theorem which sharpens the statement of Lemma 2.1 is a fundamental tool in the theory of Stokes equations. It is a simplified version of De Rham's Theorem (2.0) but it is sufficient for our purpose.

Theorem 2.3. *If* $\mathbf{f} \in H^{-1}(\Omega)^N$ *satisfies:*

$$\langle \mathbf{f}, \boldsymbol{\phi} \rangle = 0 \quad \forall \boldsymbol{\phi} \in \mathscr{V}, \tag{2.13}$$

then there exists $p \in L^2(\Omega)$ *such that:*

$$\mathbf{f} = \mathbf{grad}\, p.$$

If Ω *is connected,* p *is unique up to an additive constant.*

Proof. Assume for the moment that Ω is connected. Let us show that $\mathbf{f} = \mathbf{grad}\, p$ for p in $L^2_{\text{loc}}(\Omega)$; then Corollary 2.2 will give immediately that p is in $L^2(\Omega)$.

As Ω is a bounded and connected Lipschitz-continuous open set, there exists an *increasing* sequence $(\Omega_m)_{m \geqslant 1}$ of connected Lipschitz-continuous open sets such that

$$\bar{\Omega}_m \subset \Omega \quad \text{and} \quad \Omega = \bigcup_{m \geqslant 1} \Omega_m.$$

Let $\mathbf{u}_m \in H_0^1(\Omega)^N$ with

$$\mathbf{u}_m = \mathbf{0} \quad \text{outside } \Omega_m \quad \text{and} \quad \operatorname{div} \mathbf{u}_m = 0.$$

In order to apply Lemma 2.1, we want to regularize $\mathbf{u}_m$ in such a way that it stays divergence-free. This is achieved by the classical process of *mollifiers* (cf. for example Adams [1]):

for $\varepsilon > 0$ let ρ_ε be a regularizing sequence of $\mathscr{D}(\mathbb{R}^N)$ that vanishes for $\|x\| > \varepsilon$ and that satisfies:

$$\rho_\varepsilon(x) \geqslant 0, \qquad \int_{\mathbb{R}^N} \rho_\varepsilon(x)\, dx = 1, \qquad \lim_{\varepsilon \to 0} \rho_\varepsilon = \delta \quad \text{in } \mathscr{D}'(\mathbb{R}^N).$$

The properties of the convolution imply that:

$$\operatorname{div}(\rho_\varepsilon * \mathbf{u}_m) = \rho_\varepsilon * \operatorname{div} \mathbf{u}_m = 0,$$

$$\rho_\varepsilon * \mathbf{u}_m \in \mathscr{D}(\Omega)^N \quad \text{if } \varepsilon \text{ is sufficiently small,}$$

$$\lim_{\varepsilon \to 0} (\rho_\varepsilon * \mathbf{u}_m) = \mathbf{u}_m \quad \text{in } H^1(\Omega)^N.$$

Thus

$$\langle \mathbf{f}, \mathbf{u}_m \rangle = \lim_{\varepsilon \to 0} \langle \mathbf{f}, \rho_\varepsilon * \mathbf{u}_m \rangle = 0 \quad \text{in view of (2.13).}$$

Therefore Lemma 2.1 implies that there exists p_m in $L^2(\Omega_m)$ such that

$$f = \mathbf{grad}\, p_m \quad \text{in } \Omega_m.$$

Now, since $\mathbf{grad}\, p_m = \mathbf{grad}\, p_{m+1}$ in Ω_m then $p_m - p_{m+1}$ is constant in Ω_m. But, as p_m is unique up to an additive constant, this constant may be chosen so that

$$p_{m+1} = p_m \quad \text{in } \Omega_m \quad \forall m \geqslant 1.$$

Hence there exists $p \in L^2_{\text{loc}}(\Omega)$ such that

$$f = \mathbf{grad}\, p \quad \text{in } \Omega.$$

When Ω is not connected, we apply the above argument to each of its connected components. □

Corollary 2.5. *The space $\mathscr{V}$ is dense in V for the norm of $H_0^1(\Omega)^N$.*

Proof. The proof is based upon the following property of Banach spaces:

$$\text{(2.14)} \quad \begin{cases} \textit{A subspace } \mathscr{M} \textit{ of a Banach space } M \textit{ is dense in } M \textit{ iff every element of} \\ M' \textit{ that vanishes on } \mathscr{M} \textit{ also vanishes on } M. \end{cases}$$

First, we observe that every $\mathbf{f}$ in $H^{-1}(\Omega)^N$ that vanishes on $\mathscr{V}$ also vanishes on V since, by virtue of Theorem 2.3, $\mathbf{f}$ is necessarily of the form $\mathbf{f} = \mathbf{grad}\, p$ for some $p \in L^2(\Omega)$. But, as V is closed in $H_0^1(\Omega)^N$, V' can be identified with a subspace of $H^{-1}(\Omega)^N$ and the result follows from (2.14). □

So far, we have been mainly interested in subspaces of $H^1(\Omega)^N$; but subsequently, it will be worthwhile to use functions with less regularity. Bearing this in mind, we introduce the following spaces:

$$H(\operatorname{div}; \Omega) = \{\mathbf{v} \in L^2(\Omega)^N;\ \operatorname{div} \mathbf{v} \in L^2(\Omega)\},$$

which is clearly a Hilbert space for the norm:

$$\text{(2.15)} \qquad \|\mathbf{v}\|_{H(\operatorname{div};\Omega)} = \{\|\mathbf{v}\|_{0,\Omega}^2 + \|\operatorname{div} \mathbf{v}\|_{0,\Omega}^2\}^{1/2};$$

and

$$H_0(\operatorname{div}; \Omega) = \overline{\mathscr{D}(\Omega)^N}^{H(\operatorname{div};\Omega)}.$$

Theorem 2.4. *Let Ω be a Lipschitz-continuous open subset of $\mathbb{R}^N$ (not necessarily bounded). The space $\mathscr{D}(\bar{\Omega})^N$ is dense in $H(\text{div};\Omega)$.*

Proof. The proof is also based on property (2.14). Let l belong to $(H(\text{div};\Omega))'$, the dual space of $H(\text{div};\Omega)$. As this is a Hilbert space, we can associate with l a function $\mathbf{l}$ of $H(\text{div};\Omega)$ such that:

$$\langle l, \mathbf{u}\rangle = \sum_{i=1}^{N} (l_i, u_i) + (l_{N+1}, \operatorname{div}\mathbf{u}) \qquad \forall \mathbf{u} \in H(\text{div};\Omega),$$

where

$$l_{N+1} = \operatorname{div}\mathbf{l}.$$

Now, assume that l vanishes on $\mathscr{D}(\bar{\Omega})^N$ and let $\tilde{l}_i$ denote the extension of l_i by zero outside Ω. The above formula may be rewritten as follows:

$$\int_{\mathbb{R}^N} \left\{ \sum_{i=1}^{N} \tilde{l}_i \phi_i + \tilde{l}_{N+1} \operatorname{div}\boldsymbol{\phi} \right\} dx = 0 \qquad \forall \boldsymbol{\phi} \in \mathscr{D}(\mathbb{R}^N)^N.$$

This equality implies that, in the sense of distributions on $\mathbb{R}^N$,

$$\tilde{\mathbf{l}} = \mathbf{grad}\, \tilde{l}_{N+1}.$$

Thus $\tilde{l}_{N+1} \in H^1(\mathbb{R}^N)$, since $\tilde{\mathbf{l}} \in L^2(\mathbb{R}^N)^N$. Then, it follows from Theorem 1.2 2°) that $l_{N+1} \in H_0^1(\Omega)$. As $\mathscr{D}(\Omega)$ is dense in $H_0^1(\Omega)$, let $(\psi_\mu)_{\mu \geq 1}$ be a sequence of $\mathscr{D}(\Omega)$ that tends to l_{N+1} in $H^1(\Omega)$. Then

$$\langle l, \mathbf{u}\rangle = \lim_{\mu \to \infty} \{(\mathbf{grad}\, \psi_\mu, \mathbf{u}) + (\psi_\mu, \operatorname{div}\mathbf{u})\} = 0 \qquad \forall \mathbf{u} \in H(\text{div};\Omega).$$

Therefore, if l vanishes on $\mathscr{D}(\bar{\Omega})^N$ then l also vanishes on $H(\text{div};\Omega)$, thus proving the required density. □

The next theorems concern the normal component of boundary values of functions of $H(\text{div};\Omega)$.

Theorem 2.5. *The mapping $\gamma_n\colon \mathbf{v} \to \mathbf{v}\cdot\mathbf{n}|_\Gamma$ defined on $\mathscr{D}(\bar{\Omega})^N$ can be extended by continuity to a linear and continuous mapping, still denoted by γ_n, from $H(\text{div};\Omega)$ into $H^{-1/2}(\Gamma)$.*

Proof. Let $\phi \in \mathscr{D}(\bar{\Omega})$ and $\mathbf{v} \in \mathscr{D}(\bar{\Omega})^N$. The following Green's formula holds:

$$(v, \mathbf{grad}\,\phi) + (\operatorname{div}\mathbf{v}, \phi) = \int_\Gamma \phi \mathbf{v}\cdot\mathbf{n}\, ds.$$

As $\mathscr{D}(\bar{\Omega})$ is dense in $H^1(\Omega)$, this equality is still valid for ϕ in $H^1(\Omega)$ and $\mathbf{v}$ in $\mathscr{D}(\bar{\Omega})^N$. Therefore,

$$\left| \int_\Gamma \phi \mathbf{v}\cdot\mathbf{n}\, ds \right| \leq \|\mathbf{v}\|_{H(\text{div};\Omega)} \|\phi\|_{1,\Omega} \qquad \forall \phi \in H^1(\Omega), \qquad \forall \mathbf{v} \in \mathscr{D}(\bar{\Omega})^N.$$

Now, let μ be any element of $H^{1/2}(\Gamma)$. Then there exists an element ϕ of $H^1(\Omega)$ such that $\phi = \mu$ on Γ. Hence the above inequality implies that

$$\left|\int_\Gamma \mu \mathbf{v}\cdot\mathbf{n}\, ds\right| \leqslant \|\mathbf{v}\|_{H(\mathrm{div};\,\Omega)} \|\mu\|_{1/2,\Gamma} \qquad \forall \mu \in H^{1/2}(\Gamma), \qquad \forall \mathbf{v} \in \mathscr{D}(\bar{\Omega})^N.$$

Thus

$$\|\mathbf{v}\cdot\mathbf{n}\|_{-1/2,\Gamma} \leqslant \|\mathbf{v}\|_{H(\mathrm{div};\,\Omega)}.$$

Therefore, the linear mapping $\gamma_n\colon \mathbf{v} \to \mathbf{v}\cdot\mathbf{n}|_\Gamma$ defined on $\mathscr{D}(\bar{\Omega})^N$ is continuous for the norm of $H(\mathrm{div};\Omega)$. Since $\mathscr{D}(\bar{\Omega})^N$ is dense in $H(\mathrm{div};\Omega)$, γ_n can be extended by continuity to a mapping still called $\gamma_n \in \mathscr{L}(H(\mathrm{div};\Omega); H^{-1/2}(\Gamma))$ such that:

(2.16) $$\|\gamma_n\|_{\mathscr{L}(H(\mathrm{div};\,\Omega);\,H^{-1/2}(\Gamma))} \leqslant 1. \qquad \square$$

By extension, $\gamma_n \mathbf{v}$ is called the normal component of $\mathbf{v}$ on Γ and is denoted simply by $\mathbf{v}\cdot\mathbf{n}$.

From Theorems 2.4 and 2.5, we derive the following Green's formula:

(2.17) $$(\mathbf{v}, \mathbf{grad}\,\phi) + (\mathrm{div}\,\mathbf{v}, \boldsymbol{\phi}) = \langle \mathbf{v}\cdot\mathbf{n}, \phi\rangle_\Gamma \qquad \forall \mathbf{v} \in H(\mathrm{div};\Omega), \qquad \forall \phi \in H^1(\Omega).$$

As a consequence, we can now extend Green's formula for the Laplace operator to a wider range of functions.

Corollary 2.6. *Let $u \in H^1(\Omega)$ and $\Delta u \in L^2(\Omega)$. Then $\partial u/\partial n \in H^{-1/2}(\Gamma)$ and*

(2.18) $$(\mathbf{grad}\, u, \mathbf{grad}\, v) = -(\Delta u, v) + \langle \partial u/\partial n, v\rangle_\Gamma \qquad \forall v \in H^1(\Omega).$$

The proof is left as an exercise.

Another interesting consequence is that now we can interpret properly the variational problem (N′) of Section 1.4 and show that it is equivalent to the Neumann's problem (N). We skip the proof, as it follows immediately from (2.18).

Corollary 2.7. *Problem* (1.37) *and Problem* (1.31) (1.32) (1.33) *are equivalent.*

The next two results give further information about γ_n.

Corollary 2.8. *We have*:

$$\mathscr{R}(\gamma_n) = H^{-1/2}(\Gamma)$$

and

$$\|\gamma_n\|_{\mathscr{L}(H(\mathrm{div};\,\Omega);\,H^{-1/2}(\Gamma))} = 1.$$

Proof. Let $\mu \in H^{-1/2}(\Gamma)$; we want to exhibit an element $\mathbf{v}$ of $H(\mathrm{div};\Omega)$ such that

$$\mathbf{v}\cdot\mathbf{n} = \mu \quad \text{on } \Gamma \quad \text{and} \quad \|\mathbf{v}\cdot\mathbf{n}\|_{-1/2,\Gamma} \geqslant \|\mathbf{v}\|_{H(\mathrm{div};\,\Omega)}.$$

In view of (2.16), this will prove the corollary. To this end, consider the problem:

Find ϕ in $H^1(\Omega)$ such that

$$-\Delta\phi + \phi = 0 \quad \textit{in } \Omega,$$

$$\partial\phi/\partial n = \mu \quad \textit{on } \Gamma.$$

Unlike the Neumann's problem of Section 1.4, this problem has exactly one solution ϕ in $H^1(\Omega)$. We set $\mathbf{v} = \mathbf{grad}\,\phi$. Then $\mathbf{v} \in H(\text{div}; \Omega)$ and $\mathbf{v} \cdot \mathbf{n} = \mu$. Moreover

$$\|\phi\|_{1,\Omega}^2 = \langle \mu, \phi \rangle_\Gamma \leqslant \|\mu\|_{-1/2,\Gamma} \|\phi\|_{1,\Omega}.$$

As $\operatorname{div} \mathbf{v} = \phi$, it follows that

$$\|\mathbf{v}\|_{H(\text{div};\Omega)} \leqslant \|\mu\|_{-1/2,\Gamma} = \|\mathbf{v} \cdot \mathbf{n}\|_{-1/2,\Gamma}. \qquad \square$$

Theorem 2.6. *We have*:

$$H_0(\text{div}; \Omega) = \text{Ker}(\gamma_n) = \{\mathbf{u} \in H(\text{div}; \Omega);\ \mathbf{u} \cdot \mathbf{n}|_\Gamma = 0\}.$$

Proof. Let us establish that $\mathcal{D}(\Omega)^N$ is dense in $\text{Ker}(\gamma_n)$. Again, the proof makes use of property (2.14). Let $l \in (\text{Ker}(\gamma_n))'$ and let $\mathbf{l}$ be the element of $\text{Ker}(\gamma_n)$ associated with l by:

$$(2.19) \qquad \langle l, \mathbf{u} \rangle = \sum_{i=1}^{N} (l_i, u_i) + (l_{N+1}, \operatorname{div} \mathbf{u}) \qquad \forall \mathbf{u} \in \text{Ker}(\gamma_n).$$

Now, suppose that l vanishes on $\mathcal{D}(\Omega)^N$. This implies that

$$\mathbf{l} = \mathbf{grad}\, l_{N+1}.$$

Hence $l_{N+1} \in H^1(\Omega)$ and we can apply Green's formula (2.17) to (2.19):

$$\langle l, \mathbf{u} \rangle = \langle \mathbf{u} \cdot \mathbf{n}, l_{N+1} \rangle_\Gamma \qquad \forall \mathbf{u} \in \text{Ker}(\gamma_n).$$

Therefore l also vanishes on $\text{Ker}(\gamma_n)$. Consequently $\text{Ker}(\gamma_n) \subset H_0(\text{div}; \Omega)$.

The inclusion $H_0(\text{div}; \Omega) \subset \text{Ker}(\gamma_n)$ is an immediate consequence of Green's formula (2.17). $\square$

We end this section by examining the subspace of divergence-free functions of $H_0(\text{div}; \Omega)$:

$$H = \{\mathbf{u} \in H_0(\text{div}; \Omega);\ \operatorname{div} \mathbf{u} = 0\}.$$

Because it is a closed subspace of $L^2(\Omega)^N$, we have the decomposition:

$$L^2(\Omega)^N = H \oplus H^\perp$$

where $H^\perp$ denotes the orthogonal of H in $L^2(\Omega)^N$ for the scalar product $(.,.)$. The following theorem characterizes $H^\perp$.

Theorem 2.7. *If Ω is connected, we have:*

$$H^{\perp} = \{\mathbf{grad}\, q;\, q \in H^1(\Omega)\}.$$

Proof. For the sake of brevity, we denote the space $\{\mathbf{grad}\, q; q \in H^1(\Omega)\}$ by X. First, observe that in view of Theorem 1.9, X is a closed subspace of $L^2(\Omega)^N$. Hence, if we prove that $X^{\perp} = H$ this will imply:

$$H^{\perp} = (X^{\perp})^{\perp} = \bar{X} = X,$$

which is the required result.

First, let $\mathbf{u} \in H$. Formula (2.17) and Theorem 2.6 yield

$$(\mathbf{u}, \mathbf{grad}\, q) = 0 \qquad \forall q \in H^1(\Omega).$$

Therefore $H \subset X^{\perp}$.

Conversely, let $\mathbf{u} \in L^2(\Omega)^N$ with

$$(\mathbf{u}, \mathbf{grad}\, q) = 0 \qquad \forall q \in H^1(\Omega).$$

By choosing q in $\mathscr{D}(\Omega)$, this implies that $\operatorname{div} \mathbf{u} = 0$. Hence $\mathbf{u} \in H(\operatorname{div}; \Omega)$. Therefore we can apply formula (2.17) which gives $\mathbf{u} \cdot \mathbf{n} = 0$. In view of Theorem 2.6, this means that $\mathbf{u} \in H$. Therefore $X^{\perp} \subset H$. □

Theorem 2.8. *The space $\mathscr{V}$ is dense in H.*

Proof. This is another application of property (2.14). Let $l \in H'$ and let $\mathbf{l} \in H$ be associated with l by

$$\langle l, \mathbf{u} \rangle = (\mathbf{l}, \mathbf{u}) \qquad \forall \mathbf{u} \in H.$$

Then, if l vanishes on $\mathscr{V}$, the function $\mathbf{l}$ satisfies the hypotheses of Theorem 2.3. Hence

$$\mathbf{l} = \mathbf{grad}\, p$$

for some $p \in L^2(\Omega)$. As $\mathbf{l} \in L^2(\Omega)^N$, this implies that $p \in H^1(\Omega)$. Thus formula (2.17) yields:

$$\langle l, \mathbf{u} \rangle = -(p, \operatorname{div} \mathbf{u}) + \langle \mathbf{u} \cdot \mathbf{n}, p \rangle_{\Gamma} = 0 \qquad \forall \mathbf{u} \in H. \qquad \square$$

2.3. Some Properties of Spaces Related to the curl Operator

Unless otherwise specified, we assume in this section that the dimension N is two or three and that Ω is a *bounded* region of $\mathbb{R}^N$ with a *Lipschitz-continuous* boundary Γ. When $N = 2$, we define the curl operator for distributions ϕ of $\mathscr{D}'(\Omega)$ and $\mathbf{v}$ of $\mathscr{D}'(\Omega)^2$ by

$$\mathbf{curl}\,\phi = \left(\frac{\partial\phi}{\partial x_2}, -\frac{\partial\phi}{\partial x_1}\right),$$

$$\operatorname{curl}\mathbf{v} = \frac{\partial v_2}{\partial x_1} - \frac{\partial v_1}{\partial x_2}.$$

When $N = 3$, we define the curl of a distribution $\mathbf{v}$ of $\mathscr{D}'(\Omega)^3$ by

$$\mathbf{curl}\,\mathbf{v} = \left(\frac{\partial v_3}{\partial x_2} - \frac{\partial v_2}{\partial x_3}, \frac{\partial v_1}{\partial x_3} - \frac{\partial v_3}{\partial x_1}, \frac{\partial v_2}{\partial x_1} - \frac{\partial v_1}{\partial x_2}\right).$$

It can be easily checked that the following identities hold:

$$\operatorname{curl}(\mathbf{curl}\,\phi) = -\Delta\phi \qquad N = 2,$$

$$\left.\begin{array}{l}\mathbf{curl}(\mathbf{curl}\,\mathbf{v})\\ \mathbf{curl}(\operatorname{curl}\mathbf{v})\end{array}\right\} = -\Delta\mathbf{v} + \mathbf{grad}(\operatorname{div}\mathbf{v}). \qquad \begin{array}{l}N = 3,\\ N = 2\end{array}$$

Note that the curl of a two-dimensional vector field is a scalar. In order to avoid a multiplicity of notation, we agree nevertheless to denote it like a vector, provided there is no confusion.

To begin with, we establish an extension of the following classical *Stokes' Theorem*:

> *if a function of class $\mathscr{C}^1$ has a vanishing* **curl** *in a simply-connected region of $\mathbb{R}^N$, then this function is a gradient.*

Theorem 2.9. *Assume in addition that Ω is simply-connected. A function* $\mathbf{u}$ *of $L^2(\Omega)^N$ satisfies:*

$$\mathbf{curl}\,\mathbf{u} = \mathbf{0} \quad \text{in } \Omega$$

iff there exists a unique function $\dot{p}$ of $H^1(\Omega)/\mathbb{R}$ such that:

$$\mathbf{u} = \mathbf{grad}\,\dot{p}.$$

Proof. The proof is very similar to that of Theorem 2.3: we first prove that $\mathbf{u} = \mathbf{grad}\,p$ for some p in $L^2_{\text{loc}}(\Omega)$ and we apply Corollary 2.2 to show that $p \in L^2(\Omega)$. Obviously $p \in H^1(\Omega)$ and is uniquely defined in $H^1(\Omega)/\mathbb{R}$.

Here again, we can find an *increasing sequence* of *simply-connected*, Lipschitz-continuous open sets $(\Omega_m)_{m\geqslant 1}$ such that:

$$\bar{\Omega}_m \subset \Omega \quad \text{and} \quad \Omega = \bigcup_{m\geqslant 1} \Omega_m.$$

The idea is to regularize $\mathbf{u}$ so that its restriction to Ω_m has a vanishing **curl**. Then we shall be able to apply Stokes' Theorem to this smooth function.

For $\varepsilon > 0$, let $\rho_\varepsilon \in \mathscr{D}(\mathbb{R}^N)$ be the mollifier introduced in Theorem 2.3 and let $\tilde{\mathbf{u}}$ denote the extension of $\mathbf{u}$ by zero outside Ω. Then

$$(2.20)\qquad \begin{cases} \rho_\varepsilon * \tilde{\mathbf{u}} \in \mathscr{D}(\mathbb{R}^N)^N, \\ \lim_{\varepsilon\to 0} (\rho_\varepsilon * \tilde{\mathbf{u}}) = \tilde{\mathbf{u}} \quad \text{in } L^2(\mathbb{R}^N)^N, \\ \mathbf{curl}(\rho_\varepsilon * \tilde{\mathbf{u}}) = \rho_\varepsilon * \mathbf{curl}\, \tilde{\mathbf{u}}. \end{cases}$$

But if ε is sufficiently small,

$$\bigcup_{x\in\Omega_m} B(x;\varepsilon) \subset \Omega,$$

where $B(x;\varepsilon)$ denotes the ball with center x and radius ε. Therefore

$$\mathbf{curl}(\rho_\varepsilon * \tilde{\mathbf{u}}) = \mathbf{0} \quad \text{in } \Omega_m.$$

Hence, it follows from Stokes' Theorem that there exists a smooth function p_ε (in $H^1(\Omega_m)$ at least) such that

$$\rho_\varepsilon * \tilde{\mathbf{u}} = \mathbf{grad}\, p_\varepsilon \quad \text{in } \Omega_m.$$

Moreover, (2.20) implies that there exists $p_m \in H^1(\Omega_m)$ such that

$$\lim_{\varepsilon\to 0} p_\varepsilon = p_m \quad \text{in } H^1(\Omega_m)/\mathbb{R}$$

and

$$\mathbf{u} = \mathbf{grad}\, p_m \quad \text{in } \Omega_m.$$

The proof then ends like that of Theorem 2.3. □

With Theorem 2.7, this gives us another characterization of $H^\perp$:

Corollary 2.9. *If Ω is like in Theorem 2.9, the following identity holds in $L^2(\Omega)^N$:*

$$H^\perp = \mathrm{Ker}(\mathbf{curl}) = \{\mathbf{v} \in L^2(\Omega)^N;\ \mathbf{curl}\,\mathbf{v} = \mathbf{0}\}.$$

Remark 2.2. As a consequence of Corollary 2.9, if $\mathbf{u} \in H$ and $\mathbf{curl}\,\mathbf{u} = \mathbf{0}$ then $\mathbf{u} = \mathbf{0}$, provided of course that Ω is bounded and simply-connected.

Just like we defined the space $H(\mathrm{div};\Omega)$, we introduce the space $H(\mathbf{curl};\Omega)$:

$$H(\mathbf{curl};\Omega) = \{\mathbf{v} \in L^2(\Omega)^N;\ \mathbf{curl}\,\mathbf{v} \in L^2(\Omega)^N\},$$

a Hilbert space for the norm

$$\|\mathbf{v}\|_{H(\mathbf{curl};\Omega)} = \{\|\mathbf{v}\|_{0,\Omega}^2 + \|\mathbf{curl}\,\mathbf{v}\|_{0,\Omega}^2\}^{1/2};$$

and we set

$$H_0(\mathbf{curl};\Omega) = \overline{\mathscr{D}(\Omega)^N}^{H(\mathbf{curl};\Omega)}.$$

For these spaces, we propose to establish the analogues of Theorems 2.4, 2.5 and 2.6. We begin with two preliminary lemmas.

Lemma 2.3. *Suppose Ω is a Lipschitz-continuous open subset of $\mathbb{R}^N$ (not necessarily bounded). Then the set of functions of $H(\mathbf{curl}; \Omega)$ with compact support is dense in $H(\mathbf{curl}; \Omega)$.*

Proof. We use a classical truncation process. Let $\mathbf{u} \in H(\mathbf{curl}; \Omega)$ and let ϕ be a function of $\mathscr{C}^\infty(\mathbb{R}^N)$ such that $0 \leqslant \phi \leqslant 1$ everywhere in $\mathbb{R}^N$ and

$$\phi(x) = \begin{cases} 1 & \text{for } \|x\| \leqslant 1, \\ 0 & \text{for } \|x\| \geqslant 2. \end{cases}$$

Next, for an arbitrary real number $a > 0$, we make the change of variable:

$$\phi_a(x) = \phi(x/a).$$

As ϕ_a is smooth, $\phi_a \mathbf{u} \in H(\mathbf{curl}; \Omega)$; moreover

$$\lim_{a \to \infty} \phi_a \mathbf{u} = \mathbf{u} \quad \text{in } H(\mathbf{curl}; \Omega)$$

and the support of $\phi_a \mathbf{u}$ is compact. Hence $\phi_a \mathbf{u}$ is the desired sequence. □

Lemma 2.4. *If a function* $\mathbf{f}$ *of* $H(\mathbf{curl}; \Omega)$ *satisfies:*

$$(2.21) \qquad (\mathbf{f}, \mathbf{curl}\, \boldsymbol{\phi}) - (\mathbf{curl}\, \mathbf{f}, \boldsymbol{\phi}) = 0 \quad \forall \boldsymbol{\phi} \in \mathscr{D}(\bar{\Omega})^N,$$

then $\mathbf{f} \in H_0(\mathbf{curl}; \Omega)$.

Proof. Let $\tilde{\mathbf{f}}$ denote the extension of $\mathbf{f}$ by zero outside Ω. To begin with, observe that (2.21) implies that $\tilde{\mathbf{f}} \in H(\mathbf{curl}; \mathbb{R}^N)$. Next, in order to construct a sequence of $\mathscr{D}(\Omega)^N$ that converges to $\mathbf{f}$ in $H(\mathbf{curl}; \Omega)$, we combine the regularization techniques of Corollary 2.2 and Theorem 2.3.

1°) Suppose for the moment that Ω is strictly star-shaped. Again, we make the change of variable introduced in Corollary 2.2:

$$\tilde{\mathbf{f}}_\theta(x) = \tilde{\mathbf{f}}(x/\theta) \qquad \forall x \in \Omega_\theta = \theta\Omega,$$

but this time, we take θ in $(0, 1)$ in order to obtain a function $\tilde{\mathbf{f}}_\theta$ with compact support in Ω. Obviously, $\tilde{\mathbf{f}}_\theta \in H(\mathbf{curl}; \mathbb{R}^N)$ and

$$\lim_{\theta \to 1} \tilde{\mathbf{f}}_\theta = \tilde{\mathbf{f}} \quad \text{in } H(\mathbf{curl}; \mathbb{R}^N).$$

Hence, for $\varepsilon > 0$ sufficiently small, $\rho_\varepsilon * \tilde{\mathbf{f}}_\theta$ belongs to $\mathscr{D}(\Omega)^N$ and

$$\lim_{\varepsilon \to 0} \lim_{\theta \to 1} (\rho_\varepsilon * \tilde{\mathbf{f}}_\theta) = \tilde{\mathbf{f}} \quad \text{in } H(\mathbf{curl}; \Omega).$$

2°) In the general case, Ω can be covered by a finite family of open sets:

$$\Omega \subset \bigcup_{1 \leqslant i \leqslant q} \mathscr{O}_i$$

such that each $\Omega_i = \mathscr{O}_i \cap \Omega$ is Lipschitz-continuous, bounded and strictly star-shaped, for $1 \leqslant i \leqslant q$. Let $(\alpha_i)_{i \geqslant 1}$ be a partition of unity subordinate to the family

$(\mathcal{O}_i)_{i\geqslant 1}$ (cf. Yosida [84]), i.e.

$$\alpha_i \in \mathscr{D}(\mathcal{O}_i), \qquad 0 \leqslant \alpha_i(x) \leqslant 1 \quad \text{and} \quad \sum_{i=1}^{q} \alpha_i(x) \equiv 1 \quad \text{in } \Omega.$$

Thus

$$\tilde{\mathbf{f}} = \sum_{i=1}^{q} \alpha_i \tilde{\mathbf{f}} \quad \text{in } \mathbb{R}^N.$$

Clearly, $\alpha_i \tilde{\mathbf{f}} \in H(\mathbf{curl}; \Omega)$ and $\operatorname{supp}(\alpha_i \tilde{\mathbf{f}}) \subset \bar{\Omega}_i$. Therefore, we can finish the proof by applying the result of 1°) to each $\alpha_i \tilde{\mathbf{f}}$. □

Theorem 2.10. *Let Ω be like in Lemma 2.3. Then $\mathscr{D}(\bar{\Omega})^N$ is dense in $H(\mathbf{curl}; \Omega)$.*

The proof, which is an easy variant of Theorem 2.4, is left as an exercise.

Now, we are in a position to define the tangential trace on Γ of functions of $H(\mathbf{curl}; \Omega)$. In the case $N = 2$, we denote by $\boldsymbol{\tau}$ the unit tangent to Γ like in Figure 1—i.e. $\boldsymbol{\tau} = (-n_2, n_1)$.

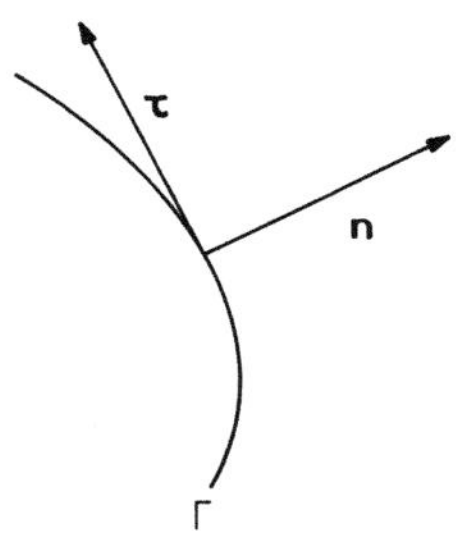

Figure 1

Theorem 2.11. *The mapping $\gamma_\tau: \mathbf{v} \to \mathbf{v}\cdot\boldsymbol{\tau}|_\Gamma$ for $N = 2$ or $\gamma_\tau: \mathbf{v} \to \mathbf{v} \times \mathbf{n}|_\Gamma$ for $N = 3$ defined on $\mathscr{D}(\bar{\Omega})^N$ can be extended by continuity to a linear and continuous mapping, still denoted by γ_τ, from $H(\mathbf{curl}; \Omega)$ into $H^{-1/2}(\Gamma)$ if $N = 2$ or $H^{-1/2}(\Gamma)^3$ if $N = 3$. Moreover, the following Green's formula holds:*

$$(2.22) \qquad \begin{aligned} &(\mathbf{curl}\,\mathbf{v}, \boldsymbol{\phi}) - (\mathbf{v}, \mathbf{curl}\,\boldsymbol{\phi}) = \langle \gamma_\tau \mathbf{v}, \boldsymbol{\phi}\rangle_\Gamma \qquad \forall \mathbf{v} \in H(\mathbf{curl}; \Omega), \\ &\forall \boldsymbol{\phi} \in H^1(\Omega)^3 \quad \text{if } N = 3 \quad \text{or} \quad \phi \in H^1(\Omega) \quad \text{if } N = 2. \end{aligned}$$

We skip the proof as it is entirely similar to that of Theorem 2.5.

To avoid a multiplicity of notations, we denote $\gamma_\tau \mathbf{v}$ by $\mathbf{v}\cdot\boldsymbol{\tau}|_\Gamma$ or $\mathbf{v} \times \mathbf{n}|_\Gamma$ according that $N = 2$ or $N = 3$.

Now we turn to $\operatorname{Ker}(\gamma_\tau)$ in $H(\mathbf{curl}; \Omega)$.

Theorem 2.12. *We have*

$$\operatorname{Ker}(\gamma_\tau) = H_0(\mathbf{curl}; \Omega).$$

Proof. By taking limits in (2.22), it is clear that $H_0(\mathbf{curl}; \Omega) \subset \operatorname{Ker}(\gamma_\tau)$. The converse is an obvious application of Lemma 2.4. □

Remark 2.3. The tangential components on Γ of a vector field $\mathbf{v}$ are defined formally by:

$$\mathbf{v}_T = \mathbf{v} - (\mathbf{v} \cdot \mathbf{n})\mathbf{n}.$$

Clearly, we have $\mathbf{v}_T \times \mathbf{n} = \mathbf{v} \times \mathbf{n}$. As a consequence, each vector field $\mathbf{v}$ of $H(\mathbf{curl}; \Omega)$ satisfies $\mathbf{v} \times \mathbf{n}|_\Gamma = \mathbf{0}$ iff its tangential components vanish on Γ. Similarly, each function q of $H^1(\Omega)$ satisfies $(\mathbf{grad}\, q) \times \mathbf{n} = \mathbf{0}$ on Γ iff q is constant on each connected component of Γ.

Remark 2.4. When $N = 2$, we observe that the function $\mathbf{v} = (v_1, v_2)$ belongs to $H(\text{curl}; \Omega)$ iff the function $\mathbf{w} = (-v_2, v_1)$ belongs to $H(\text{div}; \Omega)$. Moreover, $\mathbf{w} \cdot \mathbf{n} = -\mathbf{v} \cdot \boldsymbol{\tau}$. Hence all the properties of $H(\text{div}; \Omega)$ stated in Theorems 2.4, 2.5, 2.6 and Corollary 2.8 carry over directly to $H(\text{curl}; \Omega)$.

Remark 2.5. If $\mathbf{v} \in H_0(\mathbf{curl}; \Omega)$, then $\mathbf{curl}\, \mathbf{v} \in H$. Indeed, since $\operatorname{div} \mathbf{curl}\, \mathbf{v} = 0$, it suffices to check that $(\mathbf{curl}\, \mathbf{v}) \cdot \mathbf{n} = 0$. For ϕ in $H^2(\Omega)$, Green's formulas (2.17) and (2.22) imply that

$$\langle (\mathbf{curl}\, \mathbf{v}) \cdot \mathbf{n}, \phi \rangle_\Gamma = \int_\Omega \mathbf{curl}\, \mathbf{v} \cdot \mathbf{grad}\, \phi \, dx = \langle \mathbf{v} \times \mathbf{n}, \mathbf{grad}\, \phi \rangle_\Gamma = 0.$$

Hence the result follows from the density of $H^2(\Omega)$ in $H^1(\Omega)$.

We end this paragraph by examining the intersection of $H_0(\text{div}; \Omega)$ and $H_0(\mathbf{curl}; \Omega)$.

Lemma 2.5. *The following equality holds algebraically and topologically*:

$$H_0^1(\Omega)^N = H_0(\text{div}; \Omega) \cap H_0(\mathbf{curl}; \Omega).$$

Proof. We establish the lemma for $N = 3$; the proof is similar and simpler when $N = 2$. Let $\mathbf{v} \in H_0(\text{div}; \Omega) \cap H_0(\mathbf{curl}; \Omega)$ and let $\tilde{\mathbf{v}}$ denote the extension of $\mathbf{v}$ by zero outside Ω. From Green's formulas (2.17) and (2.22), we easily derive that $\tilde{\mathbf{v}} \in H(\text{div}; \mathbb{R}^3) \cap H(\mathbf{curl}; \mathbb{R}^3)$ with

$$\|\operatorname{div} \tilde{\mathbf{v}}\|_{0,\mathbb{R}^3} = \|\operatorname{div} \mathbf{v}\|_{0,\Omega},$$

$$\|\mathbf{curl}\, \tilde{\mathbf{v}}\|_{0,\mathbb{R}^3} = \|\mathbf{curl}\, \mathbf{v}\|_{0,\Omega},$$

$$\|\tilde{\mathbf{v}}\|_{0,\mathbb{R}^3} = \|\mathbf{v}\|_{0,\Omega}.$$

Let $\mathscr{F}\tilde{\mathbf{v}}$ denote the Fourier transform of $\tilde{\mathbf{v}}$, defined as usual by:

$$\mathscr{F}\tilde{\mathbf{v}}(\mu) = \int_{\mathbb{R}^3} e^{-2i\pi(x,\mu)}\tilde{\mathbf{v}}(x)\,dx, \tag{2.23}$$

where

$$(x, \mu) = \sum_{i=1}^{3} x_i \mu_i.$$

Owing to the properties of the Fourier transform, it is well known that $\mathscr{F}\tilde{\mathbf{v}} \in L^2(\mathbb{R}^3)$,[3]

$$\|\mathscr{F}\tilde{\mathbf{v}}\|_{0,\mathbb{R}^3} = \|\mathbf{v}\|_{0,\Omega},$$

$$\mathscr{F}(\mathbf{curl}\,\tilde{\mathbf{v}}) = 2i\pi(\mu_2\mathscr{F}\tilde{v}_3 - \mu_3\mathscr{F}\tilde{v}_2, \mu_3\mathscr{F}\tilde{v}_1 - \mu_1\mathscr{F}\tilde{v}_3, \mu_1\mathscr{F}\tilde{v}_2 - \mu_2\mathscr{F}\tilde{v}_1),$$

$$\mathscr{F}(\operatorname{div}\tilde{\mathbf{v}}) = 2i\pi(\mu_1\mathscr{F}\tilde{v}_1 + \mu_2\mathscr{F}\tilde{v}_2 + \mu_3\mathscr{F}\tilde{v}_3).$$

From these two relations, we readily derive that

$$\|\mathscr{F}(\partial\tilde{v}_i/\partial x_j)\|_{0,\mathbb{R}^3}^2 \leqslant \|\operatorname{div}\mathbf{v}\|_{0,\Omega}^2 + \|\mathbf{curl}\,\mathbf{v}\|_{0,\Omega}^2.$$

Therefore $\tilde{\mathbf{v}} \in H^1(\mathbb{R}^3)^3$. This implies that $\mathbf{v} \in H_0^1(\Omega)^3$. Furthermore,

$$\begin{aligned}\|\mathbf{v}\|_{1,\Omega} &= \|\tilde{\mathbf{v}}\|_{1,\mathbb{R}^3}\\ &\leqslant C(\|\mathbf{v}\|_{0,\Omega}^2 + \|\operatorname{div}\mathbf{v}\|_{0,\Omega}^2 + \|\mathbf{curl}\,\mathbf{v}\|_{0,\Omega}^2)^{1/2},\end{aligned}$$

with a constant C independent of Ω.

The inverse inclusion and inequality are obvious. □

The next corollary is a simple consequence of Lemma 2.5.

Corollary 2.10. *Let Ω be an arbitrary open subset of $\mathbb{R}^N$ and let $\mathbf{v}$ belong to $H(\operatorname{div};\Omega) \cap H(\mathbf{curl};\Omega)$. Then $\mathbf{v}$ belongs to $H_{\mathrm{loc}}^1(\Omega)^3$.*

Remark 2.6. It stems from the proof of Lemma 2.5 that, algebraically and topologically:

$$H^1(\mathbb{R}^N)^N = H(\operatorname{div};\mathbb{R}^N) \cap H(\mathbf{curl};\mathbb{R}^N).$$

Remark 2.7. Lemma 2.5 and Theorem 2.1 readily yield that

$$|\mathbf{v}|_{1,\Omega} \leqslant C(\|\operatorname{div}\mathbf{v}\|_{0,\Omega}^2 + \|\mathbf{curl}\,\mathbf{v}\|_{0,\Omega}^2)^{1/2} \qquad \forall \mathbf{v} \in H_0(\operatorname{div};\Omega) \cap H_0(\mathbf{curl};\Omega).$$

§ 3. A Decomposition of Vector Fields

In the preceding paragraph, we have derived the following decompositions into orthogonal subspaces:

$$L^2(\Omega)^N = H \oplus H^\perp, \qquad H_0^1(\Omega)^N = V \oplus V^\perp,$$

and we have characterized $H^{\perp}$ and $V^{\perp}$. Here, we are going to characterize H and V as spaces of **curl** of stream functions. This will show, in particular, that every function of $L^2(\Omega)^N$ is the sum of the **curl** of a stream function and a gradient.

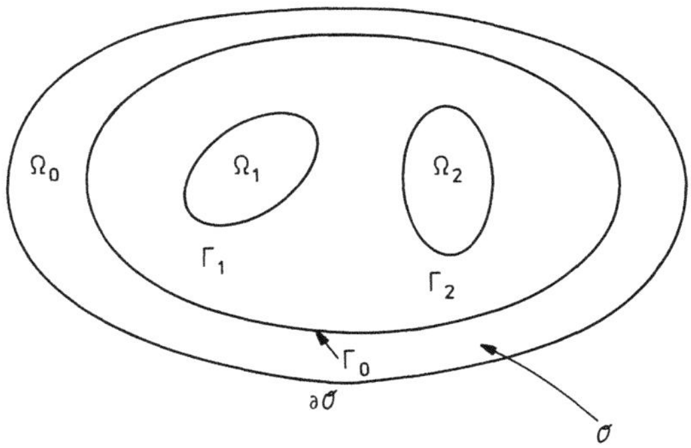

Figure 2

We shall make the following assumptions on Ω: Ω is *bounded, connected* but eventually multiply-connected, and its boundary Γ is *Lipschitz-continuous.* We shall denote by Γ_0 the exterior boundary of Ω (cf. Figure 2) and by Γ_i, $1 \leqslant i \leqslant p$, the other components of Γ. The duality between $H^{-1/2}(\Gamma_i)$ and $H^{1/2}(\Gamma_i)$ will be denoted by $\langle .,. \rangle_{\Gamma_i}$. Finally, we shall assume that the dimension N is either 2 or 3.

3.1. Decomposition of Two-Dimensional Vector Fields

Theorem 3.1. *A function* $\mathbf{v}$ *of* $L^2(\Omega)^2$ *satisfies:*

(3.1) $$\operatorname{div} \mathbf{v} = 0, \quad \langle \mathbf{v}\cdot\mathbf{n}, 1\rangle_{\Gamma_i} = 0 \quad \text{for } 0 \leqslant i \leqslant p$$

iff there exists a stream function ϕ *in* $H^1(\Omega)$ *such that:*

(3.2) $$\mathbf{v} = \mathbf{curl}\, \phi.$$

Proof. 1°) Let us show that (3.2) implies (3.1). Let $\phi \in H^1(\Omega)$ and set $\mathbf{v} = \mathbf{curl}\, \phi$. Clearly, $\operatorname{div} \mathbf{v} = 0$. Next, as $\mathscr{D}(\bar{\Omega})$ is dense in $H^1(\Omega)$, it suffices to prove that

$$\int_{\Gamma_i} \mathbf{curl}\, \phi \cdot \mathbf{n}\, ds = 0 \qquad \forall \phi \in \mathscr{D}(\bar{\Omega}), \qquad 0 \leqslant i \leqslant p.$$

This is indeed true since

$$\int_{\Gamma_i} \mathbf{curl}\, \phi \cdot \mathbf{n}\, ds = \int_{\Gamma_i} (\partial\phi/\partial\tau)\, ds = 0.$$

2°) Conversely, let $\mathbf{v}$ satisfy (3.1). The idea is to extend $\mathbf{v}$ to the whole plane so that it remains divergence-free. Then, it will be easy to construct its stream function by means of the Fourier transform.

a) We propose to use an extension that has the same normal component as $\mathbf{v}$ on Γ.

Let $\mathcal{O}$ be any bounded, smooth, simply-connected open set with $\bar{\Omega} \subset \mathcal{O}$. Then, for $p \geqslant 1$, the set $\mathcal{O} - \bar{\Omega}$ is not connected and we denote by Ω_i that component which is bounded by Γ_i, for $1 \leqslant i \leqslant p$, and Ω_0 that component bounded by Γ_0 and $\partial\mathcal{O}$ (cf. Figure 2). Consider the following problem:

Find w defined in $\mathcal{O} - \bar{\Omega}$ such that

$$\text{(N)} \quad \begin{cases} \Delta w = 0 & \textit{in } \mathcal{O} - \bar{\Omega}, \\ \partial w/\partial n = \mathbf{v}\cdot\mathbf{n} & \textit{on } \Gamma_i \quad \textit{for } 0 \leqslant i \leqslant p, \\ \partial w/\partial n = 0 & \textit{on } \partial\mathcal{O}, \end{cases}$$

where $\mathbf{n}$ denotes the unit normal to Γ_i pointing outside Ω_i, $0 \leqslant i \leqslant p$. Problem (N) consists of $p + 1$ non-homogeneous Neumann's problems, each defined in Ω_i for $0 \leqslant i \leqslant p$, like the one we analyzed in Section 1.4 since they include the compatibility conditions

$$\langle \mathbf{v}\cdot\mathbf{n}, 1\rangle_{\Gamma_i} = 0 \quad \text{for } 0 \leqslant i \leqslant p.$$

Therefore, there exists a function $w \in H^1(\mathcal{O} - \bar{\Omega})$ satisfying (N), determined uniquely up to an additive constant in each Ω_i. We set

$$\boldsymbol{\theta} = \mathbf{grad}\, w.$$

Then $\boldsymbol{\theta} \in H(\text{div}; \mathcal{O} - \bar{\Omega})$ with

$$\text{(3.3)} \quad \begin{cases} \operatorname{div} \boldsymbol{\theta} = \Delta w = 0 & \text{in } \mathcal{O} - \bar{\Omega}, \\ \boldsymbol{\theta}\cdot\mathbf{n} = \partial w/\partial n = \mathbf{v}\cdot\mathbf{n} & \text{on } \Gamma_i \quad \text{for } 0 \leqslant i \leqslant p, \\ \boldsymbol{\theta}\cdot\mathbf{n} = 0 & \text{on } \partial\mathcal{O}. \end{cases}$$

b) Now we can extend $\mathbf{v}$ as follows:

$$\tilde{\mathbf{v}} = \begin{cases} \mathbf{v} & \text{in } \Omega, \\ \boldsymbol{\theta} & \text{in } \mathcal{O} - \bar{\Omega}, \\ \mathbf{0} & \text{elsewhere.} \end{cases}$$

Clearly, $\tilde{\mathbf{v}} \in L^2(\mathbb{R}^2)^2$. Let us calculate its divergence; as a distribution, $\operatorname{div} \tilde{\mathbf{v}}$ satisfies:

$$\langle \operatorname{div} \tilde{\mathbf{v}}, \phi\rangle = -\int_{\mathbb{R}^2} \tilde{\mathbf{v}}\cdot\mathbf{grad}\,\phi\, dx \qquad \forall \phi \in \mathcal{D}(\mathbb{R}^2),$$

that is

$$\langle \operatorname{div} \tilde{\mathbf{v}}, \phi\rangle = -\int_{\Omega} \mathbf{v}\cdot\mathbf{grad}\,\phi\, dx - \sum_{i=0}^{p} \int_{\Omega_i} \boldsymbol{\theta}\cdot\mathbf{grad}\,\phi\, dx.$$

As $\mathbf{v}$ and $\boldsymbol{\theta}$ are both divergence-free, Green's formula (2.17) and (3.3) yield:

$$\langle \operatorname{div} \tilde{\mathbf{v}}, \phi \rangle = -\langle \mathbf{v} \cdot \mathbf{n}, \phi \rangle_{\Gamma} - \sum_{i=0}^{p} \langle \mathbf{v} \cdot \mathbf{n}, \phi \rangle_{\Gamma_i}.$$

In the above sum, the normal to Γ_i is directed outside Ω_i and therefore inside Ω; hence each term of this sum cancels a term of $\langle \mathbf{v} \cdot \mathbf{n}, \phi \rangle_{\Gamma}$. Therefore

$$\langle \operatorname{div} \tilde{\mathbf{v}}, \phi \rangle = 0 \qquad \forall \phi \in \mathscr{D}(\mathbb{R}^2),$$

i.e.

$$\operatorname{div} \tilde{\mathbf{v}} = 0.$$

c) As the support of $\tilde{\mathbf{v}}$ is compact, it follows that its Fourier transform $\mathscr{F}\tilde{\mathbf{v}}$ defined by (2.23) is a holomorphic function of μ. In terms of Fourier transforms, the conditions $\operatorname{div} \tilde{\mathbf{v}} = 0$ and $\tilde{\mathbf{v}} = \mathbf{curl}\, \phi$ read:

$$\mu_1 \mathscr{F}\tilde{v}_1 + \mu_2 \mathscr{F}\tilde{v}_2 = 0, \tag{3.4}$$

$$\mathscr{F}\tilde{v}_1 = 2i\pi\mu_2 \mathscr{F}\phi, \qquad \mathscr{F}\tilde{v}_2 = -2i\pi\mu_1 \mathscr{F}\phi. \tag{3.5}$$

If we take $\mathscr{F}\phi = \mathscr{F}\tilde{v}_1/(2i\pi\mu_2)$ then thanks to (3.4), both equalities of (3.5) are valid, i.e.

$$\mathscr{F}\phi = \mathscr{F}\tilde{v}_1/(2i\pi\mu_2) = -\mathscr{F}\tilde{v}_2/(2i\pi\mu_1).$$

Therefore the inverse transform of $\mathscr{F}\phi$ is the required stream function of $\tilde{\mathbf{v}}$, provided $\mathscr{F}\phi \in L^2(\mathbb{R}^2)$. But since $\mathscr{F}\tilde{v}_j \in L^2(\mathbb{R}^2)$, it suffices to show that $\mathscr{F}\phi$ is bounded in a neighborhood of the origin.

According to (3.4), we have $\mathscr{F}\tilde{v}_1(\mu_1, 0) = 0$. Hence, using the holomorphy of $\mathscr{F}\tilde{v}_1$, we obtain

$$\mathscr{F}\tilde{v}_1(\mu_1, \mu_2) = \mu_2(\partial \mathscr{F}\tilde{v}_1(\mu_1, 0)/\partial\mu_2) + O(|\mu_2|^2)$$

so that

$$\mathscr{F}\phi(\mu_1, \mu_2) = [1/(2i\pi)](\partial \mathscr{F}\tilde{v}_1(\mu_1, 0)/\partial\mu_2) + O(|\mu_2|).$$

Clearly, this implies that $\mathscr{F}\phi$ is bounded in a neighborhood of zero. □

An immediate consequence of this theorem is that when $\mathbf{v}$ belongs to $L^s(\Omega)^2$ for some real $s \geqslant 2$ (resp. $H^m(\Omega)^2$ for some integer $m \geqslant 0$) and $\mathbf{v}$ satisfies $\operatorname{div} \mathbf{v} = 0$, $\langle \mathbf{v} \cdot \mathbf{n}, 1 \rangle_{\Gamma_i} = 0$ for $0 \leqslant i \leqslant p$, then

$$\mathbf{v} = \mathbf{curl}\, \phi \quad \text{with } \phi \in W^{1,s}(\Omega) \quad (\text{resp. } \phi \in H^{m+1}(\Omega)).$$

Note that, since Ω is connected, the stream function ϕ of $\mathbf{v}$ is unique up to an additive constant. In particular, when $\mathbf{v} \in H$, its stream function ϕ satisfies $(\partial\phi/\partial\tau)|_{\Gamma_i} = 0$, that is

$$\phi|_{\Gamma_i} = \text{a constant } c_i, \quad \text{for } 0 \leqslant i \leqslant p.$$

Therefore, ϕ is uniquely determined if we fix one of these constants. For example, $\mathbf{v}$ has one and only one stream function ϕ that *vanishes* on Γ_0. This function belongs to the space

(3.6) $\Phi = \{\chi \in H^1(\Omega);\ \chi|_{\Gamma_0} = 0,\ \chi|_{\Gamma_i} = \text{an arbitrary constant for } 1 \leqslant i \leqslant p\}$,

which is a closed subspace of $H^1(\Omega)$. Moreover, $|.|_{1,\Omega}$ and $\|.\|_{1,\Omega}$ are two equivalent norms on Φ, by virtue of the following generalization of Poincaré's Theorem 1.1. The proof can be derived as an easy consequence of Theorem 2.1 2°).

Lemma 3.1. *Let Γ_0 be a portion of Γ with strictly positive measure. Then, for any real $r > 1$, $|.|_{1,r,\Omega}$ and $\|.\|_{1,r,\Omega}$ are two equivalent norms on the space*

$$\{v \in W^{1,r}(\Omega);\ v|_{\Gamma_0} = 0\}.$$

The next corollary characterizes the particular stream function that belongs to Φ as the solution of a boundary value problem. It follows readily from Theorem 3.1 and Lemma 3.1.

Corollary 3.1. *The space H has the following characterization:*

$$H = \{\mathbf{curl}\,\phi;\ \phi \in \Phi\}$$

and the mapping **curl** *is an isomorphism from Φ onto H. In addition, the stream function ϕ of $\mathbf{v}$ is the only solution of the problem:*

(3.7) $$\phi \in \Phi,\ (\mathbf{curl}\,\phi, \mathbf{curl}\,\chi) = (\mathbf{v}, \mathbf{curl}\,\chi) \quad \forall \chi \in \Phi.$$

Remark 3.1. Problem (3.7) has the following interpretation:

$$\begin{cases} -\Delta\phi = \operatorname{curl}\mathbf{v} & \text{in } \mathscr{D}'(\Omega), \\ \phi|_{\Gamma_0} = 0, \quad \phi|_{\Gamma_i} = c_i, \\ \langle \partial\phi/\partial n + \mathbf{v}\cdot\boldsymbol{\tau}, 1\rangle_{\Gamma_i} = 0 & 1 \leqslant i \leqslant p. \end{cases}$$

The first equation is obtained by choosing the test functions χ in $\mathscr{D}(\Omega)$. To derive the last boundary condition on Γ_i, we observe that (2.18) implies formally:

$$(\mathbf{curl}\,\phi, \mathbf{curl}\,\chi) = -(\Delta\phi, \chi) + \langle \partial\phi/\partial n, \chi\rangle_\Gamma \quad \forall \chi \in H^1(\Omega).$$

Likewise, (2.22) yields formally:

$$(\mathbf{v}, \mathbf{curl}\,\chi) = (\operatorname{curl}\mathbf{v}, \chi) - \langle \mathbf{v}\cdot\boldsymbol{\tau}, \chi\rangle_\Gamma \quad \forall \chi \in H^1(\Omega).$$

By comparing these two equalities with (3.7), we get:

$$\sum_{i=1}^{p} \langle \partial\phi/\partial n + \mathbf{v}\cdot\boldsymbol{\tau}, c_i\rangle_{\Gamma_i} = 0 \quad \forall c_i \in \mathbb{R}.$$

The results of Theorem 3.1 and its corollary yield the following orthogonal decomposition of $L^2(\Omega)^2$.

Theorem 3.2. *Every function $\mathbf{v}$ of $L^2(\Omega)^2$ has the following orthogonal decomposition:*

$$\mathbf{v} = \mathbf{grad}\, q + \mathbf{curl}\, \phi, \tag{3.8}$$

where $q \in H^1(\Omega)/\mathbb{R}$ *is the only solution of*

$$(\mathbf{grad}\, q, \mathbf{grad}\, \mu) = (\mathbf{v}, \mathbf{grad}\, \mu) \quad \forall \mu \in H^1(\Omega), \tag{3.9}$$

and $\phi \in \Phi$ *is the only solution of*

$$(\mathbf{curl}\, \phi, \mathbf{curl}\, \chi) = (\mathbf{v} - \mathbf{grad}\, q, \mathbf{curl}\, \chi) \quad \forall \chi \in \Phi. \tag{3.10}$$

Proof. For $\mathbf{v} \in L^2(\Omega)^2$, the Neumann's problem (3.9) has a unique solution $q \in H^1(\Omega)/\mathbb{R}$. This solution q satisfies

$$\Delta q = \operatorname{div} \mathbf{v} \quad \text{in } H^{-1}(\Omega).$$

Hence $\mathbf{v} - \mathbf{grad}\, q$ is a divergence-free vector of $H(\operatorname{div}; \Omega)$. Moreover, Green's formula (2.17) applied to (3.9) yields

$$0 = (\mathbf{v} - \mathbf{grad}\, q, \mathbf{grad}\, \mu) = \langle (\mathbf{v} - \mathbf{grad}\, q) \cdot \mathbf{n}, \mu \rangle_\Gamma \quad \forall \mu \in H^1(\Omega),$$

implying that $(\mathbf{v} - \mathbf{grad}\, q) \cdot \mathbf{n} = 0$ in $H^{-1/2}(\Gamma)$. In other words, $\mathbf{v} - \mathbf{grad}\, q \in H$. Thus, we infer from Corollary 3.1 that there exists a unique function $\phi \in \Phi$ verifying (3.8) and (3.10). □

Remark 3.2. When $\mathbf{v} \in H(\operatorname{div}; \Omega)$, problem (3.9) has the following interpretation:

$$\begin{cases} \Delta q = \operatorname{div} \mathbf{v} & \text{in } \Omega, \\ \partial q / \partial n = \mathbf{v} \cdot \mathbf{n} & \text{on } \Gamma. \end{cases}$$

If $\mathbf{v}$ belongs only to $L^2(\Omega)^2$, the last equality is formal.

The techniques of this section provide a straightforward extension of Lemma 2.2 to smoother functions.

Lemma 3.2. *Let* Ω *be a bounded and connected domain of* $\mathbb{R}^2$, *with a boundary* Γ *of class* $\mathscr{C}^{m,1}$ $(m \geqslant 2)$, *and let* $\mathbf{g}$ *be given in* $H^{m-1/2}(\Gamma)^2$ *with*

$$\int_\Gamma \mathbf{g} \cdot \mathbf{n}\, ds = 0.$$

Then, there exists $\mathbf{u} = f(\mathbf{g})$ *in* $H^m(\Omega)^2$ *such that*

$$\operatorname{div} \mathbf{u} = 0 \quad \text{in } \Omega, \qquad \mathbf{u} = \mathbf{g} \quad \text{on } \Gamma,$$

$$\|\mathbf{u}\|_{m,\Omega} \leqslant C \|\mathbf{g}\|_{m-1/2,\Gamma}.$$

Proof. Let us sketch the proof. The condition $\mathbf{u} = \mathbf{g}$ on Γ can be split into $\mathbf{u} \cdot \mathbf{n} = \mathbf{g} \cdot \mathbf{n}$ and $\mathbf{u} \cdot \boldsymbol{\tau} = \mathbf{g} \cdot \boldsymbol{\tau}$ on Γ. As it is simpler to prescribe each condition separately, we propose a function $\mathbf{u}$ of the form

$$\mathbf{u} = \mathbf{grad}\, q + \mathbf{curl}\, \phi.$$

Here q is the only solution (up to an additive constant) of the Neumann's problem:

$$\begin{cases} \Delta q = 0 & \text{in } \Omega, \\ \partial q/\partial n = \mathbf{g}\cdot\mathbf{n} & \text{on } \Gamma. \end{cases}$$

And ϕ is the unique solution of the biharmonic equation:

$$\Delta^2 \phi = 0 \quad \text{in } \Omega,$$

$$\phi = 0 \quad \text{on } \Gamma, \qquad \partial\phi/\partial n = \partial q/\partial\tau - \mathbf{g}\cdot\boldsymbol{\tau} \quad \text{on } \Gamma.$$

The reader can easily check that $\mathbf{u}$ satisfies the statement of the lemma. □

Now, we turn to the space V. According to Theorem 3.1, each function $\mathbf{v} \in V$ has a stream function $\phi \in H^1(\Omega)$, i.e.

$$\mathbf{v} = \mathbf{curl}\,\phi.$$

Again, ϕ is uniquely determined if we fix $\phi = 0$ on Γ_0. In this case, it can be readily verified that ϕ belongs to the space

$$\Psi = \{\chi \in H^2(\Omega);\ \chi|_{\Gamma_0} = 0,\ \chi|_{\Gamma_i} \text{ is constant for } 1 \leqslant i \leqslant p,\ \partial\chi/\partial n = 0 \text{ on } \Gamma\}. \tag{3.11}$$

Note that the seminorm $\|\Delta\phi\|_{0,\Omega}$ is a norm on Ψ equivalent to $\|\phi\|_{2,\Omega}$. Indeed, since $\mathbf{grad}\,\phi \in H_0^1(\Omega)^2$ when $\phi \in \Psi$, it can be easily shown like in Section 1.5 and by virtue of Lemma 3.1 that

$$\|\Delta\phi\|_{0,\Omega} = |\phi|_{2,\Omega} \geqslant C_1 |\phi|_{1,\Omega} \geqslant C_2 \|\phi\|_{0,\Omega}. \tag{3.12}$$

With this space, we can formulate for V the analogues of Corollary 3.1 and Theorem 3.2.

Corollary 3.2. *The space V verifies the identity*:

$$V = \{\mathbf{curl}\,\phi;\ \phi \in \Psi\}$$

and the mapping **curl** *is an isomorphism from Ψ onto V. Moreover, the stream function ϕ of $\mathbf{v}$ can be characterized as the only solution of*:

$$\phi \in \Psi, \qquad (\Delta\phi, \Delta\chi) = -(\operatorname{curl}\mathbf{v}, \Delta\chi) \qquad \forall \chi \in \Psi. \tag{3.13}$$

Remark 3.3. Problem (3.13) has the following interpretation:

$$\begin{cases} \Delta^2\phi = -\Delta(\operatorname{curl}\mathbf{v}) \quad \text{in } H^{-2}(\Omega), \\ \phi|_{\Gamma_0} = 0, \qquad \phi|_{\Gamma_i} = \text{a constant } c_i \quad \text{for } 1 \leqslant i \leqslant p, \\ \partial\phi/\partial n = 0 \quad \text{on } \Gamma, \\ \displaystyle\int_{\Gamma_i} \partial(\Delta\phi + \operatorname{curl}\mathbf{v})/\partial n\, ds = 0 \quad \text{for } 1 \leqslant i \leqslant p. \end{cases} \tag{3.14}$$

Indeed, by restricting χ to $\mathscr{D}(\Omega)$, we readily derive the first equality of (3.14) and by taking its scalar product with $\chi \in \Psi$, integrating by parts and comparing with (3.13) we find (formally):

$$\langle \partial(\Delta\phi + \operatorname{curl} \mathbf{v})/\partial n, \chi \rangle_\Gamma = 0 \qquad \forall \chi \in \Psi.$$

As $\chi \in \Psi$, this implies in turn (formally):

$$\int_{\Gamma_i} \partial(\Delta\phi + \operatorname{curl} \mathbf{v})/\partial n \, ds = 0 \quad \text{for } 1 \leqslant i \leqslant p.$$

Theorem 3.3. *Every function* $\mathbf{v}$ *of* $H_0^1(\Omega)^2$ *has the following orthogonal decomposition:*

(3.15) $$\mathbf{v} = (-\Delta)^{-1} \,\mathbf{grad}\, q + \mathbf{curl}\, \phi, \quad (\textit{cf. Definition } 2.1)$$

where $\phi \in \Psi$ *is the only solution of problem* (3.13) *and* q *is uniquely determined in* $L_0^2(\Omega)$ *by* (3.15).

Proof. The decomposition (3.15) follows immediately from Corollaries 2.3 and 3.2. Thus, it suffices to check that ϕ verifies (3.13). From (3.15), we infer that

$$\operatorname{curl} \mathbf{v} = -\Delta\phi + \operatorname{curl}(-\Delta)^{-1} \,\mathbf{grad}\, q.$$

Then, by multiplying both sides of this equality by $\operatorname{curl} \mathbf{w}$ for $\mathbf{w} \in \mathscr{V}$, we get

$$\begin{aligned}(\operatorname{curl} \mathbf{v}, \operatorname{curl} \mathbf{w}) &= (-\Delta\phi, \operatorname{curl} \mathbf{w}) + (\operatorname{curl}(-\Delta)^{-1} \,\mathbf{grad}\, q, \operatorname{curl} \mathbf{w}) \\ &= (-\Delta\phi, \operatorname{curl} \mathbf{w}) + ((-\Delta)^{-1} \,\mathbf{grad}\, q, \mathbf{curl}(\operatorname{curl} \mathbf{w})) \\ &= (-\Delta\phi, \operatorname{curl} \mathbf{w}) + ((-\Delta)^{-1} \,\mathbf{grad}\, q, -\Delta \mathbf{w})\end{aligned}$$

since $\operatorname{div} \mathbf{w} = 0$. Therefore

$$\begin{aligned}(\operatorname{curl} \mathbf{v}, \operatorname{curl} \mathbf{w}) &= (-\Delta\phi, \operatorname{curl} \mathbf{w}) + (\mathbf{grad}\, q, \mathbf{w}) \\ &= (-\Delta\phi, \operatorname{curl} \mathbf{w}) \qquad \forall \mathbf{w} \in \mathscr{V}.\end{aligned}$$

As $\mathscr{V}$ is dense in V, this means that

$$(-\Delta\phi, \operatorname{curl} \mathbf{w}) = (\operatorname{curl} \mathbf{v}, \operatorname{curl} \mathbf{w}) \qquad \forall \mathbf{w} \in V,$$

which is indeed equivalent to (3.13). □

3.2. Application to the Regularity of Functions of $H(\operatorname{div}; \Omega) \cap H(\mathbf{curl}; \Omega)$

We have seen in Lemma 2.5 that the functions of $H_0(\operatorname{div}; \Omega) \cap H_0(\mathbf{curl}; \Omega)$ belong to $H_0^1(\Omega)^N$ and that those of $H(\operatorname{div}; \Omega) \cap H(\mathbf{curl}; \Omega)$ belong locally to $H^1(\Omega)^N$. Nevertheless, in practice we shall often be faced with the intermediate situation of just one of these boundary conditions and the question that arises now is: what happens near the boundary Γ? If Γ is sufficiently smooth, the answer, given by Duvaut & Lions [26] and Gobert [40], is that if either $\mathbf{u} \cdot \mathbf{n} = 0$ or $\mathbf{u} \times \mathbf{n} = \mathbf{0}$

then $\mathbf{u} \in H^1(\Omega)^N$. However, their techniques do not apply when Γ has corners. It turns out that this property remains valid in this case, provided Ω has no reentrant corners (at least in two dimensions) and in two dimensions, the results of Section 3.1 supply a very simple proof.

Recall that Ω is a *connected, bounded,* open subset of $\mathbb{R}^2$. We set

$$U = H_0(\mathrm{div}; \Omega) \cap H(\mathbf{curl}; \Omega), \qquad W = H(\mathrm{div}; \Omega) \cap H_0(\mathbf{curl}; \Omega).$$

Proposition 3.1. *If Γ is of class $\mathscr{C}^{1,1}$ or if Γ is piecewise smooth with no reentrant corners, then U and W are both algebraically and topologically included in $H^1(\Omega)^2$.*

Proof. 1°) Let $\mathbf{v} \in U$; according to Theorem 3.2 and Remarks 3.1, 3.2, $\mathbf{v}$ can be split as follows:

$$\mathbf{v} = \mathbf{grad}\, q + \mathbf{curl}\, \phi, \tag{3.8}$$

where q is the solution (unique up to an additive constant) of the homogeneous Neumann's problem:

$$\begin{cases} \Delta q = \mathrm{div}\, \mathbf{v} & \text{in } \Omega, \\ \partial q/\partial n = 0 & \text{on } \Gamma \end{cases} \tag{3.16}$$

and ϕ is the solution of the non-homogeneous Dirichlet problem:

$$-\Delta \phi = \mathrm{curl}\, \mathbf{v} \quad \text{in } \Omega,$$

$$\phi|_{\Gamma_0} = 0, \qquad \phi|_{\Gamma_i} = \text{a constant } c_i,$$

where the constants c_i are determined by (3.10). As $\mathrm{div}\, \mathbf{v}$ and $\mathrm{curl}\, \mathbf{v}$ are both in $L^2(\Omega)$, and because of the hypotheses on Γ, it follows from Theorems 1.8 and 1.10 that both q and ϕ belong to $H^2(\Omega)$ and furthermore:

$$\begin{cases} \|q\|_{2,\Omega} \leqslant C_1 \|\mathrm{div}\, \mathbf{v}\|_{0,\Omega}, \\ \|\phi\|_{2,\Omega} \leqslant C_2(\|\mathrm{curl}\, \mathbf{v}\|_{0,\Omega} + \|\gamma_0 \phi\|_{3/2,\Gamma}). \end{cases} \tag{3.17}$$

Now observe that since ϕ is constant on each Γ_i and vanishes on Γ_0, we have:

$$\|\gamma_0 \phi\|_{3/2,\Gamma} \leqslant C_3 \|\phi\|_{1,\Omega} \leqslant C_4 |\phi|_{1,\Omega}.$$

In addition, by virtue of (3.10) and (3.9), we get

$$|\phi|_{1,\Omega} \leqslant \|\mathbf{v} - \mathbf{grad}\, q\|_{0,\Omega} \leqslant \|\mathbf{v}\|_{0,\Omega}.$$

Hence, collecting these inequalities and substituting into (3.8), we infer that $\mathbf{v} \in H^1(\Omega)^2$ and

$$\|\mathbf{v}\|_{1,\Omega} \leqslant C_5(\|\mathbf{v}\|_{0,\Omega} + \|\mathrm{div}\, \mathbf{v}\|_{0,\Omega} + \|\mathrm{curl}\, \mathbf{v}\|_{0,\Omega}). \tag{3.18}$$

2°) As far as the space W is concerned, we can use the simple trick of Remark 2.4: for $\mathbf{u}$ in W, the vector $\mathbf{v} = (-u_2, u_1)$ belongs to U, with $\mathrm{div}\, \mathbf{v} = -\mathrm{curl}\, \mathbf{u}$, $\mathrm{curl}\, \mathbf{v} = \mathrm{div}\, \mathbf{u}$ and $\mathbf{v} \cdot \mathbf{n} = -\mathbf{u} \cdot \boldsymbol{\tau}$. Therefore, we infer from 1°) that $\mathbf{v}$ belongs to $H^1(\Omega)^2$ and satisfies the bound (3.18); and thus the same holds for $\mathbf{u}$. $\square$

This proposition has interesting consequences and calls for some comments.

Remark 3.4. There exist polygonal domains Ω with reentrant corners for which the solution q of (3.16) is not in $H^2(\Omega)$ (cf. Grisvard [42]). Thus, it suffices to take $\phi = 0$ in (3.8) to obtain a function $\mathbf{v}$ of U that does not belong to $H^1(\Omega)^2$. Therefore, the condition on the angles of polygonal domains is in general both necessary and sufficient.

Remark 3.5. Let Ω be *simply-connected* and like in Proposition 3.1. Then, since $p = 0$, we infer from (3.17) that the following equivalence holds in U:

$$\|\mathbf{v}\|_{1,\Omega} \cong \|\operatorname{div}\mathbf{v}\|_{0,\Omega} + \|\operatorname{curl}\mathbf{v}\|_{0,\Omega}. \tag{3.19}$$

In turn, this implies that $H \cap \operatorname{Ker}(\mathbf{curl}) = \{\mathbf{0}\}$, as was previously mentioned in Remark 2.2.

On the other hand, if Ω is not simply-connected the equivalence (3.19) is no longer true for there exist non zero functions in $H \cap \operatorname{Ker}(\mathbf{curl})$.

Remark 3.6. Assume Ω is like in Proposition 3.1. Let $P \in \mathscr{L}(H^1(\Omega); H^1(\mathbb{R}^2))$ be the extension operator of Theorem 1.2, 3°). Then P applied to each component of functions of U (or W) is also an extension from U (or W) into $H^1(\mathbb{R}^2)^2$.

3.3. Decomposition of Three-Dimensional Vector Fields

Theorem 3.4. *A vector field* $\mathbf{v} \in L^2(\Omega)^3$ *satisfies*

$$\operatorname{div}\mathbf{v} = 0, \quad \langle \mathbf{v}\cdot\mathbf{n}, 1\rangle_{\Gamma_i} = 0 \quad \textit{for } 0 \leqslant i \leqslant p \tag{3.1}$$

if and only if there exists a vector potential $\boldsymbol{\phi}$ *in* $H^1(\Omega)^3$ *such that*

$$\mathbf{v} = \mathbf{curl}\,\boldsymbol{\phi}.$$

Furthermore,

$$\operatorname{div}\boldsymbol{\phi} = 0.$$

Proof. The proof is quite similar to that of Theorem 3.1.

1°) Let $\boldsymbol{\phi} \in H^1(\Omega)^3$ and $\mathbf{v} = \mathbf{curl}\,\boldsymbol{\phi}$. Then $\operatorname{div}\mathbf{v} = 0$ and we must check that $\langle \mathbf{v}\cdot\mathbf{n}, 1\rangle_{\Gamma_i} = 0$. For $0 \leqslant i \leqslant p$, let θ_i be a function of $\mathscr{D}(\mathbb{R}^3)$ satisfying:

$$0 \leqslant \theta_i(x) \leqslant 1 \quad \text{in } \mathbb{R}^3 \quad \text{and} \quad \theta_i(x) = \delta_{ij} \quad \text{in a neighborhood of } \Gamma_j.$$

We set $\mathbf{w}_i = \mathbf{curl}(\theta_i\boldsymbol{\phi})$. Obviously $\mathbf{w}_i \in L^2(\Omega)^3$ and $\operatorname{div}\mathbf{w}_i = 0$. Moreover,

$$\mathbf{w}_i\cdot\mathbf{n}|_{\Gamma_j} = 0 \quad \text{if } j \neq i, \qquad \mathbf{w}_i\cdot\mathbf{n}|_{\Gamma_i} = \mathbf{v}\cdot\mathbf{n}|_{\Gamma_i}.$$

Hence

$$\langle \mathbf{v}\cdot\mathbf{n}, 1\rangle_{\Gamma_i} = \langle \mathbf{w}_i\cdot\mathbf{n}, 1\rangle_{\Gamma} = \int_\Omega \operatorname{div}\mathbf{w}_i\,dx = 0 \quad \text{for } 0 \leqslant i \leqslant p.$$

2°) Conversely, let $\mathbf{v} \in L^2(\Omega)^3$ satisfy (3.1). As in the two-dimensional case, we extend $\mathbf{v}$ to the whole space so that the extended function $\tilde{\mathbf{v}}$ belongs to $L^2(\mathbb{R}^3)^3$, is divergence-free and has a compact support. Again, the Fourier transform of $\tilde{\mathbf{v}}$, $\mathscr{F}\tilde{\mathbf{v}}$, is holomorphic in $\mathbb{R}^3$ since the support of $\tilde{\mathbf{v}}$ is compact. In terms of Fourier transforms, the conditions $\operatorname{div} \tilde{\mathbf{v}} = 0$ and $\tilde{\mathbf{v}} = \mathbf{curl}\, \boldsymbol{\phi}$ become respectively:

$$\sum_{i=1}^{3} \mu_i \mathscr{F}\tilde{v}_i = 0, \tag{3.20}$$

and

$$\mathscr{F}\tilde{\mathbf{v}} = 2i\pi(\mu_2 \mathscr{F}\phi_3 - \mu_3 \mathscr{F}\phi_2, \mu_3 \mathscr{F}\phi_1 - \mu_1 \mathscr{F}\phi_3, \mu_1 \mathscr{F}\phi_2 - \mu_2 \mathscr{F}\phi_1). \tag{3.21}$$

Observe that if (3.20) holds then (3.21) and the following condition:

$$\sum_{i=1}^{3} \mu_i \mathscr{F}\phi_i = 0 \tag{3.22}$$

determine $\mathscr{F}\boldsymbol{\phi}$ uniquely. The unique solution of (3.20), (3.21) and (3.22) is:

$$\mathscr{F}\boldsymbol{\phi} = \frac{1}{2i\pi \|\mu\|^2}(\mu_3 \mathscr{F}\tilde{v}_2 - \mu_2 \mathscr{F}\tilde{v}_3, \mu_1 \mathscr{F}\tilde{v}_3 - \mu_3 \mathscr{F}\tilde{v}_1, \mu_2 \mathscr{F}\tilde{v}_1 - \mu_1 \mathscr{F}\tilde{v}_2).$$

Clearly, these equations imply that $\mu_j \mathscr{F}\phi_i \in L^2(\mathbb{R}^3)$ and that

$$|\mathscr{F}\phi_i| \leqslant (|\mathscr{F}\tilde{v}_j| + |\mathscr{F}\tilde{v}_k|)/(2\pi \|\mu\|).$$

Therefore, the inverse transform of $\mathscr{F}\boldsymbol{\phi}$ belongs to $H^1(\Omega)^3$, provided that $\mathscr{F}\boldsymbol{\phi}$ is bounded at the origin. First, we observe that (3.20) implies that $\mathscr{F}\tilde{\mathbf{v}}(0) = \mathbf{0}$. Hence, since $\mathscr{F}\tilde{\mathbf{v}}$ is holomorphic, it follows that, in a neighborhood of zero:

$$\mathscr{F}\tilde{\mathbf{v}}(\mu) = \sum_{j=1}^{3} \mu_j(\partial \mathscr{F}\tilde{\mathbf{v}}(0)/\partial \mu_j) + O(\|\mu\|^2).$$

Therefore $\mathscr{F}\boldsymbol{\phi}$ is bounded as μ tends to zero.

By restricting to Ω the inverse transform $\boldsymbol{\phi}$ of $\mathscr{F}\boldsymbol{\phi}$, we thus find a function $\boldsymbol{\phi}$ in $H^1(\Omega)^3$ such that $\mathbf{v} = \mathbf{curl}\, \boldsymbol{\phi}$ and moreover, $\operatorname{div} \boldsymbol{\phi} = 0$. □

Remark 3.7. The above construction shows that one can always choose a divergence-free potential vector. Of course, this is not the only possibility. For instance, we might have replaced (3.22) by:

$$\mathscr{F}\phi_3 = 0. \tag{3.22'}$$

Then (3.20), (3.21) and (3.22′) have the only solution

$$\mathscr{F}\phi_1 = \mathscr{F}\tilde{v}_2/(2i\pi\mu_3), \qquad \mathscr{F}\phi_2 = -\mathscr{F}\tilde{v}_1/(2i\pi\mu_3), \qquad \mathscr{F}\phi_3 = 0.$$

The corresponding potential vector $\boldsymbol{\phi}$ of $\mathbf{v}$ has the form $(\phi_1, \phi_2, 0)$.

Remark 3.8. Every divergence-free potential vector $\boldsymbol{\phi}$ of $\mathbf{v}$ satisfies:

$$-\Delta\boldsymbol{\phi} = \mathbf{curl}\, \mathbf{v} \quad \text{in } H^{-1}(\Omega)^3.$$

We shall study further on several boundary conditions that must be added to this equation in order to specify $\boldsymbol{\phi}$ uniquely.

Remark 3.9. A simple consequence of Theorem 3.4 is that if $\mathbf{v} \in L^2(\Omega)^3$ satisfies (3.1) and in addition $\mathbf{curl}\,\mathbf{v} = \mathbf{0}$ and $\mathbf{v} \times \mathbf{n}|_\Gamma = \mathbf{0}$ then $\mathbf{v} = \mathbf{0}$.

Remark 3.10. When Γ has only one component (i.e. $p = 0$) then (3.1) reduces to the sole condition $\operatorname{div} \mathbf{v} = 0$.

The next results give further information on the regularity of the vector potential.

Remark 3.11. When $\mathbf{v} \in H \cap L^s(\Omega)^3$ for some $s > 2$, it can be shown that the construction of Theorem 3.4 gives a vector potential $\boldsymbol{\phi}$ in $W^{1,s}(\Omega)^3$ with $\operatorname{div} \boldsymbol{\phi} = 0$.

Corollary 3.3. *Let* $\mathbf{v} \in H^1(\Omega)^3$ *satisfy* (3.1). *Then it has a divergence-free vector potential* $\boldsymbol{\phi}$ *in* $H^2(\Omega)^3$:

$$\mathbf{v} = \mathbf{curl}\,\boldsymbol{\phi}, \qquad \operatorname{div} \boldsymbol{\phi} = 0.$$

The proof is left as an exercise. (Hint: use Lemma 2.2 to extend $\mathbf{v}$ to $\mathbb{R}^3$ by matching its boundary values with that of its extension; then construct the vector potential by Fourier transforms.)

Remark 3.12. By interpolating between Theorem 3.4 and Corollary 3.3 we immediately see that if $\mathbf{v} \in H^s(\Omega)^3$, with $s \in [0, 1]$, satisfies (3.1) then it has a divergence-free vector potential $\boldsymbol{\phi}$ in $H^{s+1}(\Omega)^3$.

Remark 3.13. When Γ is of class $\mathscr{C}^{m,1}$ ($m \geqslant 2$), it can be shown that each $\mathbf{v}$ in $H^m(\Omega)^3$ satisfying (3.1) has a divergence-free vector potential $\boldsymbol{\phi}$ in $H^{m+1}(\Omega)^3$.

Now, we turn to the boundary conditions that characterize $\boldsymbol{\phi}$.

Theorem 3.5. 1°) *Let* $\mathbf{v} \in L^2(\Omega)^3$ *satisfy* (3.1); *among the vector potentials* $\boldsymbol{\phi}$ *verifying*:

$$\mathbf{curl}\,\boldsymbol{\phi} = \mathbf{v}, \qquad \operatorname{div} \boldsymbol{\phi} = 0, \tag{3.23}$$

we can choose $\boldsymbol{\phi}$ *in* $H(\mathbf{curl}; \Omega)$ *such that*:

$$\boldsymbol{\phi} \cdot \mathbf{n} = 0. \tag{3.24}$$

Moreover, when Ω *is simply-connected,* $\boldsymbol{\phi}$ *is entirely determined by* (3.23) *and* (3.24); *it can be characterized as the unique solution of the boundary value problem*:

$$\begin{cases} \boldsymbol{\phi} \in H, \\ -\Delta \boldsymbol{\phi} = \mathbf{curl}\,\mathbf{v} \quad \text{in } H^{-1}(\Omega)^3 \\ (\mathbf{curl}\,\boldsymbol{\phi} - \mathbf{v}) \cdot \mathbf{n} = 0 \quad \text{on } \Gamma. \end{cases} \tag{3.25}$$

2°) *If* Γ *is smooth* (*for instance of class* $\mathscr{C}^{1,1}$), *then* $\boldsymbol{\phi} \in H^1(\Omega)^3$.

Proof. 1°) In view of Theorem 3.4, there exists $\boldsymbol{\phi}$ in $H^1(\Omega)^3$ satisfying (3.23). Therefore, if we can find $\boldsymbol{\psi}$ in $H(\mathbf{curl};\Omega)$ such that

$$(3.26)\qquad \begin{cases} \left.\begin{aligned} \mathbf{curl}\,\boldsymbol{\psi} &= \mathbf{0} \\ \operatorname{div}\boldsymbol{\psi} &= 0 \end{aligned}\right\} \text{ in } \Omega, \\ \boldsymbol{\psi}\cdot\mathbf{n} = \boldsymbol{\phi}\cdot\mathbf{n} \quad \text{on } \Gamma, \end{cases}$$

the difference $\boldsymbol{\phi} - \boldsymbol{\psi}$ will be the desired function. Keeping this in mind, let $\chi \in H^1(\Omega)/\mathbb{R}$ be the solution of the non-homogeneous Neumann's problem:

$$(3.27)\qquad \begin{cases} \Delta\chi = 0 \quad \text{in } \Gamma, \\ \partial\chi/\partial n = \boldsymbol{\phi}\cdot\mathbf{n} \quad \text{on } \Gamma. \end{cases}$$

Then

$$\boldsymbol{\psi} = \mathbf{grad}\,\chi$$

satisfies precisely (3.26) and of course belongs to $H(\mathbf{curl};\Omega)$.

When Ω is simply-connected, the uniqueness of $\boldsymbol{\phi}$ is an obvious consequence of Remark 2.2. In addition, it is clear that $\boldsymbol{\phi}$ satisfies (3.25). Conversely (3.25) has at most one solution, for if (3.25) holds with $\mathbf{v} = \mathbf{0}$ then, on the one hand $\mathbf{curl}\,\boldsymbol{\phi} \in H$ and on the other hand $\mathbf{curl}\,\boldsymbol{\phi} \in \operatorname{Ker}(\mathbf{curl})$. Therefore $\mathbf{curl}\,\boldsymbol{\phi} = \mathbf{0}$. And since $\boldsymbol{\phi} \in H$, Remark 2.2 implies again that $\boldsymbol{\phi} = \mathbf{0}$.

2°) If Γ is of class $\mathscr{C}^{1,1}$, then $\boldsymbol{\phi}\cdot\mathbf{n} \in H^{1/2}(\Gamma)$. Therefore, Theorem 1.10 (with $k = 0$) asserts that $\chi \in H^2(\Omega)$. Hence $\mathbf{grad}\,\chi \in H^1(\Omega)^3$. □

Theorem 3.6. 1°) *Each function in* $L^2(\Omega)^3$ *that satisfies* (3.1) *has at most one divergence-free vector potential* $\boldsymbol{\phi} \in H(\mathbf{curl};\Omega)$ *that satisfies the boundary conditions:*

$$(3.28)\qquad \boldsymbol{\phi} \times \mathbf{n} = \mathbf{0} \quad \text{on } \Gamma, \qquad \langle \boldsymbol{\phi}\cdot\mathbf{n}, 1\rangle_{\Gamma_i} = 0, \qquad 0 \leqslant i \leqslant p.$$

2°) *When* Ω *is simply-connected, each* $\mathbf{v} \in H$ *has exactly one divergence-free vector potential* $\boldsymbol{\phi}$ *in* $H(\mathbf{curl};\Omega)$ *satisfying* (3.28). *It is characterized as the only solution of the boundary value problem:*

$$(3.29)\qquad \begin{cases} -\Delta\boldsymbol{\phi} = \mathbf{curl}\,\mathbf{v} \quad \text{in } H^{-1}(\Omega)^3, \\ \operatorname{div}\boldsymbol{\phi} = 0 \quad \text{in } \Omega, \\ \boldsymbol{\phi} \text{ satisfies } (3.28). \end{cases}$$

3°) *In addition to the above hypotheses, if* Γ *is* $\mathscr{C}^{1,1}$ *or if* Ω *is a convex polyhedron, then this vector potential* $\boldsymbol{\phi}$ *belongs to* $H^1(\Omega)^3$.

Proof. 1°) We proceed by contradiction. Assume that $\boldsymbol{\phi} \in H(\mathbf{curl};\Omega)$ satisfies:

$$\mathbf{curl}\,\boldsymbol{\phi} = \mathbf{0} \quad \text{in } \Omega, \qquad \boldsymbol{\phi} \times \mathbf{n} = \mathbf{0} \quad \text{on } \Gamma,$$

$$\operatorname{div}\boldsymbol{\phi} = 0 \quad \text{in } \Omega, \qquad \langle \boldsymbol{\phi}\cdot\mathbf{n}, 1\rangle_{\Gamma_i} = 0 \quad \text{for } 0 \leqslant i \leqslant p.$$

Then Remark 3.9 says that $\boldsymbol{\phi} = \mathbf{0}$.

2°) Let $\mathbf{v} \in H$ and let $\tilde{\mathbf{v}}$ denote its extension by zero outside Ω. Then $\operatorname{div} \tilde{\mathbf{v}} = 0$ in $\mathbb{R}^3$ and since the support of $\tilde{\mathbf{v}}$ is compact we can apply Theorem 3.4, i.e.: there exists $\boldsymbol{\psi} \in H^1(\mathbb{R}^3)^3$ such that

$$\tilde{\mathbf{v}} = \mathbf{curl}\, \boldsymbol{\psi}, \qquad \operatorname{div} \boldsymbol{\psi} = 0 \quad \text{in } \mathbb{R}^3.$$

The idea is to construct first a function $\mathbf{w}$ with vanishing divergence and rotation and with the same tangential components as $\boldsymbol{\psi}$. Then the difference $\boldsymbol{\psi} - \mathbf{w}$ will satisfy the first boundary condition in (3.28). The next step will consist in showing that among all such functions $\mathbf{w}$ there is one which has the same average normal component on Γ_i as $\boldsymbol{\psi}$; and therefore their difference will satisfy all the requirements.

To begin with, we observe that

$$\mathbf{curl}\, \boldsymbol{\psi} = \mathbf{0} \quad \text{in } \mathbb{R}^3 - \bar{\Omega}.$$

Now, let $\mathcal{O}$ be an open ball containing $\bar{\Omega}$ and let Ω_i denote the component of $\mathcal{O} - \bar{\Omega}$ bounded by Γ_i like in Figure 2. Since Ω is simply-connected, each Ω_i is also simply-connected (cf. for example Bernardi [8]) and therefore Theorem 2.9 implies that there exist $q_i \in H^1(\Omega_i)$ such that:

$$\boldsymbol{\psi} = \mathbf{grad}\, q_i \quad \text{in } \Omega_i \quad \text{for } 0 \leqslant i \leqslant p. \tag{3.30}$$

In addition, $q_i \in H^2(\Omega_i)$ since $\boldsymbol{\psi} \in H^1(\mathcal{O})^3$. Next let $\chi \in H^1(\Omega)$ be the solution of the non-homogeneous Dirichlet problem

$$\begin{cases} \Delta\chi = 0 & \text{in } \Omega, \\ \chi|_{\Gamma_i} = q_i & \text{for } 0 \leqslant i \leqslant p \end{cases} \tag{3.31}$$

and set

$$\mathbf{q} = (q_0, \ldots, q_p).$$

Clearly, the mapping $\mathbf{q} \to \chi$ belongs to $\mathscr{L}(\prod_{i=0}^{p} H^2(\Omega_i); H^1(\Omega))$ and the set $\{\chi, q_0, q_1, \ldots, q_p\}$ is an extension of χ in $H^1(\mathcal{O})$. Finally, we take

$$\mathbf{w} = \mathbf{grad}\, \chi;$$

we have:

$$\operatorname{div} \mathbf{w} = 0, \qquad \mathbf{curl}\, \mathbf{w} = \mathbf{0} \quad \text{in } \Omega$$

and

$$\mathbf{w} \times \mathbf{n} = \mathbf{grad}\, \chi \times \mathbf{n} = \mathbf{grad}\, q_i \times \mathbf{n} = \boldsymbol{\psi} \times \mathbf{n} \quad \text{on } \Gamma_i, \qquad 0 \leqslant i \leqslant p.$$

Hence, the function $\boldsymbol{\psi} - \mathbf{w}$ verifies (3.23) and the first condition in (3.28).

Of course, the function $\mathbf{w}$ is not unique because (3.30) fixes each q_i up to an additive constant. Our purpose now is to prove that these $p + 1$ constants may be chosen so that

$$\langle (\boldsymbol{\psi} - \mathbf{w}) \cdot \mathbf{n}, 1 \rangle_{\Gamma_i} = 0 \quad \text{for } 0 \leqslant i \leqslant p.$$

For this, let $\mathbf{q}$ denote an arbitrary (but fixed) representative of the solution of (3.30) and, just here, let $\Delta^{-1}(\mathbf{q})$ denote the corresponding solution of (3.31). For $\mathbf{c} \in \mathbb{R}^{p+1}$, we set:

$$\chi(\mathbf{c}) = \Delta^{-1}(\mathbf{q}) + \Delta^{-1}(\mathbf{c}),$$

and

$$\mathbf{w}(\mathbf{c}) = \mathbf{grad}\,\chi(\mathbf{c}) = \mathbf{grad}\,\Delta^{-1}(\mathbf{q}) + \mathbf{grad}\,\Delta^{-1}(\mathbf{c}).$$

Let us examine more closely the properties of $\mathbf{grad}\,\Delta^{-1}(\mathbf{c})$. First, we observe that the set $E = \{\mathbf{grad}\,\Delta^{-1}(\mathbf{c}); \mathbf{c} \in \mathbb{R}^{p+1}\}$ is a p-dimensional subspace of

$$F = \{\mathbf{v} \in L^2(\Omega)^N;\ \operatorname{div} \mathbf{v} = 0, \mathbf{curl}\,\mathbf{v} = \mathbf{0}, \mathbf{v} \times \mathbf{n}|_\Gamma = \mathbf{0}\}$$

because Δ^{-1} is an isomorphism and Ker(**grad**) has dimension 1. Next, we note that:

$$\langle \mathbf{v} \cdot \mathbf{n}, 1 \rangle_{\Gamma_0} = -\sum_{i=1}^{p} \langle \mathbf{v} \cdot \mathbf{n}, 1 \rangle_{\Gamma_i} \quad \text{for all divergence-free } \mathbf{v}.$$

Hence, if we prove that the mapping $\mathbf{v} \to (\langle \mathbf{v} \cdot \mathbf{n}, 1 \rangle_{\Gamma_i})_{1 \leqslant i \leqslant p}$ from E into $\mathbb{R}^p$ is injective, it will be automatically surjective since E and $\mathbb{R}^p$ have the same dimension. But it follows immediately from 1°) that this mapping is indeed injective. Therefore, we can find $\mathbf{c}$ in $\mathbb{R}^{p+1}$ such that

$$\langle \mathbf{w}(\mathbf{c}) \cdot \mathbf{n}, 1 \rangle_{\Gamma_i} = \langle \boldsymbol{\psi} \cdot \mathbf{n}, 1 \rangle_{\Gamma_i} \quad \text{for } 0 \leqslant i \leqslant p,$$

and the function $\boldsymbol{\phi} = \boldsymbol{\psi} - \mathbf{w}(\mathbf{c})$ does satisfy (3.23), (3.28) and also (3.29).

Finally, let us prove that (3.29) has at most one solution. When $\mathbf{v} = \mathbf{0}$, we have $\mathbf{curl}\,\boldsymbol{\phi} \in \operatorname{Ker}(\mathbf{curl})$ and also $\mathbf{curl}\,\boldsymbol{\phi} \in H$, owing to Remark 2.5. Hence, Remark 2.2 implies that $\mathbf{curl}\,\boldsymbol{\phi} = \mathbf{0}$. Therefore, according to Remark 3.9, $\boldsymbol{\phi} = \mathbf{0}$.

3°) Since the functions q_i belong to $H^2(\Omega_i)$, it follows from Theorem 1.8 that the solution χ of (3.31) belongs to $H^2(\Omega)$ provided either Γ is $\mathscr{C}^{1,1}$ or Ω is a convex polyhedron. Therefore $\mathbf{w} \in H^1(\Omega)^3$ and so does $\boldsymbol{\phi}$. □

Remark 3.14. When Ω is a bounded, convex polyhedron and $\mathbf{v} \in H \cap L^s(\Omega)^3$ for some $s > 2$ we derive from Theorem 3.6 and Remark 3.11 that its unique vector potential $\boldsymbol{\phi}$ satisfying

$$\operatorname{div} \boldsymbol{\phi} = 0, \qquad \boldsymbol{\phi} \times \mathbf{n} = \mathbf{0} \quad \text{on } \Gamma$$

has the extra regularity: $\boldsymbol{\phi} \in W^{1,p}(\Omega)^3$ for some p with $2 < p \leqslant s$. Indeed the regularity of $\boldsymbol{\phi}$ is determined, through Problem (3.31), by the regularity of the solution of a Dirichlet problem (for $-\Delta$) with right-hand side in $L^s(\Omega)$ (cf. Theorem 1.8).

Like in the two-dimensional case, we derive from Theorem 3.6 the following decomposition of $L^2(\Omega)^3$ and $H_0^1(\Omega)^3$.

Corollary 3.4. *Every function* $\mathbf{v}$ *of* $L^2(\Omega)^3$ *has the orthogonal decomposition*:

$$\mathbf{v} = \mathbf{grad}\, q + \mathbf{curl}\,\boldsymbol{\phi}, \tag{3.22}$$

where $q \in H^1(\Omega)/\mathbb{R}$ *is the only solution of* (3.9) *and* $\boldsymbol{\phi} \in H^1(\Omega)^3$ *satisfies* $\mathbf{curl}\,\boldsymbol{\phi} \in H$ *and* $\operatorname{div} \boldsymbol{\phi} = 0$. *In addition, if* Ω *is simply-connected, we can choose for* $\boldsymbol{\phi}$ *the only solution of* (3.29) *in* $H(\mathbf{curl}; \Omega)$.

Corollary 3.5. *Every function* $\mathbf{v}$ *of* $H_0^1(\Omega)^3$ *has the orthogonal decomposition:*

$$\mathbf{v} = (-\Delta)^{-1}\,\mathbf{grad}\, q + \mathbf{curl}\,\boldsymbol{\phi}, \tag{3.33}$$

where $\boldsymbol{\phi} \in H^2(\Omega)^3$ *with* $\operatorname{div} \boldsymbol{\phi} = 0$ *and* $\mathbf{curl}\,\boldsymbol{\phi} \in V$ *is a solution of*

$$(-\Delta\boldsymbol{\phi}, \mathbf{curl}\,\mathbf{w}) = (\mathbf{curl}\,\mathbf{v}, \mathbf{curl}\,\mathbf{w}) \quad \forall \mathbf{w} \in V, \tag{3.34}$$

and $q \in L_0^2(\Omega)$ *is uniquely determined by* (3.33). *Moreover, when* Ω *is simply-connected, problem* (3.34) *has exactly one solution* $\boldsymbol{\phi}$ *in* H *with* $\mathbf{curl}\,\boldsymbol{\phi}$ *in* V.

3.4. The Imbedding of $H(\operatorname{div}; \Omega) \cap H_0(\mathbf{curl}; \Omega)$ into $H^1(\Omega)^3$

Throughout this section, we assume that Ω is *bounded* and that *either* Γ *is* $\mathscr{C}^{1,1}$ *or* that Ω is a *convex polyhedron.* Like in Section 3.2, we set:

$$W = H(\operatorname{div}; \Omega) \cap H_0(\mathbf{curl}; \Omega), \qquad W_0 = \{\mathbf{w} \in W;\ \operatorname{div} \mathbf{w} = 0\}.$$

Lemma 3.3. *The space* W *is continuously imbedded into* $H^1(\Omega)^3$ *iff* W_0 *is also continuously imbedded into* $H^1(\Omega)^3$.

Proof. Let $\boldsymbol{\phi} \in W$ and let $q \in H^1(\Omega)$ be the solution of the Dirichlet problem

$$\Delta q = \operatorname{div} \boldsymbol{\phi} \quad \text{in } \Omega, \qquad q = 0 \quad \text{on } \Gamma.$$

Then

$$\|\mathbf{grad}\, q - \boldsymbol{\phi}\|_{0,\Omega} \leqslant \|\boldsymbol{\phi}\|_{0,\Omega}$$

and the assumptions on Ω imply that $q \in H^2(\Omega)$ with

$$\|q\|_{2,\Omega} \leqslant C_1 \|\operatorname{div} \boldsymbol{\phi}\|_{0,\Omega}.$$

Moreover, we have $(\mathbf{grad}\, q) \times \mathbf{n} = \mathbf{0}$ on Γ. Therefore the function

$$\boldsymbol{\psi} = \boldsymbol{\phi} - \mathbf{grad}\, q$$

belongs to W_0. Hence, if W_0 is continuously imbedded in $H^1(\Omega)^3$, then $\boldsymbol{\phi} \in H^1(\Omega)^3$ and

$$\begin{aligned}
\|\boldsymbol{\phi}\|_{1,\Omega} &\leqslant \|\boldsymbol{\psi}\|_{1,\Omega} + C_1 \|\operatorname{div} \boldsymbol{\phi}\|_{0,\Omega} \\
&\leqslant C_2(\|\boldsymbol{\psi}\|_{0,\Omega} + \|\mathbf{curl}\,\boldsymbol{\psi}\|_{0,\Omega}) + C_1 \|\operatorname{div} \boldsymbol{\phi}\|_{0,\Omega} \\
&\leqslant C_2(\|\boldsymbol{\phi}\|_{0,\Omega} + \|\mathbf{curl}\,\boldsymbol{\phi}\|_{0,\Omega}) + C_1 \|\operatorname{div} \boldsymbol{\phi}\|_{0,\Omega},
\end{aligned}$$

thus showing that the imbedding is also continuous. The converse is obvious. □

Lemma 3.4. *Here we assume only that Ω is bounded, Lipschitz-continuous, simply-connected and Γ has just one component ($p = 0$). Then the mapping $\boldsymbol{\phi} \to \mathbf{curl}\,\boldsymbol{\phi}$ is an isomorphism from W_0 onto H and there exist two positive constants C_1 and C_2 such that:*

$$\|\boldsymbol{\phi}\|_{0,\Omega} \leqslant C_1 \|\mathbf{curl}\,\boldsymbol{\phi}\|_{0,\Omega} \quad \forall \boldsymbol{\phi} \in W_0, \tag{3.35}$$

$$\|\boldsymbol{\phi}\|_{0,\Omega} \leqslant C_2\{\|\mathbf{curl}\,\boldsymbol{\phi}\|_{0,\Omega} + \|\operatorname{div}\boldsymbol{\phi}\|_{0,\Omega}\} \quad \forall \boldsymbol{\phi} \in W. \tag{3.36}$$

Proof. Since Ω is simply-connected and $p = 0$, Theorem 3.6 claims that each function $\mathbf{u}$ in H has exactly one vector potential $\boldsymbol{\phi}$ in W_0:

$$\mathbf{u} = \mathbf{curl}\,\boldsymbol{\phi}.$$

Therefore, the mapping **curl** is one-to-one, linear and continuous from W_0 onto H. As W_0 and H are Banach spaces, it follows that it is an isomorphism. Furthermore, the mapping $\boldsymbol{\phi} \to \|\mathbf{u}\|_{H(\operatorname{div};\Omega)} = \|\mathbf{curl}\,\boldsymbol{\phi}\|_{0,\Omega}$ is a norm on W_0 equivalent to the norm of W. This establishes (3.35).

Now, (3.36) is an easy consequence of (3.35) and the decomposition of Lemma 3.3: $\boldsymbol{\phi}$ in W can be written as

$$\boldsymbol{\phi} = \boldsymbol{\psi} + \mathbf{grad}\,q,$$

where $\boldsymbol{\psi} \in W_0$ and, by virtue of Poincaré's Theorem 1.1,

$$|q|_{1,\Omega} \leqslant C\|\operatorname{div}\boldsymbol{\phi}\|_{0,\Omega}.$$

Hence

$$\|\boldsymbol{\phi}\|_{0,\Omega} \leqslant C_1\|\mathbf{curl}\,\boldsymbol{\psi}\|_{0,\Omega} + C\|\operatorname{div}\boldsymbol{\phi}\|_{0,\Omega},$$

and the result follows since $\mathbf{curl}\,\boldsymbol{\psi} = \mathbf{curl}\,\boldsymbol{\phi}$. Note that, although we use here part of the proof of Lemma 3.3, we do not require its hypotheses on Ω for we do not ask q to belong to $H^2(\Omega)$. □

Theorem 3.7. *The space W is continuously imbedded into $H^1(\Omega)^3$.*

Proof. 1°) Assume for the moment that Ω is simply-connected and $p = 0$. According to Lemma 3.3, it is equivalent to examine W_0. By virtue of Lemma 3.4, the mapping **curl** is an isomorphism from W_0 onto H. But, since the hypotheses of Theorem 3.6, 3°) are satisfied, **curl** is also an isomorphism from $W_0 \cap H^1(\Omega)^3$ onto H. Hence, algebraically and topologically, we have

$$W_0 = W_0 \cap H^1(\Omega)^3$$

and therefore W_0 is continuously imbedded into $H^1(\Omega)^3$.

2°) When Ω is arbitrary, it is no longer convex and thus, by hypothesis, Γ must be $\mathscr{C}^{1,1}$. Now, we can easily find a finite open covering $\{\mathcal{O}_i\}_{1 \leqslant i \leqslant l}$ of $\bar{\Omega}$ such that each $\mathcal{O}_i \cap \Omega$ is simply-connected and has a connected boundary. Let $\{\alpha_i\}$ be a partition of unity subordinate to $\{\mathcal{O}_i\}$; then each $\boldsymbol{\phi}$ in W can be expressed as

$$\boldsymbol{\phi} = \sum_{i=1}^{l} \alpha_i \boldsymbol{\phi} \quad \text{in } \Omega$$

and therefore the problem reduces to the regularity of $\alpha_i \boldsymbol{\phi}$. Note that, since Γ is smooth and $\alpha_i \in \mathscr{D}(\mathcal{O}_i)$, we can consider that $\alpha_i \boldsymbol{\phi}$ is defined on a subset of $\mathcal{O}_i \cap \Omega$, say $\mathcal{O}'_i$, with a smooth boundary and the same properties as $\mathcal{O}_i \cap \Omega$. In addition, we have:

$$\alpha_i(\boldsymbol{\phi} \times \mathbf{n}) = \mathbf{0} \quad \text{on } \partial\mathcal{O}'_i$$

as either factor vanishes on each point of $\partial\mathcal{O}'_i$. Hence $\alpha_i \boldsymbol{\phi} \in W$ on $\mathcal{O}'_i$ and we infer from 1°) that $\alpha_i \boldsymbol{\phi} \in H^1(\mathcal{O}'_i)^3$ with

$$\|\alpha_i \boldsymbol{\phi}\|_1 \leqslant C_i\{\|\alpha_i \boldsymbol{\phi}\|_0 + \|\operatorname{div}(\alpha_i \boldsymbol{\phi})\|_0 + \|\mathbf{curl}(\alpha_i \boldsymbol{\phi})\|_0\},$$

where all norms are taken over $\mathcal{O}'_i$. Therefore, with the same notation,

$$\|\alpha_i \boldsymbol{\phi}\|_1 \leqslant C'_i\{\|\boldsymbol{\phi}\|_0 + \|\operatorname{div} \boldsymbol{\phi}\|_0 + \|\mathbf{curl}\, \boldsymbol{\phi}\|_0\},$$

thus proving the theorem. □

3.5. The Imbedding of $H_0(\text{div}; \Omega) \cap H(\mathbf{curl}; \Omega)$ into $H^1(\Omega)^3$

Again we keep the hypotheses of Section 3.4 and we set:

$$U = H_0(\operatorname{div}; \Omega) \cap H(\mathbf{curl}; \Omega), \qquad U_0 = \{\mathbf{u} \in U; \operatorname{div} \mathbf{u} = 0\}.$$

We begin with two results analogous to Lemmas 3.3 and 3.4.

Lemma 3.5. *The space U is continuously imbedded into $H^1(\Omega)^3$ iff U_0 is also continuously imbedded into $H^1(\Omega)^3$.*

The proof, which follows the lines of Lemma 3.3, is left to the reader.

Lemma 3.6. *Suppose that Ω is bounded, Lipschitz-continuous and simply-connected. Then the mapping $\boldsymbol{\phi} \to \mathbf{curl}\, \boldsymbol{\phi}$ is an isomorphism from the space U_0 onto the space:*

$$T = \{\mathbf{u} \in L^2(\Omega)^3; \operatorname{div} \mathbf{u} = 0, \langle \mathbf{u} \cdot \mathbf{n}, 1 \rangle_{\Gamma_i} = 0, 0 \leqslant i \leqslant p\};$$

and there exist two positive constants C_1 and C_2 such that

$$\text{(3.37)} \qquad \|\boldsymbol{\phi}\|_{0,\Omega} \leqslant C_1 \|\mathbf{curl}\, \boldsymbol{\phi}\|_{0,\Omega} \quad \forall \boldsymbol{\phi} \in U_0,$$

$$\text{(3.38)} \qquad \|\boldsymbol{\phi}\|_{0,\Omega} \leqslant C_2\{\|\mathbf{curl}\, \boldsymbol{\phi}\|_{0,\Omega} + \|\operatorname{div} \boldsymbol{\phi}\|_{0,\Omega}\} \quad \forall \boldsymbol{\phi} \in U.$$

Proof. Since Ω is simply-connected, we infer from Theorem 3.5 and the argument of Lemma 3.4 that the mapping $\boldsymbol{\phi} \to \mathbf{curl}\, \boldsymbol{\phi}$ is an isomorphism from U_0 onto T and that (3.37) holds. In turn, (3.38) follows readily from (3.37), the decomposition (3.32) and Theorem 1.9. Note that here again there is no continuity requirement on Γ, except of course the Lipschitz-continuity. □

Lemma 3.7. *If U is continuously imbedded into $H^1(\Omega)^3$, then for each $\boldsymbol{\psi} \in H^1(\Omega)^3$ with $\int_\Gamma \boldsymbol{\psi} \cdot \mathbf{n}\, ds = 0$, the solution χ of the non-homogeneous Neumann's problem:*

$$(3.39) \qquad \Delta\chi = 0 \quad \text{in } \Omega, \qquad \partial\chi/\partial n = \boldsymbol{\psi} \cdot \mathbf{n} \quad \text{on } \Gamma$$

belongs to $H^2(\Omega)/\mathbb{R}$. Conversely, when Ω is simply-connected, this condition is sufficient for the above imbedding.

Proof. 1°) Let $\boldsymbol{\psi} \in H^1(\Omega)^3$ with $\int_\Gamma \boldsymbol{\psi} \cdot \mathbf{n}\, ds = 0$; then (3.39) has a unique solution χ in $H^1(\Omega)/\mathbb{R}$ and

$$\boldsymbol{\phi} = \boldsymbol{\psi} - \mathbf{grad}\,\chi$$

belongs to U. Therefore, if $\boldsymbol{\phi}$ belongs to $H^1(\Omega)^3$, then χ is necessarily in $H^2(\Omega)/\mathbb{R}$.

2°) Conversely, we derive from the proof of Theorem 3.5 that, if $\chi \in H^2(\Omega)/\mathbb{R}$, then each function of T has a unique vector potential in $U_0 \cap H^1(\Omega)^3$. Hence Lemmas 3.5 and 3.6 imply that U is continuously imbedded in $H^1(\Omega)^3$. □

Now, when Γ is $\mathscr{C}^{1,1}$, the solution of (3.39) belongs indeed to $H^2(\Omega)/\mathbb{R}$ and therefore Lemma 3.7 guarantees that $U \hookrightarrow H^1(\Omega)^3$, provided that Ω is simply-connected. Furthermore, this last restriction can be removed by using the argument of Theorem 3.7,2°). Thus, we have established the following result.

Theorem 3.8. *If Ω is bounded and has a $\mathscr{C}^{1,1}$ boundary, then the space U is continuously imbedded in $H^1(\Omega)^3$.*

Unfortunately, when Γ is a polyhedron and $\boldsymbol{\psi}$ belongs to $H^1(\Omega)^3$, its normal component $\boldsymbol{\psi} \cdot \mathbf{n}$ is not in $H^{1/2}(\Gamma)$ and we have not been able to find in the literature any direct information about the regularity of the solution of (3.39). In this case, Lemma 3.7 seems to be of no help and we propose to establish the imbedding of U into $H^1(\Omega)^3$ by means of a different argument due to Nédélec [60]. Briefly speaking, this author uses a particular equivalence of norms which, on a smooth and convex domain Ω, holds with *constants independent* of Ω. Then, he extends this equivalence to convex polyhedrons by regularizing their boundaries. The starting point of this approach is the following technical lemma; the proof, which is too long to be included here, can be found in Grisvard [42].

Lemma 3.8. *Let Ω be a bounded open subset of $\mathbb{R}^3$ with a $\mathscr{C}^2$ boundary Γ. Every function $\boldsymbol{\phi}$ of $H^1(\Omega)^3$ with $\boldsymbol{\phi} \cdot \mathbf{n} = 0$ on Γ satisfies:*

$$(3.40) \qquad |\boldsymbol{\phi}|_{1,\Omega}^2 + \int_\Gamma (\mathscr{R}\boldsymbol{\phi}, \boldsymbol{\phi})\, ds = \|\mathbf{curl}\,\boldsymbol{\phi}\|_{0,\Omega}^2 + \|\operatorname{div} \boldsymbol{\phi}\|_{0,\Omega}^2,$$

where $\mathscr{R}$ denotes the curvature tensor on the tangent plane to Γ.

Since the curvature tensor $\mathscr{R}$ is positive definite when Ω is convex, we derive immediately the equivalence of norms:

Corollary 3.6. *Let Ω be a bounded and convex open subset of $\mathbb{R}^3$ with a $\mathscr{C}^2$ boundary Γ. The functions $\boldsymbol{\phi}$ of $U \cap H^1(\Omega)^3$ satisfy the inequality:*

$$|\boldsymbol{\phi}|_{1,\Omega}^2 \leqslant \|\mathbf{curl}\,\boldsymbol{\phi}\|_{0,\Omega}^2 + \|\operatorname{div}\boldsymbol{\phi}\|_{0,\Omega}^2. \tag{3.41}$$

Theorem 3.9. *Let Ω be a bounded and convex polyhedron. The space U is continuously imbedded in $H^1(\Omega)^3$ and the functions of U satisfy* (3.41).

Proof. It is possible to construct (cf. Grisvard [42]) an increasing sequence $(\Omega_l)_{l\geqslant 1}$ of open convex sets, each with a $\mathscr{C}^2$ boundary Γ_l and such that

$$\bar{\Omega}_l \subset \Omega \quad \forall l \geqslant 1, \quad \Omega = \bigcup_{l\geqslant 1} \Omega_l.$$

By virtue of Lemma 3.5, we can restrict our attention to U_0. Thus, let $\boldsymbol{\phi} \in U_0$, let $\chi_l \in H^1(\Omega_l)/\mathbb{R}$ denote the solution of

$$\Delta \chi_l = 0 \quad \text{in } \Omega_l, \qquad \partial\chi_l/\partial n = \boldsymbol{\phi}\cdot\mathbf{n} \quad \text{on } \Gamma_l, \tag{3.42}$$

and define $\boldsymbol{\phi}_l$ on Ω_l by

$$\boldsymbol{\phi}_l = \boldsymbol{\phi} - \mathbf{grad}\,\chi_l.$$

Note that the non-homogeneous Neumann's problem (3.42) has a unique solution χ_l, since $\operatorname{div}\boldsymbol{\phi} = 0$ in Ω. Furthermore, owing to Corollary 2.10, $\boldsymbol{\phi} \in H^1(\Omega_l)^3 \,\forall l$; and since Γ_l is smooth this, in turn, implies that $\boldsymbol{\phi}\cdot\mathbf{n} \in H^{1/2}(\Gamma_l)$. Hence $\chi_l \in H^2(\Omega_l)/\mathbb{R}$ and therefore $\boldsymbol{\phi}_l \in H^1(\Omega_l)^3$ with $\boldsymbol{\phi}_l\cdot\mathbf{n} = 0$ on Γ_l. Thus we can apply Corollary 3.6 to $\boldsymbol{\phi}_l$:

$$|\boldsymbol{\phi}_l|_{1,\Omega_l} \leqslant \|\mathbf{curl}\,\boldsymbol{\phi}_l\|_{0,\Omega_l} \leqslant \|\mathbf{curl}\,\boldsymbol{\phi}\|_{0,\Omega} \quad \forall l. \tag{3.43}$$

In addition, (3.42) is equivalent to

$$\int_{\Omega_l} \boldsymbol{\phi}_l\cdot\mathbf{grad}\,\mu\,dx = 0 \quad \forall \mu \in H^1(\Omega_l).$$

Hence, we easily derive that

$$\|\boldsymbol{\phi}_l\|_{0,\Omega_l} \leqslant \|\boldsymbol{\phi}\|_{0,\Omega}, \tag{3.44}$$

$$\|\mathbf{grad}\,\chi_l\|_{0,\Omega_l} \leqslant \|\boldsymbol{\phi}\|_{0,\Omega}, \tag{3.45}$$

$$\left|\int_{\Omega_l} \mathbf{grad}\,\chi_l\cdot\mathbf{grad}\,\mu\,dx\right| \leqslant \|\boldsymbol{\phi}\|_{0,\Omega-\Omega_l}\|\mathbf{grad}\,\mu\|_{0,\Omega-\Omega_l} \quad \forall \mu \in H^1(\Omega). \tag{3.46}$$

Let us extend $\mathbf{grad}\,\chi_l$, $\boldsymbol{\phi}_l$ and $\partial\boldsymbol{\phi}_l/\partial x_i$ by zero outside Ω_l and, as usual, let us denote the extended functions by a tilde. Then (3.45) implies that

$$\widetilde{\mathbf{grad}\,\chi_l} \to \mathbf{w} \quad \text{weakly in } L^2(\Omega)^3 \quad \text{with } \mathbf{curl}\,\mathbf{w} = \mathbf{0}.$$

Indeed, if $\boldsymbol{\lambda} \in \mathscr{D}(\Omega)^3$, there exists an integer k such that $\boldsymbol{\lambda} \in \mathscr{D}(\Omega_l)^3 \,\forall l \geqslant k$. Then

$$\int_{\Omega} \mathbf{w} \cdot \mathbf{curl}\, \boldsymbol{\lambda}\, dx = \lim_{l \to \infty} \int_{\Omega} \mathbf{grad}\, \chi_l \cdot \mathbf{curl}\, \boldsymbol{\lambda}\, dx$$

and the result follows from the fact that

$$\int_{\Omega_l} \mathbf{grad}\, \chi_l \cdot \mathbf{curl}\, \boldsymbol{\lambda}\, dx = 0 \qquad \forall l \geqslant k.$$

Since Ω is simply-connected, this means that there exists χ in $H^1(\Omega)$ such that

$$\mathbf{w} = \mathbf{grad}\, \chi.$$

In addition, (3.46) implies that $\mathbf{grad}\, \chi = \mathbf{0}$. Hence, we see that

$$\tilde{\boldsymbol{\phi}}_l \to \boldsymbol{\phi} \quad \text{weakly in } L^2(\Omega)^3.$$

Together with (3.43), this implies that

$$\partial \tilde{\boldsymbol{\phi}}_l / \partial x_i \to \partial \boldsymbol{\phi} / \partial x_i \quad \text{weakly in } L^2(\Omega)^3, \qquad 1 \leqslant i \leqslant 3.$$

Thus $\boldsymbol{\phi} \in H^1(\Omega)^3$ and we derive from (3.43) that

$$|\boldsymbol{\phi}|_{1,\Omega} \leqslant \|\mathbf{curl}\, \boldsymbol{\phi}\|_{0,\Omega}. \qquad \square$$

As an immediate consequence of this theorem and Lemma 3.7, we see on the one hand that if Ω is a bounded and convex polyhedron, the solution χ of problem (3.39) is automatically in $H^2(\Omega)$. On the other hand, if Ω is a non convex polyhedron, there are cases where the solution of (3.39) is not in $H^2(\Omega)$ and hence U is not included in $H^1(\Omega)^3$.

As an interesting consequence of Theorem 3.8, we have the following corollary established by Foias & Temam [27]:

Corollary 3.7. *Let Ω be a bounded, open region of $\mathbb{R}^3$ with a $\mathscr{C}^{m,1}$ boundary Γ ($m \geqslant 1$). We have:*

$$H^m(\Omega)^3 = \{\mathbf{v} \in L^2(\Omega)^3;\ \operatorname{div} \mathbf{v} \in H^{m-1}(\Omega),\ \mathbf{curl}\, \mathbf{v} \in H^{m-1}(\Omega)^3,\ \mathbf{v} \cdot \mathbf{n} \in H^{m-1/2}(\Gamma)\}, \tag{3.47}$$

$$\|\mathbf{v}\|_{m,\Omega} \leqslant C\{\|\mathbf{v}\|_{0,\Omega} + \|\operatorname{div} \mathbf{v}\|_{m-1,\Omega} + \|\mathbf{curl}\, \mathbf{v}\|_{m-1,\Omega} + \|\mathbf{v} \cdot \mathbf{n}\|_{m-1/2,\Gamma}\}. \tag{3.48}$$

§ 4. Analysis of an Abstract Variational Problem

In this paragraph, we construct an abstract framework well adapted to the solution of a variety of linear boundary value problems with a constraint, like the Stokes problem. Several algorithms are proposed to deal with the constraint. Although they are introduced in connection with the continuous problem, they will prove to be useful mainly for solving the discretized problems.

4.1. A General Result

Let X and M be two (real) Hilbert spaces with norms $\|\,.\,\|_X$ and $\|\,.\,\|_M$ respectively. Let X' and M' be their corresponding dual spaces and let $\|\cdot\|_{X'}$ and $\|\,.\,\|_{M'}$ denote their dual norms. As usual, we denote by $\langle .,.\rangle$ the duality pairing between the spaces X and X' or M and M'.

We introduce two bilinear continuous forms:

$$a(.,.): X \times X \to \mathbb{R}, \qquad b(.,.): X \times M \to \mathbb{R},$$

with norms

$$\|a\| = \sup_{u,v \in X, u, v \neq 0} \frac{a(u,v)}{\|u\|_X \|v\|_X}, \qquad \|b\| = \sup_{v \in X, \mu \in M, v \neq 0, \mu \neq 0} \frac{b(v,\mu)}{\|v\|_X \|\mu\|_M}.$$

Then, we consider the following variational problem, called *Problem* (*Q*):

Given $l \in X'$ *and* $\chi \in M'$, *find a pair* $(u,\lambda) \in X \times M$ *such that*

$$a(u,v) + b(v,\lambda) = \langle l, v\rangle \quad \forall v \in X, \tag{4.1}$$

$$b(u,\mu) = \langle \chi, \mu\rangle \quad \forall \mu \in M. \tag{4.2}$$

In order to study Problem (Q), we require some extra notations. With the bilinear forms $a(.,.)$ and $b(.,.)$, we associate two linear operators $A \in \mathscr{L}(X;X')$ and $B \in \mathscr{L}(X,M')$ defined by:

$$\langle Au, v\rangle = a(u,v) \quad \forall u, v \in X, \tag{4.3}$$

$$\langle Bv, \mu\rangle = b(v,\mu) \quad \forall v \in X, \quad \forall \mu \in M. \tag{4.4}$$

Let $B' \in \mathscr{L}(M;X')$ be the dual operator of B, i.e.

$$\langle B'\mu, v\rangle = \langle \mu, Bv\rangle = b(v,\mu) \quad \forall v \in X, \quad \forall \mu \in M. \tag{4.5}$$

It can be readily verified that

$$\|A\|_{\mathscr{L}(X;X')} = \|a\|, \qquad \|B\|_{\mathscr{L}(X;M')} = \|b\|. \tag{4.6}$$

With these operators, Problem (Q) may be equivalently written in the form:

Find $(u,\lambda) \in X \times M$ *satisfying*

$$Au + B'\lambda = l \quad \textit{in } X',$$

$$Bu = \chi \quad \textit{in } M'.$$

Let us now introduce the linear operator $\Phi \in \mathscr{L}(X \times M; X' \times M')$ defined by:

$$\Phi(v,\mu) = (Av + B'\mu, Bv).$$

Then, we shall say that *Problem* (Q) *is well-posed* if Φ is an isomorphism from $X \times M$ onto $X' \times M'$. Our purpose is to derive necessary and sufficient conditions for Problem (Q) to be well-posed.

We set:

$$V = \mathrm{Ker}(B),$$

and more generally, for each $\chi \in M'$, we define the affine manifold

$$V(\chi) = \{v \in X;\ Bv = \chi\}.$$

Equivalently, we have:

$$(4.7)\qquad \begin{cases} V(\chi) = \{v \in X;\ b(v,\mu) = \langle \chi, \mu\rangle \quad \forall \mu \in M\}, \\ V = V(0). \end{cases}$$

Moreover, the continuity of the operator B implies that V is a closed subspace of X.

Now, with Problem (Q), we associate the following problem, called *Problem* (P):

Find $u \in V(\chi)$ such that

$$(4.8)\qquad a(u,v) = \langle l, v\rangle \quad \forall v \in V.$$

Clearly, if $(u,\lambda) \in X \times M$ is a solution of Problem (Q), then $u \in V(\chi)$ and u is a solution of (4.8), i.e. u is a solution of Problem (P). We want to find suitable conditions which ensure that the converse of this statement holds. For this, we define the polar set V^0 of V by

$$V^0 = \{g \in X';\ \langle g, v\rangle = 0 \quad \forall v \in V\}.$$

Lemma 4.1. *The three following properties are equivalent:*

(i) *there exists a constant $\beta > 0$ such that*

$$(4.9)\qquad \inf_{\mu \in M} \sup_{v \in X} \frac{b(v,\mu)}{\|v\|_X \|\mu\|_M} \geqslant \beta;$$

(ii) *the operator B' is an isomorphism from M onto V^0 and*

$$(4.10)\qquad \|B'\mu\|_{X'} \geqslant \beta \|\mu\|_M \quad \forall \mu \in M;$$

(iii) *the operator B is an isomorphism from $V^\perp$ onto M' and*

$$(4.11)\qquad \|Bv\|_{M'} \geqslant \beta \|v\|_X \quad \forall v \in V^\perp.$$

Proof. 1°) Let us show that the properties (i) and (ii) are equivalent. By (4.5), statement (i) means that

$$\|B'\mu\|_{X'} = \sup_{v \in X, v \neq 0} \frac{\langle B'\mu, v\rangle}{\|v\|_X} \geqslant \beta \|\mu\|_M \quad \forall \mu \in M,$$

so that (4.9) is equivalent to (4.10). Hence (ii) implies (i). In order to prove that (i) implies (ii), it remains only to show that, under the condition (4.10), B' is an isomorphism from M onto V^0. Clearly, it follows from (4.10) that B' is a one-to-one operator from M onto its range $\mathscr{R}(B')$ with a continuous inverse. Hence B' is an isomorphism from M onto $\mathscr{R}(B')$ so that $\mathscr{R}(B')$ is a closed subspace of X'. Thus, we have only to prove that

$$\mathscr{R}(B') = V^0.$$

For this we apply the Closed Range Theorem of Banach (cf. Yosida [84]) which asserts that

$$\mathscr{R}(B') = (\operatorname{Ker}(B))^0 = V^0.$$

This proves part 1°).

2°) We now show that properties (ii) and (iii) are equivalent. First, we observe that V^0 may be isometrically identified with $(V^{\perp})'$. Indeed, for $v \in X$, let $v^{\perp}$ denote the orthogonal projection of v onto $V^{\perp}$. Then, with each $g \in (V^{\perp})'$, we associate the element $\tilde{g} \in X'$ defined by:

$$\langle \tilde{g}, v \rangle = \langle g, v^{\perp} \rangle \qquad \forall v \in X.$$

Obviously, $\tilde{g} \in V^0$ and it is easy to check that the correspondence $g \to \tilde{g}$ is an isometric bijection from $(V^{\perp})'$ onto V^0. This enables us to identify $(V^{\perp})'$ and V^0.

As a consequence, we have that B is an isomorphism from $V^{\perp}$ onto M' with

$$\| B^{-1} \|_{\mathscr{L}(M'; V^{\perp})} \leqslant 1/\beta$$

if and only if B' is an isomorphism from M onto $(V^{\perp})' = V^0$ with

$$\| (B')^{-1} \|_{\mathscr{L}(V^0; M)} \leqslant 1/\beta.$$

Therefore, properties (ii) and (iii) are equivalent. □

The condition (4.9), usually called an "inf-sup condition", was introduced independently by Babuska [4] and Brezzi [13].

In order to state our main result, we introduce the linear continuous operator $\pi \in \mathscr{L}(X'; V')$ by:

$$\langle \pi f, v \rangle = \langle f, v \rangle \qquad \forall f \in X', \qquad \forall v \in V.$$

Clearly, we have

$$\| \pi f \|_{V'} \leqslant \| f \|_{X'}.$$

Theorem 4.1. *Problem* (Q) *is well-posed* (*i.e. the operator Φ is an isomorphism from $X \times M$ onto $X' \times M'$*) *if and only if the following conditions hold*:

(i) *the operator πA is an isomorphism from V onto V'*;

(ii) *the bilinear form $b(.,.)$ satisfies the inf-sup condition* (4.9).

Proof. 1°) The conditions (i) and (ii) are sufficient. It follows from (4.9) and Lemma 4.1 that there exists a unique element $u_0 \in V^{\perp}$ such that

$$\begin{cases} Bu_0 = \chi, \\ \| u_0 \|_X \leqslant (1/\beta) \| \chi \|_{M'}. \end{cases}$$

Therefore, Problem (P) may be equivalently stated in the following way:

Find $w = u - u_0 \in V$ *satisfying*

$$a(w, v) = \langle l, v\rangle - a(u_0, v) \qquad \forall v \in V$$

or satisfying

$$\pi A w = \pi(l - Au_0).$$

Since πA is an isomorphism from V onto V', Problem (P) has a unique solution $u = u_0 + w \in V(\chi)$ and we have

$$\|w\|_X \leqslant C_1 \|\pi(l - Au_0)\|_{V'} \leqslant C_1 \|l - Au_0\|_{X'},$$

so that

$$\|u\|_X \leqslant C_2(\|l\|_{X'} + \|\chi\|_{M'}).$$

Now, $l - Au$ belongs to V^0. Thus, according to Lemma 4.1, there exists a unique $\lambda \in M$ such that

$$B'\lambda = l - Au$$

with

$$\|\lambda\|_M \leqslant (1/\beta)\|l - Au\|_{X'} \leqslant C_3(\|l\|_{X'} + \|\chi\|_{M'}).$$

Hence, Problem (Q) has a unique solution (u, λ) and the mapping $(l, \chi) \to (u, \lambda)$ is continuous from $X' \times M'$ onto $X \times M$. This means that Φ is an isomorphism from $X' \times M'$ onto $X \times M$.

2°) The conditions (i) and (ii) are necessary. Assume that Φ is an isomorphism from $X \times M$ onto $X' \times M'$. We first show that the inf-sup condition (4.9) holds. Let χ be in M' and set $(u, \lambda) = \Phi^{-1}(0, \chi)$. We have $Bu = \chi$ so that $\mathscr{R}(B) = M'$. Hence B is a continuous and one-to-one mapping from $V^\perp$ onto M', and therefore an isomorphism from $V^\perp$ onto M'. Thus, by virtue of Lemma 4.1, the condition (4.9) holds.

Let us next prove that πA is an isomorphism from V onto V'. We first check that the operator πA is injective on V. Indeed, let $u \in V$ satisfy $\pi A u = 0$ so that $Au \in V^0$. Since the inf-sup condition holds, it follows from Lemma 4.1 that B' is an isomorphism from M onto V^0. Hence, there exists a unique $\lambda \in M$ such that $B'\lambda = -Au$. Thus, we have $\Phi(u, \lambda) = (0, 0)$ and therefore, $u = 0$.

Now, we check that πA is surjective. Let g be in V'. By the Hahn-Banach Theorem, there exists at least one element $l \in X'$ such that $g = \pi l$. We set $(u, \lambda) = \Phi^{-1}(l, 0)$. Clearly, $u \in V$ and

$$Au + B'\lambda = l.$$

Since for all $v \in V$

$$\langle \pi B'\lambda, v\rangle = \langle B'\lambda, v\rangle = \langle \lambda, Bv\rangle = 0,$$

we have $\pi B'\lambda = 0$ so that

$$\pi A u = \pi l = g.$$

Hence, πA is a one-to-one linear continuous mapping from V onto V' and therefore an isomorphism from V onto V'. □

Remark 4.1. In the first part of the proof of Theorem 4.1, we have used the fact that Problem (P) has a unique solution as soon as πA is an isomorphism from V onto V' and the affine manifold $V(\chi)$ is non void. The inf-sup condition (4.9) implies indeed that $V(\chi)$ is non void.

Corollary 4.1. *Assume that the bilinear form* $a(.,.)$ *is* V*-elliptic, i.e. there exists a constant* $\alpha > 0$ *such that*

(4.12) $$a(v, v) \geqslant \alpha \|v\|_X^2 \quad \forall v \in V.$$

Then, Problem (Q) *is well-posed if and only if the bilinear form* $b(.,.)$ *satisfies the inf-sup condition* (4.9).

Proof. Let l be in V'. Since $a(.,.)$ is V-elliptic, we may apply the Lax & Milgram Theorem 1.7: there exists a unique $u \in V$ such that

$$a(u, v) = \langle l, v \rangle \quad \forall v \in V,$$

or equivalently

$$\pi A u = l.$$

Moreover, the mapping $l \to u$ is continuous from V' into V. Hence, πA is an isomorphism from V onto V' and the desired result follows from Theorem 4.1. □

Remark 4.2. The above analysis can be readily extended to the case where X and M are reflexive Banach spaces. The space V is unchanged but $V^{\perp}$ is replaced by the quotient space X/V. The results and proofs of this section carry over with very few modifications.

4.2. A Saddle-Point Approach

Under adequate hypotheses, it is possible to formulate Problems (P) and (Q) in terms of optimization problems.

In addition to the notations of the previous section, we introduce two quadratic functionals $J: X \to \mathbb{R}$ and $\mathscr{L}: X \times M \to \mathbb{R}$ defined by

(4.13) $$J(v) = (1/2)a(v, v) - \langle l, v \rangle$$

and

(4.14) $$\mathscr{L}(v, \mu) = J(v) + b(v, \mu) - \langle \chi, \mu \rangle.$$

Usually, J is called the energy functional associated with Problem (P) and $\mathscr{L}$ the Lagrangean functional associated with Problem (Q).

Consider the following problem, called *Problem* (L):

Find a saddle-point $(u, \lambda) \in X \times M$ of the Lagrangean functional $\mathscr{L}$ over $X \times M$, i.e. find a pair $(u, \lambda) \in X \times M$ such that

$$\mathscr{L}(u, \mu) \leqslant \mathscr{L}(u, \lambda) \leqslant \mathscr{L}(v, \lambda) \quad \forall v \in X, \quad \forall \mu \in M. \tag{4.15}$$

Theorem 4.2. *Assume the conditions* (i) *and* (ii) *of Theorem* 4.1. *Assume in addition that the bilinear form $a(.,.)$ is symmetric and semi-positive definite on X, i.e.*

$$a(v, v) \geqslant 0 \quad \forall v \in X. \tag{4.16}$$

Then Problem (L) *has a unique solution $(u, \lambda) \in X \times M$ which is precisely the solution of Problem* (Q).

Proof. The first inequality in (4.15) can be written as follows:

$$b(u, \mu - \lambda) \leqslant \langle \chi, \mu - \lambda \rangle \quad \forall \mu \in M.$$

As μ is any element of M, this amounts to

$$b(u, \mu) = \langle \chi, \mu \rangle \quad \forall \mu \in M.$$

Next, the second inequality in (4.15) is equivalent to

$$\mathscr{L}(u, \lambda) = \inf_{v \in X} \mathscr{L}(v, \lambda).$$

Since, by hypothesis, $a(.,.)$ is symmetric, we have

$$D_v \mathscr{L}(u, \lambda) \cdot v = a(u, v) + b(v, \lambda) - \langle l, v \rangle.$$

Furthermore, using (4.16), we have

$$D_{vv}^2 \mathscr{L}(u, \lambda) \cdot (v, v) = a(v, v) \geqslant 0 \quad \forall v \in X.$$

Therefore $v \to \mathscr{L}(v, \lambda)$ is a convex functional and its minimum u is characterized by the condition $D_v \mathscr{L}(u, \lambda) \cdot v = 0$, i.e.

$$a(u, v) + b(v, \lambda) = \langle l, v \rangle \quad \forall v \in X.$$

Thus, (u, λ) is a saddle-point of $\mathscr{L}$ iff it is also a solution of Problem (Q). Hence the theorem is established. □

Corollary 4.2. *Under the hypotheses of Theorem* 4.2, *the solution (u, λ) of Problem* (Q) *is characterized by*:

$$\operatorname{Min}_{v \in X} \left(\sup_{\mu \in M} \mathscr{L}(v, \mu) \right) = \mathscr{L}(u, \lambda) = \operatorname{Max}_{\mu \in M} \left(\inf_{v \in X} \mathscr{L}(v, \mu) \right), \tag{4.17}$$

where Min *instead of inf (resp.* Max *instead of sup) means that the extremum is taken.*

Proof. This is a consequence of a general optimization result: the Lagrangean functional $\mathscr{L}$ has a saddle-point (u, λ) iff

$$\operatorname*{Min}_{v \in X}\left(\sup_{\mu \in M} \mathscr{L}(v, \mu)\right) = \operatorname*{Max}_{\mu \in M}\left(\inf_{v \in X} \mathscr{L}(v, \mu)\right) \tag{4.17'}$$

and this quantity is equal to $\mathscr{L}(u, \lambda)$.

Let us recall the proof for the reader's convenience. Setting

$$\phi(v) = \sup_{\mu \in M} \mathscr{L}(v, \mu), \qquad \phi^*(\mu) = \inf_{v \in X} \mathscr{L}(v, \mu),$$

we first notice that

$$\sup_{\mu \in M} \phi^*(\mu) \leqslant \inf_{v \in X} \phi(v). \tag{4.18}$$

In fact, since

$$\mathscr{L}(v, \mu) \leqslant \phi(v) \quad \forall v \in X, \quad \forall \mu \in M,$$

we get

$$\phi^*(\mu) \leqslant \inf_{v \in X} \phi(v) \quad \forall \mu \in M,$$

so that (4.18) follows.

Now let (u, λ) be a saddle-point of $\mathscr{L}$. This means that

$$\phi(u) = \mathscr{L}(u, \lambda) = \phi^*(\lambda).$$

Hence

$$\inf_{v \in X} \phi(v) \leqslant \phi(u) = \phi^*(\lambda) \leqslant \sup_{\mu \in M} \phi^*(\mu). \tag{4.19}$$

With (4.18) this implies that

$$\inf_{v \in X} \phi(v) = \phi(u) = \mathscr{L}(u, \lambda) = \phi^*(\lambda) = \sup_{\mu \in M} \phi^*(\mu),$$

which in particular implies (4.17').

Conversely, let (4.17') hold and let u realize the minimum of ϕ and λ the maximum of ϕ^*. Then (4.17') says that

$$\phi(u) = \phi^*(\lambda).$$

In addition, the definition of ϕ and ϕ^* implies that

$$\phi^*(\lambda) \leqslant \mathscr{L}(u, \lambda) \leqslant \phi(u);$$

thus

$$\phi(u) = \mathscr{L}(u, \lambda) = \phi^*(\lambda),$$

meaning that (u, λ) is a saddle-point of $\mathscr{L}$.

Finally, this optimization result and the uniqueness of the solution of (Q) yield the characterization (4.17). $\square$

Note that

$$\sup_{\mu \in M} \mathscr{L}(v, \mu) = \begin{cases} \infty & \text{if } v \notin V(\chi), \\ J(v) & \text{if } v \in V(\chi). \end{cases}$$

Hence, we have

$$\inf_{v \in X} \sup_{\mu \in M} \mathscr{L}(v, \mu) = \inf_{v \in V(\chi)} J(v).$$

Since $b(u, \mu) = \langle \chi, \mu \rangle$, the first equality (4.17) becomes

(4.20) $$J(u) = \operatorname*{Min}_{v \in V(\chi)} J(v).$$

Therefore, the solution u of Problem (P) is characterized as the solution of the constrained optimization problem (4.20).

On the other hand, let us consider the dual problem of (4.20):

$$\operatorname*{Max}_{\mu \in M} \inf_{v \in X} \mathscr{L}(v, \mu).$$

Assume in addition to the hypotheses of Theorem 4.2 that the bilinear form $a(.,.)$ is X-elliptic, i.e.

(4.21) $$a(v, v) \geqslant \alpha \| v \|_X^2 \quad \forall v \in X, \quad \alpha > 0.$$

Now, for any $\mu \in M$, we define $u(\mu) \in X$ to be the solution of the optimization problem:

$$\mathscr{L}(u(\mu), \mu) = \inf_{v \in X} \mathscr{L}(v, \mu).$$

We have seen in the proof of Theorem 4.2 that this problem is equivalent to the equation:

(4.22) $$a(u(\mu), v) = \langle l, v \rangle - b(v, \mu) \quad \forall v \in X.$$

By (4.21), problem (4.22) has indeed a unique solution $u(\mu) \in X$. Since $u(\lambda) = u$, we note that

$$\mathscr{L}(u(\lambda), \lambda) = \operatorname*{Max}_{\mu \in M} \mathscr{L}(u(\mu), \mu).$$

Let us give a simple expression for $\mathscr{L}(u(\mu), \mu)$. We have

$$\mathscr{L}(u(\mu), \mu) = (1/2) a(u(\mu), u(\mu)) - \langle l, u(\mu) \rangle + b(u(\mu), \mu) - \langle \chi, \mu \rangle,$$

so that by (4.22)

$$\mathscr{L}(u(\mu), \mu) = -(1/2) a(u(\mu), u(\mu)) - \langle \chi, \mu \rangle.$$

Hence, setting

(4.23) $$K(\mu) = (1/2) a(u(\mu), u(\mu)) + \langle \chi, \mu \rangle,$$

the dual problem of (4.20) becomes

$$K(\lambda) = \underset{\mu \in M}{\text{Min}}\, K(\mu). \tag{4.24}$$

Therefore, the second argument λ of the solution of Problem (Q) is characterized as the solution of the unconstrained optimization problem (4.24).

4.3. Approximation by Regularization or Penalty

We want to introduce a perturbed form of Problem (Q) which may be easier to solve in practice. In addition to $a(.,.)$ and $b(.,.)$, we consider a third continuous bilinear form

$$c(.,.): M \times M \to \mathbb{R},$$

with norm

$$\|c\| = \sup_{\lambda, \mu \in M, \lambda, \mu \neq 0} \frac{c(\lambda, \mu)}{\|\lambda\|_M \|\mu\|_M}.$$

We assume that the form $c(.,.)$ is M-elliptic in the sense that there exists a constant $\gamma > 0$ such that:

$$c(\mu, \mu) \geqslant \gamma \|\mu\|_M^2 \quad \forall \mu \in M. \tag{4.25}$$

With the form $c(.,.)$, we associate the operator $C \in \mathscr{L}(M; M')$ defined by

$$\langle C\lambda, \mu\rangle = c(\lambda, \mu) \quad \forall \lambda, \mu \in M. \tag{4.26}$$

Let $\varepsilon > 0$ be a parameter which will *tend to zero*. We consider the following problem, called *Problem* (Q^ε):

Find a pair $(u^\varepsilon, \lambda^\varepsilon) \in X \times M$ *such that*

$$a(u^\varepsilon, v) + b(v, \lambda^\varepsilon) = \langle l, v\rangle \quad \forall v \in X, \tag{4.27}$$

$$-\varepsilon c(\lambda^\varepsilon, \mu) + b(u^\varepsilon, \mu) = \langle \chi, \mu\rangle \quad \forall \mu \in M. \tag{4.28}$$

Since the operator C is nonsingular, equation (4.28) is equivalent to

$$\lambda^\varepsilon = (1/\varepsilon) C^{-1}(Bu^\varepsilon - \chi). \tag{4.29}$$

Hence, by replacing λ^ε by its value in (4.27), we obtain the following problem, called *Problem* (P^ε), which is obviously equivalent to Problem (Q^ε):

Find $u^\varepsilon \in X$ *such that*

$$a(u^\varepsilon, v) + (1/\varepsilon)\langle C^{-1} Bu^\varepsilon, Bv\rangle = \langle l, v\rangle + (1/\varepsilon)\langle C^{-1}\chi, Bv\rangle \quad \forall v \in X. \tag{4.30}$$

Theorem 4.3. *Assume the hypotheses* (4.9) *and* (4.25). *Assume in addition that there exists a constant* $\alpha > 0$ *such that*

$$a(v, v) + \langle C^{-1} Bv, Bv\rangle \geqslant \alpha \|v\|_X^2 \quad \forall v \in X. \tag{4.31}$$

Then Problems (Q) *and* (Q^ε) *for* $\varepsilon \leqslant 1$ *both have one and only one solution* (u, λ) *and* $(u^\varepsilon, \lambda^\varepsilon)$ *in* $X \times M$. *Moreover, the following error bound holds for* $\varepsilon \leqslant \varepsilon_0$ *small enough*

$$\|u^\varepsilon - u\|_X + \|\lambda^\varepsilon - \lambda\|_M \leqslant K\varepsilon(\|l\|_{X'} + \|\chi\|_{M'}), \tag{4.32}$$

where the constant K *depends only upon* α, β, $\|a\|$, $\|b\|$ *and* $\|c\|$.

Proof. Hypothesis (4.31) implies that the bilinear form $a(.,.)$ is V-elliptic. Hence, by Corollary 4.1, Problem (Q) has a unique solution $(u, \lambda) \in X \times M$.

Now, it follows from (4.25) and (4.31) that the bilinear form

$$u, v \to a(u, v) + (1/\varepsilon)\langle C^{-1}Bu, Bv\rangle$$

is X-elliptic when $\varepsilon \leqslant 1$. Hence Problem (P^ε) has exactly one solution u^ε in X. Therefore, if we define λ^ε by (4.29) then $(u^\varepsilon, \lambda^\varepsilon)$ is the only solution of Problem (Q^ε).

It remains to establish the bound (4.32). From (4.1) and (4.27), (4.2) and (4.28), we get:

$$\begin{cases} a(u^\varepsilon - u, v) + b(v, \lambda^\varepsilon - \lambda) = 0 & \forall v \in X, \\ -\varepsilon c(\lambda^\varepsilon, \mu) + b(u^\varepsilon - u, \mu) = 0 & \forall \mu \in M. \end{cases} \tag{4.33}$$

The first equation (4.33) together with (4.9) yields:

$$\beta\|\lambda^\varepsilon - \lambda\|_M \leqslant \sup_{v \in X} \frac{b(v, \lambda^\varepsilon - \lambda)}{\|v\|_X} \leqslant \|a\|\,\|u^\varepsilon - u\|_X$$

whence

$$\|\lambda^\varepsilon - \lambda\|_M \leqslant (\|a\|/\beta)\,\|u^\varepsilon - u\|_X. \tag{4.34}$$

Next, by taking $v = u^\varepsilon - u$ and $\mu = \lambda^\varepsilon - \lambda$ in (4.33) we find:

$$a(u^\varepsilon - u, u^\varepsilon - u) = -\varepsilon c(\lambda, \lambda^\varepsilon - \lambda) - \varepsilon c(\lambda^\varepsilon - \lambda, \lambda^\varepsilon - \lambda) \leqslant -\varepsilon c(\lambda, \lambda^\varepsilon - \lambda).$$

Then (4.34) gives

$$a(u^\varepsilon - u, u^\varepsilon - u) \leqslant \varepsilon(\|a\|\,\|c\|/\beta)\,\|\lambda\|_M\|u^\varepsilon - u\|_X.$$

Besides that, we have by the second equation (4.33):

$$B(u^\varepsilon - u) = \varepsilon C\lambda^\varepsilon.$$

Therefore, we obtain

$$\langle C^{-1}B(u^\varepsilon - u), B(u^\varepsilon - u)\rangle = \varepsilon^2\langle C\lambda^\varepsilon, \lambda^\varepsilon\rangle = \varepsilon^2 c(\lambda^\varepsilon, \lambda^\varepsilon) \leqslant \varepsilon^2\|c\|\,\|\lambda^\varepsilon\|_M^2,$$

and by (4.34)

$$\langle C^{-1}B(u^\varepsilon - u), B(u^\varepsilon - u)\rangle \leqslant \varepsilon^2\|c\|\,\{(\|a\|/\beta)\,\|u^\varepsilon - u\|_X + \|\lambda\|_M\}^2.$$

Hence, Hypothesis (4.31) yields an inequality of the form

$$\alpha x^2 \leqslant \varepsilon^2 C_1(\|\lambda\|_M + x)^2 + \varepsilon C_2\|\lambda\|_M x,$$

where x stands for $\|u^\varepsilon - u\|_X$. If ε is sufficiently small, this amounts to

$$\|u^\varepsilon - u\|_X \leqslant \varepsilon C_3 \|\lambda\|_M.$$

This, together with (4.34), proves the bound (4.32). □

Remark 4.3. Problem (Q^ε) has a unique solution even when the inf-sup condition (4.9) does not hold. Hence, in (4.28), $-\varepsilon c(\lambda^\varepsilon, \mu)$ plays the role of a regularizing term: Problem (Q^ε) appears to be a regularized version of Problem (Q).

Remark 4.4. When the bilinear forms $a(.,.)$ and $c(.,.)$ are symmetric, so is the bilinear form $u, v \to a(u, v) + (1/\varepsilon)\langle C^{-1}Bu, Bv\rangle$. In this case, setting

$$J_\varepsilon(v) = J(v) + \frac{1}{2\varepsilon}\langle C^{-1}(Bv - \chi), Bv - \chi\rangle$$

and using the ellipticity property (4.31) with $\varepsilon \leqslant 1$, we obtain that Problem (P^ε) amounts to find $u^\varepsilon \in X$ such that

$$J_\varepsilon(u^\varepsilon) = \inf_{v \in X} J_\varepsilon(v).$$

Hence, we have approximated the constrained optimization problem (4.20) for the energy functional J by an unconstrained one for the penalized functional J_ε where the expression $\langle C^{-1}(Bv - \chi), Bv - \chi\rangle/(2\varepsilon)$ is a penalty term corresponding to the constraint $Bv = \chi$. Thus, Problem (P^ε) is a penalized version of Problem (P).

In fact, we can derive a more precise result than Theorem 4.3, namely an asymptotic expansion of $(u^\varepsilon, \lambda^\varepsilon)$ in powers of ε. We define by induction a sequence $(u_n, \lambda_n) \in X \times M$ in the following way. We set $\lambda_0 = \lambda$ and, for any integer $n \geqslant 1$, the pair $(u_n, \lambda_n) \in X \times M$ is the solution of:

$$\text{(4.35)} \qquad \begin{cases} a(u_n, v) + b(v, \lambda_n) = 0 & \forall v \in X, \\ b(u_n, \mu) = c(\lambda_{n-1}, \mu) & \forall \mu \in M. \end{cases}$$

Knowing λ_{n-1}, it follows from (4.9), (4.31) and Corollary 4.1, that the equations (4.35) define indeed a unique pair $(u_n, \lambda_n) \in X \times M$.

Theorem 4.4. *Assume the hypotheses of Theorem 4.3. Then, we have for all integers $N \geqslant 1$ and for $\varepsilon \leqslant \varepsilon_0$ small enough*

$$\left\| u^\varepsilon - u - \sum_{n=1}^{N} \varepsilon^n u_n \right\|_X + \left\| \lambda^\varepsilon - \lambda - \sum_{n=1}^{N} \varepsilon^n \lambda_n \right\|_M \leqslant K_N \varepsilon^{N+1}(\|l\|_{X'} + \|\chi\|_{M'}),$$

(4.36)

where the constant K_N depends only upon N, α, β, $\|a\|$, $\|b\|$ and $\|c\|$.

Proof. The argument of the proof is very similar to that of Theorem 4.3. We set:

$$w^\varepsilon = u^\varepsilon - u - \sum_{n=1}^{N} \varepsilon^n u_n, \qquad \rho^\varepsilon = \lambda^\varepsilon - \lambda - \sum_{n=1}^{N} \varepsilon^n u_n.$$

Using (4.1), (4.2), (4.27), (4.28) and (4.35), we get:

$$\text{(4.37)} \qquad \begin{cases} a(w^\varepsilon, v) + b(v, \rho^\varepsilon) = 0 & \forall v \in X, \\ -\varepsilon c(\rho^\varepsilon, \mu) + b(w^\varepsilon, \mu) = \varepsilon^{N+1} c(\lambda_N, \mu) & \forall \mu \in M. \end{cases}$$

The first equation (4.37) together with (4.9) yields

$$\text{(4.38)} \qquad \|\rho^\varepsilon\|_M \leqslant (\|a\|/\beta) \|w^\varepsilon\|_X.$$

Next, by taking $v = w^\varepsilon$ and $\mu = \rho^\varepsilon$ in (4.37), we obtain

$$a(w^\varepsilon, w^\varepsilon) = -\varepsilon c(\rho^\varepsilon, \rho^\varepsilon) - \varepsilon^{N+1} c(\lambda_N, \rho^\varepsilon) \leqslant -\varepsilon^{N+1} c(\lambda_N, \rho^\varepsilon)$$

so that

$$a(w^\varepsilon, w^\varepsilon) \leqslant \varepsilon^{N+1} (\|a\| \, \|c\|/\beta) \|\lambda_N\|_M \|w^\varepsilon\|_X.$$

On the other hand, the second equation (4.37) gives

$$Bw^\varepsilon = \varepsilon C(\rho^\varepsilon + \varepsilon^N \lambda_N)$$

so that

$$\begin{aligned} \langle C^{-1} Bw^\varepsilon, Bw^\varepsilon \rangle &= \varepsilon^2 c(\rho^\varepsilon + \varepsilon^N \lambda_N, \rho^\varepsilon + \varepsilon^N \lambda_N) \\ &\leqslant 2\varepsilon^2 \|c\| (\|\rho^\varepsilon\|_M^2 + \varepsilon^{2N} \|\lambda_N\|_M^2) \end{aligned}$$

and by (4.38)

$$\langle C^{-1} Bw^\varepsilon, Bw^\varepsilon \rangle \leqslant 2\varepsilon^2 \|c\| \{(\|a\|/\beta)^2 \|w^\varepsilon\|_X^2 + \varepsilon^{2N} \|\lambda_N\|_M^2\}.$$

Therefore, using (4.31), we obtain an inequality of the form

$$\alpha \|w^\varepsilon\|_X^2 \leqslant C_1 \varepsilon^2 \|w^\varepsilon\|_X^2 + C_2 \varepsilon^{N+1} \|\lambda_N\|_M \|w^\varepsilon\|_X + C_3 \varepsilon^{2N+2} \|\lambda_N\|_M^2$$

which implies for $\varepsilon \leqslant \varepsilon_0$ small enough

$$\text{(4.39)} \qquad \|w^\varepsilon\|_X \leqslant C_4 \varepsilon^{N+1} \|\lambda_N\|_M.$$

Now, it follows from (4.35) and Corollary 4.1 that

$$\text{(4.40)} \qquad \|\lambda_N\|_M \leqslant C_5 \|\lambda\|_M \leqslant C_6 (\|l\|_{X'} + \|\chi\|_{M'}),$$

with constants C_5 and C_6 of the form C^N. Thus the desired inequality (4.36) is a consequence of (4.38), (4.39) and (4.40). □

4.4. Iterative Methods of Gradient Type

The purpose of this section is to derive convenient iterative methods to solve Problem (Q). To this end we propose a duality approach based on the optimization formulations (4.20) and (4.24). Assume for the moment that the bilinear form

$b(.,.)$ satisfies the *inf-sup* condition (4.9) and that the bilinear form $a(.,.)$ is *symmetric* and *X-elliptic* in the sense of (4.21). Then according to Corollary 4.1, Problem (Q) is well-posed and Theorem 4.2 claims that its unique solution (u, λ) is characterized as the unique saddle-point of the Lagrangean functional $\mathscr{L}(v, \mu)$ defined by (4.14). Furthermore, u is the unique solution of the *constrained* optimization problem (4.20) while λ is the unique solution of the *unconstrained* optimization problem (4.24). It appears clearly that this last problem is the easiest to solve and that methods of descent should be particularly well-adapted to its solution. Here we shall concentrate on a general descent algorithm and apply it to two classes of gradient methods: the *simple gradient* (also called *Uzawa's Algorithm*) and the *conjugate-gradient* methods.

Keeping in mind the above considerations, let us first generalize Theorem 4.2 and show that (u, λ) can be characterized as the unique saddle-point of a parametrized family of Lagrangean functionals. Following the preceding section, we introduce a continuous, bilinear form $c(.,.)\colon M \times M \to \mathbb{R}$, *M-elliptic* in the sense of (4.25) and *symmetric*. Likewise, we define by (4.26) the corresponding operator $C \in \mathscr{L}(M; M')$. Now, for all $r \geqslant 0$ we introduce the *augmented energy functional* $J_r\colon X \to \mathbb{R}$ and the *augmented Lagrangean functional* $\mathscr{L}_r\colon X \times M \to \mathbb{R}$ defined respectively by:

$$J_r(v) = J(v) + (r/2)\langle C^{-1}(Bv - \chi), Bv - \chi\rangle, \tag{4.41}$$

$$\mathscr{L}_r(v, \mu) = J_r(v) + b(v, \mu) - \langle \chi, \mu\rangle. \tag{4.42}$$

Since

$$J_r(v) = J(v) \qquad \forall v \in V(\chi),$$

u is still characterized by:

$$J_r(u) = \inf_{v \in V(\chi)} J_r(v). \tag{4.43}$$

On the other hand, we have the following result.

Theorem 4.5. *Suppose that* (4.9), (4.12) *and* (4.25) *hold and in addition assume that the bilinear forms* $a(.,.)$ *and* $c(.,.)$ *are symmetric with*

$$a(v, v) + r\langle C^{-1} Bv, Bv\rangle \geqslant 0 \qquad \forall v \in X. \tag{4.44}$$

Then the solution (u, λ) *of Problem* (Q) *is the unique saddle-point of the augmented Lagrangean functional* $\mathscr{L}_r$, *or equivalently is characterized by:*

$$\operatorname*{Min}_{v \in X}\Big(\sup_{\mu \in M} \mathscr{L}_r(v, \mu)\Big) = \mathscr{L}_r(u, \lambda) = \operatorname*{Max}_{\mu \in M}\Big(\inf_{v \in X} \mathscr{L}_r(v, \mu)\Big). \tag{4.45}$$

Proof. Taking into account the symmetry of $a(.,.)$ and $c(.,.)$ we get:

$$D_v \mathscr{L}_r(u, \lambda)\cdot v = a(u, v) - \langle l, v\rangle + r\langle C^{-1}(Bu - \chi), Bv\rangle + b(v, \lambda),$$

$$D^2_{vv} \mathscr{L}_r(u, \lambda)\cdot (v, v) = a(v, v) + r\langle C^{-1} Bv, Bv\rangle.$$

Then the proof follows the lines of Theorem 4.2 and Corollary 4.2. □

Now we turn to the dual problem of (4.43):

$$\operatorname*{Max}_{\mu \in M} \inf_{v \in X} \mathscr{L}_r(v, \mu).$$

Assuming that the bilinear form $a_r(.,.)$ defined by:

$$a_r(u, v) = a(u, v) + r\langle C^{-1}Bu, Bv\rangle$$

is X-elliptic, i.e. assuming there exists $\alpha_r > 0$ such that

$$a_r(v, v) \geqslant \alpha_r \|v\|_X^2 \quad \forall v \in X, \tag{4.46}$$

then it is clear that the optimization problem:

$$\mathscr{L}_r(u_r(\mu), \mu) = \inf_{v \in X} \mathscr{L}_r(v, \mu)$$

has a unique solution $u_r(\mu) \in X$. Indeed, according to the above proof, this problem amounts to solve:

$$a_r(u_r(\mu), v) = \langle l, v\rangle + rb(v, C^{-1}\chi) - b(v, \mu) \quad \forall v \in X. \tag{4.47}$$

Here again, we have $u_r(\lambda) = u$; hence

$$\mathscr{L}_r(u_r(\lambda), \lambda) = \operatorname*{Max}_{\mu \in M} \mathscr{L}_r(u_r(\mu), \mu). \tag{4.48}$$

Like in Section 4.2 we easily derive the expression:

$$\mathscr{L}_r(u_r(\mu), \mu) = -(1/2)a_r(u_r(\mu), u_r(\mu)) - \langle \chi, \mu\rangle + (r/2)\langle C^{-1}\chi, \chi\rangle.$$

Thus, setting

$$K_r(\mu) = (1/2)a_r(u_r(\mu), u_r(\mu)) + \langle \chi, \mu\rangle, \tag{4.49}$$

we obtain

$$\mathscr{L}_r(u_r(\mu), \mu) = -K_r(\mu) + (r/2)\langle C^{-1}\chi, \chi\rangle.$$

From this equality and (4.48), it follows that λ satisfies:

$$K_r(\lambda) = \operatorname*{Min}_{\mu \in M} K_r(\mu). \tag{4.50}$$

The following corollary summarizes this result.

Corollary 4.3. *Under the hypotheses of Theorem* 4.5 *and the ellipticity condition* (4.46), $\lambda \in M$ *is characterized as the unique solution of the unconstrained optimization problem* (4.50).

Remark 4.5. The mapping $\mu \to u_r(\mu)$ defined by (4.47) is affine from M into X.

In order to adapt methods of descent to solve (4.50), we must compute the first two derivatives of the quadratic functional K_r with respect to μ. First of all, we observe from (4.47) that the derivative of u_r:

$$Du_r = Du_r(\mu) \in \mathscr{L}(M; X)$$

is independent of μ. More precisely, (4.47) yields:

$$Du_r = -A_r^{-1}B'$$

where $A_r \in \mathscr{L}(X; X')$ denotes the operator associated with the bilinear form $a_r(.,.)$:

$$\langle A_r u, v\rangle = a_r(u, v) \qquad \forall u, v \in X.$$

Hence, differentiating (4.49), we find:

$$DK_r(\mu)\cdot v = \langle A_r Du_r \cdot v, u_r(\mu)\rangle + \langle \chi, v\rangle = \langle \chi - Bu_r(\mu), v\rangle.$$

Therefore

$$DK_r(\mu) = \chi - Bu_r(\mu) \in M' \tag{4.51}$$

and

$$D^2 K_r(\mu) = D^2 K_r = -BDu_r = BA_r^{-1}B' \in \mathscr{L}(M; M'). \tag{4.52}$$

Next, let us equip the Hilbert space M with a scalar product. As the bilinear form $c(.,.)$ is symmetric and M-elliptic we can choose it for scalar product. With respect to c, the gradient $g_r(\mu)$ of K_r at the point μ of M is defined by:

$$c(g_r(\mu), v) = DK_r(\mu)\cdot v.$$

In other words,

$$g_r(\mu) = C^{-1}(\chi - Bu_r(\mu)). \tag{4.53}$$

Now, let (ω^m) be an arbitrary sequence of elements of M and consider the following method of descent:

starting with an initial guess $\lambda^0 \in M$, construct the sequence $(\lambda^m) \subset M$ by:

$$\lambda^{m+1} = \lambda^m - \rho^m \omega^m \qquad m \geqslant 0, \tag{4.54}$$

where the parameter ρ^m is chosen so that

$$K_r(\lambda^m - \rho^m \omega^m) = \inf_{\rho \in \mathbb{R}} K_r(\lambda^m - \rho\omega^m).$$

This optimal ρ^m is given by:

$$\rho^m = DK_r(\lambda^m)\cdot \omega^m / D^2 K_r \cdot (\omega^m, \omega^m) \tag{4.55}$$

and of course u^m is related to λ^m by (cf. (4.47)):

$$u^m = A_r^{-1}(l + rB'C^{-1}\chi - B'\lambda^m) = u^m(\lambda^m). \tag{4.56}$$

Note that this iterative method decouples the computation of u^m and λ^m. The sequence (ω^m) is not specified for the moment. We shall see further on that proper choices of (ω^m) lead to attractive gradient methods.

From a practical point of view, the algorithm (4.54), (4.55), (4.56) can be organized more efficiently. For the sake of simplicity, we set:

$$g^m = g_r(\lambda^m), \qquad z^m = A_r^{-1} B' \omega^m.$$

From (4.52), (4.53) and (4.55), we easily derive:

$$\rho^m = c(g^m, \omega^m)/b(z^m, \omega^m).$$

In addition, by subtracting two consecutive values of (4.56) and using (4.54) we get:

$$u^{m+1} - u^m = \rho^m z^m.$$

Hence our general descent algorithm has the following procedure:

1°) Given an initial guess $\lambda^0 \in M$ compute $u^0 \in X$ by:

$$A_r u^0 = l + B'(rC^{-1}\chi - \lambda^0);$$

2°) For $m \geqslant 0$, given $\omega^m \in M$ and knowing $(u^m, \lambda^m) \in X \times M$, determine $(z^m, g^m) \in X \times M$, $\rho^m \in \mathbb{R}$ and the pair $(u^{m+1}, \lambda^{m+1}) \in X \times M$ by:

$$(4.57) \qquad \begin{cases} Cg^m = \chi - Bu^m, \\ A_r z^m = B'\omega^m, \\ \rho^m = c(g^m, \omega^m)/b(z^m, \omega^m), \\ \lambda^{m+1} = \lambda^m - \rho^m \omega^m, \\ u^{m+1} = u^m + \rho^m z^m. \end{cases}$$

At each iteration, we have to solve two linear problems: one with the operator C and one with the operator A_r. In practice, the operator C is easily invertible—for instance, in the case of the Stokes problem, one usually chooses the L^2-inner product for $c(.,.)$ so that C is the identity operator—. Hence the major computational effort is concentrated on z^m.

This method of descent has the following convergence result:

Theorem 4.6. *Assume that the form $b(.,.)$ satisfies the inf-sup condition* (4.9) *and that $a_r(.,.)$ and $c(.,.)$ are both symmetric and satisfy the ellipticity conditions* (4.46) *and* (4.25) *respectively. Then for every starting value $\lambda^0 \in M$ and every sequence $(\omega^m) \subset M$, the iterative scheme* (4.57) *defines a unique sequence (u^m, λ^m) in $X \times M$. In addition, if the sequences (g^m) and (ω^m) satisfy for all $m \geqslant 0$:*

$$(4.58) \qquad c(g^m, \omega^m) \geqslant \alpha \|g^m\|_M \|\omega^m\|_M \quad \alpha > 0 \ \textit{independent of } m,$$

then the method of descent converges:

$$\lim_{m \to \infty} \{\|u^m - u\|_X + \|\lambda^m - \lambda\|_M\} = 0.$$

Proof. Clearly the first two equations in (4.57) have a unique solution because of the ellipticity of $a_r(.,.)$ and $c(.,.)$. Furthermore, (4.52) yields:

$$D^2K_r\cdot(\mu,\mu) = \langle A_r^{-1}B'\mu, B'\mu\rangle.$$

Hence the inf-sup condition (4.9) implies that

$$\delta_r\|\mu\|_M^2 \leqslant D^2K_r\cdot(\mu,\mu) \leqslant \tau_r\|\mu\|_M^2 \quad \forall\mu\in M, \quad \delta_r, \tau_r > 0. \tag{4.59}$$

As a consequence, (4.57) defines a unique ρ^m (provided of course $\omega^m \neq 0$).

Now we turn to the convergence of the algorithm. First, we have by construction:

$$DK_r(\lambda^{m+1})\cdot\omega^m = 0.$$

Since K_r is quadratic this implies that:

$$K_r(\lambda^m) - K_r(\lambda^{m+1}) = (1/2)D^2K_r\cdot(\lambda^m - \lambda^{m+1}, \lambda^m - \lambda^{m+1}).$$

Therefore, in view of (4.59) we get:

$$K_r(\lambda^m) - K_r(\lambda^{m+1}) \geqslant (1/2)\delta_r\|\lambda^m - \lambda^{m+1}\|_M^2.$$

But by construction, the sequence $K_r(\lambda^m)$ is monotone decreasing and bounded below by $K_r(\lambda)$. Hence it is convergent and therefore

$$\lim_{m\to\infty}\|\lambda^m - \lambda^{m+1}\|_M = 0.$$

According to (4.54), this means that

$$\lim_{m\to\infty}\rho^m\|\omega^m\|_M = 0.$$

But (4.59) and the hypothesis (4.58) imply that (ρ^m) is bounded below by:

$$\rho^m \geqslant (\alpha/\tau_r)(\|g^m\|_M/\|\omega^m\|_M).$$

As a consequence,

$$\rho^m\|\omega^m\|_M \geqslant (\alpha/\tau_r)\|g^m\|_M$$

and thus

$$\lim_{m\to\infty} g^m = 0,$$

i.e.

$$\lim_{m\to\infty} DK_r(\lambda^m) = 0 \quad \text{in } M'. \tag{4.60}$$

Besides that, since $DK_r(\lambda) = 0$ we can write:

$$DK_r(\lambda^m) = DK_r(\lambda^m) - DK_r(\lambda) = D^2K_r\cdot(\lambda^m - \lambda),$$

$$D^2K_r\cdot(\lambda^m - \lambda, \lambda^m - \lambda) = DK_r(\lambda^m)\cdot(\lambda^m - \lambda).$$

By virtue of (4.59) this gives:

$$\delta_r\|\lambda^m - \lambda\|_M \leqslant \|DK_r(\lambda^m)\|_{M'}.$$

Therefore the convergence (4.60) implies that

$$\lim_{m\to\infty} \lambda^m = \lambda \quad \text{in } M$$

which in turn implies the convergence of u^m by taking limits in (4.56). □

Remark 4.6. The convergence criterion (4.58) implies that the direction of descent ω^m is never asymptotically orthogonal to the gradient g^m.

As a first application, the *simple gradient algorithm with optimal parameter* is obtained by a descent along the gradient:

$$\omega^m = g^m. \tag{4.61}$$

Thus the general step of this algorithm is:

For $m \geqslant 0$, knowing $(u^m, \lambda^m) \in X \times M$ determine $(z^m, g^m) \in X \times M$, $\rho^m \in \mathbb{R}$ and the pair $(u^{m+1}, \lambda^{m+1}) \in X \times M$ by:

$$\begin{cases} Cg^m = \chi - Bu^m, \\ A_r z^m = B' g^m, \\ \rho^m = c(g^m, g^m)/b(z^m, g^m), \\ \lambda^{m+1} = \lambda^m - \rho^m g^m, \\ u^{m+1} = u^m + \rho^m z^m. \end{cases} \tag{4.62}$$

Here the criterion (4.58) is obviously satisfied with $\alpha = 1$; thus we have:

Corollary 4.4. *Let $b(.,.)$, $a_r(.,.)$ and $c(.,.)$ satisfy the assumptions of Theorem* 4.6. *Then the simple gradient algorithm with optimal parameter* (4.62) *is convergent.*

At this stage, it is interesting to point out that the simple gradient algorithm is still convergent provided the parameters ρ^m lie in an appropriate interval, even though these parameters are *not optimal.* Let us stress the fact that the convergence of the resulting scheme *does not require the symmetry of $a(.,.)$*. The method of proof is altogether different from that of Theorem 4.6.

Theorem 4.7. *We retain the hypotheses of Corollary* 4.4 *but we do not necessarily assume that the form $a(.,.)$ is symmetric. Then the algorithm:*

$$\lambda^{m+1} = \lambda^m - \rho^m g^m \tag{4.54'}$$

$$a_r(u^m, v) = \langle l, v\rangle + rb(v, C^{-1}\chi) - b(v, \lambda^m) \qquad \forall v \in X \tag{4.56}$$

is convergent for every initial value $\lambda^0 \in M$ and every choice of ρ^m satisfying:

$$0 < \inf_m \rho^m \leqslant \sup_m \rho^m < 2\gamma\tilde{\alpha}_r \tag{4.63}$$

where

$$\tilde{\alpha}_r = \inf_{v \in X} [a_r(v, v)/\| Bv \|_{M'}^2]. \tag{4.46'}$$

Proof. We set:

$$e^m = u^m - u, \qquad \nu^m = \lambda^m - \lambda.$$

Then (4.1), (4.2) and (4.56) yield:

$$a_r(e^m, v) + b(v, \nu^m) = 0 \qquad \forall v \in X,$$

$$-c(\nu^{m+1} - \nu^m, \mu) + \rho^m b(e^m, \mu) = 0 \qquad \forall \mu \in M.$$

By taking $v = e^m$ and $\mu = \nu^{m+1}$ in the above equations, we get

$$c(\nu^{m+1} - \nu^m, \nu^{m+1}) = \rho^m b(e^m, \nu^{m+1}) = \rho^m\{b(e^m, \nu^{m+1} - \nu^m) - a_r(e^m, e^m)\}.$$

As the bilinear form $c(.,.)$ is symmetric, we may write

$$2c(\nu^{m+1} - \nu^m, \nu^{m+1}) = c(\nu^{m+1}) - c(\nu^m) + c(\nu^{m+1} - \nu^m),$$

where $c(\mu)$ stands for $c(\mu, \mu)$. Therefore, using (4.25), we obtain:

$$c(\nu^{m+1}) - c(\nu^m) + \gamma \| \nu^{m+1} - \nu^m \|_M^2 + 2\rho^m a_r(e^m, e^m)$$

$$\leqslant 2\rho^m \| Be^m \|_{M'} \| \nu^{m+1} - \nu^m \|_M.$$

By virtue of the inequality $2ab \leqslant \gamma a^2 + (1/\gamma)b^2$ and (4.46′) we find

$$2\rho^m \| Be^m \|_{M'} \| \nu^{m+1} - \nu^m \|_M \leqslant \gamma \| \nu^{m+1} - \nu^m \|_M^2 + (\rho^m)^2 a_r(e^m, e^m)/(\tilde{\alpha}_r \gamma).$$

Hence, applying (4.46), this yields:

$$c(\nu^{m+1}) - c(\nu^m) + \rho^m\{2 - \rho^m/(\tilde{\alpha}_r \gamma)\} \alpha_r \| e^m \|_X^2 \leqslant 0.$$

Since ρ^m satisfies (4.63), there exists a constant $\delta > 0$, independent of m, such that

$$2\tilde{\alpha}_r - \rho^m/\gamma \geqslant \delta.$$

Thus, we obtain:

$$c(\nu^{m+1}) - c(\nu^m) + \rho^m \delta(\alpha_r/\tilde{\alpha}_r) \| e^m \|_X^2 \leqslant 0. \tag{4.64}$$

On the other hand, the inf-sup condition (4.9) gives:

$$\| \nu^m \|_M \leqslant (1/\beta) \sup_{v \in X} \frac{b(v, \nu^m)}{\| v \|_X} = (1/\beta) \sup_{v \in X} \frac{a_r(e^m, v)}{\| v \|_X} \leqslant (\| a_r \|/\beta) \| e^m \|_X,$$

where $\| a_r \|$ denotes the norm of the form $a_r(.,.)$. Therefore,

$$\| e^m \|_X^2 \geqslant \beta^2 c(\nu^m)/(\| c \| \, \| a_r \|^2)$$

and (4.64) implies:

$$c(\nu^{m+1}) \leqslant c(\nu^m)\{1 - (\alpha_r/\tilde{\alpha}_r)(\rho^m \delta \beta^2)/(\| c \| \, \| a_r \|^2)\}.$$

Let us put $\rho = \inf_m \rho^m$; according to (4.63), $\rho > 0$. Therefore,

$$(4.65)\qquad c(v^m) \leqslant \theta^m c(v^0), \qquad 0 \leqslant \theta = 1 - (\alpha_r/\tilde{\alpha}_r)(\rho\delta\beta^2)/(\|c\|\,\|a_r\|^2) < 1.$$

Hence the sequence $(c(v^m))$ converges to zero and

$$\lim_{m\to\infty} \|\lambda^m - \lambda\|_M = 0.$$

In turn, this implies the convergence of (u^m). □

Remark 4.7. We infer from (4.65) that each iteration reduces the norm of the error: $(c(v^m))^{1/2}$ by a factor which is at least $\theta^{1/2} < 1$. Therefore, the convergence of the iterative method (4.54′) (4.56) is linear.

Remark 4.8. The simple gradient algorithm with constant parameter ρ instead of ρ^m is also widely used. Of course, fixing ρ once and for all reduces the volume of computation, but this also reduces the rate of convergence of the method. Both theoretical and practical arguments lead us to choose r and ρ according to the following criteria:

i) take r as large as possible, for the conditioning of A_r increases with r;

ii) take

$$\rho = (r\gamma^2)/\|c\|^2.$$

Indeed, we have

$$\langle C^{-1}\chi, \chi\rangle \geqslant (\gamma/\|c\|^2)\|\chi\|^2_{M'};$$

when $a(.,.)$ is semi-positive definite on X (cf. (4.16)), this implies:

$$a_r(v, v) \geqslant r\langle C^{-1}Bv, Bv\rangle \geqslant (\gamma r/\|c\|^2)\|Bv\|^2_{M'},$$

whence

$$\tilde{\alpha}_r \geqslant \gamma r/\|c\|^2.$$

Therefore the above ρ satisfies (4.63).

More specifically, when $M = M'$ and $C = \mathrm{Id}$, we can take $\rho = r$.

Now, we turn to the *conjugate-gradient method*. Instead of updating λ^m along the direction of the gradient (i.e. along the direction of steepest descent) like in (4.61), we propose the conjugate direction to that of the preceding descent with respect to the hypersurface K_r = constant. This amounts to take:

$$(4.66)\qquad \omega^m = g^m + \sigma^m \omega^{m-1},$$

where the real parameter σ^m is chosen so that

$$D^2 K_r \cdot (\omega^m, \omega^{m-1}) = 0$$

i.e.

(4.67) $$\sigma^m = -D^2K_r\cdot(g^m,\omega^{m-1})/D^2K_r\cdot(\omega^{m-1},\omega^{m-1}).$$

In fact, ρ^m and σ^m have a more amenable expression. Recall the classical theorem (that can be found in every text concerned with conjugate-gradient methods):

Theorem 4.8. *We have the following orthogonality relations:*

$$D^2K_r\cdot(\omega^i,\omega^j)=0 \quad i\neq j,$$
$$c(g^i,g^j)=0 \quad i\neq j,$$
$$c(g^i,\omega^j)=0 \quad i>j.$$

As a consequence, we can simplify (4.55) and (4.67):

Corollary 4.5. *The parameters ρ^m and σ^m verify:*

(4.68) $$\begin{cases}\sigma^m = c(g^m,g^m)/c(g^{m-1},g^{m-1}),\\ \rho^m = c(g^m,g^m)/D^2K_r\cdot(\omega^m,g^m).\end{cases}$$

Proof. To begin with, recall that

$$c(g^m,v)=DK_r(\lambda^m)\cdot v \quad \forall v\in M.$$

Next, like in Theorem 4.6 we obtain:

$$DK_r(\lambda^{m+1})-DK_r(\lambda^m)=D^2K_r\cdot(\lambda^{m+1}-\lambda^m)=-\rho^m D^2K_r\cdot\omega^m,$$

i.e.

(4.69) $$c(g^{m+1}-g^m,v)=-\rho^m D^2K_r\cdot(\omega^m,v) \quad \forall v\in M.$$

On the other hand since D^2K_r is self-adjoint (cf. (4.52)), we have:

$$\begin{aligned}\sigma^m &= -[\rho^{m-1}D^2K_r\cdot(\omega^{m-1},g^m)]/[\rho^{m-1}D^2K_r\cdot(\omega^{m-1},\omega^{m-1})]\\ &= -c(g^m-g^{m-1},g^m)/c(g^m-g^{m-1},\omega^{m-1}).\end{aligned}$$

Then Theorem 4.8 and (4.66) yield:

$$\begin{aligned}\sigma^m &= c(g^m,g^m)/c(g^{m-1},\omega^{m-1})\\ &= c(g^m,g^m)/c(g^{m-1},g^{m-1}).\end{aligned}$$

Finally, we readily derive the desired expression for ρ^m by taking $v=g^m$ in (4.69). □

With this corollary, we can organize more efficiently the general step of the *conjugate-gradient algorithm*:

For $m\geqslant 0$, knowing $(u^m,\lambda^m)\in X\times M$, compute $g^m,\omega^m\in M$, $z^m\in X$, $\rho^m,\sigma^m\in\mathbb{R}$ and the pair $(u^{m+1},\lambda^{m+1})\in X\times M$ by:

$$(4.70)\quad \begin{cases} Cg^m = \chi - Bu^m, \\ \left.\begin{aligned} \sigma^m &= c(g^m, g^m)/c(g^{m-1}, g^{m-1}) \\ \omega^m &= g^m + \sigma^m \omega^{m-1} \end{aligned}\right\} \begin{array}{l} \text{only if } m \geqslant 1, \\ \omega^0 = g^0 \text{ otherwise,} \end{array} \\ A_r z^m = B'\omega^m, \\ \rho^m = c(g^m, g^m)/b(z^m, g^m), \\ \lambda^{m+1} = \lambda^m - \rho^m \omega^m, \\ u^{m+1} = u^m + \rho^m z^m. \end{cases}$$

Note that the volume of computation per iteration is slightly larger than that of the simple-gradient algorithm; but the bulk of the computation still lies in the inversion of A_r.

Finally, it is easy to prove the convergence of this scheme.

Corollary 4.6. *Under the hypotheses of Corollary* 4.4, *the conjugate-gradient algorithm* (4.70) *is convergent.*

Proof. Observe that the convergence criterion (4.58) is equivalent to:

$$(4.71)\qquad \rho^m \geqslant \alpha' \|g^m\|_M / \|\omega^m\|_M \quad \text{for some constant } \alpha' > 0.$$

Indeed, in view of (4.57) and the expression of z^m we have

$$\begin{aligned} \rho^m &= c(g^m, \omega^m)/b(z^m, \omega^m) \\ &= c(g^m, \omega^m)/D^2 K_r \cdot (\omega^m, \omega^m). \end{aligned}$$

Hence, (4.58) and (4.59) imply (4.71) with $\alpha' = \alpha/\tau_r$; conversely (4.71) and (4.59) yield (4.58) with $\alpha = \alpha'\delta_r$.

Next, note that

$$D^2 K_r \cdot (\omega^m, g^m) = D^2 K_r \cdot (\omega^m, \omega^m) > 0 \qquad \forall m.$$

Therefore (4.71) is an immediate consequence of (4.68). Hence the scheme is convergent. □

§ 5. The Stokes Equations

Let us consider the Navier-Stokes equations describing the N-dimensional motion of an incompressible viscous fluid:

$$(5.1)\qquad \rho\left(\frac{\partial u_i}{\partial t} + \sum_{j=1}^{N} u_j \frac{\partial u_i}{\partial x_j}\right) - \sum_{j=1}^{N} \frac{\partial \sigma_{ij}}{\partial x_j} = \rho f_i, \qquad 1 \leqslant i \leqslant N,$$

$$(5.2)\qquad \operatorname{div} \mathbf{u} = \sum_{i=1}^{N} D_{ii}(\mathbf{u}) = 0 \quad \text{(incompressibility condition),}$$

where

$$\sigma_{ij} = -P\delta_{ij} + 2\mu D_{ij}(\mathbf{u}), \qquad 1 \leqslant i, j \leqslant N, \tag{5.3}$$

and

$$D_{ij}(\mathbf{u}) = (1/2)(\partial u_i/\partial x_j + \partial u_j/\partial x_i).$$

In these equations, $\mathbf{u} = (u_1, \ldots, u_N)$ is the velocity of the fluid, ρ is its density (assumed to be constant), $\mu > 0$ is its viscosity (also assumed to be constant) and P is its pressure; (σ_{ij}) is the stress tensor and $\mathbf{f} = (f_1, \ldots, f_N)$ represents a density of body forces per unit mass (gravity for instance).

As usual, we set

$$p = P/\rho, \qquad \nu = \mu/\rho. \tag{5.4}$$

Here, p is the kinematic pressure and ν is the kinematic viscosity but for the sake of simplicity, they will be called in the sequel pressure and viscosity. With these notations, the Navier-Stokes equations become

$$\begin{aligned} &\frac{\partial u_i}{\partial t} + \sum_{j=1}^{N} u_j \frac{\partial u_i}{\partial x_j} - 2\nu \sum_{j=1}^{N} \frac{\partial D_{ij}(\mathbf{u})}{\partial x_j} + \frac{\partial p}{\partial x_i} = f_i, \qquad 1 \leqslant i \leqslant N, \\ &\operatorname{div} \mathbf{u} = 0. \end{aligned} \tag{5.5}$$

Note that, when $\operatorname{div} \mathbf{u} = 0$, the following identity holds

$$\sum_{j=1}^{N} \frac{\partial D_{ij}(\mathbf{u})}{\partial x_j} = (1/2) \sum_{j=1}^{N} \left(\frac{\partial^2 u_i}{\partial x_j^2} + \frac{\partial^2 u_j}{\partial x_i \partial x_j} \right) = (1/2)\Delta u_i \tag{5.6}$$

so that (5.5) can be written more conveniently

$$\begin{cases} \dfrac{\partial \mathbf{u}}{\partial t} + \displaystyle\sum_{j=1}^{N} u_j \frac{\partial \mathbf{u}}{\partial x_j} - \nu \Delta \mathbf{u} + \mathbf{grad}\, p = \mathbf{f}, \\ \operatorname{div} \mathbf{u} = 0. \end{cases} \tag{5.7}$$

Another canonical way of writing the Navier-Stokes equations consists in introducing a parameter (the Reynolds number) which measures the effect of viscosity on the flow. For a given problem, let L be a characteristic length and U a characteristic velocity. This determines a characteristic time $T = L/U$. Then, we introduce the dimensionless quantities

$$x' = x/L, \qquad \mathbf{u}' = \mathbf{u}/U, \qquad t' = t/T.$$

Using this change of variables, it is easy to check that the Navier-Stokes equations become (with obvious notations):

$$\begin{cases} \dfrac{\partial \mathbf{u}'}{\partial t'} + \displaystyle\sum_{j=1}^{N} u_j' \frac{\partial \mathbf{u}'}{\partial x_j'} - \frac{\nu}{LU} \Delta' \mathbf{u}' + \mathbf{grad}'\, p' = \mathbf{f}' \\ \operatorname{div}_{x'} \mathbf{u}' = 0, \end{cases}$$

where

$$p' = P/(\rho U^2), \qquad \mathbf{f}' = L\mathbf{f}/U^2.$$

Now, if we define the Reynolds number Re to be the dimensionless number

$$\mathrm{Re} = LU/\nu,$$

we find that the Navier-Stokes equations may be written in dimensionless variables

$$(5.8)\qquad \begin{cases} \dfrac{\partial \mathbf{u}}{\partial t} + \sum_{j=1}^{N} u_j \dfrac{\partial \mathbf{u}}{\partial x_j} - \dfrac{1}{\mathrm{Re}} \Delta \mathbf{u} + \mathbf{grad}\, p = \mathbf{f}, \\ \operatorname{div} \mathbf{u} = 0. \end{cases}$$

We obtain again the equations (5.7) with ν replaced by $1/Re$.

For the time being, we introduce two simplifications in the equations (5.5) or (5.7). We only consider the steady-state (or stationary) case, that is $\partial \mathbf{u}/\partial t = \mathbf{0}$, and furthermore we assume that the velocity $\mathbf{u}$ is sufficiently small for ignoring the nonlinear convection terms $u_j(\partial u_i/\partial x_j)$. Thus we are led to the Stokes equations

$$(5.9)\qquad \begin{cases} -2\nu \sum_{j=1}^{N} \dfrac{\partial D_{ij}(\mathbf{u})}{\partial x_j} + \dfrac{\partial p}{\partial x_i} = f_i, \qquad 1 \leqslant i \leqslant N, \\ \operatorname{div} \mathbf{u} = 0, \end{cases}$$

which can be written more conveniently

$$(5.10)\qquad \begin{cases} -\nu \Delta \mathbf{u} + \mathbf{grad}\, p = \mathbf{f}, \\ \operatorname{div} \mathbf{u} = 0. \end{cases}$$

The Stokes equations are linear but nevertheless they deserve special attention because of the incompressibility condition $\operatorname{div} \mathbf{u} = 0$. In this paragraph, we shall establish the existence and uniqueness of the solution of the Stokes equations and we shall derive several variational formulations that will be used later on for approximation purposes.

5.1. The Dirichlet Problem in the Velocity-Pressure Formulation

In order to get a well-posed problem for the Stokes equations (5.10), we have to complete them with appropriate boundary conditions. We begin with the Dirichlet boundary conditions.

Theorem 5.1. *Let Ω be a bounded and connected open subset of $\mathbb{R}^N$ with a Lipschitz-continuous boundary Γ. Given $\mathbf{f} \in H^{-1}(\Omega)^N$ and $\mathbf{g} \in H^{1/2}(\Gamma)^N$ such that*

$$(5.11)\qquad \int_\Gamma \mathbf{g} \cdot \mathbf{n}\, ds = 0,$$

there exists a unique pair $(\mathbf{u}, p) \in H^1(\Omega)^N \times L_0^2(\Omega)$ *solution of the equations*

$$(5.12) \qquad \begin{cases} -\nu \Delta \mathbf{u} + \mathbf{grad}\, p = \mathbf{f} & \text{in } \Omega, \\ \operatorname{div} \mathbf{u} = 0 & \text{in } \Omega, \\ \mathbf{u} = \mathbf{g} & \text{on } \Gamma. \end{cases}$$

Proof. By virtue of (5.11) and Lemma 2.2, there exists a function $\mathbf{u}_0 \in H^1(\Omega)^N$ such that

$$\operatorname{div} \mathbf{u}_0 = 0 \quad \text{in } \Omega, \qquad \mathbf{u}_0 = \mathbf{g} \quad \text{on } \Gamma.$$

Now, let us put Problem (5.12) into the framework of Paragraph 4. We set:

$$X = H_0^1(\Omega)^N, \qquad M = L_0^2(\Omega)$$

$$\text{with norms } \|\cdot\|_X = |\cdot|_{1,\Omega}, \qquad \|\cdot\|_M = \|\cdot\|_{0,\Omega},$$

$$(5.13) \qquad a(\mathbf{u}, \mathbf{v}) = \nu \sum_{i,j=1}^{N} \left(\frac{\partial u_i}{\partial x_j}, \frac{\partial v_i}{\partial x_j} \right) = \nu(\mathbf{grad}\, \mathbf{u}, \mathbf{grad}\, \mathbf{v}),$$

$$b(\mathbf{v}, q) = -(q, \operatorname{div} \mathbf{v}),$$

$$\langle \mathbf{l}, \mathbf{v} \rangle = \langle \mathbf{f}, \mathbf{v} \rangle - a(\mathbf{u}_0, \mathbf{v}), \qquad \chi = 0.$$

Then

$$V = \{\mathbf{v} \in H_0^1(\Omega)^N;\ \operatorname{div} \mathbf{v} = 0\}.$$

We must check that the form $a(.,.)$ is V-elliptic and the form $b(.,.)$ satisfies the inf-sup condition (4.9). On the one hand, the ellipticity property is obvious since

$$a(\mathbf{v}, \mathbf{v}) = \nu |\mathbf{v}|_{1,\Omega}^2.$$

On the other hand, the inf-sup condition says that

$$(5.14) \qquad \sup_{\mathbf{v} \in H_0^1(\Omega)^N} \frac{(q, \operatorname{div} \mathbf{v})}{|\mathbf{v}|_{1,\Omega}} \geqslant \beta \|q\|_{0,\Omega} \qquad \forall q \in L_0^2(\Omega).$$

Let $q \in L_0^2(\Omega)$; by virtue of Corollary 2.4, there exists a unique function $\mathbf{v} \in V^\perp$ such that

$$\operatorname{div} \mathbf{v} = q, \qquad |\mathbf{v}|_{1,\Omega} \leqslant C \|q\|_{0,\Omega}.$$

Hence

$$\frac{(q, \operatorname{div} \mathbf{v})}{|\mathbf{v}|_{1,\Omega}} = \frac{\|q\|_{0,\Omega}^2}{|\mathbf{v}|_{1,\Omega}} \geqslant (1/C) \|q\|_{0,\Omega}$$

from which (5.14) follows with $\beta = 1/C$.

We are now in a position to apply Corollary 4.1: there exists a unique pair of functions $(\mathbf{w}, p) \in H_0^1(\Omega)^N \times L_0^2(\Omega)$ such that

$$\begin{cases} a(\mathbf{w}, \mathbf{v}) + b(\mathbf{v}, p) = \langle \mathbf{l}, \mathbf{v} \rangle & \forall \mathbf{v} \in H_0^1(\Omega)^N, \\ b(\mathbf{w}, q) = 0 & \forall q \in L_0^2(\Omega). \end{cases}$$

Equivalently $(\mathbf{u} = \mathbf{u}_0 + \mathbf{w}, p) \in [\mathbf{u}_0 + H_0^1(\Omega)^N] \times L_0^2(\Omega)$ is the solution of the equations

$$\begin{cases} \nu(\mathbf{grad\, u}, \mathbf{grad\, v}) - (p, \operatorname{div} \mathbf{v}) = \langle \mathbf{f}, \mathbf{v} \rangle & \forall \mathbf{v} \in H_0^1(\Omega)^N, \\ (q, \operatorname{div} \mathbf{u}) = 0 & \forall q \in L_0^2(\Omega). \end{cases}$$

Applying again Corollary 2.4, this last equation is equivalent to $\operatorname{div} \mathbf{u} = 0$. Moreover, $\mathbf{u} \in \mathbf{u}_0 + H_0^1(\Omega)^N$ if and only if

$$\mathbf{u} \in H^1(\Omega)^N, \qquad \mathbf{u}|_\Gamma = \mathbf{g}.$$

Hence, there exists a unique pair $(\mathbf{u}, p) \in H^1(\Omega)^N \times L_0^2(\Omega)$ such that

$$\begin{cases} \nu(\mathbf{grad\, u}, \mathbf{grad\, v}) - (p, \operatorname{div} \mathbf{v}) = \langle \mathbf{f}, \mathbf{v} \rangle & \forall \mathbf{v} \in H_0^1(\Omega)^N, \\ \operatorname{div} \mathbf{u} = 0, \\ \mathbf{u}|_\Gamma = \mathbf{g}. \end{cases}$$

Now, using classical arguments, it is easy to show that this last problem is equivalent to Problem (5.12). □

Remark 5.1. The choice $M = L_0^2(\Omega)$ in (5.13) is only a matter of convenience and we can just as well take $M = L^2(\Omega)/\mathbb{R}$. On the other hand, we can also choose in the above proof

$$a(\mathbf{u}, \mathbf{v}) = 2\nu \sum_{i,j=1}^{N} \int_\Omega D_{ij}(\mathbf{u}) D_{ij}(\mathbf{v})\, dx.$$

This is a consequence of the following identity:

$$\sum_{i,j=1}^{N} \int_\Omega D_{ij}(\mathbf{u}) D_{ij}(\mathbf{v})\, dx = \frac{1}{2} \sum_{i,j=1}^{N} \int_\Omega \frac{\partial u_i}{\partial x_j} \frac{\partial v_i}{\partial x_j}\, dx \tag{5.15}$$

which holds *for all* $\mathbf{u} \in H^1(\Omega)^N$ *with* $\operatorname{div} \mathbf{u} = 0$ *and all* $\mathbf{v} \in H_0^1(\Omega)^N$. In fact, using the symmetry of the operator D_{ij} with respect to i and j, we have

$$\sum_{i,j=1}^{N} \int_\Omega D_{ij}(\mathbf{u}) D_{ij}(\mathbf{v})\, dx = \sum_{i,j=1}^{N} \int_\Omega D_{ij}(\mathbf{u}) \frac{\partial v_i}{\partial x_j}\, dx.$$

Moreover, we get

$$\sum_{i,j=1}^{N} \int_\Omega \frac{\partial u_j}{\partial x_i} \frac{\partial v_i}{\partial x_j}\, dx = -\sum_{i=1}^{N} \left\langle \frac{\partial \operatorname{div} \mathbf{u}}{\partial x_i}, v_i \right\rangle = 0,$$

from which (5.15) follows.

Remark 5.2. Problem (5.12) has the following variational formulations of (Q) and (P) types respectively:

Find a pair $(\mathbf{u}, p) \in H^1(\Omega)^N \times L_0^2(\Omega)$ *such that*

$$(5.16)\qquad \begin{cases} a(\mathbf{u}, \mathbf{v}) - (p, \operatorname{div} \mathbf{v}) = \langle \mathbf{f}, \mathbf{v} \rangle & \forall \mathbf{v} \in H_0^1(\Omega)^N, \\ (q, \operatorname{div} \mathbf{u}) = 0 & \forall q \in L_0^2(\Omega), \\ \mathbf{u} = \mathbf{g} & \text{on } \Gamma, \end{cases}$$

and

Find $\mathbf{u} \in H^1(\Omega)^N$ *such that*

$$(5.17)\qquad \begin{cases} a(\mathbf{u}, \mathbf{v}) = \langle \mathbf{f}, \mathbf{v} \rangle & \forall \mathbf{v} \in V, \\ \operatorname{div} \mathbf{u} = 0 & \text{in } \Omega, \\ \mathbf{u} = \mathbf{g} & \text{on } \Gamma, \end{cases}$$

where

$$a(\mathbf{u}, \mathbf{v}) = \begin{cases} \nu(\mathbf{grad}\, \mathbf{u}, \mathbf{grad}\, \mathbf{v}), \\ 2\nu \sum_{i,j=1}^{N} (D_{ij}(\mathbf{u}), D_{ij}(\mathbf{v})). \end{cases}$$

Remark 5.3. Corollary 4.1 yields the bound:

$$\|\mathbf{u}\|_{1,\Omega} + \|p\|_{0,\Omega} \leqslant C(\|\mathbf{f}\|_{-1,\Omega} + \|\mathbf{u}_0\|_{1,\Omega})$$

for all functions $\mathbf{u}_0$ in $H^1(\Omega)^N$ satisfying $\operatorname{div} \mathbf{u}_0 = 0$, $\mathbf{u}_0|_\Gamma = \mathbf{g}$. By taking the infimum with respect to $\mathbf{u}_0$, this becomes

$$\|\mathbf{u}\|_{1,\Omega} + \|p\|_{0,\Omega} \leqslant C(\|\mathbf{f}\|_{-1,\Omega} + \|\mathbf{g}\|_{1/2,\Gamma}).$$

Problem (5.12) can also be expressed as a saddle-point problem. With the above notations, we set

$$(5.18)\qquad J(\mathbf{v}) = (1/2)a(\mathbf{v}, \mathbf{v}) - \langle \mathbf{f}, \mathbf{v} \rangle,$$

$$(5.19)\qquad \mathscr{L}(\mathbf{v}, q) = J(\mathbf{v}) - (q, \operatorname{div} \mathbf{v}).$$

Theorem 5.2. *Under the hypotheses of Theorem* 5.1, *the solution* $(\mathbf{u}, p)$ *of* (5.12) *is characterized by*:

$$(5.20)\qquad \begin{cases} \mathscr{L}(\mathbf{u}, p) = \operatorname*{Min}_{\substack{\mathbf{v} \in H^1(\Omega)^N \\ \mathbf{v}|_\Gamma = \mathbf{g}}} \sup_{q \in L_0^2(\Omega)} \mathscr{L}(\mathbf{v}, q) \\ \qquad = \operatorname*{Max}_{q \in L_0^2(\Omega)} \inf_{\substack{\mathbf{v} \in H^1(\Omega)^N \\ \mathbf{v}|_\Gamma = \mathbf{g}}} \mathscr{L}(\mathbf{v}, q). \end{cases}$$

Proof. We use the notations (5.13). Since the form $a(.,.)$ is symmetric and $H_0^1(\Omega)^N$ elliptic and the form $b(.,.)$ satisfies the inf-sup condition (5.14), we may apply Theorem 4.2. We find that the pair $(\mathbf{u} - \mathbf{u}_0, p)$ is the unique saddle-point of the Lagrangean functional

$$\mathscr{L}_{\mathbf{u}_0}(\mathbf{v}, q) = (1/2)a(\mathbf{v}, \mathbf{v}) - \langle \mathbf{f}, \mathbf{v} \rangle + a(\mathbf{u}_0, \mathbf{v}) - (q, \operatorname{div} \mathbf{v})$$

over the product space $H_0^1(\Omega)^N \times L_0^2(\Omega)$. By taking into account that $\operatorname{div} \mathbf{u}_0 = \operatorname{div} \mathbf{u} = 0$, this exactly means that

$$\mathscr{L}(\mathbf{u}, q) = \mathscr{L}(\mathbf{u}, p) \leqslant \mathscr{L}(\mathbf{v} + \mathbf{u}_0, p) \quad \forall \mathbf{v} \in H_0^1(\Omega)^N, \quad \forall q \in L_0^2(\Omega)$$

and therefore that $(\mathbf{u}, p)$ is the unique saddle-point of the functional $\mathscr{L}$ over $[\mathbf{u}_0 + H_0^1(\Omega)^N] \times L_0^2(\Omega)$.

Now, using

$$\mathbf{u}_0 + H_0^1(\Omega)^N = \{\mathbf{v} \in H^1(\Omega)^N; \mathbf{v}|_\Gamma = \mathbf{g}\}$$

and arguing as in the proof of Corollary 4.2, we obtain the characterization (5.20). □

By adapting the arguments of Paragraph 4.2 to the present situation, we find that $\mathbf{u}$ is characterized by

(5.21) $$J(\mathbf{u}) = \inf_{\substack{\mathbf{v} \in H^1(\Omega)^N \\ \operatorname{div} \mathbf{v} = 0,\, \mathbf{v}|_\Gamma = \mathbf{g}}} J(\mathbf{v}).$$

On the other hand, with any $q \in L_0^2(\Omega)$, we associate the solution $\mathbf{u}(q) \in H^1(\Omega)^N$ of the boundary value problem

(5.22) $$\begin{cases} -\nu \Delta \mathbf{u}(q) = \mathbf{f} - \mathbf{grad}\, q & \text{in } \Omega, \\ \mathbf{u} = \mathbf{g} & \text{on } \Gamma. \end{cases}$$

Then setting

(5.23) $$K(q) = (1/2)a(\mathbf{w}(q), \mathbf{w}(q))$$

where $\mathbf{w}(q) = \mathbf{u}(q) - \mathbf{u}_0$ with $\operatorname{div} \mathbf{u}_0 = 0$, $\mathbf{u}_0|_\Gamma = \mathbf{g}$, we find that the function $p \in L_0^2(\Omega)$ is characterized by

(5.24) $$K(p) = \operatorname*{Min}_{q \in L_0^2(\Omega)} K(q).$$

In order to eliminate the pressure, we can use the regularization method or penalty method introduced in Section 4.3. We consider the following problem:

Given $\varepsilon > 0$, *find* $(\mathbf{u}^\varepsilon, p^\varepsilon) \in H^1(\Omega)^N \times L_0^2(\Omega)$ *such that*

(5.25) $$\begin{cases} -\nu \Delta \mathbf{u}^\varepsilon + \mathbf{grad}\, p^\varepsilon = \mathbf{f} & \textit{in } \Omega, \\ p^\varepsilon = -(1/\varepsilon) \operatorname{div} \mathbf{u}^\varepsilon & \textit{in } \Omega, \\ \mathbf{u}^\varepsilon = \mathbf{g} & \textit{on } \Gamma. \end{cases}$$

By eliminating p^ε, we get an equivalent second order elliptic problem in $\mathbf{u}^\varepsilon$:

(5.26) $$\begin{cases} -\nu \Delta \mathbf{u}^\varepsilon - (1/\varepsilon)\, \mathbf{grad} \operatorname{div} \mathbf{u}^\varepsilon = \mathbf{f} & \textit{in } \Omega, \\ \mathbf{u}^\varepsilon = \mathbf{g} & \textit{on } \Omega. \end{cases}$$

Theorem 5.3. *Assume the hypotheses of Theorem* 5.1. *Then, there exists a unique pair of functions* $(\mathbf{u}^\varepsilon, p^\varepsilon) \in H^1(\Omega)^N \times L_0^2(\Omega)$ *solution of the equations* (5.25). *Moreover, we get the estimate:*

$$\|\mathbf{u}^\varepsilon - \mathbf{u}\|_{1,\Omega} + \|p^\varepsilon - p\|_{0,\Omega} \leqslant C\varepsilon(\|\mathbf{f}\|_{-1,\Omega} + \|\mathbf{g}\|_{1/2,\Gamma}), \tag{5.27}$$

where the constant C *is independent of* ε, $\mathbf{f}$ *and* $\mathbf{g}$.

Proof. Like in Theorem 5.1, we easily obtain the existence and uniqueness of the solution pair $(\mathbf{u}^\varepsilon, p^\varepsilon)$ of Problem (5.25). Next, considering that the difference $\mathbf{u}^\varepsilon - \mathbf{u}$ vanishes on Γ, the argument of Theorem 4.3 immediately gives:

$$\|\mathbf{u}^\varepsilon - \mathbf{u}\|_{1,\Omega} + \|p^\varepsilon - p\|_{0,\Omega} \leqslant C_1\varepsilon\|p\|_{0,\Omega},$$

with a constant C_1 that is independent of ε. Hence (5.27) follows from Remark 5.3. □

Similarly, we can also apply Theorem 4.4. We obtain that the problems

$$\begin{cases} -\nu\Delta\mathbf{u}_n + \mathbf{grad}\, p_n = \mathbf{0} & \text{in } \Omega, \\ \operatorname{div}\mathbf{u}_n = -p_{n-1} & \text{in } \Omega, \\ \mathbf{u}_n = \mathbf{0} & \text{on } \Gamma, \end{cases} \tag{5.28}$$

starting with $p_0 = p$, uniquely define by induction a sequence $(\mathbf{u}_n, p_n)$ in $H_0^1(\Omega)^N \times L_0^2(\Omega)$. Furthermore, we get for all $M \geqslant 1$ and ε small enough

$$\begin{cases} \left\|\mathbf{u}^\varepsilon - \mathbf{u} - \sum_{n=1}^{M} \varepsilon^n \mathbf{u}_n\right\|_{1,\Omega} + \left\|p^\varepsilon - p - \sum_{n=1}^{M} \varepsilon^n p_n\right\|_{0,\Omega} \\ \leqslant C_M \varepsilon^{M+1}(\|\mathbf{f}\|_{-1,\Omega} + \|\mathbf{g}\|_{1/2,\Gamma}), \end{cases} \tag{5.29}$$

where C_M is a constant independent of ε, $\mathbf{f}$ and $\mathbf{g}$.

Remark 5.4. If we choose

$$a(\mathbf{u}, \mathbf{v}) = 2\nu \sum_{i,j=1}^{N} \int_\Omega D_{ij}(\mathbf{u})D_{ij}(\mathbf{v})\, dx,$$

we obtain a different regularized problem:

$$\begin{cases} -2\nu \sum_{j=1}^{N} \dfrac{\partial D_{ij}(\mathbf{u}^\varepsilon)}{\partial x_j} + \dfrac{\partial p^\varepsilon}{\partial x_i} = f_i \quad \text{in } \Omega, \quad 1 \leqslant i \leqslant N, \\ p^\varepsilon = -(1/\varepsilon)\operatorname{div}\mathbf{u}^\varepsilon \quad \text{in } \Omega, \\ \mathbf{u}^\varepsilon = \mathbf{g} \quad \text{on } \Gamma. \end{cases}$$

By eliminating p^ε, we get the penalized problem

$$\begin{cases} -\nu\Delta\mathbf{u}^\varepsilon - (\nu + 1/\varepsilon)\,\mathbf{grad}\operatorname{div}\mathbf{u}^\varepsilon = \mathbf{f} & \text{in } \Omega, \\ \mathbf{u}^\varepsilon = \mathbf{g} & \text{on } \Gamma. \end{cases}$$

Now, in order to apply Theorem 4.3, we need here to check the assumption (4.31), namely:

$$(5.30)\quad \begin{cases} 2\nu \sum_{i,j=1}^{N} \|D_{ij}(\mathbf{v})\|_{0,\Omega}^2 + \|\operatorname{div} \mathbf{v}\|_{0,\Omega}^2 \\ \geqslant \alpha_0 |\mathbf{v}|_{1,\Omega}^2 \qquad \forall \mathbf{v} \in H_0^1(\Omega)^N, \qquad \alpha_0 > 0. \end{cases}$$

But Remark 5.1 establishes that

$$\begin{aligned} \sum_{i,j=1}^{N} (D_{ij}(\mathbf{u}), D_{ij}(\boldsymbol{\phi})) &= -(1/2)\{\langle \Delta \mathbf{u}, \boldsymbol{\phi}\rangle + \langle \mathbf{grad}(\operatorname{div} \mathbf{u}), \boldsymbol{\phi}\rangle\} \\ &= (1/2)\{(\mathbf{grad}\, \mathbf{u}, \mathbf{grad}\, \boldsymbol{\phi}) + (\operatorname{div} \mathbf{u}, \operatorname{div} \boldsymbol{\phi})\} \end{aligned}$$

for all $\mathbf{u} \in H^1(\Omega)^N$, for all $\boldsymbol{\phi} \in \mathscr{D}(\Omega)^N$. Hence

$$(5.31)\qquad 2\nu \sum_{i,j=1}^{N} \|D_{ij}(\mathbf{v})\|_{0,\Omega}^2 = \nu(|\mathbf{v}|_{1,\Omega}^2 + \|\operatorname{div} \mathbf{v}\|_{0,\Omega}^2).$$

Therefore the analogue of Theorem 5.3 holds in that case.

Remark 5.5. The inequality (5.31) is also related to the classical *Korn's Inequality*:

$$(5.31')\quad \sum_{i,j=1}^{N} \|D_{ij}(\mathbf{v})\|_{0,\Omega}^2 + \|\mathbf{v}\|_{0,\Omega}^2 \geqslant \alpha_1 \|\mathbf{v}\|_{1,\Omega}^2 \qquad \forall \mathbf{v} \in H^1(\Omega)^N, \qquad \alpha_1 > 0.$$

For the sake of completeness, here is a concise proof of (5.31′) given by Duvaut & Lions [26] in the case of a bounded, Lipschitz-continuous open subset of $\mathbb{R}^N$.

Observe that formally the following identity holds:

$$\frac{\partial^2 v_i}{\partial x_j \partial x_k} = \frac{\partial(D_{ik}(\mathbf{v}))}{\partial x_j} + \frac{\partial(D_{ij}(\mathbf{v}))}{\partial x_k} - \frac{\partial(D_{jk}(\mathbf{v}))}{\partial x_i}, \qquad 1 \leqslant i, j, k \leqslant N.$$

Hence for $\mathbf{v}$ in $H^1(\Omega)^N$ we have

$$\begin{aligned} \|\mathbf{grad}(\partial v_i/\partial x_j)\|_{-1,\Omega} &\leqslant C_1 \sum_{i,j=1}^{N} \|\mathbf{grad}(D_{ij}(\mathbf{v}))\|_{-1,\Omega} \\ &\leqslant C_2 \sum_{i,j=1}^{N} \|D_{ij}(\mathbf{v})\|_{0,\Omega}. \end{aligned}$$

Therefore, applying Theorem 2.2 to $\mathbf{grad}\, v_i$ yields immediately (5.31).

Now, let us turn to the gradient methods to solve Problem (5.12). Again, we take

$$c(p, q) = (p, q),$$

so that

$$a_r(\mathbf{u}, \mathbf{v}) = \nu(\mathbf{grad}\, \mathbf{u}, \mathbf{grad}\, \mathbf{v}) + r(\operatorname{div} \mathbf{u}, \operatorname{div} \mathbf{v})$$

which is clearly elliptic on $H_0^1(\Omega)^N$. Thus, defining $\mathbf{u}_r(q)$ by:

$$\begin{cases} a_r(\mathbf{u}_r(q), \mathbf{v}) = \langle \mathbf{f}, \mathbf{v} \rangle + (\operatorname{div} \mathbf{v}, q) & \forall \mathbf{v} \in H_0^1(\Omega)^N, \\ \mathbf{u}_r(q) = \mathbf{g} \quad \text{on } \Gamma, \end{cases}$$

we find in view of (5.23):

$$\begin{aligned} a_r(D\mathbf{u}_r \cdot \mu, \mathbf{v}) &= (\operatorname{div} \mathbf{v}, \mu) \qquad \forall \mathbf{v} \in H_0^1(\Omega)^N, \\ D\mathbf{u}_r \cdot \mu &= \mathbf{0} \quad \text{on } \Gamma, \\ DK_r(q) &= \operatorname{div} \mathbf{u}_r(q), \\ D^2 K_r \cdot \mu &= \operatorname{div}(D\mathbf{u}_r \cdot \mu). \end{aligned}$$

Hence the *simple gradient algorithm with optimal parameter* reads as follows:

1°) Given $p^0 \in L_0^2(\Omega)$, solve the non-homogeneous elliptic boundary value problem:

$$\begin{cases} -(\nu\Delta + r\,\mathbf{grad}\operatorname{div})\mathbf{u}^0 = \mathbf{f} - \mathbf{grad}\, p^0 & \text{in } \Omega, \\ \mathbf{u}^0 = \mathbf{g} \quad \text{on } \Gamma. \end{cases}$$

2°) For $m \geqslant 0$, knowing $(\mathbf{u}^m, p^m)$ in $H^1(\Omega)^N \times L_0^2(\Omega)$ with $\mathbf{u}^m|_\Gamma = \mathbf{g}$, solve the homogeneous problem:

$$\begin{cases} -(\nu\Delta + r\,\mathbf{grad}\operatorname{div})\mathbf{z}^m = \mathbf{grad}\operatorname{div} \mathbf{u}^m & \text{in } \Omega, \\ \mathbf{z}^m = \mathbf{0} \quad \text{on } \Gamma; \end{cases}$$

then compute $\rho^m \in \mathbb{R}$ and the next pair $(\mathbf{u}^{m+1}, p^{m+1}) \in H^1(\Omega)^N \times L_0^2(\Omega)$ by:

$$\begin{aligned} \rho^m &= -\|\operatorname{div} \mathbf{u}^m\|_{0,\Omega}^2 / (\operatorname{div} \mathbf{z}^m, \operatorname{div} \mathbf{u}^m), \\ p^{m+1} &= p^m - \rho^m \operatorname{div} \mathbf{u}^m, \\ \mathbf{u}^{m+1} &= \mathbf{u}^m + \rho^m \mathbf{z}^m. \end{aligned}$$

It can be readily checked that the statement of Theorem 4.6 holds, namely the above scheme is always convergent. This is still true when ρ^m is not the optimal parameter, provided that

$$0 < \inf_m \rho^m \leqslant \sup_m \rho^m < 2\tilde{\alpha}_r.$$

A simple calculation shows that

$$\tilde{\alpha}_r \leqslant r + \nu/N.$$

The *conjugate-gradient algorithm* has the same starting procedure *n*°1 while step *n*°2 is replaced by:

2°) For $m \geqslant 0$, knowing $(\mathbf{u}^m, p^m)$ in $H^1(\Omega)^N \times L_0^2(\Omega)$ with $\mathbf{u}^m|_\Gamma = \mathbf{g}$, compute $\sigma^m \in \mathbb{R}$ and $\omega^m \in L_0^2(\Omega)$ by:

$$\left.\begin{aligned} \sigma^m &= \|\operatorname{div} \mathbf{u}^m\|_{0,\Omega}^2 / \|\operatorname{div} \mathbf{u}^{m-1}\|_{0,\Omega}^2, \\ \omega^m &= \operatorname{div} \mathbf{u}^m + \sigma^m \omega^{m-1}, \end{aligned}\right\} \quad \begin{aligned} &\text{only if } m \geqslant 1, \\ &\omega^0 = \operatorname{div} \mathbf{u}^0 \text{ otherwise,} \end{aligned}$$

solve the homogeneous boundary value problem:

$$\begin{cases} -(\nu\Delta + r\,\mathbf{grad}\,\mathrm{div})\mathbf{z}^m = \mathbf{grad}\,\omega^m & \text{in } \Omega, \\ \mathbf{z}^m = \mathbf{0} & \text{on } \Gamma, \end{cases}$$

and compute $\rho^m \in \mathbb{R}$ and the next pair $(\mathbf{u}^{m+1}, p^{m+1}) \in H^1(\Omega)^N \times L_0^2(\Omega)$ by:

$$\rho^m = -\frac{\|\mathrm{div}\,\mathbf{u}^m\|_{0,\Omega}^2}{(\mathrm{div}\,\mathbf{z}^m, \mathrm{div}\,\mathbf{u}^m)},$$

$$p^{m+1} = p^m - \rho^m \omega^m,$$

$$\mathbf{u}^{m+1} = \mathbf{u}^m + \rho^m \mathbf{z}^m.$$

Here again, Corollary 4.6 guarantees that the conjugate-gradient algorithm is convergent.

We end this section with a theorem due to Cattabriga [18] concerning the existence and regularity of the solution of the Stokes problem (5.12) in more general Sobolev spaces when the boundary Γ is sufficiently smooth.

Theorem 5.4. *In addition to the hypotheses of Theorem* 5.1, *suppose that the boundary Γ is of class $\mathscr{C}^{\max(2,m+2)}$, $\mathbf{f} \in W^{m,r}(\Omega)^N$ and $\mathbf{g} \in W^{m+2-1/r,r}(\Gamma)^N$ for some integer $m \geqslant -1$ and some real r with $1 < r < \infty$. Then, Problem* (5.12) *has a unique solution $(\mathbf{u}, p) \in W^{m+2,r}(\Omega)^N \times W^{m+1,r}(\Omega)$ with $\int_\Omega p\,dx = 0$ and there exists a constant C independent of* **f** *and* **g** *such that*

$$\|\mathbf{u}\|_{m+2,r,\Omega} + \|p\|_{m+1,r,\Omega} \leqslant C(\|\mathbf{f}\|_{m,r,\Omega} + \|\mathbf{g}\|_{m+2-1/r,r,\Gamma}). \tag{5.32}$$

Remark 5.6. In the case of a polyhedral domain Ω, the regularity properties of the solution of Problem (5.12) are clearly weaker. In particular, when $N = 2$ and Ω is a convex polygon, the conclusions of the theorem are still valid for $m \leqslant 0$ and $1 < r \leqslant 2$ (cf. Grisvard [43]).

5.2. The Stream Function Formulation of the Dirichlet Problem in Two Dimensions

Let us again denote by Γ_i, $0 \leqslant i \leqslant p$, the components of the boundary Γ like in Figure 2. Now, instead of (5.11), we assume the stronger condition

$$\int_{\Gamma_i} \mathbf{g} \cdot \mathbf{n}\,ds = 0, \qquad 0 \leqslant i \leqslant p. \tag{5.33}$$

Then, according to Theorems 3.1 and 3.4, the velocity field **u** given by Theorem 5.1 may be expressed as the **curl** of a stream function ψ ($N = 2$) or a vector potential $\boldsymbol{\psi}$ ($N = 3$). We are going to show that this stream function or this vector potential can be characterized as the solution of a biharmonic problem in Ω.

We begin with the case $N = 2$. Then, the stream function ψ is unique up to an additive constant. But, as $\psi \in H^2(\Omega) \subsetneq \mathscr{C}^0(\bar{\Omega})$, ψ can be uniquely determined by fixing its value in one point of $\bar{\Omega}$. At first, we set $\psi(x_0) = 0$, where x_0 is an arbitrary point of Γ_0 (for instance!). Next, we choose a function χ (in $H^{3/2}(\Gamma)$) which satisfies

$$\partial\chi/\partial\tau = \mathbf{g}\cdot\mathbf{n} \quad \text{on } \Gamma, \qquad \chi(x_0) = 0. \tag{5.34}$$

Since $\partial\phi/\partial\tau = \mathbf{g}\cdot\mathbf{n}$ on Γ, it follows that

$$\psi = \begin{cases} \chi & \text{on } \Gamma_0, \\ \chi + c_i & \text{on } \Gamma_i, \qquad 1 \leqslant i \leqslant p, \end{cases}$$

where the constants c_i are fixed but unknown.

Theorem 5.5. *Let $N = 2$ and let the hypotheses of Theorem* 5.1 *be satisfied together with* (5.33). *Then, the associated stream function of* $\mathbf{u}$ *may be characterized as the unique function $\psi \in H^2(\Omega)$ solution of the equations*

$$\nu(\Delta\psi, \Delta\phi) = \langle \mathbf{f}, \mathbf{curl}\,\phi\rangle \quad \forall \phi \in \Psi \quad (cf.\ (3.11)), \tag{5.35}$$

$$\begin{cases} \psi = \chi \quad \text{on } \Gamma_0, \qquad \psi = \chi + c_i \quad \text{on } \Gamma_i, \qquad 1 \leqslant i \leqslant p, \\ \partial\psi/\partial n = -\mathbf{g}\cdot\boldsymbol{\tau} \quad \text{on } \Gamma, \end{cases} \tag{5.36}$$

where χ is chosen according to (5.34).

Proof. We have already shown that the velocity field $\mathbf{u}$ is the unique solution of (5.17). Now, the function $\mathbf{u}$ satisfies the conditions

$$\operatorname{div}\mathbf{u} = 0 \quad \text{in } \Omega, \qquad \mathbf{u} = \mathbf{g} \quad \text{on } \Gamma$$

if and only if $\mathbf{u} = \mathbf{curl}\,\psi$ where the stream function $\psi \in H^2(\Omega)$ satisfies the boundary conditions (5.36). Besides that, according to Corollary 3.2, $\mathbf{v} \in V$ if and only if $\mathbf{v} = \mathbf{curl}\,\phi$ with $\phi \in \Psi$. Thus, the theorem will be proved if we show that

$$(\mathbf{grad}\,\mathbf{u}, \mathbf{grad}\,\mathbf{v}) = (\Delta\psi, \Delta\phi) \qquad \forall \mathbf{v} = \mathbf{curl}\,\phi, \qquad \phi \in \Psi. \tag{5.37}$$

Here, we use the identities stated at the beginning of Section 2.3. First, we have

$$(\Delta\psi, \Delta\phi) = (\operatorname{curl}(\mathbf{curl}\,\psi), \operatorname{curl}(\mathbf{curl}\,\phi)) = (\operatorname{curl}\mathbf{u}, \operatorname{curl}\mathbf{v}).$$

Next, we take $\mathbf{v}$ in $\mathscr{V} = \{\mathbf{v} \in \mathscr{D}(\Omega)^2; \operatorname{div}\mathbf{v} = 0\}$ and recall that according to Corollary 2.5, $\mathscr{V}$ is dense in V. In the sense of distributions, we have:

$$\begin{aligned} (\operatorname{curl}\mathbf{u}, \operatorname{curl}\mathbf{v}) &= \langle \mathbf{curl}(\operatorname{curl}\mathbf{u}), \mathbf{v}\rangle \\ &= \langle -\Delta\mathbf{u}, \mathbf{v}\rangle \quad \text{because} \quad \operatorname{div}\mathbf{u} = 0, \\ &= (\mathbf{grad}\,\mathbf{u}, \mathbf{grad}\,\mathbf{v}). \end{aligned}$$

Thus the density of $\mathscr{V}$ in V yields (5.37). □

It remains to interpret Problem (5.35) (5.36). By applying (formally) Green's formula, we can easily show that ψ is the only solution of the boundary value problem:

$$\nu \Delta^2 \psi = \operatorname{curl} \mathbf{f},$$

$$\psi = \chi \quad \text{on } \Gamma_0, \qquad \psi = c_i + \chi \quad \text{on } \Gamma_i, \qquad 1 \leqslant i \leqslant p,$$

$$\partial\psi/\partial n = -\mathbf{g} \cdot \boldsymbol{\tau} \quad \text{on } \Gamma,$$

$$\int_{\Gamma_i} (\nu \partial(\Delta\psi)/\partial n - \mathbf{f} \cdot \boldsymbol{\tau})\, ds = 0, \qquad 1 \leqslant i \leqslant p.$$

Note that this last condition makes sense if $\mathbf{f} \in H(\operatorname{curl}; \Omega)$.

5.3. The Three-Dimensional Case

Now consider the case $N = 3$. To simplify the discussion, we first examine the case of a homogeneous boundary condition $\mathbf{g} = \mathbf{0}$. We know from Section 3.3 that the velocity field $\mathbf{u}$ can be expressed as:

$$\mathbf{u} = \mathbf{curl}\, \boldsymbol{\psi} \quad \text{with} \quad \operatorname{div} \boldsymbol{\psi} = 0 \quad \text{in } \Omega.$$

When Ω is simply-connected, the vector potential $\boldsymbol{\psi}$ is uniquely determined by either:

(5.38) $$\boldsymbol{\psi} \cdot \mathbf{n} = 0 \quad \text{on } \Gamma$$

or

(5.39) $$\boldsymbol{\psi} \times \mathbf{n} = \mathbf{0} \quad \text{on } \Gamma \quad \text{and} \quad \int_{\Gamma_i} \boldsymbol{\psi} \cdot \mathbf{n}\, ds = 0 \qquad 0 \leqslant i \leqslant p.$$

As $\mathbf{u}$ vanishes on Γ, we can add the boundary condition:

(5.40) $$\mathbf{curl}\, \boldsymbol{\psi} = \mathbf{0} \quad \text{on } \Gamma$$

but observe that this condition reduces to

(5.40′) $$(\mathbf{curl}\, \boldsymbol{\psi}) \times \mathbf{n} = \mathbf{0} \quad \text{on } \Gamma$$

when $\boldsymbol{\psi}$ satisfies the first of (5.39) (cf. Remark 2.5). As far as the regularity of $\boldsymbol{\psi}$ is concerned, all we want is that $\boldsymbol{\psi}$ satisfy a biharmonic problem. Thus, it seems reasonable to ask that $\Delta \boldsymbol{\psi}$ belong to $L^2(\Omega)^3$. In view of the identity

(5.41) $$-\Delta \boldsymbol{\theta} = \mathbf{curl}\, \mathbf{curl}\, \boldsymbol{\theta} - \mathbf{grad}(\operatorname{div} \boldsymbol{\theta})$$

and the fact that $\mathbf{curl}\, \boldsymbol{\psi} \in H^1(\Omega)^3$ we see that this amounts to ask that $\operatorname{div} \boldsymbol{\psi} \in H^1(\Omega)$. Of course, this is always the case when $\boldsymbol{\psi}$ is divergence-free, but in practice a divergence-free condition is unattractive and we are going to relax it entirely.

At this stage, it is important to point out that the vector potential $\boldsymbol{\psi}$ has no straightforward characterization on a multiply-connected domain Ω. For this reason, we shall only discuss briefly this last situation and restrict ourselves mainly to simply-connected regions.

Let us begin with vector potentials that satisfy (5.39). In the light of the above considerations, we introduce the space

$$\Psi = \Big\{ \boldsymbol{\phi} \in L^2(\Omega)^3;\ \operatorname{div} \boldsymbol{\phi} \in H^1(\Omega),\ \mathbf{curl}\, \boldsymbol{\phi} \in H_0^1(\Omega)^3,\ \boldsymbol{\phi} \times \mathbf{n}|_\Gamma = \mathbf{0},$$

$$\int_{\Gamma_i} \boldsymbol{\phi} \cdot \mathbf{n}\, ds = 0,\ 0 \leqslant i \leqslant p \Big\},$$

normed by

$$\|\boldsymbol{\phi}\| = \{ \|\boldsymbol{\phi}\|_{0,\Omega}^2 + \|\operatorname{div} \boldsymbol{\phi}\|_{1,\Omega}^2 + \|\mathbf{curl}\, \boldsymbol{\phi}\|_{1,\Omega}^2 \}^{1/2}.$$

Note that the functions of Ψ are not divergence-free but satisfy $\int_\Omega \operatorname{div} \boldsymbol{\phi}\, dx = 0$.

Now, let $\mathbf{u}$ be the solution of Problem (5.12) with $\mathbf{g} = \mathbf{0}$. As mentioned above, when Ω is simply-connected, $\mathbf{u}$ has a unique divergence-free vector potential $\boldsymbol{\psi}$ in Ψ:

(5.42)
$$\begin{cases} -\nu \Delta(\mathbf{curl}\, \boldsymbol{\psi}) + \mathbf{grad}\, p = \mathbf{f} & \text{in } H^{-1}(\Omega)^3, \\ \operatorname{div} \boldsymbol{\psi} = 0 \quad \text{in } \Omega, & \boldsymbol{\psi} \in \Psi. \end{cases}$$

As $\operatorname{div}(\mathbf{curl}\, \boldsymbol{\psi}) = 0$, (5.42) reduces to:

$$\nu\, \mathbf{curl}\, \mathbf{curl}(\mathbf{curl}\, \boldsymbol{\psi}) + \mathbf{grad}\, p = \mathbf{f}.$$

Let us multiply both sides of this equation with $\mathbf{curl}\, \boldsymbol{\phi}$ for $\boldsymbol{\phi}$ in Ψ. Then

$$\langle \mathbf{curl}\, \mathbf{curl}(\mathbf{curl}\, \boldsymbol{\psi}), \mathbf{curl}\, \boldsymbol{\phi} \rangle = (\mathbf{curl}(\mathbf{curl}\, \boldsymbol{\psi}), \mathbf{curl}(\mathbf{curl}\, \boldsymbol{\phi}))$$

since $\mathbf{curl}\, \boldsymbol{\phi} \in H_0^1(\Omega)^3$. Thus it suffices to show that

(5.43)
$$(\mathbf{curl}\, \mathbf{curl}\, \boldsymbol{\psi}, \mathbf{grad}\, \operatorname{div} \boldsymbol{\phi}) = 0$$

in order to obtain that

$$-\langle \Delta(\mathbf{curl}\, \boldsymbol{\psi}), \mathbf{curl}\, \boldsymbol{\phi} \rangle = (\Delta \boldsymbol{\psi}, \Delta \boldsymbol{\phi}) \qquad \forall \boldsymbol{\phi} \in \Psi.$$

But (5.43) holds as soon as $\mathbf{curl}\, \boldsymbol{\psi} \in H_0^1(\Omega)^3$. Hence we have established that $\boldsymbol{\psi}$ is a solution of the biharmonic problem:

Find $\boldsymbol{\psi}$ in Ψ such that

(5.44)
$$\nu(\Delta \boldsymbol{\psi}, \Delta \boldsymbol{\phi}) = \langle \mathbf{f}, \mathbf{curl}\, \boldsymbol{\phi} \rangle \qquad \forall \boldsymbol{\phi} \in \Psi.$$

Conversely, it is easy to prove that (5.44) has at most one solution. Indeed, if $\boldsymbol{\psi} \in \Psi$ satisfies $\Delta \boldsymbol{\psi} = \mathbf{0}$ then $\mathbf{curl}\, \boldsymbol{\psi} = \mathbf{0}$ in Ω, for $\mathbf{curl}\, \boldsymbol{\psi}$ is the solution of the Dirichlet problem:

$$\Delta(\mathbf{curl}\, \boldsymbol{\psi}) = \mathbf{0} \quad \text{in } \Omega, \qquad \mathbf{curl}\, \boldsymbol{\psi}|_\Gamma = \mathbf{0}.$$

In turn, this implies that $\mathbf{grad}(\operatorname{div} \boldsymbol{\psi}) = \mathbf{0}$. Hence $\operatorname{div} \boldsymbol{\psi} = 0$, since $\int_\Omega \operatorname{div} \boldsymbol{\psi}\, dx = 0$. Therefore $\boldsymbol{\psi} = \mathbf{0}$ (cf. Remark 3.9). Thus we have proved the following result:

Lemma 5.1. *When Ω is simply-connected and satisfies the hypotheses of Theorem* 5.1, *the biharmonic problem* (5.44) *has a unique solution $\boldsymbol{\psi}$. Furthermore* $\operatorname{div} \boldsymbol{\psi} = 0$ *and $\mathbf{curl}\, \boldsymbol{\psi}$ is the unique solution of the Stokes problem* (5.12) *with* $\mathbf{g} = \mathbf{0}$.

Remark 5.7. The velocity vector $\mathbf{u}$ has a vector potential that satisfies (5.39) only if Ω is simply-connected. However, it is worth mentioning that Problem (5.44) is well-posed even when Ω is multiply-connected. In fact, we can easily prove the following equivalence of norms.

Lemma 5.2. *In addition to the hypotheses of Theorem* 5.1, *assume that either Γ is $\mathscr{C}^{1,1}$ or that Ω is simply-connected. Then the mapping $\boldsymbol{\psi} \to \|\Delta \boldsymbol{\psi}\|_{0,\Omega}$ is a norm on Ψ equivalent to $\|\boldsymbol{\psi}\|$.*

Proof. First, observe that $\|\boldsymbol{\psi}\|$ is equivalent on Ψ to:

$$\{\|\boldsymbol{\psi}\|_{0,\Omega}^2 + \|\operatorname{div}\boldsymbol{\psi}\|_{1,\Omega}^2 + |\mathbf{curl}\,\boldsymbol{\psi}|_{1,\Omega}^2\}^{1/2}$$

by virtue of Theorem 1.1. Next, remark that (5.41) is simply the orthogonal decomposition of the vector $\Delta\boldsymbol{\theta}$. This orthogonality yields:

$$\|\Delta\boldsymbol{\psi}\|_{0,\Omega}^2 = \|\mathbf{curl}\,\mathbf{curl}\,\boldsymbol{\psi}\|_{0,\Omega}^2 + |\operatorname{div}\boldsymbol{\psi}|_{1,\Omega}^2.$$

But since $\mathbf{curl}\,\boldsymbol{\psi} \in H_0^1(\Omega)^3$, we infer from Remark 2.7 that

$$|\mathbf{curl}\,\boldsymbol{\psi}|_{1,\Omega} \cong \|\mathbf{curl}\,\mathbf{curl}\,\boldsymbol{\psi}\|_{0,\Omega}.$$

Besides that, the fact that $\int_\Omega \operatorname{div}\boldsymbol{\psi}\,dx = 0$ implies that:

$$\|\operatorname{div}\boldsymbol{\psi}\|_{1,\Omega} \cong |\operatorname{div}\boldsymbol{\psi}|_{1,\Omega}.$$

Indeed, a straightforward application of Theorem 2.1,3°) yields:

$$\|v\|_{1,\Omega} \cong \left(|v|_{1,\Omega}^2 + \left|\int_\Omega v\,dx\right|^2\right)^{1/2} \qquad \forall v \in H^1(\Omega).$$

Hence it remains to establish that

$$\|\boldsymbol{\psi}\|_{0,\Omega} \leqslant C\{\|\operatorname{div}\boldsymbol{\psi}\|_{0,\Omega}^2 + \|\mathbf{curl}\,\boldsymbol{\psi}\|_{0,\Omega}^2\}^{1/2} \qquad \forall \boldsymbol{\psi} \in \Psi. \tag{5.45}$$

When Ω is simply-connected, this is proved by Lemma 3.4. Indeed, even if Γ has several components, its argument shows that:

$$\|\boldsymbol{\psi}\|_{0,\Omega} \leqslant C\|\mathbf{curl}\,\boldsymbol{\psi}\|_{0,\Omega} \qquad \forall \boldsymbol{\psi} \in \Psi \quad \text{with} \quad \operatorname{div}\boldsymbol{\psi} = 0.$$

And if $\operatorname{div}\boldsymbol{\psi} \neq 0$, we can always obtain a divergence-free function in Ψ by considering the difference $\boldsymbol{\psi} - \mathbf{w}$ where $\mathbf{w} \in H_0^1(\Omega)^3$ satisfies (cf. Corollary 2.4):

$$\operatorname{div}\mathbf{w} = \operatorname{div}\boldsymbol{\psi}, \qquad |\mathbf{w}|_{1,\Omega} \leqslant C\|\operatorname{div}\boldsymbol{\psi}\|_{0,\Omega}.$$

When Γ is $\mathscr{C}^{1,1}$, (5.45) is an easy consequence of Theorems 3.7, 2.1 and Remark 3.9. □

Remark 5.8. It can also be proved (cf. Dominguez [24]) that when Ω is multiply-connected, the solution of Problem (5.44) is not the potential of the original Stokes problem.

Remark 5.9. Bendali, Dominguez & Gallic [6] have shown that Problem (5.44) has the following interpretation:

$$(5.46)\qquad \begin{cases} \nu\Delta^2\boldsymbol{\psi} = \mathbf{curl}\,\mathbf{f} \quad \text{in } H^{-2}(\Omega)^3, \\ \operatorname{div}\boldsymbol{\psi} = 0 \quad \text{in } \Omega, \\ (\mathbf{curl}\,\boldsymbol{\psi}) \times \mathbf{n}|_\Gamma = \mathbf{0}, \qquad \boldsymbol{\psi} \times \mathbf{n}|_\Gamma = \mathbf{0}, \\ \displaystyle\int_{\Gamma_i} \boldsymbol{\psi}\cdot\mathbf{n}\,ds = 0, \qquad 0 \leqslant i \leqslant p. \end{cases}$$

A very similar analysis can be applied to vector potentials that satisfy the boundary condition (5.38). Here we choose the space:

$$\Psi_1 = \{\boldsymbol{\phi} \in L^2(\Omega)^3;\ \operatorname{div}\boldsymbol{\phi} \in H^1(\Omega),\ \mathbf{curl}\,\boldsymbol{\phi} \in H_0^1(\Omega)^3,\ \boldsymbol{\phi}\cdot\mathbf{n}|_\Gamma = 0\}$$

equipped with the same norm as Ψ. Then we have the analogues of Lemmas 5.1 and 5.2. More precisely, we can first prove an equivalence of norms:

Lemma 5.3. *When Ω is like in Lemma* 5.1, *the mapping $\boldsymbol{\psi} \to \|\Delta\boldsymbol{\psi}\|_{0,\Omega}$ is a norm on Ψ_1 equivalent to $\|\boldsymbol{\psi}\|$.*

Owing to Lemma 5.3, the biharmonic problem:

Find $\boldsymbol{\psi}$ in Ψ_1 such that

$$(5.47)\qquad \nu(\Delta\boldsymbol{\psi}, \Delta\boldsymbol{\phi}) = \langle \mathbf{f}, \mathbf{curl}\,\boldsymbol{\phi}\rangle \qquad \forall\boldsymbol{\phi} \in \Psi_1,$$

has a unique solution $\boldsymbol{\psi}$ in Ψ_1. It is easy to check that $\boldsymbol{\psi}$ is the vector potential of $\mathbf{u}$ and hence $\operatorname{div}\boldsymbol{\psi} = 0$.

Lemma 5.4. *Assume Ω is like in Lemma* 5.1. *The biharmonic problem* (5.47) *has a unique solution $\boldsymbol{\psi}$ in Ψ_1 and* $\operatorname{div}\boldsymbol{\psi} = 0$, $\mathbf{curl}\,\boldsymbol{\psi} = \mathbf{u}$ *where* $\mathbf{u}$ *is the solution of the homogeneous Stokes problem* (5.12).

Remark 5.10. Problem (5.47) has the following interpretation:

$$(5.48)\qquad \begin{cases} \nu\Delta^2\boldsymbol{\psi} = \mathbf{curl}\,\mathbf{f} \quad \text{in } H^{-2}(\Omega)^3, \\ \operatorname{div}\boldsymbol{\psi} = 0 \quad \text{in } \Omega, \\ \mathbf{curl}\,\boldsymbol{\psi}|_\Gamma = \mathbf{0}, \qquad \boldsymbol{\psi}\cdot\mathbf{n}|_\Gamma = 0. \end{cases}$$

Finally, let us turn to the non-homogeneous Stokes problem. On the one hand, we must assume that the boundary value $\mathbf{g}$ satisfies:

$$(5.49)\qquad \int_{\Gamma_i} \mathbf{g}\cdot\mathbf{n}\,ds = 0, \qquad 0 \leqslant i \leqslant p,$$

so that $\mathbf{u}$ has indeed a vector potential $\boldsymbol{\psi}$. On the other hand, $\boldsymbol{\psi}$ cannot satisfy the boundary condition (5.39) because it implies $(\mathbf{curl}\,\boldsymbol{\psi})\cdot\mathbf{n}|_\Gamma = 0$. However, it is possible to prescribe condition (5.38). Thus a reasonable choice for the space of vector potentials is:

$$\Psi_2 = \{\boldsymbol{\phi} \in L^2(\Omega)^3;\ \operatorname{div}\boldsymbol{\phi} \in H^1(\Omega),\ \mathbf{curl}\,\boldsymbol{\phi} \in H^1(\Omega)^3,\ \boldsymbol{\phi}\cdot\mathbf{n}|_\Gamma = 0\}.$$

Likewise, let us take Ψ_1 as space of test functions. Then consider the biharmonic problem:

Find $\boldsymbol{\psi}$ *in* Ψ_2 *with* $\mathbf{curl}\,\boldsymbol{\psi}|_\Gamma = \mathbf{g}$ *such that*

(5.50) $$\nu(\Delta\boldsymbol{\psi}, \Delta\boldsymbol{\phi}) = \langle \mathbf{f}, \mathbf{curl}\,\boldsymbol{\phi}\rangle \quad \forall\boldsymbol{\phi} \in \Psi_1.$$

It is easy to check that this problem has a unique solution but the relation between $\boldsymbol{\psi}$ and $\mathbf{u}$ is not altogether trivial because (5.43) no longer holds for all $\boldsymbol{\phi}$ in Ψ_1. However, observe that $\boldsymbol{\psi}$ and the divergence-free potential $\boldsymbol{\psi}_0$ of $\mathbf{u}$ in $\Psi_1(\mathbf{curl}\,\boldsymbol{\psi}_0 = \mathbf{u}, \operatorname{div}\boldsymbol{\psi}_0 = 0, \boldsymbol{\psi}_0\cdot\mathbf{n}|_\Gamma = 0)$ both satisfy:

$$\nu(\mathbf{curl}\,\mathbf{curl}\,\boldsymbol{\psi}, \mathbf{curl}\,\mathbf{curl}\,\boldsymbol{\phi}) = \langle \mathbf{f}, \mathbf{curl}\,\boldsymbol{\phi}\rangle \quad \forall\boldsymbol{\phi} \in \Psi_1 \quad \text{with} \quad \operatorname{div}\boldsymbol{\phi} = 0.$$

Thus, setting

$$\mathbf{w} = \mathbf{curl}(\boldsymbol{\psi} - \boldsymbol{\psi}_0) \in V,$$

we derive $\mathbf{curl}\,\mathbf{w} = \mathbf{0}$ and hence $\mathbf{w} = \mathbf{0}$. Therefore $\boldsymbol{\psi}$ is a vector potential of $\mathbf{u}$, but it is not necessarily divergence-free.

Theorem 5.6. *Let* Ω *be like in Lemma* 5.1 *and let* $\mathbf{g}$ *satisfy* (5.49). *The biharmonic problem* (5.50) *has a unique solution* $\boldsymbol{\psi}$ *in* Ψ_2 *and* $\mathbf{curl}\,\boldsymbol{\psi}$ *is the unique solution of the non-homogeneous Stokes problem* (5.12).

Remark 5.11. The divergence of $\boldsymbol{\psi}$ is the solution λ of the Neumann's problem:

$$\Delta\lambda = 0 \quad \text{in } \Omega, \qquad \int_\Omega \lambda\,dx = 0,$$

$$\partial\lambda/\partial n = (\mathbf{curl}\,\mathbf{u})\cdot\mathbf{n} \quad \text{on } \Gamma.$$

Therefore $\operatorname{div}\boldsymbol{\psi} = 0$ iff $(\mathbf{curl}\,\mathbf{u})\cdot\mathbf{n}|_\Gamma = 0$.

Remark 5.12. If we want a biharmonic problem whose solution is $\boldsymbol{\psi}_0$, we must include the constraint $(\operatorname{div}\boldsymbol{\phi})|_\Gamma = 0$ in the spaces Ψ_1 and Ψ_2.

Remark 5.13. It is also possible to prescribe the condition $(\mathbf{curl}\,\boldsymbol{\psi})|_\Gamma = \mathbf{g}$, by setting on the one hand

$$(\mathbf{curl}\,\boldsymbol{\psi}) \times \mathbf{n} = \mathbf{g} \times \mathbf{n} \quad \text{on } \Gamma,$$

and on the other hand

$$\mathrm{div}_\Gamma(\boldsymbol{\psi} \times \mathbf{n}) = \mathbf{g}\cdot\mathbf{n},$$

$$\mathrm{curl}_\Gamma(\boldsymbol{\psi} \times \mathbf{n}) = 0,$$

where the subscript Γ indicates that the operators div and curl are surface operators. We refer to Roux [69] for more details.

Appendix A. Results of Standard Finite Element Approximation

This short chapter gathers most of the properties of the classical finite element approximation that will be required subsequently. The more familiar results are stated without proof; for detailed proofs and further material, the reader can refer to the very complete texts of Ciarlet [19] and Strang & Fix [77]. In addition, we include a brief mention of new or nonstandard material, developed among others by Clément [21], Bernardi [9], Scott [73], Lenoir [50], that will also be very useful later on.

A.1. Triangular Finite Elements

Definition A.1. For each integer $k \geqslant 0$, we denote by P_k the space of all polynomials defined on $\mathbb{R}^N$, of degree less than or equal to k.

Recall that an N-simplex of $\mathbb{R}^N$ is the convex hull κ of $N + 1$ points a_j, $1 \leqslant j \leqslant N + 1$, called the vertices of κ, which are not all located in a single hyperplane. For instance, a 2-simplex is a non degenerate triangle and a 3-simplex is a non degenerate tetrahedron. The size and shape of an N-simplex κ are specified by two quantities:

$$h_\kappa = \text{diameter of } \kappa$$

and

$$\rho_\kappa = \sup\{\text{diameter of } B;\ B \text{ is a ball contained in } \kappa\}.$$

In addition, the regularity of κ is measured by the ratio

$$\sigma_\kappa = h_\kappa/\rho_\kappa.$$

We denote by $\hat{\kappa}$ the reference unit simplex in the $(\hat{x}_1, \ldots, \hat{x}_N)$ space with vertices $\hat{a}_j = (\delta_{ij})_{1 \leqslant i \leqslant N}$, for $1 \leqslant j \leqslant N$, and $\hat{a}_{N+1} = (0)_{1 \leqslant i \leqslant N}$. If κ is an N-simplex with vertices a_j, $1 \leqslant j \leqslant N + 1$, there exists exactly one affine mapping

$$F_\kappa(\hat{x}) = B_\kappa \hat{x} + b_\kappa \tag{A.1}$$

that maps $\hat{\kappa}$ onto κ with $F_\kappa(\hat{a}_i) = a_i$ for $1 \leqslant i \leqslant N + 1$ (cf. Figure 3). Furthermore,

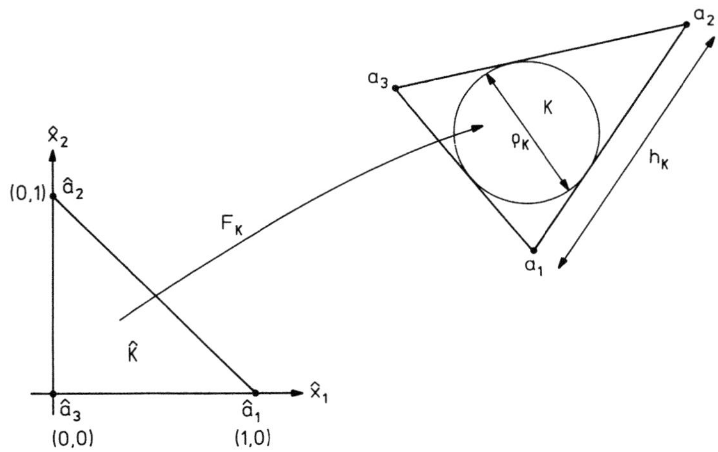

Figure 3. Triangle K and its reference unit triangle $\hat{K}$ $(N = 2)$

it can be easily shown that the matrix B_κ is nonsingular and satisfies the following bounds:

(A.2) $$\| B_\kappa \| \leqslant h_\kappa / \rho_{\hat{\kappa}}, \qquad \| B_\kappa^{-1} \| \leqslant h_{\hat{\kappa}} / \rho_\kappa,$$

where $\| . \|$ stands for both the Euclidean norm of $\mathbb{R}^N$ and its subordinate matrix norm. In addition, owing that

(A.3) $$|\det(B_\kappa)| = \text{meas}(\kappa)/\text{meas}(\hat{\kappa}),$$

there exist two positive constants $C_1(N)$, $C_2(N)$ depending only upon N, such that

(A.4) $$C_2(N)\rho_\kappa^N \leqslant |\det(B_\kappa)| \leqslant C_1(N) h_\kappa^N.$$

According to convenience, we shall sometimes replace the Euclidean coordinates of the point x of $\mathbb{R}^N$ by its barycentric coordinates, $\lambda_i = \lambda_i(x)$, with respect to the vertices a_i, $1 \leqslant i \leqslant N + 1$, defined by:

(A.5) $$\lambda_i \in P_1, \quad \lambda_i(a_j) = \delta_{ij} \quad \text{for } 1 \leqslant i, j \leqslant N + 1.$$

The barycentric coordinates satisfy the following useful identities in $\mathbb{R}^N$:

(A.6) $$\sum_{i=1}^{N+1} \lambda_i = 1, \quad p = \sum_{i=1}^{N+1} p(a_i)\lambda_i \quad \forall p \in P_1.$$

Moreover, it can be easily checked that

$$\kappa = \{x \in \mathbb{R}^N;\ 0 \leqslant \lambda_i(x) \leqslant 1,\ 1 \leqslant i \leqslant N + 1\}.$$

As mentioned above, the mapping $x = F_\kappa(\hat{x})$ establishes a one-to-one correspondence between $\hat{\kappa}$ and κ. The composition with F_κ:

$$(v\colon \kappa \to \mathbb{R}) \to (\hat{v} = v \circ F_\kappa\colon \hat{\kappa} \to \mathbb{R})$$

and with F_κ^{-1}:

$$(\hat{v}: \hat{\kappa} \to \mathbb{R}) \to (v = \hat{v} \circ F_\kappa^{-1}: \kappa \to \mathbb{R})$$

are of constant use because they enable us to work exclusively on the reference element $\hat{\kappa}$. The effects of this change of variable are described in the next lemma.

Lemma A.1. *For each integer $m \geqslant 0$ and for all real p with $1 \leqslant p \leqslant \infty$, the mapping $v \to \hat{v} = v \circ F_\kappa$ is an isomorphism from $W^{m,p}(\kappa)$ onto $W^{m,p}(\hat{\kappa})$ and the following bounds hold:*

$$\text{(A.7)} \qquad |\hat{v}|_{m,p,\hat{\kappa}} \leqslant C_1 \| B_\kappa \|^m |\det(B_\kappa)|^{-1/p} |v|_{m,p,\kappa} \quad \forall v \in W^{m,p}(\kappa),$$

$$\text{(A.8)} \qquad |v|_{m,p,\kappa} \leqslant C_2 \| B_\kappa^{-1} \|^m |\det(B_\kappa)|^{1/p} |\hat{v}|_{m,p,\hat{\kappa}} \quad \forall \hat{v} \in W^{m,p}(\hat{\kappa}).$$

Note that the gradient of a function and the unit normal vector have the simple expressions (cf. for instance Babuska & *al* [5]):

$$\text{(A.9)} \qquad \mathbf{grad}_{\hat{x}}\, \hat{v}(\hat{x}) = (B_\kappa^T \,\mathbf{grad}_x\, v) \circ F_\kappa(\hat{x}),$$

$$\text{(A.10)} \qquad \hat{\mathbf{n}}(\hat{x}) = [(B_\kappa^T \mathbf{n}) / \| B_\kappa^T \mathbf{n} \|] \circ F_\kappa(\hat{x}),$$

where $\mathbf{n}$ (resp. $\hat{\mathbf{n}}$) denotes the unit exterior normal to κ (resp. $\hat{\kappa}$).

The following theorem (which is an extension of Theorem I.1.9) and its consequences are fundamental tools of the finite element theory.

Theorem A.1. *For each integer $k \geqslant 0$ and real p with $1 \leqslant p \leqslant \infty$, there exists a constant $\hat{C} > 0$, depending only on k, p and $\hat{\kappa}$, such that:*

$$\text{(A.11)} \qquad \inf_{t \in P_k} \| \hat{u} + t \|_{k+1,p,\hat{\kappa}} \leqslant \hat{C} |\hat{u}|_{k+1,p,\hat{\kappa}} \quad \forall \hat{u} \in W^{k+1,p}(\hat{\kappa})/P_k.$$

Corollary A.1. *Let $k \geqslant 0$, $m \geqslant 0$ be integers and $p \geqslant 1$, $q \geqslant 1$ be reals such that*

$$W^{k+1,p}(\hat{\kappa}) \subsetneq W^{m,q}(\hat{\kappa}).$$

Let $\hat{\pi} \in \mathscr{L}(W^{k+1,p}(\hat{\kappa}); W^{m,q}(\hat{\kappa}))$ satisfy:

$$\hat{\pi} t = t \quad \forall t \in P_k.$$

Then there exists a constant $\hat{C} > 0$ depending on k, m, p, q, $\hat{\kappa}$ and $\hat{\pi}$ only, such that

$$\text{(A.12)} \qquad \| \hat{v} - \hat{\pi}\hat{v} \|_{m,q,\hat{\kappa}} \leqslant \hat{C} |\hat{v}|_{k+1,p,\hat{\kappa}} \quad \forall \hat{v} \in W^{k+1,p}(\hat{\kappa}).$$

When combined with Lemma A.1, (A.2) and (A.3), Corollary A.1 yields:

Corollary A.2. *Let k, m, p, q and $\hat{\pi}$ be like in Corollary A.1. Let κ be an N-simplex of $\mathbb{R}^N$ and let the operator $\pi \in \mathscr{L}(W^{k+1,p}(\kappa); W^{m,q}(\kappa))$ be defined by:*

$$\text{(A.13)} \qquad (\pi v) \circ F_\kappa = \hat{\pi}(v \circ F_\kappa) \quad (\text{i.e. } \widehat{\pi v} = \hat{\pi}\hat{v}).$$

Then there exists a constant $C > 0$, depending only on k, m, p, q, $\hat{\kappa}$ and $\hat{\pi}$ such that *for all* $v \in W^{k+1,p}(\kappa)$:

(A.14) $$|v - \pi v|_{m,q,\kappa} \leqslant C\sigma_\kappa^m(\text{meas}(\kappa))^{1/q-1/p} h_\kappa^{k+1-m}|v|_{k+1,p,\kappa}.$$

We shall assume henceforth that the *dimension N equals two*, although some results will be stated in the N-dimensional case. In addition, to simplify the discussion we shall assume that Ω is a bounded domain with a *polygonal* boundary Γ. The technical difficulties inherent to curved boundaries can be conveniently handled with the general elements introduced by Bernardi [9], but there is no space to discuss them here.

For each $h > 0$, let $\mathscr{T}_h$ be a *triangulation of* $\bar{\Omega}$ made of *closed* triangles κ with diameters bounded by h. In other words,

$$\bar{\Omega} = \bigcup_{\kappa \in \mathscr{T}_h} \kappa, \qquad h_\kappa \leqslant h,$$

where any two triangles are either disjoint or share exactly either one side or one vertex.

Definitions A.2. 1°) A family $\mathscr{T}_h$ of triangulations of $\bar{\Omega}$ is said to be regular as h tends to zero if there exists a number $\sigma > 0$, independent of h and κ, such that

(A.15) $$\sigma_\kappa \leqslant \sigma \qquad \forall \kappa \in \mathscr{T}_h.$$

2°) In addition, $\mathscr{T}_h$ is said to be uniformly regular (or quasi-uniform) as h tends to zero if there exists another constant $\tau > 0$ such that

(A.16) $$\tau h \leqslant h_\kappa \leqslant \sigma \rho_\kappa \qquad \forall \kappa \in \mathscr{T}_h.$$

Remark A.1. When $\mathscr{T}_h$ is regular, the error estimate (A.14) simplifies to:

(A.17) $$|v - \pi v|_{m,q,\kappa} \leqslant C h_\kappa^{k+1-m+N(1/q-1/p)}|v|_{k+1,p,\kappa} \qquad \forall v \in W^{k+1,p}(\kappa),$$

where the constant C is independent of v, h and κ. Because of the inclusion

$$W^{k+1,p}(\hat{\kappa}) \subsetneq W^{m,q}(\hat{\kappa}),$$

we have:

$$k + 1 - m + N(1/q - 1/p) \geqslant 0$$

and therefore the above factor h_κ can be bounded by h.

Now, we fix the integer $k \geqslant 1$ and we introduce the standard finite element spaces:

(A.18) $$\Theta_h = \{\theta_h \in \mathscr{C}^0(\bar{\Omega});\ \theta_h|_\kappa \in P_k \quad \forall \kappa \in \mathscr{T}_h\},$$
$$\Phi_h = \Theta_h \cap H_0^1(\Omega).$$

Note that they are both finite-dimensional subspaces of $W^{1,\infty}(\Omega)$; but their dimensions tend to infinity as the approximation parameter h tends to zero. Next, we define an interpolation operator. Of course, the simplest choice consists in interpolating the functions at an appropriate set of points of the N-simplex κ such as the principal lattice of order k:

(A.19)
$$\Sigma_\kappa = \left\{ x = \sum_{j=1}^{N+1} \lambda_j a_j;\ \sum_{j=1}^{N+1} \lambda_j = 1,\ \lambda_j \in \{0, 1/k, \ldots, (k-1)/k, 1\},\ 1 \leqslant j \leqslant N+1 \right\}$$

(cf. Figure 4 for $k = 3$). This yields a widely used interpolation operator I_h with well known properties, valid in all dimensions N.

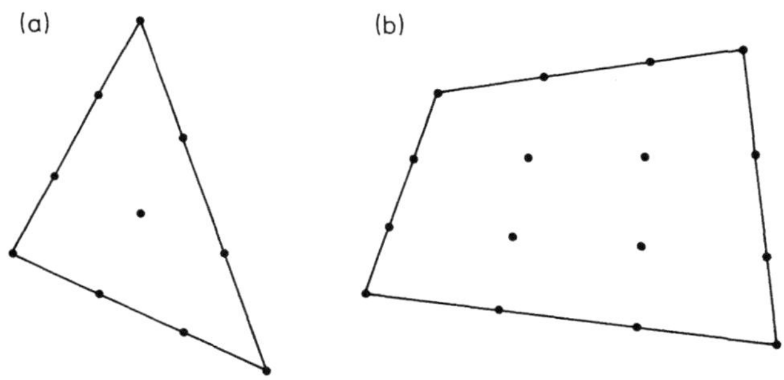

Figure 4. Principal lattice of order 3 for (a) triangle (b) quadrilateral

Lemma A.2. *Let $\mathcal{T}_h$ be a regular triangulation of $\bar{\Omega}$. For real $p > N/2$, the interpolation operator $I_h \in \mathcal{L}(W^{2,p}(\Omega); \Theta_h) \cap \mathcal{L}(W^{2,p}(\Omega) \cap W_0^{1,p}(\Omega); \Phi_h)$ defined in each element κ by:*

(A.20) $$I_h v|_\kappa \in P_k, \quad I_h v(x) = v(x) \quad \forall x \in \Sigma_\kappa,$$

satisfies the following error estimate for all integers m and real s with $0 \leqslant m \leqslant s + 1$, $1 \leqslant s \leqslant k$:

(A.21a) $$|v - I_h v|_{m,p,\Omega} \leqslant C_1 h^{s+1-m} |v|_{s+1,p,\Omega} \quad \forall v \in W^{s+1,p}(\Omega).$$

Moreover, $I_h \in \mathcal{L}(W^{1,q}(\Omega); \Theta_h) \cap \mathcal{L}(W_0^{1,q}(\Omega); \Phi_h)$ for all real $q > N$ and

(A.21b) $$|v - I_h v|_{m,q,\Omega} \leqslant C_2 h^{1-m} |v|_{1,q,\Omega} \quad \forall v \in W^{1,q}(\Omega), \quad m = 0, 1.$$

Both constants C_1 and C_2 are positive and independent of h and v.

However, in the forthcoming applications it will sometimes be handy to work with a slightly different interpolant, given here when $N = 2$.

Lemma A.3. *Let $p \in \mathbb{R}$ with $p > 1$; for each $v \in W^{2,p}(\kappa)$, there exists exactly one polynomial $\tilde{I}_\kappa v \in P_k$ such that:*

$$(A.22)\quad \begin{cases} \tilde{I}_\kappa v(a_i) = v(a_i) \quad 1 \leqslant i \leqslant 3, \\ \text{if } k \geqslant 2 \quad \int_{\kappa'} (\tilde{I}_\kappa v - v) f\, ds = 0 \quad \forall f \in P_{k-2}(\kappa'), \quad \forall \text{sides } \kappa' \text{ of } \kappa, \\ \text{if } k \geqslant 3 \quad \int_{\kappa} (\tilde{I}_\kappa v - v) f\, dx = 0 \quad \forall f \in P_{k-3}(\kappa). \end{cases}$$

Proof. First, we remark that (A.22) consists of $(1/2)(k + 1)(k + 2)$ equations, which is precisely the dimension of P_k. Hence (A.22) is a square system of linear equations and it suffices to prove that its solution is unique. Thus, we assume that $p \in P_k$ satisfies:

$$\text{(i)} \qquad p(a_i) = 0 \quad 1 \leqslant i \leqslant 3,$$

$$\text{(ii)} \qquad \int_{\kappa'} p(s) f(s)\, ds = 0 \quad \forall f \in P_{k-2}(\kappa'), \quad \forall \text{sides } \kappa' \text{ of } \kappa,$$

$$\text{(iii)} \qquad \int_{\kappa} p(x) f(x)\, dx = 0 \quad \forall f \in P_{k-3}(\kappa).$$

Suppose for the moment that $k \geqslant 3$. Note that the restriction of p to each κ' is a polynomial of degree k of the single variable s. Then it follows from (i) and (ii) that:

$$0 = \int_{\kappa'} p(s)(d^2 p(s)/ds^2)\, ds = -\int_{\kappa'} (dp(s)/ds)^2\, ds.$$

Hence $p = 0$ on $\partial\kappa$ and p can be expressed in terms of barycentric coordinates as:

$$p = \lambda_1 \lambda_2 \lambda_3 q, \quad \text{where } q \in P_{k-3}.$$

Then (iii) implies that

$$\int_{\kappa} \lambda_1 \lambda_2 \lambda_3 q^2(x)\, dx = 0,$$

i.e. $q = 0$ since the integrand is non negative. Hence $p = 0$ on κ.

The case $k \leqslant 2$ is trivial. □

It is a matter of routine to check that $\tilde{I}_\kappa \in \mathscr{L}(W^{2,p}(\kappa); P_k)$, the value of $\tilde{I}_\kappa v$ on each side κ' of κ depends only on the value of v on κ' and of course

$$\tilde{I}_\kappa p = p \quad \forall p \in P_k.$$

Furthermore, the operator $\tilde{I}_\kappa$ satisfies the fundamental relation (A.13):

$$\widehat{\tilde{I}_\kappa v} = \tilde{I}_{\hat{\kappa}} \hat{v}.$$

By collecting these properties and using Corollary A.2 and Remark A.1 we derive:

Lemma A.4. *Let $\mathscr{T}_h$ be a regular triangulation of $\bar{\Omega}$. For real $p > 1$, the interpolation operator $\tilde{I}_h \in \mathscr{L}(W^{2,p}(\Omega); \Theta_h) \cap \mathscr{L}(W^{2,p}(\Omega) \cap W_0^{1,p}(\Omega); \Phi_h)$ defined in each element κ by:*

$$\tilde{I}_h v|_\kappa = \tilde{I}_\kappa v \qquad \forall \kappa \in \mathscr{T}_h,$$

satisfies the following error estimate for all integers m and real s with $0 \leqslant m \leqslant s+1$, $1 \leqslant s \leqslant k$:

$$|v - \tilde{I}_h v|_{m,p,\Omega} \leqslant C h^{s+1-m} |v|_{s+1,p,\Omega} \qquad \forall v \in W^{s+1,p}(\Omega), \tag{A.23}$$

where the constant $C > 0$ is independent of h and v.

Apart from these two interpolants, the projection operator will play a fundamental part in the subsequent theory. To be specific, for real $p \geqslant 1$, let

$$\mathring{P}_h \in \mathscr{L}(W_0^{1,p}(\Omega); \Phi_h), \qquad P_h \in \mathscr{L}(W^{1,p}(\Omega); \Theta_h)$$

be defined respectively by

$$(\mathbf{grad}(\mathring{P}_h v - v), \mathbf{grad}\, \phi_h) = 0 \quad \forall \phi_h \in \Phi_h, \qquad \forall v \in W_0^{1,p}(\Omega); \tag{A.24}$$

$$\left\{\begin{array}{ll} (\mathbf{grad}(P_h v - v), \mathbf{grad}\, \theta_h) = 0 \quad \forall \theta_h \in \Theta_h, & \\ (P_h v - v, 1) = 0. & \end{array}\right. \qquad \forall v \in W^{1,p}(\Omega). \tag{A.25}$$

Note that $\mathring{P}_h v$ can also be interpreted as the finite element solution of the homogeneous Dirichlet problem:

$$-\Delta v = f \quad \text{in } \Omega, \qquad v = 0 \quad \text{on } \Gamma;$$

and $P_h v$ as the finite element solution of the non-homogeneous Neumann problem:

$$-\Delta v = f \quad \text{in } \Omega, \qquad \partial v / \partial n = g \quad \text{on } \Gamma,$$

with the compatibility condition

$$\int_\Omega f\, dx = \langle g, 1 \rangle_\Gamma,$$

its uniqueness proceeding from the second condition of (A.25). Now, since $\mathring{P}_h$ and P_h are projections for the seminorm of $H^1(\Omega)$, it is easy to prove error estimates in the L^2 and H^1 norms, provided $v \in H^1(\Omega)$; but it is much more difficult to establish optimal L^p and $W^{1,p}$ estimates. The following theorem is the achievement of many years of work contributed by several mathematicians; cf. for example Douglas, Dupont & Whalbin [25], Nitsche [62], Rannacher & Scott [66] and Scott [73]. Although it is by no means standard, its proof is omitted as it is far beyond the scope of this book.

Theorem A.2. *Assume that Ω is a convex polygon. Let $\mathscr{T}_h$ be a uniformly regular triangulation of $\bar{\Omega}$ and let the reals s and p be such that $0 \leqslant s \leqslant k$ and $1 \leqslant p \leqslant \infty$. For $k \geqslant 2$ or for $k = 1$ and $2 \leqslant p < \infty$, there exists a constant $C > 0$, independent of h, such that the projection P_h (resp. $\mathring{P}_h$) satisfies the error estimate:*

$$\begin{aligned}&\|v - P_h v\|_{0,p,\Omega} + h|v - P_h v|_{1,p,\Omega} \leqslant C h^{s+1} \|v\|_{s+1,p,\Omega}\\ &\forall v \in W^{s+1,p}(\Omega) \quad (\textit{resp.}\ \forall v \in W^{s+1,p}(\Omega) \cap W_0^{1,p}(\Omega)).\end{aligned} \tag{A.26}$$

When $k = 1$ and $p \in [1, 2)$, the estimate (A.26) *holds with the additional factor*

$$|\ln h|^{2/p-1}$$

in the right-hand side. When $k = 1$ and $p = \infty$, the L^∞-estimate in (A.26) *becomes*

$$\|v - P_h v\|_{0,\infty,\Omega} \leqslant C|\ln h| h^{s+1} \|v\|_{s+1,\infty,\Omega}, \tag{A.27}$$

while the $W^{1,\infty}$-estimate is unchanged.

The proof of Theorem A.2 relies on the fact that the Laplacian operator is an isomorphism from $W_0^{1,p}(\Omega) \cap W^{2,p}(\Omega)$ onto $L^p(\Omega)$ for all $p \in (1, 2 + \varepsilon)$, for some $\varepsilon > 0$. According to Remark I.1.2, this isomorphism holds on a smooth domain or on a convex polygon. This accounts for the convexity hypothesis in the statement of Theorem A.2. Let us point out that the occurrence of the logarithmic factor is a well-known phenomenon in L^∞-estimates. Finally, as mentioned above, (A.26) can be established directly when $p = 2$ and for all $k \geqslant 1$; if s is an integer, this yields the familiar estimate for P_h (resp. $\mathring{P}_h$):

$$\begin{aligned}&\|v - P_h v\|_{0,\Omega} + h|v - P_h v|_{1,\Omega} \leqslant C h^{m+1} |v|_{m+1,\Omega}\\ &\forall v \in H^{m+1}(\Omega) \quad (\text{resp.}\ \forall v \in H^{m+1}(\Omega) \cap H_0^1(\Omega)).\end{aligned} \tag{A.28}$$

Remark A.2. When the function v is not continuous, neither its interpolant $I_h v$ nor $\tilde{I}_h v$ are defined. If Ω is convex and the triangulation uniformly regular, they may be conveniently replaced by the projection $P_h v$ (or $\mathring{P}_h v$ if v vanishes on Γ). When Ω is not convex or a uniformly regular triangulation is not available, other interpolants obtained by *local regularization* may be used. They also have the advantage, over the projection, of being defined locally and not globally. The reader will find some information about this technique in Section A.3.

We shall also use subsequently a *local L^2 projection.* To be precise, for $v \in L^2(\Omega)$ and each integer $k \geqslant 0$, we define:

$$\rho_h v|_\kappa \in P_k, \qquad \int_\kappa (\rho_h v - v) f \, dx = 0 \qquad \forall f \in P_k, \qquad \forall \kappa \in \mathscr{T}_h. \tag{A.29}$$

Clearly the operator ρ_h satisfies the hypotheses of Corollary A.2 and we have the following result valid in all dimensions N.

Lemma A.5. *Let $v \in H^s(\Omega)$ for some real $s \in [0, k + 1]$. The L^2 projection ρ_h satisfies the error estimate*:

$$\|v - \rho_h v\|_{0,\Omega} \leqslant C h^s |v|_{s,\Omega}, \tag{A.30}$$

with a constant $C > 0$ independent of h and v.

Note that this lemma requires no regularity of the triangulation.

We finish this section with a number of inverse inequalities satisfied, for arbitrary N, by the functions of Θ_h.

Lemma A.6. *Let r and p be reals with $1 \leqslant r, p \leqslant \infty$. Under the assumption that $\mathcal{T}_h$ is regular if $1/r - 1/p \geqslant 1/N$ or $\mathcal{T}_h$ is uniformly regular otherwise, there exists a constant $C > 0$ independent of h and κ such that:*

(A.31) $$|v|_{1,r,\kappa} \leqslant Ch^{N(1/r-1/p)-1} \|v\|_{0,p,\kappa} \quad \forall \kappa \in \mathcal{T}_h \quad \forall v \in \Theta_h.$$

Proof. Let $v \in \Theta_h$; owing to (A.8), (A.2) and (A.3) we have:

$$|v|_{1,r,\kappa} \leqslant C_1(1/\rho_\kappa)(\text{meas}(\kappa))^{1/r}|\hat{v}|_{1,r,\hat{\kappa}}.$$

But since $\hat{v}$ belongs to the finite-dimensional space P_k on $\hat{\kappa}$, we have

$$|\hat{v}|_{1,r,\hat{\kappa}} \leqslant C_2 \|\hat{v}\|_{0,p,\hat{\kappa}}$$

where the constant C_2 depends only on r, p, k, N and the geometry of $\hat{\kappa}$. Then applying (A.7) and (A.3) we get:

(A.32) $$|v|_{1,r,\kappa} \leqslant C_3(1/\rho_\kappa)(\text{meas}(\kappa))^{1/r-1/p} \|v\|_{0,p,\kappa}.$$

In view of (A.4), if $N(1/r - 1/p) \geqslant 1$ the right-hand side of (A.32) involves a positive power of h_κ and hence the regularity of $\mathcal{T}_h$ is sufficient to yield (A.31). Otherwise, the uniform regularity of $\mathcal{T}_h$ is necessary to draw the same conclusion. □

Corollary A.3. *Let r and p be like in Lemma A.6 and assume $\mathcal{T}_h$ is a uniformly regular triangulation of $\bar{\Omega}$. There exists a constant $C > 0$ independent of h such that:*

(A.33) $$|v|_{1,r,\Omega} \leqslant Ch^{-1+\text{Min}(0,N/r-N/p)} \|v\|_{0,p,\Omega} \quad \forall v \in \Theta_h.$$

Proof. If $r \geqslant p$ we derive (A.33) from (A.31) and Jensen's inequality:

(A.34) $$\left(\sum_{i=1}^{I} |a_i|^r\right)^{1/r} \leqslant \left(\sum_{i=1}^{I} |a_i|^p\right)^{1/p}$$

which holds for every finite sum.

If $r < p$, we use the discrete version of Hölder's inequality:

(A.35) $$\left(\sum_{i=1}^{I} |a_i|^r |b_i|^{1-r/p}\right)^{1/r} \leqslant \left(\sum_{i=1}^{I} |a_i|^p\right)^{1/p} \left(\sum_{i=1}^{I} |b_i|\right)^{1/r-1/p},$$

with $|a_i| = \|v\|_{0,p,\kappa}$, $b_i = \text{meas}(\kappa)$ and the summation runs over all κ of $\mathcal{T}_h$. Thus the uniform regularity of $\mathcal{T}_h$ and (A.32) yield:

$$|v|_{1,r,\Omega} \leqslant C_3[\sigma/(\tau h)](\text{meas}(\Omega))^{1/r-1/p} \|v\|_{0,p,\Omega}.$$ □

Lemma A.7. *Let r and p be like in Lemma A.6 and let m be a non-negative integer. If $r \leqslant p$ or if the triangulation $\mathcal{T}_h$ is uniformly regular and $r > p$, there exists a* *constant* $C > 0$, *independent of* h, *such that*:

(A.36) $$|v|_{m,r,\Omega} \leqslant C h^{\mathrm{Min}(0, N/r - N/p)} |v|_{m,p,\Omega} \qquad \forall v \in \Theta_h.$$

Proof. Clearly, (A.36) is obvious when $r \leqslant p$. When $r > p$, the proof is quite similar to that of (A.33) and is left to the reader. □

A.2. Quadrilateral Finite Elements

This section is devoted exclusively to plane polygonal domains. Unless otherwise specified, the notation is that of the preceding section. In addition, since several properties of triangular finite elements are still valid for quadrilateral finite elements, we shall focus our attention on those properties which are specific to quadrilaterals.

Definition A.3. For each integer $k \geqslant 0$, we denote by Q_k the space of all polynomials in the reference space $(\hat{x}_1, \hat{x}_2)$ of the form:

$$q(\hat{x}) = \sum c_{ij} \hat{x}_1^i \hat{x}_2^j,$$

where the sum ranges over all integers i and j such that $0 \leqslant i, j \leqslant k$.

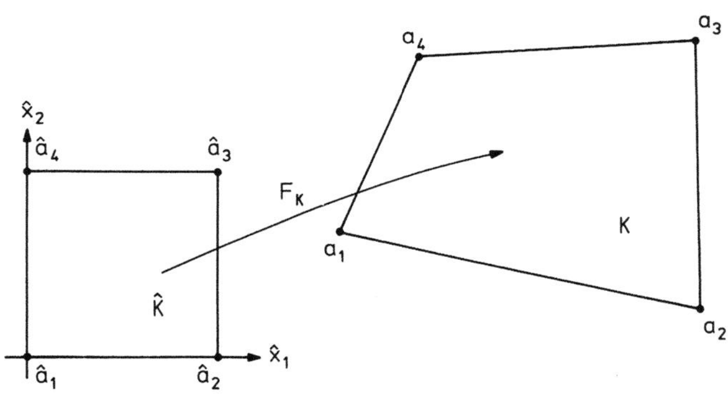

Figure 5. Quadrilateral K and its reference unit square $\hat{K}$

Of course Q_0 coincides with P_0 but for all $k \geqslant 1$ we have the strict inclusion

$$P_k \subset Q_k.$$

Let $\hat{\kappa}$ denote the reference unit square $[0, 1] \times [0, 1]$ in the $(\hat{x}_1, \hat{x}_2)$ reference space, with vertices denoted by $\hat{a}_j$, $1 \leqslant j \leqslant 4$, like in Figure 5. With the notations of this figure, for each convex quadrilateral κ with vertices a_j, there exists exactly one invertible mapping $F_\kappa \in Q_1^2$ that maps $\hat{\kappa}$ onto κ and is such that

$$F_\kappa(\hat{a}_j) = a_j, \qquad 1 \leqslant j \leqslant 4.$$

We denote by S_i the subtriangle of κ with vertices a_{i-1}, a_i and a_{i+1} (where of course a_0 coincides with a_4). Note that unless κ is a parallelogram, the mapping F_κ is not affine, but nevertheless it maps the sides of $\hat{\kappa}$ onto the corresponding sides of κ and in fact the restriction of F_κ to the sides of $\hat{\kappa}$ is affine.

For each $\hat{x}$, let $DF_\kappa(\hat{x}) = [\partial F_i(\hat{x})/\partial \hat{x}_j]_{i,j} \in \mathscr{L}(\mathbb{R}^2; \mathbb{R}^2)$ denote the derivative of F_κ at the point $\hat{x}$ with the following norm:

$$\|DF_\kappa\|_{\infty,\hat{\kappa}} = \sup_{\hat{x} \in \hat{\kappa}} \|DF_\kappa(\hat{x})\|,$$

where as usual $\|.\|$ stands for the subordinate Euclidean norm, and let J_F denote the Jacobian of F_κ, i.e.

$$J_F(\hat{x}) = \det(DF_\kappa(\hat{x})).$$

Then, if F_κ^{-1} denotes the inverse of F_κ with Jacobian $J_{F^{-1}}$, we have:

$$D(F_\kappa^{-1}) \circ F_\kappa = (DF_\kappa^T)^{-1}, \qquad J_{F^{-1}} \circ F_\kappa = 1/J_F. \tag{A.37}$$

Note again that when F_κ is affine, $DF_\kappa = B_\kappa$, $D(F_\kappa^{-1}) = (B_\kappa^T)^{-1}$ and $J_F = \det(B_\kappa)$. Of course, since J_F is not constant the simple relation (A.3) does not hold here but a plain calculation shows that J_F belongs to P_1. Hence, its extrema are attained at the vertices of $\hat{\kappa}$. Thus it is easy to check that

$$\begin{cases} \operatorname*{Max}_{\hat{x} \in \hat{\kappa}} J_F(\hat{x}) = 2 \operatorname*{Max}_{1 \leqslant i \leqslant 4} \operatorname{meas}(S_i), \\ \operatorname*{Min}_{\hat{x} \in \hat{\kappa}} J_F(\hat{x}) = 2 \operatorname*{Min}_{1 \leqslant i \leqslant 4} \operatorname{meas}(S_i), \end{cases} \tag{A.38}$$

and observe that

$$J_F(\hat{x}) \geqslant 0$$

because κ is convex. Furthermore,

$$J_F(1/2, 1/2) = \operatorname{meas}(\kappa).$$

Hence, a glance at (A.37) and (A.38) shows that a convenient choice of parameters to describe the geometry of κ is:

$$h_\kappa = \text{diameter of } \kappa,$$

$$\rho_\kappa = 2 \operatorname*{Min}_{1 \leqslant i \leqslant 4} \{\text{diameter of circle inscribed in } S_i\}.$$

This choice leads to the following upper bounds:

$$\begin{cases} \|DF_\kappa\|_{\infty,\hat{\kappa}} \leqslant C_1 h_\kappa, & \|J_F\|_{\infty,\hat{\kappa}} \leqslant C_2 h_\kappa^2 \\ \|DF_\kappa^{-1}\|_{\infty,\kappa} \leqslant C_3 (h_\kappa/\rho_\kappa^2), & \|J_{F^{-1}}\|_{\infty,\kappa} \leqslant C_4 (1/\rho_\kappa^2), \end{cases} \tag{A.39}$$

where all constants involved are independent of the geometry of κ.

We use the same concept for the triangulation $\mathscr{T}_h$ except that here it is made of *convex quadrilaterals* instead of triangles and with the above parameters, we retain *Definition* A.2 *for a regular and uniformly regular triangulation.* Note that the regularity of the triangulation guarantees that no subtriangle of κ has zero measure, i.e. κ does not degenerate into a triangle.

Next, we turn to the finite element spaces. It stems from the geometry of quadrilaterals that the basis functions here are not polynomials of P_k but rather are images of polynomials of Q_k. To be specific, for each integer $k \geqslant 0$, and for each element κ of $\mathscr{T}_h$, we introduce the finite-dimensional space:

$$Q_k(\kappa) = \{q = \hat{q} \circ F_\kappa^{-1};\ \hat{q} \in Q_k\} \tag{A.40}$$

and observe that $Q_k(\kappa) \subset \mathscr{C}^\infty(\kappa)$, but (unless κ is a parallelogram) $Q_k(\kappa)$ is *not* a space of polynomials. Then, in order to study the approximation error in such a space, we must modify slightly the statements of Theorem A.1 and its corollaries. This is achieved by a change of seminorm: for each integer $k \geqslant 0$ and real $p > 0$, we set

$$[v]_{k,p,\Omega} = (\|\partial^k v/\partial x_1^k\|_{0,p,\Omega}^p + \|\partial^k v/\partial x_2^k\|_{0,p,\Omega}^p)^{1/p}. \tag{A.41}$$

Clearly, $[\cdot]_{0,p,\Omega} = \|\cdot\|_{0,p,\Omega}$ and $[\cdot]_{1,p,\Omega} = |\cdot|_{1,p,\Omega}$; in addition, this seminorm has the following important property proved by Aronszajn & Smith (cf. Smith [74]):

Lemma A.8. *Let Ω be a bounded domain of $\mathbb{R}^2$ with a Lipschitz-continuous boundary. For each integer $k \geqslant 0$ and real $p > 0$ there exists a positive constant C such that*

$$\|v\|_{k,p,\Omega} \leqslant C\{\|v\|_{0,p,\Omega} + [v]_{k,p,\Omega}\} \quad \forall v \in W^{k,p}(\Omega). \tag{A.42}$$

Now, Theorem A.1 and its first corollary have the following counterpart:

Theorem A.3. *Let $k \geqslant 0$, $m \geqslant 0$ be integers and $p \geqslant 1$, $q \geqslant 1$ be reals such that*

$$W^{k+1,p}(\hat{\kappa}) \subsetneq W^{m,q}(\hat{\kappa}).$$

Let $\hat{\pi} \in \mathscr{L}(W^{k+1,p}(\hat{\kappa}); W^{m,q}(\hat{\kappa}))$ satisfy

$$\hat{\pi} t = t \quad \forall t \in Q_k.$$

Then there exists a positive constant $\hat{C}$ that depends only on k, m, p, q, $\hat{\kappa}$ and $\hat{\pi}$, such that

$$|\hat{v} - \hat{\pi}\hat{v}|_{m,q,\hat{\kappa}} \leqslant \hat{C}[\hat{v}]_{k+1,p,\hat{\kappa}} \quad \forall \hat{v} \in W^{k+1,p}(\hat{\kappa}). \tag{A.43}$$

Thus, in order to estimate the interpolation error in $Q_k(\kappa)$, we must derive an upper bound for $|v|_{m,q,\kappa}$ in terms of $\hat{v}$ and for $[\hat{v}]_{k+1,p,\hat{\kappa}}$ in terms of v. This is the object of the next lemma (compare with Lemma A.1). The material for the proof can be found in Cartan [17].

Lemma A.9. *Let κ be a convex quadrilateral. For each integer $m \geqslant 0$ and real $p \in [1, \infty]$ there exist positive constants C_1, C_2 and C_3 independent of the geometry* *of* κ *such that the following upper bounds hold for all* $v \in W^{m,p}(\kappa)$:

$$\text{(A.44)} \quad \begin{cases} |v|_{0,p,\kappa} \leqslant C_1 h_\kappa^{2/p} |\hat{v}|_{0,p,\hat{\kappa}}, \\ |v|_{m,p,\kappa} \leqslant C_2 (h_\kappa/\rho_\kappa)^{3m-2} (h_\kappa^{2/p}/\rho_\kappa^m) \left\{ \sum_{l=1}^{m} |\hat{v}|_{l,p,\hat{\kappa}}^p \right\}^{1/p}, \quad m \geqslant 1, \end{cases}$$

$$\text{(A.45)} \quad [\hat{v}]_{m,p,\hat{\kappa}} \leqslant C_3 (h_\kappa^m/\rho_\kappa^{2/p}) |v|_{m,p,\kappa}, \quad m \geqslant 0.$$

Then, we have the following analogue of Corollary A.2 (compare with Remark A.1).

Corollary A.5. *Let* κ *be a convex quadrilateral and let* k, m, p, q *and* $\hat{\pi}$ *be like in Theorem* A.3. *We define the operator* $\pi \in \mathscr{L}(W^{k+1,p}(\kappa); W^{m,q}(\kappa))$ *by*:

$$\widehat{\pi v} = \hat{\pi} \hat{v}.$$

Then there exists a positive constant C, *independent of the geometry of* κ, *such that*:

$$\text{(A.46)} \quad \begin{cases} |v - \pi v|_{0,q,\kappa} \leqslant C \sigma_\kappa^{2/p} h_\kappa^{k+1+2/q-2/p} |v|_{k+1,p,\kappa}, \\ |v - \pi v|_{m,q,\kappa} \leqslant C \sigma_\kappa^{4m-2+2/p} h_\kappa^{k+1-m+2/q-2/p} |v|_{k+1,p,\kappa} \quad m \geqslant 1 \\ \qquad \forall v \in W^{k+1,p}(\kappa). \end{cases}$$

When κ *belongs to a regular triangulation* $\mathscr{T}_h$, *these two inequalities reduce to*:

$$\text{(A.47)} \quad |v - \pi v|_{m,q,\kappa} \leqslant C h^{k+1-m+2/q-2/p} |v|_{k+1,p,\kappa}.$$

Remark A.3. For all $m \geqslant 2$, a simple exercise on the differentiation of compound functions shows that, unlike $[\hat{v}]_{m,p,\hat{\kappa}}$, the full seminorm $|\hat{v}|_{m,p,\hat{\kappa}}$ satisfies a very poor upper bound in terms of $\|v\|_{m,p,\kappa}$. The trouble arises from the cross derivatives $\partial^m \hat{v}/\partial \hat{x}_1^i \, \partial \hat{x}_2^{m-i}$ for $1 \leqslant i \leqslant m-1$, because they involve (in particular) the factor $\partial^2 F_\kappa/\partial \hat{x}_1 \partial \hat{x}_2$ instead of products of the form $(\partial F_\kappa/\partial \hat{x}_k)(\partial F_\kappa/\partial \hat{x}_l)$. And, unless κ is nearly a parallelogram, $\partial^2 F_\kappa/\partial \hat{x}_1 \partial \hat{x}_2$ is only of the order of h_κ whereas $(\partial F_\kappa/\partial \hat{x}_k)(\partial F_\kappa/\partial \hat{x}_l)$ is of the order of h_κ^2. This is why the seminorm $[\,.\,]_{m,p,\Omega}$ plays a vital part in deriving accurate error estimates for quadrilateral finite elements.

Remark A.4. Observe that the exponent of the regularity factor σ_κ is much higher in (A.46) than in (A.17). This comes from the fact that the transformation F_κ is not affine.

Like in the preceding section, we fix an integer $k \geqslant 1$ and we introduce similar finite element spaces:

$$\Theta_h = \{\theta_h \in \mathscr{C}^0(\bar{\Omega});\ \theta_h|_\kappa \in Q_k(\kappa) \quad \forall \kappa \in \mathscr{T}_h\},$$

$$\Phi_h = \Theta_h \cap H_0^1(\Omega).$$

We obtain the analogue of the interpolant I_h by interpolating first the function values at the principal lattice of order k of $\hat{\kappa}$ (cf. Figure 4):

$$\Sigma_{\hat{\kappa}} = \{(i/k, j/k);\ 0 \leqslant i,j \leqslant k\},$$

i.e. for each $\hat{v} \in \mathscr{C}^0(\hat{\kappa})$, we set

$$\text{(A.48)} \qquad \hat{I}\hat{v} \in Q_k, \quad \hat{I}\hat{v}(\hat{x}) = \hat{v}(\hat{x}) \quad \forall \hat{x} \in \Sigma_{\hat{\kappa}}.$$

Then, if $\mathscr{T}_h$ is regular the interpolation operator I_h defined on each κ of $\mathscr{T}_h$ by

$$I_h v = \hat{I}\hat{v} \qquad \forall v \in \mathscr{C}^0(\bar{\Omega})$$

satisfies the conclusions of Lemma A.2, namely:

$$\text{(A.49)} \qquad \begin{aligned} &|v - I_h v|_{m,p,\Omega} \leqslant C h^{s+1-m} |v|_{s+1,p,\Omega} \qquad \forall v \in W^{s+1,p}(\Omega), \\ &\forall m \in \mathbb{N}, \quad \forall s, p \in \mathbb{R} \quad \text{with } p > 1, \quad 0 \leqslant m \leqslant s+1, \quad 1 \leqslant s \leqslant k; \end{aligned}$$

and

$$\text{(A.50)} \quad |v - I_h v|_{m,q,\Omega} \leqslant C h^{1-m} |v|_{1,q,\Omega} \qquad \forall v \in W^{1,q}(\Omega), \quad q > 2, \quad m = 0, 1.$$

Similarly, we define the interpolation operator $\tilde{I}_{\hat{\kappa}}$ on $W^{2,p}(\hat{\kappa})$ for $p > 1$ by:

$$\tilde{I}_{\hat{\kappa}}\hat{v} \in Q_k,$$

$$\tilde{I}_{\hat{\kappa}}\hat{v}(\hat{a}_i) = \hat{v}(\hat{a}_i) \qquad 1 \leqslant i \leqslant 4,$$

$$\text{if } k \geqslant 2 \quad \begin{cases} \displaystyle\int_{\hat{\kappa}'} (\tilde{I}_{\hat{\kappa}}\hat{v} - \hat{v}) f \, d\hat{s} = 0 \qquad \forall f \in P_{k-2}(\hat{\kappa}'), \qquad \forall \text{sides } \hat{\kappa}' \text{ of } \hat{\kappa}, \\ \displaystyle\int_{\hat{\kappa}} (\tilde{I}_{\hat{\kappa}}\hat{v} - \hat{v}) f \, d\hat{x} = 0 \qquad \forall f \in Q_{k-2}(\hat{\kappa}). \end{cases}$$

Following the lines of Lemma A.3, it can be easily proved that this system of linear equations defines $\tilde{I}_{\hat{\kappa}}$ uniquely. Then, if $\mathscr{T}_h$ is regular, the operator $\tilde{I}_h$ defined in each κ of $\mathscr{T}_h$ by

$$\widehat{\tilde{I}_h v} = \tilde{I}_{\hat{\kappa}}\hat{v} \qquad \forall v \in W^{2,p}(\Omega) \quad \text{with } p > 1$$

satisfies the statement of Lemma A.4.

Likewise, the projections P_h and $\mathring{P}_h$ can be extended to the case of quadrilaterals and it can be shown that they have the same properties as in the triangular case. Similarly, the L^2 projection ρ_h defined by:

$$\text{(A.51)} \qquad \begin{cases} \left.\begin{array}{l} \hat{\rho}\hat{v} \in Q_k \\ \displaystyle\int_{\hat{\kappa}} (\hat{\rho}\hat{v} - \hat{v}) f \, d\hat{x} = 0 \qquad \forall f \in Q_k \end{array}\right\} \text{ on } \hat{\kappa} \\ \widehat{\rho_h v} = \hat{\rho}\hat{v} \qquad \forall \kappa \in \mathscr{T}_h \qquad \forall v \in L^2(\Omega) \end{cases}$$

satisfies the statement of Lemma A.5 *provided the triangulation is regular.*

Finally, Lemma A.6 and Corollary A.3 remain valid in the case of quadrilateral finite elements whereas in general Lemma A.7 holds here only when $m \leqslant 1$. Nevertheless these two values of m are sufficient for subsequent applications.

Remark A.5. It is sometimes very useful to construct triangulations which contain triangles as well as quadrilaterals. All the results which are valid for *both triangles and quadrilaterals* can be applied to these triangulations.

A.3. Interpolation of Discontinuous Functions

The interpolation of L^1 functions was first analyzed by Clément [21] who proposed a local regularization operator. This technique was later generalized by Bernardi [9] who introduced more general finite elements in order to interpolate essentially functions that were either discontinuous or that were defined on domains with curved boundaries. We have no space here to describe "curved" finite elements but we shall discuss briefly the interpolation of functions that are not supposed to be continuous. For the sake of simplicity, *we shall restrict ourselves to the finite element space Θ_h defined by* (A.18) *on triangular finite elements.* The forthcoming results are valid for higher dimensions and quadrilateral finite elements and also for spaces with other boundary conditions, like for example $\Theta_h \cap H_0^1(\Omega)$.

Let $\mathscr{T}_h$ be the triangulation of $\bar{\Omega}$ corresponding to Θ_h and let

$$\Sigma_h = \bigcup_{\kappa \in \mathscr{T}_h} \Sigma_\kappa = \{a_i;\ 1 \leqslant i \leqslant N_h\},$$

all the points a_i being distinct. For each integer i, $1 \leqslant i \leqslant N_h$, there exists exactly one function $\theta_i \in \Theta_h$ such that

$$\theta_i(a_j) = \delta_{ij}, \qquad 1 \leqslant j \leqslant N_h.$$

The set $\{\theta_i; 1 \leqslant i \leqslant N_h\}$ is a basis of Θ_h. Now, for each i, let

$$\text{(A.52)} \qquad \begin{cases} \Delta_i = \bigcup \{\kappa \in \mathscr{T}_h;\ \operatorname{supp}(\theta_i) \cap \kappa \neq \phi\}, \\ = \bigcup \{\kappa \in \mathscr{T}_h;\ a_i \in \kappa\}. \end{cases}$$

If the triangulation $\mathscr{T}_h$ is regular, it can be shown that on the one hand, the number of triangles κ in Δ_i is bounded by a constant, say M, independent of h and i; on the other hand, the number of macro-elements Δ_i containing a given triangle κ is also bounded by a constant independent of κ and h. In addition, for any pair of triangles κ and κ' in the same macro-element Δ_i we have:

$$h_\kappa \leqslant C h_{\kappa'} \quad \text{with a constant } C \text{ independent of } h \text{ and } i.$$

Thus, the macro-elements Δ_i can only assume a *finite number of different configurations.* To each configuration, there corresponds a reference region $\hat{\Delta}_i$, contained in the unit disc$\{\hat{x}; \|\hat{x}\| \leqslant 1\}$, and composed of at most M equal triangles with a common vertex at $\hat{x} = 0$ (see the example of Figure 6). Furthermore, there exists a continuous and invertible mapping F_{Δ_i} from $\hat{\Delta}_i$ onto Δ_i such that

$$F_{\Delta_i}|_{\hat{\kappa}_j} \quad \text{is an } \textit{affine} \text{ mapping from } \hat{\kappa}_j \text{ onto } \kappa_j$$

for all $\hat{\kappa}_j$ contained in $\hat{\Delta}_i$.

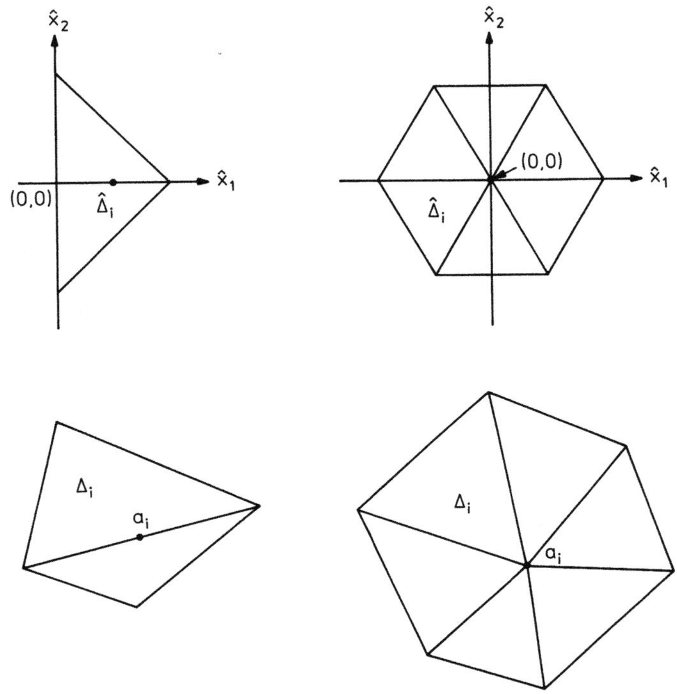

Figure 6. Examples of regions Δ_i and their reference regions $\hat{\Delta}_i$

Now, we are in a position to define an interpolation operator $R_h \in \mathscr{L}(L^1(\Omega); \Theta_h)$ by a local L^2 projection. Let Δ_i be any macro-element, $\hat{\Delta}_i$ its reference region and $\hat{v} \in L^1(\hat{\Delta}_i)$; let $\hat{R}_i \hat{v}$ be the unique polynomial of $P_k(\hat{\Delta}_i)$ determined by:

(A.53) $$\int_{\hat{\Delta}_i} (\hat{R}_i \hat{v} - \hat{v}) p \, d\hat{x} = 0 \qquad \forall p \in P_k(\hat{\Delta}_i).$$

Then, for $v \in L^1(\Omega)$ we define $R_h v \in \Theta_h$ by

(A.54) $$R_h v = \sum_{i=1}^{N_h} \hat{R}_i (v \circ F_{\Delta_i})(\hat{a}_i) \theta_i,$$

where $\hat{a}_i = F_{\Delta_i}^{-1}(a_i)$. It is proved in Clément [21] and Bernardi (loc. cit.) that this operator R_h enjoys basically the same interpolation properties as I_h and $\tilde{I}_h$, with the advantage that it works on rough functions. The major local interpolation result is:

Theorem A.4. *Let $\mathcal{T}_h$ be a regular triangulation of $\bar{\Omega}$, κ an arbitrary triangle of $\mathcal{T}_h$, l and m two positive integers with $0 \leqslant l \leqslant k + 1$ and p and q two real numbers in $[1, \infty]$ such that:*

$$W^{l,p}(\kappa) \hookrightarrow W^{m,q}(\kappa).$$

Let Δ denote the union of all macro-elements Δ_i containing κ. Then for all functions $v \in W^{l,p}(\Delta)$ we have the estimate:

$$\| v - R_h v \|_{m,q,\kappa} \leqslant C h_\kappa^{l-m+2(1/q-1/p)} \| v \|_{l,p,\Delta}, \tag{A.55}$$

with a constant C independent of κ, h and v.

Chapter II. Numerical Solution of the Stokes Problem in the Primitive Variables

§ 1. General Approximation

The abstract problem discussed in Chapter I, § 4 lends itself readily to a straightforward approximation that converges under reasonable assumptions with an error proportional to the approximation error of the spaces involved. When applied to the Stokes problem, this approach yields a conforming approximation of the velocity and pressure, although the approximate velocity field is (in general) not exactly divergence-free. The wide range of finite element methods developped in the remainder of the chapter are all founded on the material of this paragraph. Non-conforming methods can also be put into this framework (cf. Zine [85]) but for the sake of conciseness we have skipped them entirely.

1.1. An Abstract Approximation Result

This section is devoted to the approximation of the abstract variational problem analyzed in § 4 of Chapter I. We keep here the same notation and we put the problem in exactly the same situation. Recall that our *Problem* (Q) reads:

Given l in X' and χ in M', find a pair (u, λ) in $X \times M$ such that

(1.1) $$a(u, v) + b(v, \lambda) = \langle l, v \rangle \quad \forall v \in X,$$

(1.2) $$b(u, \mu) = \langle \chi, \mu \rangle \quad \forall \mu \in M.$$

Here, X and M are two real Hilbert spaces and $a(.,.)$ and $b(.,.)$ are two continuous bilinear forms defined respectively on $X \times X$ and $X \times M$. With the form $b(.,.)$ we associate the linear operators B and B' defined by:

$$\langle Bv, \mu \rangle = \langle v, B'\mu \rangle = b(v, \mu) \quad \forall v \in X, \quad \forall \mu \in M$$

and we set

$$V(\chi) = \{v \in X;\ Bv = \chi\},$$
$$V = V(0) = \mathrm{Ker}(B).$$

Recall the *Problem* (P) associated with Problem (Q):

Given l in X' and χ in M', find u in $V(\chi)$ such that:

$$a(u,v) = \langle l, v\rangle \quad \forall v \in V. \tag{1.3}$$

We retain the two hypotheses which guarantee that Problems (Q) and (P) are equivalent and have a unique solution (cf. Theorem I.4.1 and its Corollary):

there exists a constant $\alpha > 0$ such that

$$a(v,v) \geqslant \alpha \| v \|_X^2 \quad \forall v \in V; \tag{1.4}$$

there exists a constant $\beta > 0$ such that

$$\inf_{\mu \in M} \sup_{v \in X} \frac{b(v,\mu)}{\| v \|_X \| \mu \|_M} \geqslant \beta. \tag{1.5}$$

Let h denote a discretization parameter tending to zero and, for each h, let X_h and M_h be two *finite-dimensional* spaces such that:

$$X_h \subset X, \qquad M_h \subset M.$$

Let X_h' and M_h' denote their dual spaces with the dual norms:

$$\| l_h \|_{X_h'} = \sup_{v_h \in X_h} \frac{\langle l_h, v_h\rangle}{\| v_h \|_X}, \qquad \| \chi_h \|_{M_h'} = \sup_{\mu_h \in M_h} \frac{\langle \chi_h, \mu_h\rangle}{\| \mu_h \|_M}.$$

Clearly,

$$\| l \|_{X_h'} \leqslant \| l \|_{X'}, \qquad \| \chi \|_{M_h'} \leqslant \| \chi \|_{M'} \qquad \forall l \in X', \qquad \forall \chi \in M'.$$

Like in the continuous case, we associate with $a(.,.)$ and $b(.,.)$ the operators $A_h \in \mathscr{L}(X; X_h')$, $B_h \in \mathscr{L}(X; M_h')$ and $B_h' \in \mathscr{L}(M; X_h')$ defined by:

$$\langle A_h u, v_h\rangle = a(u, v_h) \quad \forall v_h \in X_h, \quad \forall u \in X,$$

$$\langle B_h v, \mu_h\rangle = b(v, \mu_h) \quad \forall \mu_h \in M_h, \quad \forall v \in X,$$

$$\langle v_h, B_h' \mu\rangle = b(v_h, \mu) \quad \forall v_h \in X_h, \quad \forall \mu \in M.$$

Strictly speaking, B_h' is not the dual operator of B_h but if B_h is restricted to X_h and B_h' to M_h then B_h and B_h' are indeed dual operators. In addition, we obviously have:

$$\| B_h v \|_{M_h'} \leqslant \| B v \|_{M'} \qquad \forall v \in X$$

with similar inequalities for $\| A_h v \|_{X_h'}$ and $\| B_h' \mu \|_{X_h'}$.

For each $\chi \in M'$, we define the finite-dimensional analogue of $V(\chi)$:

$$V_h(\chi) = \{ v_h \in X_h;\ b(v_h, \mu_h) = \langle \chi, \mu_h\rangle \quad \forall \mu_h \in M_h \}$$

and we set

$$V_h = V_h(0) = \operatorname{Ker}(B_h) \cap X_h;$$

i.e.

(1.6) $$V_h = \{v_h \in X_h;\ b(v_h, \mu_h) = 0 \quad \forall \mu_h \in M_h\}.$$

Right away we remark that generally $V_h \not\subset V$ and $V_h(\chi) \not\subset V(\chi)$ because M_h is a proper subspace of M.

Now we approximate Problem (Q) by *Problem* (Q_h):

Find a pair (u_h, λ_h) *in* $X_h \times M_h$ *satisfying*:

(1.7) $$a(u_h, v_h) + b(v_h, \lambda_h) = \langle l, v_h \rangle \qquad \forall v_h \in X_h,$$

(1.8) $$b(u_h, \mu_h) = \langle \chi, \mu_h \rangle \qquad \forall \mu_h \in M_h$$

and we associate with (Q_h) the following *Problem* (P_h):

Find $u_h \in V_h(\chi)$ *such that*:

(1.9) $$a(u_h, v_h) = \langle l, v_h \rangle \qquad \forall v_h \in V_h.$$

As $V_h \not\subset V$, Problem (P_h) may be viewed as an *external approximation* of Problem (P). Here again, the first component u_h of any solution (u_h, λ_h) of Problem (Q_h) is also a solution of Problem (P_h). The converse is proved as part of the next theorem.

Theorem 1.1. 1°) *Assume that the following conditions hold*:

(i) $V_h(\chi)$ *is not empty*;

(ii) *there exists a constant* $\alpha^* > 0$ *such that*:

(1.10) $$a(v_h, v_h) \geqslant \alpha^* \|v_h\|_X^2 \qquad \forall v_h \in V_h.$$

Then Problem (P_h) *has a unique solution* $u_h \in V_h(\chi)$ *and there exists a constant* C_1 *depending only upon* α^*, $\|a\|$ *and* $\|b\|$ *such that the "error bound" holds*:

(1.11) $$\|u - u_h\|_X \leqslant C_1 \left\{ \inf_{v_h \in V_h(\chi)} \|u - v_h\|_X + \inf_{\mu_h \in M_h} \|\lambda - \mu_h\|_M \right\}.$$

2°) *Assume that hypothesis* (ii) *holds and, in addition, that*:

(iii) *there exists a constant* $\beta^* > 0$ *such that*

(1.12) $$\sup_{v_h \in X_h} \frac{b(v_h, \mu_h)}{\|v_h\|_X} \geqslant \beta^* \|\mu_h\|_M \qquad \forall \mu_h \in M_h.$$

Then $V_h(\chi) \neq \phi$ *and there exists a unique* λ_h *in* M_h *such that* (u_h, λ_h) *is the only solution of Problem* (Q_h). *Furthermore, there exists a constant* C_2 *depending only upon* α^*, β^*, $\|a\|$ *and* $\|b\|$ *such that*:

(1.13) $$\|u - u_h\|_X + \|\lambda - \lambda_h\|_M \leqslant C_2 \left\{ \inf_{v_h \in X_h} \|u - v_h\|_X + \inf_{\mu_h \in M_h} \|\lambda - \mu_h\|_M \right\}.$$

Proof. 1°) As $V_h(\chi)$ is not empty, we choose a u_h^0 in $V_h(\chi)$ and we solve the problem:

Find z_h *in* V_h *such that*

$$a(z_h, v_h) = \langle l, v_h \rangle - a(u_h^0, v_h) \qquad \forall v_h \in V_h.$$

From (1.10), this problem has a unique solution z_h and therefore

$$u_h = z_h + u_h^0$$

is the only solution of Problem (P_h).

Now let w_h be an arbitrary element of $V_h(\chi)$; then $v_h = u_h - w_h \in V_h$ and

$$a(v_h, v_h) = \langle l, v_h \rangle - a(w_h, v_h). \tag{1.14}$$

Since $v_h \in X_h$, we can take $v = v_h$ in (1.1) and substitute in (1.14). This yields:

$$a(v_h, v_h) = a(u - w_h, v_h) + b(v_h, \lambda).$$

Moreover, since $v_h \in V_h$, we have $b(v_h, \mu_h) = 0\ \forall \mu_h \in M_h$. Hence

$$a(v_h, v_h) = a(u - w_h, v_h) + b(v_h, \lambda - \mu_h) \qquad \forall \mu_h \in M_h. \tag{1.15}$$

The V_h-ellipticity of a and the continuity of a and b yield:

$$\|v_h\|_X \leqslant (\|a\|\,\|u - w_h\|_X + \|b\|\,\|\lambda - \mu_h\|_M)/\alpha^*.$$

Therefore

$$\|u - u_h\|_X \leqslant \left(1 + \frac{\|a\|}{\alpha^*}\right)\|u - w_h\|_X + \frac{\|b\|}{\alpha^*}\|\lambda - \mu_h\|_M$$

$$\forall w_h \in V_h(\chi) \qquad \forall \mu_h \in M_h.$$

This yields (1.11) with $C_1 = \max\left(1 + \frac{\|a\|}{\alpha^*}, \frac{\|b\|}{\alpha^*}\right)$.

2°) Let us apply Lemma I.4.1 to the particular case of X_h and M_h. Hypothesis (iii) implies that B_h is an isomorphism from $V_h^\perp$ (taken in X_h) onto M_h'. Therefore $V_h(\chi)$ is not empty and according to n° 1 Problem (P_h) has a unique solution u_h. Furthermore, it follows from Corollary I.4.1 that there exists a unique λ_h in M_h such that the pair (u_h, λ_h) is the only solution of Problem (Q_h).

To derive the error bound (1.13) we shall first prove that

$$\inf_{w_h \in V_h(\chi)} \|u - w_h\|_X \leqslant \left(1 + \frac{\|b\|}{\beta^*}\right) \inf_{v_h \in X_h} \|u - v_h\|_X. \tag{1.16}$$

Let v_h be an arbitrary element of X_h; like above, there exists a unique z_h in $V_h^\perp$ such that

$$B_h z_h = B_h(u - v_h)$$

and

$$\|z_h\|_X \leqslant \frac{1}{\beta^*}\|B_h(u - v_h)\|_{M_h'} \leqslant \frac{1}{\beta^*}\|b\|\,\|u - v_h\|_X.$$

Thus, if we set $w_h = z_h + v_h$, then

$$b(w_h, \mu_h) = b(u - v_h, \mu_h) + b(v_h, \mu_h) = \langle \chi, \mu_h \rangle \qquad \forall \mu_h \in M_h$$

and therefore $w_h \in V_h(\chi)$. Furthermore

$$\|u - w_h\|_X \leqslant \|u - v_h\|_X + \|z_h\|_X \leqslant \left(1 + \frac{\|b\|}{\beta^*}\right)\|u - v_h\|_X.$$

As v_h is arbitrary, this implies (1.16).

It remains to evaluate $\|\lambda - \lambda_h\|_M$. From (1.1) and (1.7) we derive that:

$$b(v_h, \lambda_h - \mu_h) = a(u - u_h, v_h) + b(v_h, \lambda - \mu_h) \qquad \forall v_h \in X_h, \qquad \forall \mu_h \in M_h.$$

Therefore hypothesis (1.12) yields:

$$\|\lambda_h - \mu_h\|_M \leqslant \frac{1}{\beta^*} \sup_{v_h \in X_h} \frac{1}{\|v_h\|_X}\{a(u - u_h, v_h) + b(v_h, \lambda - \mu_h)\}$$
$$\leqslant \frac{1}{\beta^*}\{\|a\|\,\|u - u_h\|_X + \|b\|\,\|\lambda - \mu_h\|_M\}.$$

Hence

$$\text{(1.17)} \quad \|\lambda - \lambda_h\|_M \leqslant \frac{1}{\beta^*}\left\{\|a\|\,\|u - u_h\|_X + (\beta^* + \|b\|) \inf_{\mu_h \in M_h} \|\lambda - \mu_h\|_M\right\}.$$

Then the bound (1.13) follows immediately from (1.11), (1.16) and (1.17). □

Remark 1.1. The bound (1.11) can be slightly improved without making use of the inf-sup condition (1.12). Indeed, by applying (1.10) to (1.15) we obtain:

$$\|v_h\|_X \leqslant \frac{1}{\alpha^*}\left(\|a\|\,\|u - w_h\|_X + \sup_{v_h \in V_h} \frac{b(v_h, \lambda - \mu_h)}{\|v_h\|_X}\right).$$

Therefore

$$\|u - u_h\|_X \leqslant \left(1 + \frac{\|a\|}{\alpha^*}\right) \inf_{w_h \in V_h(\chi)} \|u - w_h\|_X + \frac{1}{\alpha^*} \inf_{\mu_h \in M_h} \sup_{v_h \in V_h} \frac{b(v_h, \lambda - \mu_h)}{\|v_h\|_X}.$$

(1.18)

Note that the expression

$$\inf_{\mu_h \in M_h} \sup_{v_h \in V_h} \frac{b(v_h, \lambda - \mu_h)}{\|v_h\|_X}$$

takes into account the fact that $V_h \not\subset V$: it vanishes when $V_h \subset V$.

Remark 1.2. If besides Hypotheses (1.10) and (1.12) we assume that the bilinear form $a(.,.)$ is symmetric and semi-positive definite on X_h, then we can relate Problems (P_h) and (Q_h) to optimization problems. As in the continuous case, and with the same notations, it can be shown that the solution u_h of (P_h) is characterized by:

$$J(u_h) = \inf_{v_h \in V_h(\chi)} J(v_h),$$

while the solution (u_h, p_h) of (Q_h) is characterized by:

$$\mathscr{L}(u_h, p_h) = \operatorname*{Min}_{v_h \in X_h} \sup_{q_h \in M_h} \mathscr{L}(v_h, q_h) = \operatorname*{Max}_{q_h \in M_h} \inf_{v_h \in X_h} \mathscr{L}(v_h, q_h).$$

Remark 1.3. From the argument of Theorem 1.1, we readily derive that if Hypotheses (1.10) and (1.12) hold then the solution (u_h, λ_h) is bounded as follows:

$$\|u_h\|_X \leqslant \frac{1}{\alpha^*}\left\{\|l\|_{X'} + \frac{1}{\beta^*}(\alpha^* + \|a\|)\|\chi\|_{M'}\right\},$$

$$\|\lambda_h\|_M \leqslant \frac{1}{\beta^*}\{\|l\|_{X'} + \|a\|\,\|u_h\|_X\}.$$

Observe that the bilinear form $a(.,.)$ is V_h-elliptic as soon as $a(v_h, v_h) > 0$ for all $v_h \neq 0$. Similarly, the bilinear form $b(.,.)$ satisfies the discrete inf-sup condition (1.12) provided $\operatorname{Ker}(B_h') \cap M_h = \{0\}$. But of course in either case the constants α^* and β^* will generally depend upon h. Now, in order to derive optimal error bounds in Theorem 1.1, it is clear that both constants α^* and β^* must be independent of h. And since usually $V_h \not\subset V$, the V-ellipticity of $a(.,.)$ does not necessarily carry over to V_h. As a consequence, hypothesis (1.10) must be checked in each particular case; but for the applications we have in mind, this is not a major obstacle. On the other hand, the discrete inf-sup condition (1.12) which acts as a uniform compatibility condition between X_h and M_h is much more delicate to check. The following lemma due to Fortin [28] establishes a useful criterion for (1.12).

Lemma 1.1. *The inf-sup condition* (1.12) *holds with a constant* $\beta^* > 0$ *independent of* h *if and only if there exists an operator* $\Pi_h \in \mathscr{L}(X; X_h)$ *satisfying*:

$$b(v - \Pi_h v, \mu_h) = 0 \quad \forall \mu_h \in M_h, \quad \forall v \in X \tag{1.19}$$

and

$$\|\Pi_h v\|_X \leqslant C\|v\|_X \quad \forall v \in X \tag{1.20}$$

with a constant $C > 0$ *independent of* h.

Proof. Assume that such an operator Π_h exists; then we have for all $\mu_h \in M_h$:

$$\sup_{v_h \in X_h} \frac{b(v_h, \mu_h)}{\|v_h\|_X} \geqslant \sup_{v \in X} \frac{b(\Pi_h v, \mu_h)}{\|\Pi_h v\|_X} = \sup_{v \in X} \frac{b(v, \mu_h)}{\|\Pi_h v\|_X},$$

owing to (1.19). Thus (1.20) and (1.5) imply that

$$\sup_{v_h \in X_h} \frac{b(v_h, \mu_h)}{\|v_h\|_X} \geqslant \frac{1}{C} \sup_{v \in X} \frac{b(v, \mu_h)}{\|v\|_X} \geqslant \frac{\beta}{C}\|\mu_h\|_M,$$

and (1.12) follows with $\beta^* = \beta/C$.

Conversely, suppose that (1.12) holds with a constant $\beta^* > 0$ independent of h. Then for each v in X there exists a unique element $\Pi_h v$ of $V_h^\perp$ such that

$$b(\Pi_h v, \mu_h) = b(v, \mu_h) \quad \forall \mu_h \in M_h$$

and

$$\| \Pi_h v \|_X \leqslant \frac{1}{\beta^*} \| B_h v \|_{M_h'} \leqslant \frac{\|b\|}{\beta^*} \| v \|_X.$$

Clearly $\Pi_h \in \mathscr{L}(X; X_h)$ and satisfies (1.20) with $C = \|b\|/\beta^*$. □

In practice, the construction of Π_h is by no means easy. The reader will find in Section 1.4 how to establish the inf-sup condition in a number of cases without constructing Π_h explicitly.

Remark 1.4. Another useful way of writing the inf-sup condition (1.12) is:

$$\begin{cases} \text{for each } \mu_h \in M_h \text{ there exists a } v_h \text{ in } X_h \text{ (unique in } V_h^\perp\text{) such that:} \\ \qquad b(v_h, \mu_h) = \|\mu_h\|_M^2, \qquad \|v_h\|_X \leqslant \dfrac{1}{\beta^*} \|\mu_h\|_M. \end{cases}$$

This result, which is also valid in the continuous case, uses explicitly the fact that $\| . \|_M$ is a Hilbert norm.

Remark 1.5. In the particular case where the bilinear form $a(.,.)$ coincides with the scalar product $((.,.))_X$ associated with the Hilbert norm $\| . \|_X$, formula (1.17) simplifies to:

$$(1.17') \quad \|\lambda - \lambda_h\|_M \leqslant \frac{1}{\beta^*} \left\{ \inf_{w_h \in V_h(\chi)} \|u - w_h\|_X + (\beta^* + \|b\|) \inf_{\mu_h \in M_h} \|\lambda - \mu_h\|_M \right\}.$$

Indeed, we have:

$$b(v_h, \lambda_h - \mu_h) = ((u - w_h, v_h))_X + b(v_h, \lambda - \mu_h)$$
$$\forall v_h \in V_h^\perp, \quad \forall w_h \in V_h(\chi), \quad \forall \mu_h \in M_h$$

and the v_h (in $V_h^\perp$) of Remark 1.4 gives (1.17′).

Theorem 1.1 readily yields the following general convergence results.

Corollary 1.1. *Assume that the following hypotheses hold:*

1°) *the form* $a(.,.)$ *satisfies* (1.10) *with a constant* $\alpha^* > 0$ *independent of* h;

2°) *there exist a dense subvariety* $\mathscr{V}(\chi)$ *of* $V(\chi)$, *a dense subspace* $\mathscr{M}$ *of* M *and two mappings* $r_h\colon \mathscr{V}(\chi) \to V_h(\chi)$ *and* $\rho_h\colon \mathscr{M} \to M_h$ *with*:

$$\lim_{h \to 0} \|r_h v - v\|_X = 0 \quad \forall v \in \mathscr{V}(\chi),$$

$$\lim_{h \to 0} \|\rho_h \mu - \mu\|_M = 0 \quad \forall \mu \in \mathscr{M}.$$

Then

$$\lim_{h\to 0} \|u - u_h\|_X = 0.$$

Corollary 1.2. *We retain the above hypotheses on* $a(.,.)$ *and* M *and we assume that* $b(.,.)$ *satisfies a uniform inf-sup condition* (1.12). *If there exists a dense subspace* $\mathscr{X}$ *of* X *and a mapping* $r_h: \mathscr{X} \to X_h$ *satisfying*:

$$\lim_{h\to 0} \|r_h v - v\|_X = 0 \qquad \forall v \in \mathscr{X}$$

then

$$\lim_{h\to 0} \{\|u - u_h\|_X + \|\lambda - \lambda_h\|_M\} = 0.$$

Now, let us extend the classical duality argument of Aubin [3] and Nitsche [61] to the case of Problems (P) and (P_h). For this, we introduce a Hilbert space H with scalar product $(.,.)$ and associated norm $|.|$ such that

$$X \subset H \text{ with continuous imbedding and } X \text{ is dense in } H.$$

We identify H with its dual space H' for the scalar product $(.,.)$. Therefore, H can be identified with a subspace of X':

$$H \subset X' \text{ with continuous and dense imbedding.}$$

In order to evaluate $|u - u_h|$, we introduce for each g in H the unique solution pair (φ_g, ξ_g) of the dual problem:

$$(1.21) \qquad \begin{cases} a(v, \varphi_g) + b(v, \xi_g) = (g, v) & \forall v \in X \\ \quad b(\varphi_g, \mu) = 0 & \forall \mu \in M. \end{cases}$$

Theorem 1.2. *Assume that Problem* (P_h) *has a unique solution* u_h. *Then there exists a constant* C, *depending only upon* $\|a\|$ *and* $\|b\|$, *such that*:

$$(1.22) \qquad \begin{aligned} |u - u_h| \leqslant C&\left\{\|u - u_h\|_X + \inf_{\mu_h \in M_h} \|\lambda - \mu_h\|_M\right\} \\ &\times \sup_{g\in H} \frac{1}{|g|}\left\{\inf_{\varphi_h \in V_h} \|\varphi_g - \varphi_h\|_X + \inf_{\xi_h \in M_h} \|\xi_g - \xi_h\|_M\right\}. \end{aligned}$$

Proof. On the one hand, we have:

$$|u - u_h| = \sup_{g\in H} \frac{(g, u - u_h)}{|g|}.$$

On the other hand by choosing $v = u - u_h$ in (1.21), we get

$$(1.23) \qquad (g, u - u_h) = a(u - u_h, \varphi_g) + b(u - u_h, \xi_g).$$

Then taking into account (1.1) and (1.6) we find:

$$a(u - u_h, \varphi_h) = -b(\varphi_h, \lambda) = -b(\varphi_h, \lambda - \mu_h) \qquad \forall \varphi_h \in V_h, \qquad \forall \mu_h \in M_h.$$

Besides that, as $\varphi_g \in V$, we have

$$b(\varphi_g, \lambda - \mu_h) = 0 \qquad \forall \mu_h \in M_h$$

and as $u \in V(\chi)$ and $u_h \in V_h(\chi)$, we also have:

$$b(u - u_h, \xi_h) = 0 \qquad \forall \xi_h \in M_h.$$

When substituted into (1.23), these three equalities yield:

$$(g, u - u_h) = a(u - u_h, \varphi_g - \varphi_h) + b(\varphi_g - \varphi_h, \lambda - \mu_h) + b(u - u_h, \xi_g - \xi_h)$$
$$\forall \varphi_h \in V_h, \qquad \forall \mu_h, \xi_h \in M_h.$$

Hence

$$|(g, u - u_h)| \leqslant C\{\|u - u_h\|_X + \|\lambda - \mu_h\|_M\}\{\|\varphi_g - \varphi_h\|_X + \|\xi_g - \xi_h\|_M\}$$
$$\forall \mu_h, \xi_h \in M_h, \qquad \forall \varphi_h \in V_h$$

where $C = \max(\|a\|, \|b\|)$. □

Remark 1.6. When Problem (Q_h) has a solution (u_h, λ_h) a straightforward modification of the above argument shows that:

$$|u - u_h| \leqslant C\{\|u - u_h\|_X + \|\lambda - \lambda_h\|_M\}$$
$$\times \sup_{g \in H} \frac{1}{|g|}\left\{\inf_{\varphi_h \in X_h} \|\varphi_g - \varphi_h\|_X + \inf_{\xi_h \in M_h} \|\xi_g - \xi_h\|_M\right\},$$

with the constant C of (1.22).

1.2. Decoupling the Computation of u_h and λ_h

In this short section, we propose to apply the technique of Sections I.4.3 and I.4.4 to dissociate the computation of λ_h from that of u_h. These methods are often used in practice.

Let us consider first the regularization procedure of Section I.4.3. Recall that we require a continuous, bilinear form $c(.,.)$ on $M_h \times M_h$ which is supposed to be M_h-*elliptic*, i.e. there exists a constant $\gamma^* > 0$ such that:

(1.24)
$$c(\mu_h, \mu_h) \geqslant \gamma^* \|\mu_h\|_M^2 \qquad \forall \mu_h \in M_h.$$

With the form $c(.,.)$ we associate as usual the operator $C_h \in \mathscr{L}(M_h; M_h')$ by:

$$\langle C_h \mu_h, \nu_h \rangle = c(\mu_h, \nu_h) \qquad \forall \mu_h, \nu_h \in M_h.$$

Like in the continuous case, for each $\varepsilon > 0$ we introduce the *Problem* (Q_h^ε):

Find a pair $(u_h^\varepsilon, \lambda_h^\varepsilon) \in X_h \times M_h$ *such that*

(1.25)
$$\begin{cases} a(u_h^\varepsilon, v_h) + b(v_h, \lambda_h^\varepsilon) = \langle l, v_h \rangle & \forall v_h \in X_h, \\ -\varepsilon c(\lambda_h^\varepsilon, \mu_h) + b(u_h^\varepsilon, \mu_h) = \langle \chi, \mu_h \rangle & \forall \mu_h \in M_h. \end{cases}$$

By virtue of (1.24), the operator C_h is non-singular and therefore λ_h^ε can be eliminated from the above equations. Thus Problem $(\mathrm{Q}_h^\varepsilon)$ is equivalent to the following *Problem* $(\mathrm{P}_h^\varepsilon)$:

Find $u_h^\varepsilon \in X_h$ *satisfying*:

(1.26) $$a(u_h^\varepsilon, v_h) + \frac{1}{\varepsilon}\langle C_h^{-1} B_h u_h^\varepsilon, B_h v_h\rangle = \langle l, v_h\rangle + \frac{1}{\varepsilon}\langle C_h^{-1}\chi, B_h v_h\rangle \quad \forall v_h \in X_h,$$

where $C_h^{-1} \in \mathscr{L}(M_h'; M_h)$ denotes the inverse of C_h.

Clearly, the situation here is exactly that of Section I.4.3, with the operators B and C replaced by B_h and C_h. Hence the statement of Theorem I.4.3 is valid for Problems $(\mathrm{P}_h^\varepsilon)$ and $(\mathrm{Q}_h^\varepsilon)$:

Theorem 1.3. *In addition to* (1.12) *and* (1.24), *assume that there exists a constant* $\alpha^* > 0$ *such that*:

(1.27) $$a(v_h, v_h) + \langle C_h^{-1} B_h v_h, B_h v_h\rangle \geqslant \alpha^* \| v_h \|_X^2 \quad \forall v_h \in X_h.$$

Then Problems (Q_h) *and* $(\mathrm{Q}_h^\varepsilon)$ *for* $\varepsilon \leqslant 1$ *have both a unique solution* (u_h, λ_h) *and* $(u_h^\varepsilon, \lambda_h^\varepsilon)$ *in* $X_h \times M_h$. *Moreover, for all* $\varepsilon \leqslant \varepsilon_0$ *small enough we have the following error bound*:

(1.28) $$\| u_h^\varepsilon - u_h \|_X + \| \lambda_h^\varepsilon - \lambda_h \|_M \leqslant K^* \varepsilon (\| l \|_{X'} + \| \chi \|_{M'}),$$

where the constant K^* *depends only upon* α^*, β^*, $\| a \|$, $\| b \|$ *and* $\| c \|$.

Likewise, we can refine (1.28) and obtain an asymptotic expansion for $(u_h^\varepsilon, \lambda_h^\varepsilon)$ in powers of ε. Again, we introduce the sequence of solutions $(u_h^n, \lambda_h^n) \in X_h \times M_h$ of the problems:

(1.29) $$\begin{cases} a(u_h^n, v_h) + b(v_h, \lambda_h^n) = 0 & \forall v_h \in X_h, \\ b(u_h^n, \mu_h) = c(\lambda_h^{n-1}, \mu_h) & \forall \mu_h \in M_h, \end{cases}$$

starting with $\lambda_h^0 = \lambda_h$. We have the analogue of Theorem I.4.4:

Theorem 1.4. *Under the hypotheses of Theorem* 1.3, *we have for all integers* $N \geqslant 1$ *and for* $\varepsilon \leqslant \varepsilon_0$ *small enough*:

(1.30) $$\begin{gathered} \left\| u_h^\varepsilon - u_h - \sum_{n=1}^N \varepsilon^n u_h^n \right\|_X + \left\| \lambda_h^\varepsilon - \lambda_h - \sum_{n=1}^N \varepsilon^n \lambda_h^n \right\|_M \\ \leqslant K_N^* \varepsilon^{N+1} (\| l \|_{X'} + \| \chi \|_{M'}), \end{gathered}$$

where the constant K_N^* *depends only upon* N, α^*, β^*, $\| a \|$, $\| b \|$ *and* $\| c \|$.

Now, we turn to the gradient algorithms of Section I.4.4. With the above notations, we set for each real parameter $r \geqslant 0$:

(1.31) $$a_r^h(u, v) = a(u, v) + r\langle C_h^{-1} B_h u, B_h v\rangle \quad \forall u, v \in X.$$

In addition to the ellipticity of $c(.,.)$, assume that the form $a_r^h(.,.)$ is X_h-elliptic: there exists a constant $\alpha^* > 0$ such that

$$a_r^h(v_h, v_h) \geqslant \alpha^* \|v_h\|_X^2 \qquad \forall v_h \in X_h. \tag{1.32}$$

Then the *simple gradient algorithm with optimal parameter* has the following discrete version:

1°) Given an initial guess $\lambda_h^0 \in M_h$, compute the solution $u_h^0 \in X_h$ of the problem

$$a_r^h(u_h^0, v_h) = \langle l, v_h\rangle + b(v_h, rC_h^{-1}\chi - \lambda_h^0) \qquad \forall v_h \in X_h;$$

2°) For $m \geqslant 0$, knowing $(u_h^m, \lambda_h^m) \in X_h \times M_h$, determine $(z_h^m, g_h^m) \in X_h \times M_h$, $\rho_h^m \in \mathbb{R}$ and the pair $(u_h^{m+1}, \lambda_h^{m+1}) \in X_h \times M_h$ by:

$$(1.33)\quad \begin{cases} \text{(a)} & \begin{cases} c(g_h^m, \mu_h) = \langle \chi, \mu_h\rangle - b(u_h^m, \mu_h) & \forall \mu_h \in M_h, \\ a_r^h(z_h^m, v_h) = b(v_h, g_h^m) & \forall v_h \in X_h, \end{cases} \\ \text{(b)} & \rho_h^m = \dfrac{c(g_h^m, g_h^m)}{b(z_h^m, g_h^m)}, \\ \text{(c)} & \begin{cases} \lambda_h^{m+1} = \lambda_h^m - \rho_h^m g_h^m, \\ u_h^{m+1} = u_h^m + \rho_h^m z_h^m. \end{cases} \end{cases}$$

Needless to say, the above scheme is a gradient algorithm only when the bilinear forms $a(.,.)$ and $c(.,.)$ are symmetric. Then the following result is a direct consequence of Corollary I.4.4.

Theorem 1.5. *Suppose the bilinear forms $a_r^h(.,.)$, $b(.,.)$ and $c(.,.)$ satisfy respectively* (1.32), (1.12) *and* (1.24) *and assume that $a(.,.)$ and $c(.,.)$ are symmetric. Then the simple gradient algorithm* (1.33) *is convergent for every choice of the starting value $\lambda_h^0 \in M_h$:*

$$\lim_{m\to\infty} \{\|u_h^m - u_h\|_X + \|\lambda_h^m - \lambda_h\|_M\} = 0.$$

Like in Section I.4.4, observe that the simple gradient algorithm can converge without optimal parameters. In that case, the bilinear form $a(.,.)$ need not be symmetric and we have the analogue of Theorem I.4.7:

Theorem 1.6. *We retain all hypotheses of Theorem* 1.5 *except the symmetry assumption on $a(.,.)$. Then the algorithm* (1.33a) (1.33c) *is convergent for every choice of the initial guess $\lambda_h^0 \in M_h$ and every sequence of numbers (ρ_h^m) in the range*:

$$0 < \inf_m \rho_h^m \leqslant \sup_m \rho_h^m < 2\gamma^* \tilde{\alpha}_r^*$$

where

$$\tilde{\alpha}_r^* = \inf_{v_h \in X_h} (a_r^h(v_h, v_h)/\|B_h v_h\|_{M_h'}^2).$$

By now, it stands out clearly that the discrete conjugate-gradient algorithm is entirely similar to the scheme (I.4.70). Let us describe it for the sake of completeness:

1°) Starting from an initial guess $\lambda_h^0 \in M_h$, compute the solution $u_h^0 \in X_h$ of the problem:

$$a_r^h(u_h^0, v_h) = \langle l, v_h \rangle + b(v_h, rC_h^{-1}\chi - \lambda_h^0) \qquad \forall v_h \in X_h;$$

2°) For $m \geqslant 0$, knowing $(u_h^m, \lambda_h^m) \in X_h \times M_h$ compute g_h^m, $\omega_h^m \in M_h$, $z_h^m \in X_h$, ρ_h^m, $\sigma_h^m \in \mathbb{R}$ and the pair $(u_h^{m+1}, \lambda_h^{m+1}) \in X_h \times M_h$ by:

$$(1.34) \qquad \begin{cases} c(g_h^m, \mu_h) = \langle \chi_h, \mu_h \rangle - b(u_h^m, \mu_h) \qquad \forall \mu_h \in M_h, \\ \left.\begin{aligned} \sigma_h^m &= \dfrac{c(g_h^m, g_h^m)}{c(g_h^{m-1}, g_h^{m-1})} \\ \omega_h^m &= g_h^m + \sigma_h^m \omega_h^{m-1} \end{aligned}\right\} \begin{array}{l} \text{only if } m \geqslant 1 \\ \omega_h^0 = g_h^0 \text{ otherwise,} \end{array} \\ a_r^h(z_h^m, v_h) = b(v_h, \omega_h^m), \\ \rho_h^m = \dfrac{c(g_h^m, g_h^m)}{b(z_h^m, g_h^m)}, \\ \lambda_h^{m+1} = \lambda_h^m - \rho_h^m \omega_h^m, \\ u_h^{m+1} = u_h^m + \rho_h^m z_h^m. \end{cases}$$

Theorem 1.7. *The conjugate-gradient algorithm converges with the hypotheses of Theorem* 1.5.

1.3. Application to the Homogeneous Stokes Problem

For the sake of simplicity, we focus our attention on homogeneous boundary conditions. Let Ω be a bounded, connected, open subset of $\mathbb{R}^N$ with a Lipschitz-continuous boundary Γ and let $\mathbf{f}$ be a given function of $H^{-1}(\Omega)^N$. Recall that the homogeneous Stokes equations:

$$(1.35) \qquad \begin{cases} \textit{Find } (\mathbf{u}, p) \textit{ in } H_0^1(\Omega)^N \times L_0^2(\Omega) \textit{ such that} \\ \left.\begin{aligned} -\nu\Delta\mathbf{u} + \mathbf{grad}\, p &= \mathbf{f} \\ \operatorname{div} \mathbf{u} &= 0 \end{aligned}\right\} \textit{ in } \Omega, \end{cases}$$

has a unique solution. Moreover, setting either

$$(1.36) \quad \text{or} \quad \begin{array}{l} \text{(a)} \\ \\ \text{(b)} \end{array} \begin{cases} a(\mathbf{u}, \mathbf{v}) = 2\nu \displaystyle\sum_{i,j=1}^{N} (D_{ij}(\mathbf{u}), D_{ij}(\mathbf{v})) \\ a(\mathbf{u}, \mathbf{v}) = \nu(\mathbf{grad}\, \mathbf{u}, \mathbf{grad}\, \mathbf{v}), \end{cases}$$

we know that (1.35) is equivalent to the variational formulation:

Find a pair $(\mathbf{u}, p) \in H_0^1(\Omega)^N \times L_0^2(\Omega)$ *such that*

$$(1.37)\qquad \begin{cases} a(\mathbf{u}, \mathbf{v}) - (p, \operatorname{div} \mathbf{v}) = \langle \mathbf{f}, \mathbf{v} \rangle \quad \forall \mathbf{v} \in H_0^1(\Omega)^N, \\ (q, \operatorname{div} \mathbf{u}) = 0 \quad \forall q \in L_0^2(\Omega). \end{cases}$$

With the following substitutions:

$$X = H_0^1(\Omega)^N, \quad M = L_0^2(\Omega), \quad \|\cdot\|_X = |\cdot|_{1,\Omega}, \quad \|\cdot\|_M = \|\cdot\|_{0,\Omega},$$

$$b(\mathbf{v}, q) = -(q, \operatorname{div} \mathbf{v}),$$

$$\chi = 0, \quad \mathbf{l} = \mathbf{f},$$

this is exactly the problem studied in Section I.5.1.

Now, for each h let W_h and Q_h be two finite-dimensional spaces such that

$$W_h \subset H^1(\Omega)^N, \quad Q_h \subset L^2(\Omega)$$

and throughout this section we assume that Q_h *contains the constant functions.* We set:

$$(1.38)\qquad \begin{cases} X_h = W_h \cap H_0^1(\Omega)^N = \{\mathbf{v}_h \in W_h;\ \mathbf{v}_{h|_\Gamma} = \mathbf{0}\}, \\ M_h = Q_h \cap L_0^2(\Omega) = \left\{q_h \in Q_h;\ \int_\Omega q_h\, dx = 0\right\}. \end{cases}$$

With these spaces, Problem (1.37) is approximated by:

Find a pair $(\mathbf{u}_h, p_h) \in X_h \times M_h$ *such that:*

$$(1.39)\qquad \begin{cases} a(\mathbf{u}_h, \mathbf{v}_h) - (p_h, \operatorname{div} \mathbf{v}_h) = \langle \mathbf{f}, \mathbf{v}_h \rangle \quad \forall \mathbf{v}_h \in X_h, \\ (q_h, \operatorname{div} \mathbf{u}_h) = 0 \quad \forall q_h \in M_h. \end{cases}$$

As $\operatorname{div} \mathbf{u}_h \in L_0^2(\Omega)$, observe that the second equation in (1.39) is equivalent to

$$(q_h, \operatorname{div} \mathbf{u}_h) = 0 \quad \forall q_h \in Q_h.$$

In view of this remark, the corresponding space V_h is given by:

$$V_h = \{\mathbf{v}_h \in X_h;\ (q_h, \operatorname{div} \mathbf{v}_h) = 0 \quad \forall q_h \in Q_h\}.$$

Hence the *Problem* (P_h) associated with (1.39) is:

Find $\mathbf{u}_h \in V_h$ *satisfying*

$$(1.40)\qquad a(\mathbf{u}_h, \mathbf{v}_h) = \langle \mathbf{f}, \mathbf{v}_h \rangle \quad \forall \mathbf{v}_h \in V_h.$$

Remark 1.7. As mentioned in the preceding section, V_h is generally not included in V: $\{\mathbf{v} \in H_0^1(\Omega)^N; \operatorname{div} \mathbf{v} = 0\}$; this will be the case in all the examples of this chapter. Thus the functions of V_h are not divergence-free but satisfy only

$$\rho_h(\operatorname{div} \mathbf{v}_h) = 0,$$

where ρ_h is the orthogonal projection of $L^2(\Omega)$ onto Q_h. As a consequence, the expressions (1.36a) and (1.36b) for the bilinear form $a(.,.)$ in (1.39) do not yield equivalent formulations.

In order to study Problem (1.40) we relate the continuous and discrete spaces by the following hypotheses:

Hypothesis H1 (Approximation property of X_h). *There exist an operator* $r_h \in \mathscr{L}(H^2(\Omega)^N; W_h) \cap \mathscr{L}((H^2(\Omega) \cap H_0^1(\Omega))^N; X_h)$ *and an integer* l *such that:*

$$(1.41) \qquad \|\mathbf{v} - r_h \mathbf{v}\|_{1,\Omega} \leqslant C h^m \|\mathbf{v}\|_{m+1,\Omega} \qquad \forall \mathbf{v} \in H^{m+1}(\Omega)^N, \qquad 1 \leqslant m \leqslant l.$$

Hypothesis H2 (Approximation property of Q_h). *There exist an operator* $S_h \in \mathscr{L}(L^2(\Omega); Q_h)$ *such that:*

$$(1.42) \qquad \|q - S_h q\|_{0,\Omega} \leqslant C h^m \|q\|_{m,\Omega} \qquad \forall q \in H^m(\Omega), \qquad 0 \leqslant m \leqslant l.$$

Hypothesis H3 (Uniform inf-sup condition). *For each* $q_h \in M_h$ *there exists a* $\mathbf{v}_h \in X_h$ *such that*

$$(1.43) \qquad (q_h, \operatorname{div} \mathbf{v}_h) = \|q_h\|_{0,\Omega}^2$$

$$(1.44) \qquad |\mathbf{v}_h|_{1,\Omega} \leqslant C \|q_h\|_{0,\Omega}$$

with a constant $C > 0$ *independent of* h, q_h *and* $\mathbf{v}_h$.

Recall that according to Remark 1.4 the statement of Hypothesis H3 is equivalent to the inf-sup condition (1.12) with $\beta^* = 1/C$.

Theorem 1.8. *Under Hypotheses* H1, H2 *and* H3, *Problem* (1.39) *has a unique solution* $(\mathbf{u}_h, p_h) \in V_h \times M_h$ *and* $\mathbf{u}_h$ *is also the only solution of Problem* (1.40). *In addition,* $(\mathbf{u}_h, p_h)$ *tends to the solution* $(\mathbf{u}, p)$ *of Problem* (1.35):

$$(1.45) \qquad \lim_{h \to 0} \{|\mathbf{u}_h - \mathbf{u}|_{1,\Omega} + \|p_h - p\|_{0,\Omega}\} = 0.$$

Furthermore, when $(\mathbf{u}, p)$ *belongs to* $H^{m+1}(\Omega)^N \times (H^m(\Omega) \cap L_0^2(\Omega))$ *for some integer* m *with* $1 \leqslant m \leqslant l$, *we have the error bound*:

$$(1.46) \qquad |\mathbf{u} - \mathbf{u}_h|_{1,\Omega} + \|p - p_h\|_{0,\Omega} \leqslant C h^m \{\|\mathbf{u}\|_{m+1,\Omega} + \|p\|_{m,\Omega}\}.$$

Proof. Let us apply Theorem 1.1. Owing to Hypothesis H3, the pair of spaces (X_h, M_h) satisfies a uniform inf-sup condition; therefore it suffices to check the ellipticity of $a(.,.)$ in order to obtain that Problem (1.39) has a unique solution. When $a(.,.)$ is defined by (1.36b), we have:

$$a(\mathbf{v}, \mathbf{v}) = \nu |\mathbf{v}|_{1,\Omega}^2 \qquad \forall \mathbf{v} \in H^1(\Omega)^N$$

and when $a(.,.)$ is defined by (1.36a) we use Korn's inequality (cf. (I.5.31))

$$a(\mathbf{v}, \mathbf{v}) \geqslant \nu |\mathbf{v}|_{1,\Omega}^2 \qquad \forall \mathbf{v} \in H_0^1(\Omega)^N.$$

Thus in both cases the form $a(.,.)$ is X-elliptic. Therefore Theorem 1.1 implies that Problem (1.39) has a unique solution $(\mathbf{u}_h, p_h) \in X_h \times M_h$, where $\mathbf{u}_h$ is the unique solution of Problem (1.40), and we have:

$$(1.47)\quad |\mathbf{u}-\mathbf{u}_h|_{1,\Omega} + \|p-p_h\|_{0,\Omega} \leqslant C_1 \left\{ \inf_{\mathbf{v}_h \in X_h} |\mathbf{u}-\mathbf{v}_h|_{1,\Omega} + \inf_{q_h \in M_h} \|p-q_h\|_{0,\Omega} \right\}$$

with a constant C_1 independent of h.

Now, observe that if s_h does not map $L_0^2(\Omega)$ onto M_h we can replace it by:

$$\bar{s}_h q = s_h q - \frac{1}{\operatorname{meas}(\Omega)} \int_\Omega s_h q \, dx;$$

then $\bar{s}_h \in \mathscr{L}(L^2(\Omega); M_h)$ and $\|\bar{s}_h q - q\|_{0,\Omega} \leqslant \|s_h q - q\|_{0,\Omega}\ \forall q \in L_0^2(\Omega)$. Thus if $p \in H^m(\Omega) \cap L_0^2(\Omega)$, Hypothesis H2 gives:

$$\inf_{q_h \in M_h} \|p-q_h\|_{0,\Omega} \leqslant \|p - s_h p\|_{0,\Omega} \leqslant C_2 h^m \|p\|_{m,\Omega}.$$

Likewise, if $\mathbf{u} \in H^{m+1}(\Omega)^N \cap V$, Hypothesis H1 yields:

$$\inf_{\mathbf{v}_h \in X_h} |\mathbf{u}-\mathbf{v}_h|_{1,\Omega} \leqslant |\mathbf{u} - r_h \mathbf{u}|_{1,\Omega} \leqslant C_3 h^m \|\mathbf{u}\|_{m+1,\Omega}.$$

These inequalities and (1.47) imply (1.46).

It remains to establish the limit (1.45). For this we make use of Corollary 1.2 and the above considerations. We know that $H^2(\Omega) \cap H_0^1(\Omega)$ is dense in $H_0^1(\Omega)$ and that $H^1(\Omega) \cap L_0^2(\Omega)$ is dense in $L_0^2(\Omega)$. Hence it suffices to work with the above operators r_h and s_h to obtain (1.45). □

Of course, (1.46) implies that $\|\mathbf{u}-\mathbf{u}_h\|_{0,\Omega} = O(h^m)$, but it is possible to refine this estimate by making use of Theorem 1.2. Here we take

$$H = L^2(\Omega)^N.$$

Then Problem (1.21) is the homogeneous Stokes Problem:

$$(1.48)\quad \begin{cases} \textit{Find a pair } (\boldsymbol{\varphi}, \xi) \textit{ in } H_0^1(\Omega)^N \times L_0^2(\Omega) \textit{ such that:} \\ \left.\begin{aligned} -\nu\Delta\boldsymbol{\varphi} + \mathbf{grad}\,\xi &= \mathbf{g} \\ \operatorname{div}\boldsymbol{\varphi} &= 0 \end{aligned}\right\} \quad \textit{in } \Omega, \textit{ where } \mathbf{g} \in L^2(\Omega)^N. \end{cases}$$

We shall require below the following concept of regularity for this problem.

Definition 1.1. We say that Problem (1.48) is *regular* if the mapping

$$(\boldsymbol{\varphi}, \xi) \mapsto -\nu\Delta\boldsymbol{\varphi} + \mathbf{grad}\,\xi$$

is an *isomorphism* from $[H^2(\Omega)^N \cap V] \times [H^1(\Omega) \cap L_0^2(\Omega)]$ onto $L^2(\Omega)^N$.

This definition means that $\boldsymbol{\varphi}$ belongs to $H^2(\Omega)^N$ and ξ to $H^1(\Omega)$ whenever the right-hand side $\mathbf{g}$ belongs to $L^2(\Omega)^N$ and

$$(1.49)\quad \|\boldsymbol{\varphi}\|_{2,\Omega} + \|\xi\|_{1,\Omega} \leqslant C\|\mathbf{g}\|_{0,\Omega}.$$

Observe that in view of Theorem I.5.4, Problem (1.48) is regular as soon as the boundary Γ of Ω is of class $\mathscr{C}^2$. When Γ is only Lipschitz-continuous—and subsequently Γ will be a polygonal line—Remark I.5.6 asserts that Problem (1.48) is regular provided Ω is a *plane, bounded* and *convex polygon.*

Theorem 1.9. *Assume that Hypotheses* H1, H2 *and* H3 *are satisfied and that Problem* (1.48) *is regular. Then, if the solution* $(\mathbf{u}, p)$ *of the Stokes Problem* (1.35) *belongs to* $H^{m+1}(\Omega)^N \times H^m(\Omega) \cap L_0^2(\Omega)$ *for some integer* m *with* $1 \leqslant m \leqslant l$, *we have the following error bound*:

$$\|\mathbf{u} - \mathbf{u}_h\|_{0,\Omega} \leqslant C h^{m+1}(\|\mathbf{u}\|_{m+1,\Omega} + \|p\|_{m,\Omega}). \tag{1.50}$$

Proof. According to Theorem 1.2 and Remark 1.6, we have:

$$\begin{aligned} \|\mathbf{u} - \mathbf{u}_h\|_{0,\Omega} &\leqslant C_1\{|\mathbf{u} - \mathbf{u}_h|_{1,\Omega} + \|p - p_h\|_{0,\Omega}\} \\ &\times \sup_{\mathbf{g} \in L^2(\Omega)^N} \frac{1}{\|\mathbf{g}\|_{0,\Omega}} \left\{ \inf_{\boldsymbol{\varphi}_h \in X_h} |\boldsymbol{\varphi} - \boldsymbol{\varphi}_h|_{1,\Omega} + \inf_{\xi_h \in M_h} \|\xi - \xi_h\|_{0,\Omega} \right\}. \end{aligned} \tag{1.51}$$

Since Problem (1.48) is regular, $\boldsymbol{\varphi} \in H^2(\Omega)^N \cap V$ and $\xi \in H^1(\Omega) \cap L_0^2(\Omega)$. Then the extra power of h in (1.50) follows from Hypotheses H1 and H2 with $m = 1$ and (1.46) substituted into (1.51).

Thus Hypothesis H1 with $m = 1$ and Hypothesis H2 with $m = 1$ yield:

$$\inf_{\boldsymbol{\varphi}_h \in V_h} |\boldsymbol{\varphi} - \boldsymbol{\varphi}_h|_{1,\Omega} \leqslant C_3 h |\boldsymbol{\varphi}|_{2,\Omega}, \qquad \inf_{\xi_h \in M_h} \|\xi - \xi_h\|_{0,\Omega} \leqslant C_4 h |\xi|_{1,\Omega}.$$

Therefore combining (1.49) and (1.51) we obtain:

$$\|\mathbf{u} - \mathbf{u}_h\|_{0,\Omega} \leqslant C_5 h \left\{ |\mathbf{u} - \mathbf{u}_h|_{1,\Omega} + \inf_{q_h \in M_h} \|p - q_h\|_{0,\Omega} \right\}$$

and (1.50) follows from (1.46) and (1.42). □

As mentioned in the previous section, the verification of Hypothesis H3 is bound to be often quite intricate. In fact, the choice of the spaces X_h and M_h is generally dictated by the fact (conjectured or established) that H3 is satisfied. The reader will find in the next section how to construct such pairs of spaces.

We finish this section with a brief survey of the iterative methods proposed in Section 1.2 to decouple the computation of $\mathbf{u}_h$ from that of p_h. We choose the scalar product of $L^2(\Omega)$ for the bilinear form $c(.,.)$:

$$c(p, q) = \int_\Omega p(x) q(x)\, dx.$$

Then, the *penalized version* of Problem (1.40) becomes:

$$\begin{cases} \textit{Find a function } \mathbf{u}_h^\varepsilon \in X_h \textit{ such that} \\ a(\mathbf{u}_h^\varepsilon, \mathbf{v}_h) + \dfrac{1}{\varepsilon}(\rho_h(\operatorname{div} \mathbf{u}_h^\varepsilon), \rho_h(\operatorname{div} \mathbf{v}_h)) = \langle \mathbf{f}, \mathbf{v}_h \rangle \quad \forall \mathbf{v}_h \in X_h, \end{cases} \tag{1.52}$$

where ρ_h is the operator of orthogonal projection of $L^2(\Omega)$ onto Q_h. Clearly, Problem (1.52) dissociates the computation of $\mathbf{u}_h^\varepsilon$ from that of p_h^ε since here p_h^ε is given explicitly by

$$p_h^\varepsilon = -\frac{1}{\varepsilon}\rho_h(\operatorname{div}\mathbf{u}_h^\varepsilon).$$

But, of course, this problem offers a practical interest only if the calculation of $\rho_h(\operatorname{div}\mathbf{v}_h)$ is simple. This will be precisely the case of nearly all methods discussed in this chapter because the functions of M_h will be piecewise discontinuous and ρ_h will be a local operator.

As far as the convergence of $\mathbf{u}_h^\varepsilon$ is concerned, a straightforward application of Theorem 1.3 gives the following result.

Theorem 1.10. *Problem* (1.52) *has a unique solution* $\mathbf{u}_h^\varepsilon$ *for all* $\varepsilon > 0$. *Moreover, under Hypothesis* H3, *we have for all* $\varepsilon \leqslant \varepsilon_0$ *sufficiently small:*

$$|\mathbf{u}_h - \mathbf{u}_h^\varepsilon|_{1,\Omega} + \left\| p_h + \frac{1}{\varepsilon}\rho_h(\operatorname{div}\mathbf{u}_h^\varepsilon)\right\|_{0,\Omega} \leqslant C\varepsilon\|\mathbf{f}\|_{-1,\Omega}$$

with a constant $C > 0$ *independent of* h *and* ε.

Similarly, $\mathbf{u}_h^\varepsilon$ and p_h^ε can be expanded in powers of ε. Starting with $p_h^0 = p_h$, we define the sequence $(\mathbf{u}_h^n, p_h^n) \in X_h \times M_h$, solution of

$$\text{(1.53)} \qquad \begin{cases} a(\mathbf{u}_h^n, \mathbf{v}_h) - (\operatorname{div}\mathbf{v}_h, p_h^n) = 0 & \forall \mathbf{v}_h \in X_h, \\ (\operatorname{div}\mathbf{u}_h^n, q_h) = -(p_h^{n-1}, q_h) & \forall q_h \in Q_h. \end{cases}$$

Then Theorem 1.4 yields the following asymptotic expansion:

Theorem 1.11. *Under Hypothesis* H3, *we have for all integers* $M \geqslant 1$ *and all* $\varepsilon \leqslant \varepsilon_0$ *sufficiently small:*

$$\left|\mathbf{u}_h^\varepsilon - \mathbf{u}_h - \sum_{n=1}^{M}\varepsilon^n\mathbf{u}_h^n\right|_{1,\Omega} + \left\|\frac{1}{\varepsilon}\rho_h(\operatorname{div}\mathbf{u}_h^\varepsilon) + p_h + \sum_{n=1}^{M}\varepsilon^n p_h^n\right\|_{0,\Omega}$$
$$\leqslant K_M\varepsilon^{M+1}\|\mathbf{f}\|_{-1,\Omega},$$

with a constant K_M *independent of* h *and* ε.

Now, let us discuss the gradient algorithms. With the above choice of $c(.,.)$, the bilinear form $a_r^h(.,.)$ reads:

$$a_r^h(\mathbf{u},\mathbf{v}) = a(\mathbf{u},\mathbf{v}) + r(\rho_h(\operatorname{div}\mathbf{u}), \rho_h(\operatorname{div}\mathbf{v})).$$

It is $H_0^1(\Omega)^N$-elliptic and obviously symmetric. Therefore, the algorithms described by (1.33) and (1.34) are genuine gradient algorithms. The formulas for the *simple gradient algorithm with optimal parameter* are:

1°) Predict the initial value $p_h^0 \in M_h$ and compute the solution $\mathbf{u}_h^0 \in X_h$ of:

$$a_r^h(\mathbf{u}_h^0, \mathbf{v}_h) = (p_h^0, \operatorname{div} \mathbf{v}_h) + \langle \mathbf{f}, \mathbf{v}_h \rangle \qquad \forall \mathbf{v}_h \in X_h;$$

2°) For $m \geqslant 0$, knowing $(\mathbf{u}_h^m, p_h^m)$ determine $\mathbf{z}_h^m \in X_h$, $\mu_h^m \in \mathbb{R}$ and the pair $(\mathbf{u}_h^{m+1}, p_h^{m+1}) \in X_h \times M_h$ by:

$$a_r^h(\mathbf{z}_h^m, \mathbf{v}_h) = -(\rho_h(\operatorname{div} \mathbf{u}_h^m), \rho_h(\operatorname{div} \mathbf{v}_h)) \qquad \forall \mathbf{v}_h \in X_h;$$

$$\mu_h^m = -\frac{\| \rho_h(\operatorname{div} \mathbf{u}_h^m) \|_{0,\Omega}^2}{(\rho_h(\operatorname{div} \mathbf{u}_h^m), \rho_h(\operatorname{div} \mathbf{z}_h^m))},$$

$$p_h^{m+1} = p_h^m - \mu_h^m \rho_h(\operatorname{div} \mathbf{u}_h^m),$$

$$\mathbf{u}_h^{m+1} = \mathbf{u}_h^m + \mu_h^m \mathbf{z}_h^m.$$

The *conjugate-gradient algorithm* initializes $\mathbf{u}_h^0$ like above and replaces step n° 2 by:

2°) For $m \geqslant 0$, knowing $(\mathbf{u}_h^m, p_h^m) \in X_h \times M_h$ compute $(\mathbf{z}_h^m, \omega_h^m) \in X_h \times M_h$, $(\mu_h^m, \sigma_h^m) \in \mathbb{R} \times \mathbb{R}$ and the pair $(\mathbf{u}_h^{m+1}, p_h^{m+1}) \in X_h \times M_h$ by:

$$\left.\begin{aligned} \sigma_h^m &= \frac{\| \rho_h(\operatorname{div} \mathbf{u}_h^m) \|_{0,\Omega}^2}{\| \rho_h(\operatorname{div} \mathbf{u}_h^{m-1}) \|_{0,\Omega}^2} \\ \omega_h^m &= \rho_h(\operatorname{div} \mathbf{u}_h^m) + \sigma_h^m \omega_h^{m-1}, \end{aligned}\right\} \quad \begin{aligned} &\text{only if } m \geqslant 1, \\ &\omega_h^0 = \rho_h(\operatorname{div} \mathbf{u}_h^0), \end{aligned}$$

$$a_r^h(\mathbf{z}_h^m, \mathbf{v}_h) = -(\omega_h^m, \operatorname{div} \mathbf{v}_h) \qquad \forall \mathbf{v}_h \in X_h,$$

$$\mu_h^m = -\frac{\| \rho_h(\operatorname{div} \mathbf{u}_h^m) \|_{0,\Omega}^2}{(\rho_h(\operatorname{div} \mathbf{u}_h^m), \rho_h(\operatorname{div} \mathbf{z}_h^m))},$$

$$p_h^{m+1} = p_h^m - \mu_h^m \omega_h^m,$$

$$\mathbf{u}_h^{m+1} = \mathbf{u}_h^m + \mu_h^m \mathbf{z}_h^m.$$

It follows from Theorems 1.5 and 1.7 that both gradient algorithms are *convergent* provided the Hypothesis H3 holds. Furthermore, the simple-gradient algorithm converges for any choice of the parameters μ_h^m such that:

$$0 < \inf_m \mu_h^m \leqslant \sup_m \mu_h^m < 2\left(\frac{\nu}{N} + r\right).$$

1.4. Checking the inf-sup Condition

This short section is dedicated to the construction of pairs of spaces (X_h, M_h) that satisfy uniformly the inf-sup condition (1.12). The underlying idea due to Boland & Nicolaïdes [11], is that if (1.12) holds uniformly for a pair of spaces $(\bar{X}_h, \bar{M}_h)$ then one can generate a whole family of pairs of spaces that also satisfy (1.12) uniformly provided they satisfy a *local* inf-sup condition. In other words, the global condition (1.12) can be reduced to a local condition, which is of course much easier to check.

Suppose Ω is partitioned into a finite number of disjoint, Lipschitz-continuous open subsets Ω_r with boundary Γ_r:

$$\bar{\Omega} = \bigcup_{r=1}^{R} \bar{\Omega}_r.$$

Let X_h and M_h be defined by (1.38) with $\mathbb{R} \subset Q_h$. For $1 \leqslant r \leqslant R$ we set:

$$\text{(1.54)} \qquad \begin{cases} X_h(\Omega_r) = \{\mathbf{v} \in X_h;\ \mathbf{v} = \mathbf{0} \text{ in } \Omega - \Omega_r\}, \\ Q_h(\Omega_r) = \{q_{|\Omega_r};\ q \in Q_h\}, \\ M_h(\Omega_r) = Q_h(\Omega_r) \cap L_0^2(\Omega_r), \end{cases}$$

$$\text{(1.55)} \qquad \overline{M}_h = \{q \in L_0^2(\Omega);\ q_{|\Omega_r} \text{ is constant, } 1 \leqslant r \leqslant R\}.$$

Note that the functions of $X_h(\Omega_r)$ belong to $H_0^1(\Omega_r)^N$. We introduce as an assumption the following concept of uniform, local inf-sup condition with respect to this partition:

Hypothesis H4. *There exists a constant $\lambda^* > 0$, independent of h and r, such that:*

$$\text{(1.56)} \qquad \sup_{\mathbf{v}_h \in X_h(\Omega_r)} \left[\frac{\int_{\Omega_r} q_h \operatorname{div} \mathbf{v}_h \, dx}{|\mathbf{v}_h|_{1,\Omega_r}}\right] \geqslant \lambda^* \|q_h\|_{0,\Omega_r} \qquad \forall q_h \in M_h(\Omega_r), \qquad 1 \leqslant r \leqslant R.$$

Let us establish the salient result of this section.

Theorem 1.12. *Let the pair of spaces (X_h, M_h) defined by* (1.38) *satisfy Hypothesis* H4. *If there exists a subspace $\overline{X}_h$ of X_h such that the pair $(\overline{X}_h, \overline{M}_h)$ satisfies the inf-sup condition* (1.12) *with a constant $\bar{\beta}$ independent of h, then (X_h, M_h) also satisfies* (1.12) *with a constant β^* independent of h.*

Proof. From the definition (1.54) we derive immediately the *orthogonal* decomposition of $Q_h(\Omega_r)$:

$$Q_h(\Omega_r) = M_h(\Omega_r) \oplus \mathbb{R}.$$

Thus each function $q_h \in M_h$ can be split as follows:

$$q_h = \tilde{q}_h + \bar{q}_h,$$

where

$$\bar{q}_{h|\Omega_r} = \frac{1}{\operatorname{meas}(\Omega_r)} \int_{\Omega_r} q_h \, dx$$

and $\tilde{q}_r = \tilde{q}_{h|\Omega_r} \in M_h(\Omega_r)$. Observe that $\bar{q}_h \in \overline{M}_h$ and that the orthogonality of the decomposition implies:

$$\text{(1.57)} \qquad \|q_h\|_{0,\Omega}^2 = \|\tilde{q}_h\|_{0,\Omega}^2 + \|\bar{q}_h\|_{0,\Omega}^2.$$

Now, owing to Hypothesis H4 and Remark 1.4 there exists a function $\tilde{\mathbf{v}}_r \in X_h(\Omega_r)$ such that

$$(1.58)\qquad \begin{cases} \displaystyle\int_{\Omega_r} \tilde{q}_r \operatorname{div} \tilde{\mathbf{v}}_r \, dx = \|\tilde{q}_r\|_{0,\Omega_r}^2, \\ |\tilde{\mathbf{v}}_r|_{1,\Omega_r} \leqslant \dfrac{1}{\lambda^*} \|\tilde{q}_r\|_{0,\Omega_r}. \end{cases}$$

Similarly, since the pair $(\bar{X}_h, \bar{M}_h)$ satisfies (1.12) there exists a function $\bar{\mathbf{v}}_h \in \bar{X}_h$ such that

$$(1.59)\qquad \begin{cases} \displaystyle\int_{\Omega} \bar{q}_h \operatorname{div} \bar{\mathbf{v}}_h \, dx = \|\bar{q}_h\|_{0,\Omega}^2, \\ |\bar{\mathbf{v}}_h|_{1,\Omega} \leqslant \dfrac{1}{\beta} \|\bar{q}_h\|_{0,\Omega}. \end{cases}$$

Let $\tilde{\mathbf{v}}_h$ be the function of X_h defined by:

$$\tilde{\mathbf{v}}_{h|\Omega_r} = \tilde{\mathbf{v}}_r.$$

We propose to associate with q_h the function $\mathbf{v}_h \in X_h$:

$$\mathbf{v}_h = \tilde{\mathbf{v}}_h + \alpha \bar{\mathbf{v}}_h,$$

for some $\alpha > 0$ and we hope to adjust the parameter α so that the pair $(\mathbf{v}_h, q_h)$ verifies the inf-sup condition.

Let us evaluate $(q_h, \operatorname{div} \mathbf{v}_h)$. We have:

$$(q_h, \operatorname{div} \mathbf{v}_h) = (\tilde{q}_h, \operatorname{div} \tilde{\mathbf{v}}_h) + (\bar{q}_h, \operatorname{div} \tilde{\mathbf{v}}_h) + \alpha(\tilde{q}_h, \operatorname{div} \bar{\mathbf{v}}_h) + \alpha(\bar{q}_h, \operatorname{div} \bar{\mathbf{v}}_h).$$

Now,

$$(\bar{q}_h, \operatorname{div} \tilde{\mathbf{v}}_h) = 0 \quad \text{since } \tilde{\mathbf{v}}_r \text{ vanishes on } \Gamma_r,$$

$$(\tilde{q}_h, \operatorname{div} \tilde{\mathbf{v}}_h) = \|\tilde{q}_h\|_{0,\Omega}^2,$$

$$(\bar{q}_h, \operatorname{div} \bar{\mathbf{v}}_h) = \|\bar{q}_h\|_{0,\Omega}^2,$$

by virtue of (1.58) and (1.59) respectively and

$$(\tilde{q}_h, \operatorname{div} \bar{\mathbf{v}}_h) \leqslant \frac{\sqrt{N}}{\beta} \|\tilde{q}_h\|_{0,\Omega} \|\bar{q}_h\|_{0,\Omega} \quad \text{in view of (1.59).}$$

Hence, collecting these results we obtain:

$$(q_h, \operatorname{div} \mathbf{v}_h) \geqslant \|\tilde{q}_h\|_{0,\Omega}^2 + \alpha \|\bar{q}_h\|_{0,\Omega}^2 - \alpha \frac{\sqrt{N}}{\beta} \|\tilde{q}_h\|_{0,\Omega} \|\bar{q}_h\|_{0,\Omega}.$$

Then, the inequality:

$$\|\tilde{q}_h\|_{0,\Omega} \|\bar{q}_h\|_{0,\Omega} \leqslant \varepsilon \|\tilde{q}_h\|_{0,\Omega}^2 + \frac{1}{4\varepsilon} \|\bar{q}_h\|_{0,\Omega}^2$$

yields

$$(q_h, \operatorname{div} \mathbf{v}_h) \geqslant \left(1 - \frac{\alpha\varepsilon\sqrt{N}}{\beta}\right) \|\tilde{q}_h\|_{0,\Omega}^2 + \left(1 - \frac{\sqrt{N}}{4\varepsilon\beta}\right) \|\bar{q}_h\|_{0,\Omega}^2$$

for arbitrary $\varepsilon > 0$. With the choice

$$\varepsilon = \frac{1}{2} \frac{\bar{\beta}}{\alpha\sqrt{N}},$$

this becomes:

$$(q_h, \operatorname{div} \mathbf{v}_h) \geqslant \frac{1}{2} \|\tilde{q}_h\|_{0,\Omega}^2 + \alpha\left(1 - \frac{\alpha N}{2\bar{\beta}^2}\right) \|\bar{q}_h\|_{0,\Omega}^2.$$

Let us choose for example

$$\alpha = \frac{\bar{\beta}^2}{N}.$$

Then (1.57) implies:

$$(q_h, \operatorname{div} \mathbf{v}_h) \geqslant \min\left(\frac{1}{2}, \frac{\alpha}{2}\right) \|q_h\|_{0,\Omega}^2. \tag{1.60}$$

Finally, we have:

$$\begin{aligned} |\mathbf{v}_h|_{1,\Omega} &\leqslant |\tilde{\mathbf{v}}_h|_{1,\Omega} + \alpha |\bar{\mathbf{v}}_h|_{1,\Omega} \\ &\leqslant \frac{1}{\lambda^*} \|\tilde{q}_h\|_{0,\Omega} + \frac{\alpha}{\bar{\beta}} \|\bar{q}_h\|_{0,\Omega} \\ &\leqslant \left[\left\{\frac{1}{\lambda^*}\right\}^2 + \left\{\frac{\alpha}{\bar{\beta}}\right\}^2\right]^{1/2} \|q_h\|_{0,\Omega} \end{aligned}$$

by virtue of (1.57), (1.58) and (1.59). Combined with (1.60), this last inequality gives the expected inf-sup condition:

$$\frac{(q_h, \operatorname{div} \mathbf{v}_h)}{|\mathbf{v}_h|_{1,\Omega}} \geqslant \beta^* \|q_h\|_{0,\Omega}$$

with a constant β^* that depends solely upon N, $\bar{\beta}$ and λ^*. □

The reader can also refer to Stenberg [75] for a related approach.

§ 2. Simplicial Finite Element Methods Using Discontinuous Pressures

The methods discussed in this paragraph are essentially the oldest finite element methods developed to solve the Stokes and Navier-Stokes problem. Since their first publication by Fortin [29] and by Crouzeix & Raviart [23], they have been substantially simplified and generalized by a number of authors. We shall study here contributions by Bernardi & Raugel [10], Fortin [30], Johnson & Pitkäranta [46], Mansfield [56] and Boland & Nicolaides [12].

The situation and notations are those of Section 1.3.

2.1. A First Order Approximation on Triangular Elements

Throughout this section, we assume that Ω is a *bounded polygon* in $\mathbb{R}^2$ so that it can be entirely triangulated. For each $h > 0$, $\mathscr{T}_h$ is a triangulation of $\bar{\Omega}$ made of triangles κ with diameters bound by h:

$$\bar{\Omega} = \bigcup_{\kappa \in \mathscr{T}_h} \kappa.$$

By now, the reader is well aware that the choice of compatible spaces X_h and M_h, whatever their accuracy, is crucial for the success of the approximation. For example, the most straightforward choice of first-order spaces is:

$$W_h = \{w \in \mathscr{C}^0(\bar{\Omega})^2;\ w|_\kappa \in P_1^2 \quad \forall \kappa \in \mathscr{T}_h\},$$

$$X_h = W_h \cap H_0^1(\Omega)^2,$$

$$Q_h = \{q \in L^2(\Omega);\ q|_\kappa \in P_0 \quad \forall \kappa \in \mathscr{T}_h\},$$

$$M_h = Q_h \cap L_0^2(\Omega).$$

But, more often than not, this choice leads to $V_h = \{0\}$ as can be checked in the simple example of Figure 7. Here, X_h has only two degrees of freedom, while the definition of V_h requires five independent conditions. Hence, $V_h = \{0\}$.

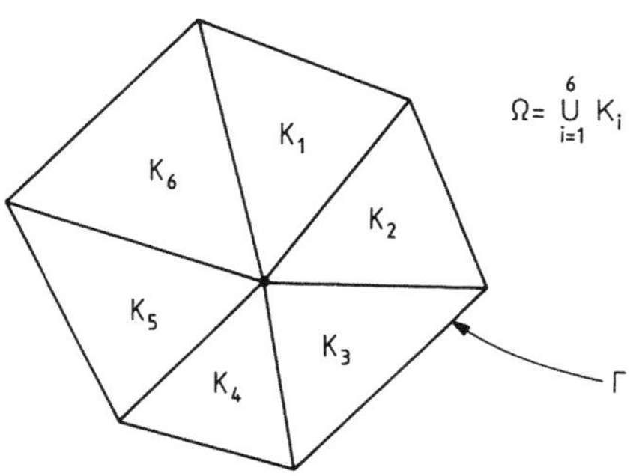

Figure 7. Example of incompatible pair of spaces (X_h, M_h)

This example illustrates the fact that the space W_h needs more degrees of freedom in order to generate a sufficiently large space V_h. Fortin [29] suggested first to take piecewise polynomials of degree two for the functions of W_h. For a long time this was believed to be the simplest velocity space to associate with the above pressure space Q_n. However, following a further idea of Fortin [30], Bernardi & Raugel [10] recently analyzed an intermediate velocity space which involves less degrees of freedom and lends itself more readily to higher order extensions. We shall study this element in detail.

Let κ be an arbitrary triangle of $\mathscr{T}_h$ with vertices a_1, a_2, a_3 like in Figure 8. We denote by f_i the side opposite a_i and by $\mathbf{n}_i$ and $\mathbf{t}_i$ the unit outward normal and unit tangent to f_i. Our aim is a function $\mathbf{w}$ with quadratic components in κ and affine tangential components on each side f_i of κ:

$$\mathbf{w}|_\kappa \in P_2^2, \qquad \mathbf{w}\cdot\mathbf{t}_i|_{f_i} \in P_1.$$

This can be achieved by splitting $\mathbf{w}$:

$$\mathbf{w} = \mathbf{w}_1 + \mathbf{w}_2$$

into $\mathbf{w}_1 \in P_1^2$ and $\mathbf{w}_2 \in P_2^2$ such that $\mathbf{w}_2 \cdot \mathbf{t}_i|_{f_i} = 0$.

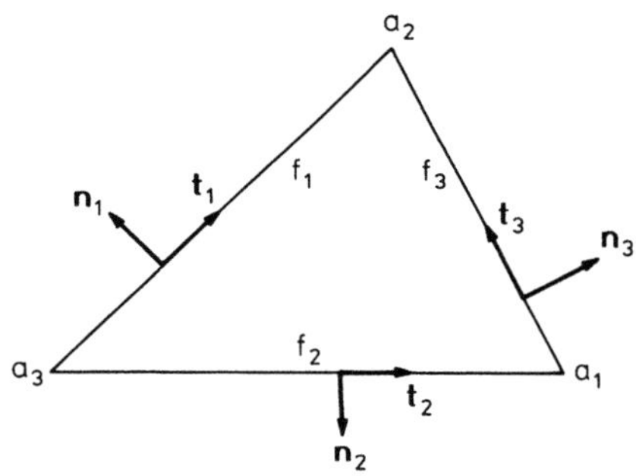

Figure 8

As an example, observe that the function

$$\mathbf{p}_1 = \mathbf{n}_1 \lambda_2 \lambda_3$$

vanishes on the sides f_2 and f_3 and satisfies $\mathbf{p}_1 \cdot \mathbf{t}_1|_{f_1} = 0$. Generalizing this remark, we set:

$$\text{(2.1)} \qquad \mathbf{p}_1 = \mathbf{n}_1 \lambda_2 \lambda_3, \qquad \mathbf{p}_2 = \mathbf{n}_2 \lambda_3 \lambda_1, \qquad \mathbf{p}_3 = \mathbf{n}_3 \lambda_1 \lambda_2,$$

and we take the velocities $\mathbf{w}$ in the polynomial subspace of P_2^2:

$$\text{(2.2)} \qquad \mathscr{P}_1(\kappa) = P_1^2 \oplus \operatorname{span}\{\mathbf{p}_1, \mathbf{p}_2, \mathbf{p}_3\}.$$

Hence, we propose the following pair of spaces:

$$\text{(2.3)} \qquad \begin{cases} W_h = \{\mathbf{w} \in \mathscr{C}^0(\bar{\Omega})^2;\ \mathbf{w}|_\kappa \in \mathscr{P}_1(\kappa), \quad \forall \kappa \in \mathscr{T}_h\}, \\ \bar{X}_h = W_h \cap H_0^1(\Omega)^2, \end{cases}$$

$$\left\{\begin{aligned} Q_h &= \{q \in L^2(\Omega);\ q|_\kappa \in P_0,\quad \forall \kappa \in \mathcal{T}_h\},\\ \overline{M}_h &= Q_h \cap L_0^2(\Omega). \end{aligned}\right. \tag{2.4}$$

(The bars on top of X_h and M_h are put to match the notation of Section 1.4.)

Remark 2.1. Let $\mathbf{p} \in \mathscr{P}_1(\kappa)$ and let

$$I_\kappa \mathbf{p} = \sum_{i=1}^{3} \mathbf{p}(a_i)\lambda_i$$

denote its standard interpolant on P_1^2. Then since $\mathbf{p}_i(a_j) = \mathbf{0}$ for $1 \leqslant i, j \leqslant 3$, $\mathbf{p}$ has the form:

$$\mathbf{p} = I_\kappa \mathbf{p} + \sum_{i=1}^{3} \alpha_i \mathbf{p}_i \quad \text{with } \alpha_i \in \mathbb{R}.$$

In addition, since $\mathbf{p}_i|_{f_j} = \mathbf{0}$ for $i \neq j$, we immediately derive that

$$\alpha_i \mathbf{p}_i|_{f_i} = (\mathbf{p} - I_\kappa \mathbf{p})|_{f_i}.$$

This remark suggests to choose the following degrees of freedom for the functions of $\mathscr{P}_1(\kappa)$: the values of $\mathbf{p}$ at the vertices a_i of κ and the flux of $\mathbf{p}$ through each side f_i of κ. The next lemma checks the $\mathscr{P}_1(\kappa)$-unisolvence of these degrees of freedom.

Lemma 2.1. *A polynomial* $\mathbf{p}$ *of* $\mathscr{P}_1(\kappa)$ *is uniquely determined by*:

$$\left\{\begin{aligned} &\mathbf{p}(a_i), && 1 \leqslant i \leqslant 3\\ &\int_{f_i} \mathbf{p}\cdot\mathbf{n}_i\, ds, && 1 \leqslant i \leqslant 3. \end{aligned}\right. \tag{2.5}$$

Moreover, on any side $f_i = [a_k, a_l]$ *of* κ, $\mathbf{p}$ *depends only upon the degrees of freedom defined on that side, namely*: $\mathbf{p}(a_k)$, $\mathbf{p}(a_l)$ *and* $\int_{f_i} \mathbf{p}\cdot\mathbf{n}_i\, ds$.

Proof. First, observe that (2.5) involves nine linear conditions and that $\mathbf{p}$ has nine coefficients. Thus it suffices to prove that if all the degrees of freedom in (2.5) vanish then $\mathbf{p} = \mathbf{0}$.

In view of Remark 2.1, we have $I_\kappa \mathbf{p} = \mathbf{0}$. Likewise, $\int_{f_i} \alpha_i \mathbf{p}_i \cdot \mathbf{n}_i\, ds = 0$; i.e. $\int_{f_i} \alpha_i \lambda_j \lambda_k\, ds = 0$. Hence $\alpha_i = 0$ for $1 \leqslant i \leqslant 3$ and therefore $\mathbf{p} = \mathbf{0}$.

Similarly, if $\mathbf{p}(a_k) = \mathbf{p}(a_l) = \mathbf{0}$ and $\int_{f_i} \mathbf{p}\cdot\mathbf{n}_i\, ds = 0$ we readily find that $\mathbf{p}|_{f_i} = \mathbf{0}$. □

Lemma 2.1 leads to the following interpolation operator:
for $\mathbf{v} \in \mathscr{C}^0(\bar{\Omega})^2$, let $r_\kappa \mathbf{v}$ be the unique polynomial of $\mathscr{P}_1(\kappa)$ defined by:

$$\left\{\begin{aligned} &r_\kappa \mathbf{v}(a_i) = \mathbf{v}(a_i) && 1 \leqslant i \leqslant 3,\\ &\int_{f_i} (r_\kappa \mathbf{v} - \mathbf{v})\cdot\mathbf{n}_i\, ds = 0 && 1 \leqslant i \leqslant 3, \end{aligned}\right. \tag{2.6}$$

and let
$$r_h\mathbf{v}|_\kappa = r_\kappa\mathbf{v} \qquad \forall\kappa\in\mathscr{T}_h.$$
It follows from Lemma 2.1 that
$$r_h\in\mathscr{L}(\mathscr{C}^0(\bar{\Omega})^2;\, W_h)\cap\mathscr{L}((\mathscr{C}^0(\bar{\Omega})\cap H_0^1(\Omega))^2;\, \bar{X}_h).$$
However, if we want to check the inf-sup condition—i.e. Hypothesis H3—it is convenient, in vew of Lemma 1.1, to work right away with the appropriate operator π_h. At first sight, it appears that the above operator r_h would do the trick since
$$\int_\kappa \operatorname{div}(\mathbf{v} - r_\kappa\mathbf{v})\,dx = \sum_{i=1}^{3}\int_{f_i}(\mathbf{v} - r_\kappa\mathbf{v})\cdot\mathbf{n}_i\,ds = 0.$$
But strictly speaking, r_h does not satisfy H3 because it is defined on $H^2(\Omega)^2$ instead of $H^1(\Omega)^2$. Thus we must replace r_h by another operator which does not involve the values of $\mathbf{v}$ at the vertices of κ. The easiest way of turning the difficulty consists in replacing the values of $\mathbf{v}$ by those of the projection $P_h\mathbf{v}$ on X_h. From a practical point of view, this global regularization is not satisfactory because the corresponding proof requires the uniform regularity of the triangulation (cf. Theorem A.2 and Girault & Raviart [32]). This additional requirement, which stems from the global regularization and not from the above approximation, can be released by using instead the local regularization operator in Section A.3.

Thus, following Bernardi & Raugel [10], with each $\mathbf{v}\in H_0^1(\Omega)^2$ we associate the function $\mathbf{w}_h = R_h\mathbf{v}\in\Phi_h^2$ where
$$\Phi_h = \{\phi\in\mathscr{C}^0(\bar{\Omega});\, \phi|_\kappa\in P_1 \quad \forall\kappa\in\mathscr{T}_h\}\cap H_0^1(\Omega)$$
and R_h is defined by (A.53), (A.54). Then we define the operator $\pi_h\in\mathscr{L}(H_0^1(\Omega)^2;\, \bar{X}_h)$ by:

(2.7)
$$\begin{cases} \pi_h\mathbf{v}(a) = R_h\mathbf{v}(a) & \forall \text{node } a \text{ of } \mathscr{T}_h,\\ \displaystyle\int_f (\pi_h\mathbf{v} - \mathbf{v})\cdot\mathbf{n}\,ds = 0 & \forall \text{side } f \text{ of } \mathscr{T}_h.\end{cases}$$

Now assume that the family of triangulations $\mathscr{T}_h$ is regular in the sense of Definition A.2, i.e.

(2.8)
$$h_\kappa/\rho_\kappa \leqslant \sigma \qquad \forall\kappa\in\mathscr{T}_h, \qquad \sigma \text{ independent of } h.$$

The next lemma shows that π_h satisfies Hypothesis H1 with $l = 1$ as well as Hypothesis H3.

Lemma 2.2. *If the triangulation $\mathscr{T}_h$ is regular, then*

(2.9)
$$|\mathbf{v} - \pi_h\mathbf{v}|_{m,\Omega} \leqslant Ch^{k-m}|\mathbf{v}|_{k,\Omega} \qquad \forall\mathbf{v}\in H^k(\Omega)^2$$

for $m = 0$ or 1 and $k = 1$ or 2, with a positive constant C independent of h and $\mathbf{v}$.

In addition, whatever the triangulation we have

$$\int_\Omega \operatorname{div}(\mathbf{v} - \pi_h \mathbf{v}) q \, dx = 0 \quad \forall q \in Q_h. \tag{2.10}$$

Proof. As mentioned above, the second equation in (2.7) implies directly (2.10).

Let κ be an arbitrary triangle of $\mathscr{T}_h$. We infer from Remark 2.1 and the first equation in (2.7) that

$$\pi_h \mathbf{v}|_\kappa = R_h \mathbf{v}|_\kappa + \sum_{i=1}^{3} \alpha_i \mathbf{p}_i$$

where

$$\alpha_i = \left[\int_{f_i} (\mathbf{v} - R_h \mathbf{v}) \cdot \mathbf{n}_i \, ds\right] \Big/ \left[\int_{f_i} \lambda_j \lambda_k \, ds\right].$$

First, according to Theorem A.4, the operator R_h has the following local interpolation error:

$$\|\mathbf{v} - R_h \mathbf{v}\|_{0,\kappa} + h_\kappa |\mathbf{v} - R_h \mathbf{v}|_{1,\kappa} \leqslant C_1 h_\kappa^k |\mathbf{v}|_{k,\Delta_\kappa} \quad \forall \mathbf{v} \in H^k(\Omega)^2, \tag{2.11}$$

$k = 1$ or 2, where Δ_κ denotes the union of all elements of $\mathscr{T}_h$ that share at least a vertex with κ.

Next, formula (A.8) implies that

$$\begin{cases} |\mathbf{p}_i|_{m,\kappa} \leqslant C_2 |\det(B_\kappa)|^{1/2} \|B_\kappa^{-1}\|^m |\hat{\lambda}_j \hat{\lambda}_k|_{m,\hat{\kappa}} \\ \qquad\quad \leqslant C_3 |\det(B_\kappa)|^{1/2} \|B_\kappa^{-1}\|^m, \end{cases} \tag{2.12}$$

since $|\hat{\lambda}_j \hat{\lambda}_k|_{m,\hat{\kappa}}$ is a constant independent of h and κ. Similarly,

$$\int_{f_i} \lambda_j \lambda_k \, ds = [\operatorname{meas}(f_i)/\operatorname{meas}(\hat{f}_i)] \int_{\hat{f}_i} \hat{\lambda}_j \hat{\lambda}_k \, d\hat{s}.$$

Finally,

$$\left| \int_{f_i} (\mathbf{v} - R_h \mathbf{v}) \cdot \mathbf{n}_i \, ds \right| \leqslant [\operatorname{meas}(f_i)/\operatorname{meas}(\hat{f}_i)] \int_{\hat{f}_i} \|\hat{\mathbf{v}} - \widehat{R_h \mathbf{v}}\| \, d\hat{s}$$

and

$$\int_{\hat{f}_i} \|\hat{\mathbf{v}} - \widehat{R_h \mathbf{v}}\| \, d\hat{s} \leqslant C_4 \|\hat{\mathbf{v}} - \widehat{R_h \mathbf{v}}\|_{1,\hat{\kappa}}$$

by the trace Theorem I.1.5. But we infer from (A.7) and (A.2) that:

$$\|\hat{\mathbf{v}} - \widehat{R_h \mathbf{v}}\|_{1,\hat{\kappa}} \leqslant C_5 |\det(B_\kappa)|^{-1/2} \{ \|\mathbf{v} - R_h \mathbf{v}\|_{0,\kappa}^2 + h_\kappa^2 |\mathbf{v} - R_h \mathbf{v}|_{1,\kappa}^2 \}^{1/2}.$$

Thus (2.11) yields:

$$|\alpha_i| \leqslant C_6 |\det(B_\kappa)|^{-1/2} h_\kappa^k |\mathbf{v}|_{k,\Delta_\kappa}.$$

Combined with (2.12), (A.2) and (2.8), this inequality gives:

$$\left|\sum_{i=1}^{3} \alpha_i \mathbf{p}_i\right|_{m,\kappa} \leqslant C_7 \sigma^m h_\kappa^{k-m} |\mathbf{v}|_{k,\Delta_\kappa}.$$

Then with another application of (2.11), we derive:

$$|\mathbf{v} - \pi_h \mathbf{v}|_{m,\kappa} \leqslant C_8 \sigma^m h^{k-m} |\mathbf{v}|_{k,\Delta_\kappa}, \qquad m = 0, 1, \qquad k = 1, 2.$$

But the regularity of $\mathscr{T}_h$ implies that the maximum number of occurrences of a given triangle κ in the sets Δ_κ is bounded by a fixed constant M independent of h and κ. Hence

$$\left(\sum_{\kappa \in \mathscr{T}_h} |\mathbf{v}|_{k,\Delta_\kappa}^2\right)^{1/2} \leqslant M |\mathbf{v}|_{k,\Omega}$$

and therefore

$$|\mathbf{v} - \pi_h \mathbf{v}|_{m,\Omega} \leqslant C_8 M \sigma^m h^{k-m} |\mathbf{v}|_{k,\Omega}.$$

In particular, when $k = m = 1$ this establishes the bound (1.20) of Lemma 1.1:

$$|\pi_h \mathbf{v}|_{1,\Omega} \leqslant (1 + C_8 M \sigma) |\mathbf{v}|_{1,\Omega} \qquad \forall \mathbf{v} \in H_0^1(\Omega)^2.$$

And when $k = 2$ and $m = 1$ this yields Hypothesis H1 with $l = 1$. □

Finally, Hypothesis H2 is a direct consequence of Lemma A.5 for the L^2-projection ρ_h onto Q_h:

$$\|q - \rho_h q\|_{0,\Omega} \leqslant Ch|q|_{1,\Omega} \qquad \forall q \in H^1(\Omega). \tag{2.13}$$

Since all the assumptions of Theorems 1.8 and 1.9 are satisfied, the following convergence result is established for our first-order scheme.

Theorem 2.1. *Suppose Ω is a bounded, plane polygon. Let the solution* $(\mathbf{u}, p)$ *of the Stokes problem satisfy*:

$$\mathbf{u} \in [H^2(\Omega) \cap H_0^1(\Omega)]^2, \qquad p \in H^1(\Omega) \cap L_0^2(\Omega)$$

and let the spaces $\bar{X}_h$ and $\bar{M}_h$ be defined by (2.3) *and* (2.4) *respectively. If the family of triangulations $\mathscr{T}_h$ is regular then the solution* $(\mathbf{u}_h, p_h)$ *of* (1.39) *satisfies the error estimate*:

$$|\mathbf{u} - \mathbf{u}_h|_{1,\Omega} + \|p - p_h\|_{0,\Omega} \leqslant C_1 h(|\mathbf{u}|_{2,\Omega} + |p|_{1,\Omega}). \tag{2.14}$$

In addition, if Ω is convex, we have

$$\|\mathbf{u} - \mathbf{u}_h\|_{0,\Omega} \leqslant C_2 h^2(|\mathbf{u}|_{2,\Omega} + |p|_{1,\Omega}). \tag{2.15}$$

Remark 2.2. Both upper bounds are stated in terms of seminorms of $\mathbf{u}$ and p whereas the estimates of Theorems 1.8 and 1.9 involve full norms in their right-hand sides. This slight refinement is due to the fact that both (2.11) and (2.13) are formulated with seminorms only.

2.2. Higher-Order Approximation on Triangular Elements

The finite element spaces discussed in this section were introduced by Crouzeix & Raviart [23] and Mansfield [56]. They are a direct generalization of the space $\mathscr{P}_1(\kappa)$. The reader will perhaps find them easy to grasp, but originally their analysis was far from trivial on account of the inf-sup condition. Fortunately, the material of Section 1.4 has considerably reduced this difficulty.

In this section, we assume that Ω is a *bounded, plane polygon.* Let us fix an integer $l \geqslant 2$. Let $\mathscr{T}_h$ be a triangulation of $\bar{\Omega}$ and let κ be any triangle of $\mathscr{T}_h$. Here again, we want to construct velocities with components that are polynomials of degree $l+1$ and tangential components of degree l on each side of κ. But now, the higher degree of polynomials provide a simple answer to this problem for it suffices that all terms of degree $l+1$ vanish on the sides of κ; in other words they must have the common factor $\lambda_1\lambda_2\lambda_3$. More precisely, let us denote by $\tilde{P}_k$ the space of homogeneous polynomials of degree k:

$$\tilde{P}_k = \operatorname{span}\{x_1^i x_2^{k-i};\ 0 \leqslant i \leqslant k\}.$$

Then we take the velocities $\mathbf{w}$ in the polynomial subspace of P_{l+1}^2:

$$\mathscr{P}_l(\kappa) = [P_l \oplus \{\lambda_1\lambda_2\lambda_3\tilde{P}_{l-2}\}]^2 \tag{2.16}$$

and the pressures q in P_{l-1}. This leads to the following choice of spaces:

$$\begin{cases} W_h = \{\mathbf{w}\in\mathscr{C}^0(\bar{\Omega})^2;\ \mathbf{w}|_\kappa\in\mathscr{P}_l(\kappa)\quad\forall\kappa\in\mathscr{T}_h\}, \\ X_h = W_h\cap H_0^1(\Omega)^2, \end{cases} \tag{2.17}$$

$$\begin{cases} Q_h = \{q\in L^2(\Omega);\ q|_\kappa\in P_{l-1}\quad\forall\kappa\in\mathscr{T}_h\}, \\ M_h = Q_h\cap L_0^2(\Omega). \end{cases} \tag{2.18}$$

The choice of degrees of freedom for the velocities $\mathbf{w}$ is no longer dictated by the fulfillment of the inf-sup condition. Therefore we can simply take the values of $\mathbf{w}$ associated with P_l on the sides of κ and the derivatives corresponding to P_{l-2} at the center of κ, i.e.:

(2.19)
- (a) $\mathbf{w}(a)$ on all points a of $\Sigma_\kappa \cap f_i$, $1 \leqslant i \leqslant 3$, where Σ_κ denotes the principal lattice of order l (cf. (A.19)),
- (b) $\partial^k \mathbf{w}(a_\kappa)/\partial x_1^i \partial x_2^{k-i}$ on the center a_κ of κ, $0 \leqslant i \leqslant k$, $0 \leqslant k \leqslant l-2$.

From a theoretical point of view, derivatives are seldom an attractive choice because they require a lot of regularity; they can be replaced by the interior moments:

$$\text{(2.19)}\quad \text{(c)}\qquad \int_\kappa \mathbf{w}\cdot\mathbf{q}\,dx \quad \forall\mathbf{q}\in P_{l-2}^2.$$

In particular, when $l = 2$, (2.19a) (2.19b) or (2.19c) become:

(a) $$\begin{cases} \mathbf{w}(a_i) \text{ on the three vertices } a_i \text{ of } \kappa, \\ \mathbf{w}(a_{ij}) \text{ on the midpoint } a_{ij} \text{ of the segment } [a_i, a_j],\ 1 \leqslant i < j \leqslant 3, \end{cases}$$

(b) $$\mathbf{w}(a_\kappa),$$

or

(c) $$\int_\kappa \mathbf{w}\, dx.$$

Note that all degrees of freedom (2.19) are defined *separately* for each component of $\mathbf{w}$. The next lemma checks their unisolvence.

Lemma 2.3. *A polynomial* $\mathbf{p}$ *of* $\mathscr{P}_l(\kappa)$ *is uniquely determined by the* $l(l+5)$ *degrees of freedom* (2.19a) (2.19b) *or* (2.19a) (2.19c). *In addition, the restriction of* $\mathbf{p}$ *to any side* f_i *of* κ *depends only upon the degrees of freedom defined on that side.*

Proof. As the dimension of $\mathscr{P}_l(\kappa)$ is $l(l+5)$ it suffices to show that $\mathbf{p} = \mathbf{0}$ when all its degrees of freedom vanish.

Let p be any component of $\mathbf{p}$. If its degrees of freedom are zero on any side f_i of κ, we readily obtain that p vanishes identically on that side because p reduces to a polynomial of degree l on each side of κ. Hence, if the degrees of freedom (2.19a) are zero, we have

$$p = \lambda_1 \lambda_2 \lambda_3 q \quad \text{with } q \text{ in } P_{l-2}.$$

Then, if all interior degrees of freedom vanish we find in the case of (2.19c):

$$\int_\kappa \lambda_1 \lambda_2 \lambda_3 q^2\, dx = 0$$

and in the case of (2.19b):

$$\partial^k q(a_\kappa)/\partial x_1^i \partial x_2^{k-i} = 0 \qquad 0 \leqslant i \leqslant k, \qquad 0 \leqslant k \leqslant l-2.$$

Each of these equalities implies that $q = 0$ in κ. □

Thus (2.19) yields two interpolation operators:

$r_\kappa \mathbf{v}$(resp. $r'_\kappa \mathbf{v}$): the unique polynomial of $\mathscr{P}_l(\kappa)$ that has the same degrees of freedom (2.19a) (2.19c) (resp. (2.19a) (2.19b)) as the function $\mathbf{v}$ on κ. Then we define r_h by:

$$r_h \mathbf{v}|_\kappa = r_\kappa \mathbf{v} \qquad \forall \kappa \in \mathscr{T}_h$$

and similarly for r'_h. Lemma 2.3 implies that:

$$r'_h \in \mathscr{L}(\mathscr{C}^{l-2}(\bar{\Omega})^2; W_h) \cap \mathscr{L}([\mathscr{C}^{l-2}(\bar{\Omega}) \cap H_0^1(\Omega)]^2; X_h),$$
$$r_h \in \mathscr{L}(\mathscr{C}^0(\bar{\Omega})^2; W_h) \cap \mathscr{L}([\mathscr{C}^0(\bar{\Omega}) \cap H_0^1(\Omega)]^2; X_h).$$

Furthermore, it is easy to check that both operators r_κ and r'_κ are *invariant under affine transformations*:

$$\widehat{r_\kappa v} = r_{\hat\kappa}\hat v, \qquad \widehat{r'_\kappa v} = r'_{\hat\kappa}\hat v.$$

Therefore, since P_l^2 is invariant under both r_κ and r'_κ, we can apply immediately Corollary A.2 and derive the following interpolation result:

Lemma 2.4. *If the triangulation $\mathcal{T}_h$ is regular then the above operator r_h satisfies*:

$$(2.20)\qquad |\mathbf{v} - r_h\mathbf{v}|_{m,\Omega} \leqslant Ch^{k+1-m}|\mathbf{v}|_{k+1,\Omega} \qquad \forall \mathbf{v}\in H^{k+1}(\Omega)^2, \qquad m = 0 \text{ or } 1$$

with the integer $k\in[1, l]$ and a positive constant C that depends upon k, m, l and Ω but is independent of h and $\mathbf{v}$.

Likewise, the operator r'_h satisfies the bound (2.20) *with $k = l$ or $l-1$.*

Remark 2.3. This lemma's proof is trivial because the particularly simple interpolants r_κ and r'_κ are preserved by affine transformations.

Now let us examine the inf-sup condition. First, observe that owing to Lemma 2.2 the pair of spaces $(\bar X_h, \bar M_h)$ defined by (2.3) (2.4) satisfies a uniform inf-sup condition. Next, the space $\bar X_h$ is obviously contained in the space X_h defined by (2.17). Therefore, according to Theorem 1.12 if the pair (X_h, M_h) given by (2.17) (2.18) satisfies the local inf-sup condition stated in Hypothesis H4 then it will also satisfy globally a uniform inf-sup condition. Now, in order to check Hypothesis H4, one must first choose an appropriate partition $\{\Omega_r; 1 \leqslant r \leqslant R\}$ of Ω. It is quite clear that this partition must bear some relation with the triangulation $\mathcal{T}_h$. Let us try the easiest guess which consists in taking for partition the triangulation itself:

$$\Omega_r = \kappa, \qquad \kappa\in\mathcal{T}_h.$$

Thus,

$$(2.21)\qquad \begin{cases} X_h(\kappa) = \{\mathbf{v}\in\mathcal{P}_l(\kappa);\ \mathbf{v}|_{\partial\kappa} = \mathbf{O}\},\\ \quad M_h(\kappa) = P_{l-1}\cap L_0^2(\kappa).\end{cases}$$

The next theorem establishes that this pair of spaces does indeed satisfy H4.

Theorem 2.2. *Assume that $\mathcal{T}_h$ is a regular triangulation of $\bar\Omega$. There exists a constant $\lambda^* > 0$, independent of h and κ, such that*

$$(2.22)\qquad \sup_{\mathbf{v}\in X_h(\kappa)}\left\{\left(\int_\kappa q \operatorname{div}\mathbf{v}\,dx\right)\Big/|\mathbf{v}|_{1,\kappa}\right\} \geqslant \lambda^*\|q\|_{0,\kappa} \qquad \forall q\in M_h(\kappa).$$

Proof. Let $q\in M_h(\kappa)$; we must construct a vector $\mathbf{v}$ in $X_h(\kappa)$ such that

$$(2.23)\qquad \left(\int_\kappa q\operatorname{div}\mathbf{v}\,dx\right)\Big/|\mathbf{v}|_{1,\kappa} \geqslant \lambda^*\|q\|_{0,\kappa}.$$

First, as $\mathbf{v}$ vanishes on the boundary of κ, we can write:

$$\int_\kappa q \operatorname{div} \mathbf{v}\, dx = -\int_\kappa \mathbf{v} \cdot \mathbf{grad}\, q\, dx.$$

Then since $\mathbf{grad}\, q \in P_{l-2}^2$, we can try:

$$\mathbf{v} = -\lambda_1 \lambda_2 \lambda_3 \,\mathbf{grad}\, q$$

which belongs indeed to $\mathscr{P}_l(\kappa)$ and vanishes on $\partial\kappa$. With this choice,

$$\int_\kappa q \operatorname{div} \mathbf{v}\, dx = \int_\kappa \lambda_1 \lambda_2 \lambda_3 \sum_{i=1}^{2} (\partial q/\partial x_i)^2\, dx. \tag{2.24}$$

Next, a straightforward calculation shows that all polynomials ϕ of (say) P_l satisfy:

$$\int_\kappa \lambda_1 \lambda_2 \lambda_3 \phi^2\, dx \geqslant C_1 \|\phi\|_{0,\kappa}^2 \tag{2.25}$$

with a constant $C_1 > 0$ independent of κ, h and ϕ. Indeed, in terms of the reference triangle $\hat{\kappa}$ we have:

$$\int_\kappa \lambda_1 \lambda_2 \lambda_3 \phi^2\, dx = [\operatorname{meas}(\kappa)/\operatorname{meas}(\hat{\kappa})] \int_{\hat{\kappa}} \hat{\lambda}_1 \hat{\lambda}_2 \hat{\lambda}_3 \hat{\phi}^2\, d\hat{x}.$$

But the mapping

$$\hat{\phi} \to \left(\int_{\hat{\kappa}} \hat{\lambda}_1 \hat{\lambda}_2 \hat{\lambda}_3 \hat{\phi}^2\, d\hat{x} \right)^{1/2}$$

is a norm on P_l, equivalent to the $L^2(\hat{\kappa})$-norm. Hence

$$\int_{\hat{\kappa}} \hat{\lambda}_1 \hat{\lambda}_2 \hat{\lambda}_3 \hat{\phi}^2\, d\hat{x} \geqslant \hat{C} \|\hat{\phi}\|_{0,\hat{\kappa}}^2 = \hat{C}[\operatorname{meas}(\hat{\kappa})/\operatorname{meas}(\kappa)] \|\phi\|_{0,\kappa}^2.$$

This yields (2.25) with the equivalence constant $\hat{C}$. Therefore, combining (2.24) and (2.25) we get:

$$\int_\kappa q \operatorname{div} \mathbf{v}\, dx \geqslant \hat{C} |q|_{1,\kappa}^2.$$

Finally, using the argument of Lemma A.6 (cf. formula (A.32)) we get:

$$|\mathbf{v}|_{1,\kappa} \leqslant (C_2/\rho_\kappa) \|\mathbf{v}\|_{0,\kappa}$$

with a constant C_2 independent of h, κ and $\mathbf{v}$. But

$$\|\mathbf{v}\|_{0,\kappa} = \|\lambda_1 \lambda_2 \lambda_3 \,\mathbf{grad}\, q\|_{0,\kappa} \leqslant |q|_{1,\kappa}.$$

Hence

$$|\mathbf{v}|_{1,\kappa} \leqslant (C_2/\rho_\kappa) |q|_{1,\kappa}$$

and consequently,

$$\left(\int_\kappa q \operatorname{div} \mathbf{v}\, dx\right) \Big/ |\mathbf{v}|_{1,\kappa} \geqslant (\hat{C}/C_2)\rho_\kappa |q|_{1,\kappa}.$$

Therefore the theorem is established provided we show that

$$\|q\|_{0,\kappa} \leqslant C_3 h_\kappa |q|_{1,\kappa} \qquad \forall q \in M_h(\kappa)$$

with a constant $C_3 > 0$ independent of κ, h and q. This will be the object of the next lemma. Assuming this result, we immediately derive that

$$\rho_\kappa |q|_{1,\kappa} \geqslant (1/C_3)(\rho_\kappa/h_\kappa)\|q\|_{0,\kappa} \geqslant [1/(\sigma C_3)]\|q\|_{0,\kappa}$$

on account of the regularity of $\mathscr{T}_h$. □

Lemma 2.5. *Let κ be an N-simplex of $\mathbb{R}^N$. There exists a constant $C > 0$, independent of h and κ such that the following inequality holds for all functions q in $H^1(\kappa) \cap L_0^2(\kappa)$:*

(2.26) $$\|q\|_{0,\kappa} \leqslant C h_\kappa |q|_{1,\kappa}.$$

Proof. Recall that

$$\|q\|_{0,\kappa} = \inf_{c \in \mathbb{R}} \|q + c\|_{0,\kappa} \qquad \forall q \in L_0^2(\kappa).$$

Therefore

$$\|q\|_{0,\kappa} = |\det(B_\kappa)|^{1/2} \inf_{c \in \mathbb{R}} \|\hat{q} + c\|_{0,\hat{\kappa}}$$
$$\leqslant |\det(B_\kappa)|^{1/2} \|\hat{q}\|_{H^1(\hat{\kappa})/\mathbb{R}}.$$

Then the equivalence Theorem I.1.9 yields:

$$\|q\|_{0,\kappa} \leqslant \hat{C}_1 |\det(B_\kappa)|^{1/2} |\hat{q}|_{1,\hat{\kappa}} \leqslant \hat{C}_2 h_\kappa |q|_{1,\kappa}$$

in view of (A.7) and (A.2). □

Remark 2.4. We know from Theorem I.1.9 that

$$\|q\|_{0,\kappa} \leqslant C(\kappa)|q|_{1,\kappa} \qquad \forall q \in H^1(\kappa) \cap L_0^2(\kappa),$$

but the constant $C(\kappa)$ depends on κ and Lemma 2.5 precises this dependence.

Remark 2.5. Observe that the crucial idea in the proof of Theorem 2.2 is that

$$\left(\prod_{1 \leqslant i \leqslant N+1} \lambda_i\right) \mathbf{grad}\, q \quad \text{belongs to } X_h(\kappa)$$

whenever q belongs to $M_h(\kappa)$. We shall see later on that this fundamental property applies as well to elements in three dimensions.

As mentioned above, Theorems 1.12 and 2.2 imply the uniform inf-sup condition (1.12):

Lemma 2.6. *Let $\mathscr{T}_h$ be a regular triangulation of $\bar{\Omega}$ and let the spaces X_h and M_h be defined by* (2.17) (2.18). *Then there exists a constant $\beta^* > 0$ independent of h such that*:

$$\sup_{\mathbf{v}_h \in X_h} \left\{ \left(\int_\Omega q_h \operatorname{div} \mathbf{v}_h \, dx \right) \Big/ |\mathbf{v}_h|_{1,\Omega} \right\} \geqslant \beta^* \| q_h \|_{0,\Omega} \qquad \forall q_h \in M_h.$$

Again, Hypothesis H2 follows immediately from Lemma A.5 for the L^2-projection ρ_h onto Q_h:

(2.27) $$\| q - \rho_h q \|_{0,\Omega} \leqslant C h^l |q|_{l,\Omega} \qquad \forall q \in H^l(\Omega).$$

Thus Theorems 1.8 and 1.9 imply the convergence and error estimates for the higher-order schemes.

Theorem 2.3. *Assume that Ω is a bounded, plane polygon. Let the solution* $(\mathbf{u}, p)$ *of the Stokes problem satisfy*:

$$\mathbf{u} \in [H^{k+1}(\Omega) \cap H_0^1(\Omega)]^2, \qquad p \in H^k(\Omega) \cap L_0^2(\Omega)$$

for some integer k with $1 \leqslant k \leqslant l$ where the integer $l \geqslant 2$; and let the spaces X_h and M_h be defined respectively by (2.17) *and* (2.18). *If $\mathscr{T}_h$ is a regular family of triangulations of $\bar{\Omega}$ then the solution* $(\mathbf{u}_h, p_h)$ *of* (1.39) *satisfies*:

(2.28) $$|\mathbf{u} - \mathbf{u}_h|_{1,\Omega} + \| p - p_h \|_{0,\Omega} \leqslant C_1 h^k (|\mathbf{u}|_{k+1,\Omega} + |p|_{k,\Omega}).$$

If in addition Ω is convex, we have the L^2-estimate:

(2.29) $$\| \mathbf{u} - \mathbf{u}_h \|_{0,\Omega} \leqslant C_2 h^{k+1} (|\mathbf{u}|_{k+1,\Omega} + |p|_{k,\Omega}).$$

Remark 2.6. Of course, according to (1.45) the hypotheses of Theorem 2.1 on Ω and $\mathscr{T}_h$ guarantee the convergence of $\mathbf{u}_h$ and p_h without regularity assumption on the exact solution $\mathbf{u}$ and p.

2.3. The Three-Dimensional Case: First and Higher-Order Schemes

In this section we assume that Ω is a *bounded polyhedron* of $\mathbb{R}^3$ and $\mathscr{T}_h$ is a triangulation of $\bar{\Omega}$ that consists of tetrahedra κ with diameters bounded by h. If κ is a tetrahedron with vertices a_1, a_2, a_3, a_4 like in Figure 9 we denote by F_i the face opposite a_i, $\mathbf{n}_i$ its outward unit normal and e_{ij} the edge $[a_i, a_j]$.

We shall develop the first and second-order schemes separately, since they are particular cases. The *first-order scheme* is a very straightforward extension of the two-dimensional scheme discussed in Section 2.1 and we shall skim through it rapidly. Let κ be an arbitrary tetrahedron; what we want is a velocity vector $\mathbf{w}$, with affine tangential components on the faces of κ, that is compatible

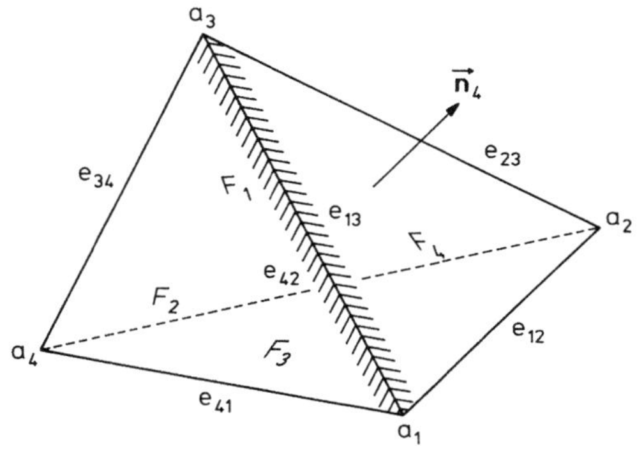

Figure 9

with constant pressures in κ. Formulas (2.1) and (2.2) suggest to take $\mathbf{w}$ in the space $\mathscr{P}_1(\kappa)$ with:

$$(2.30)\quad \begin{cases} \mathscr{P}_1(\kappa) = P_1^3 \oplus \operatorname{span}\{\mathbf{p}_1, \mathbf{p}_2, \mathbf{p}_3, \mathbf{p}_4\} \subset P_3^3 \\ \text{where} \\ \mathbf{p}_1 = \mathbf{n}_1\lambda_2\lambda_3\lambda_4, \quad \mathbf{p}_2 = \mathbf{n}_2\lambda_3\lambda_4\lambda_1, \quad \mathbf{p}_3 = \mathbf{n}_3\lambda_4\lambda_1\lambda_2, \quad \mathbf{p}_4 = \mathbf{n}_4\lambda_1\lambda_2\lambda_3. \end{cases}$$

Note that $\mathbf{p}_i$ vanishes on all faces F_j with $j \neq i$ and obviously $\mathbf{p}_i \times \mathbf{n}_i = \mathbf{0}$ so that

$$\mathbf{p}_i \times \mathbf{n}|_{\partial\kappa} = \mathbf{0} \qquad 1 \leqslant i \leqslant 4.$$

As far as the degrees of freedom of $\mathbf{w}$ are concerned we can easily take the values of $\mathbf{w}$ at the vertices a_i of κ and its flux through each face F_i. The argument of Lemma 2.1 shows that these 16 degrees of freedom are $\mathscr{P}_1(\kappa)$-unisolvent:

Lemma 2.7. *A polynomial* $\mathbf{p}$ *of* $\mathscr{P}_1(\kappa)$ *is uniquely determined by the* 16 *values:*

$$(2.31)\quad \begin{cases} \mathbf{p}(a_i) \qquad 1 \leqslant i \leqslant 4, \\ \displaystyle\int_{F_i} \mathbf{p}\cdot\mathbf{n}_i\, ds \qquad 1 \leqslant i \leqslant 4. \end{cases}$$

In addition, on any face F_i *of* κ, $\mathbf{p}$ *depends only upon the degrees of freedom defined on that face.*

The corresponding velocity and pressure spaces are:

$$(2.32)\quad \begin{cases} W_h = \{\mathbf{w} \in \mathscr{C}^0(\bar{\Omega})^3;\ \mathbf{w}|_\kappa \in \mathscr{P}_1(\kappa) \quad \forall \kappa \in \mathscr{T}_h\}, \\ X_h = W_h \cap H_0^1(\Omega)^3, \end{cases}$$

$$(2.33)\quad \begin{cases} Q_h = \{q \in L^2(\Omega);\ q|_\kappa \in P_0 \quad \forall \kappa \in \mathscr{T}_h\}, \\ M_h = Q_h \cap L_0^2(\Omega). \end{cases}$$

By virtue of Lemma 2.7, we might again define an interpolation operator r_κ on $\mathscr{P}_1(\kappa)$ by:

$$r_\kappa \mathbf{v}(a_i) = \mathbf{v}(a_i), \qquad \int_{F_i} (r_\kappa \mathbf{v} - \mathbf{v}) \cdot \mathbf{n}_j \, ds = 0, \qquad 1 \leqslant i, j \leqslant 4.$$

But like in the two-dimensional case, this operator will not satisfy Hypothesis H3 because it is not defined on $H^1(\Omega)^3$. Therefore, we propose to replace the values of $\mathbf{v}$ by those of the local regularization operator in $\mathbb{R}^3$ developed by Bernardi [9] that generalizes the two dimensional operator R_h of Section A.3. There is no space here to give a detailed description of this operator (also denoted by R_h). All we need to know is that $R_h \in \mathscr{L}(H_0^1(\Omega)^3; \Phi_h^3)$ and that R_h satisfies (2.11) when the triangulation $\mathscr{T}_h$ is regular. Then we define the operator $\pi_h \in \mathscr{L}(H_0^1(\Omega)^3; \bar{X}_h)$ by:

$$\text{(2.34)} \qquad \begin{cases} \pi_h \mathbf{v}(a) = R_h \mathbf{v}(a) & \forall \text{node } a \text{ of } \mathscr{T}_h, \\ \displaystyle\int_F (\pi_h \mathbf{v} - \mathbf{v}) \cdot \mathbf{n} \, ds = 0 & \forall \text{face } F \text{ of } \mathscr{T}_h. \end{cases}$$

With very minor modifications, the proof of Lemma 2.2 can be adapted to show that π_h satisfies Hypothesis H1 with $l = 1$ and also Hypothesis H3:

Lemma 2.8. *The operator π_h defined by* (2.34) *satisfies*:

$$\int_\Omega \operatorname{div}(\mathbf{v} - \pi_h \mathbf{v}) q \, dx = 0 \qquad \forall q \in Q_h.$$

Moreover, if the triangulation $\mathscr{T}_h$ is regular, π_h has the error bound:

$$\text{(2.35)} \qquad |\mathbf{v} - \pi_h \mathbf{v}|_{m,\Omega} \leqslant C h^{k-m} |\mathbf{v}|_{k,\Omega} \qquad \forall \mathbf{v} \in H^k(\Omega)^3,$$

with $m = 0$ or 1 and $k = 1$ or 2.

Finally, applying Theorems 1.8 and 1.9 we obtain the expected estimate for this first-order scheme:

Theorem 2.4. *Let Ω be a bounded polyhedron in $\mathbb{R}^3$ and let the solution $(\mathbf{u}, p)$ of the Stokes problem satisfy*:

$$\mathbf{u} \in [H^2(\Omega) \cap H_0^1(\Omega)]^3, \qquad p \in H^1(\Omega) \cap L_0^2(\Omega).$$

If the triangulation $\mathscr{T}_h$ is regular then the solution $(\mathbf{u}_h, p_h)$ of (1.39) *with the spaces $\bar{X}_h$ and $\bar{M}_h$ defined respectively by* (2.32) *and* (2.33) *satisfies the estimate*:

$$\text{(2.36)} \qquad |\mathbf{u} - \mathbf{u}_h|_{1,\Omega} + \|p - p_h\|_{0,\Omega} \leqslant C_1 h (|\mathbf{u}|_{2,\Omega} + |p|_{1,\Omega}).$$

Moreover, if the Stokes Problem (1.48) *is regular, we have the L^2-estimate*:

$$\text{(2.37)} \qquad \|\mathbf{u} - \mathbf{u}_h\|_{0,\Omega} \leqslant C_2 h^2 (|\mathbf{u}|_{2,\Omega} + |p|_{1,\Omega}).$$

Now we turn to the *second-order scheme*. As expected we wish to construct a velocity vector $\mathbf{w}$, with quadratic tangential components on the boundary of κ, which is compatible with affine pressures in κ. In the light of the corresponding scheme in $\mathbb{R}^2$, one is tempted to take $\mathbf{w}$ in the space $\{P_2 \oplus (\lambda_1\lambda_2\lambda_3\lambda_4 P_0)\}^3$. Unfortunately, this space's dimension is too small to meet the requirements of the inf-sup condition with affine pressures. Following Fortin [30] and Bernardi & Raugel [10], the best we can do is add to this space the cubics $\mathbf{p}_i$ of (2.30); thus we shall take $\mathbf{w}$ in the subspace of P_4^3:

$$\mathscr{P}_2(\kappa) = \{P_2 \oplus (\lambda_1\lambda_2\lambda_3\lambda_4 P_0)\}^3 \oplus \operatorname{span}\{\mathbf{p}_1, \mathbf{p}_2, \mathbf{p}_3, \mathbf{p}_4\}. \tag{2.38}$$

Right away, observe that $\mathscr{P}_1(\kappa) \subset \mathscr{P}_2(\kappa)$; therefore we can try as much as possible to apply the material of Section 1.4 in order to establish the inf-sup condition. This means that the inf-sup condition imposes no constraint on the degrees of freedom of the velocity $\mathbf{w}$ and thus we can choose the most convenient ones. Now, the degrees of freedom naturally attached to P_2 are:

$w(a_i)$ $1 \leqslant i \leqslant 4$ and $w(a_{ij})$ where a_{ij} is the midpoint of e_{ij}, $1 \leqslant i < j \leqslant 4$;

those corresponding to $\{\mathbf{p}_i;\ 1 \leqslant i \leqslant 4\}$ are:

$$\int_{F_i} \mathbf{w}\cdot\mathbf{n}_i\, ds \qquad 1 \leqslant i \leqslant 4,$$

and the simplest one corresponding to the "bubble function" $\lambda_1\lambda_2\lambda_3\lambda_4$ is

$$w(a_\kappa) \quad \text{where } a_\kappa \text{ is the center of } \kappa.$$

The next lemma shows that this set of moments is $\mathscr{P}_2(\kappa)$-unisolvent.

Lemma 2.9. *A polynomial $\mathbf{p}$ of $\mathscr{P}_2(\kappa)$ is uniquely determined by the 37 values:*

$$\begin{cases} \mathbf{p}(a_i) \quad 1 \leqslant i \leqslant 4, \quad \mathbf{p}(a_{ij}) \quad 1 \leqslant i < j \leqslant 4, \quad \mathbf{p}(a_\kappa), \\ \displaystyle\int_{F_i} \mathbf{p}\cdot\mathbf{n}_i\, ds \quad 1 \leqslant i \leqslant 4. \end{cases} \tag{2.39}$$

Moreover the restriction of $\mathbf{p}$ to any face F_i of κ depends exclusively upon its degrees of freedom on that face.

Proof. A polynomial of $\mathscr{P}_2(\kappa)$ has 37 coefficients which is precisely the number of degrees of freedom defined by (2.39). Therefore it suffices to prove that zero moments generate only the zero polynomial. Now, on any edge e_{ij} a component p of $\mathbf{p}$ reduces to a quadratic function of one variable. Therefore $p(a_i) = p(a_j) = p(a_{ij}) = 0$ imply that $p|_{e_{ij}} = 0$. Thus if the degrees of freedom of $\mathbf{p}$ vanish on the boundary of any face F_i then necessarily $\mathbf{p}|_{F_i} = c\mathbf{p}_i$ and if $\int_{F_i} \mathbf{p}\cdot\mathbf{n}_i\, ds = 0$ then $c = 0$.

Finally, if $\mathbf{p}|_{\partial\kappa} = \mathbf{0}$ then each component p of $\mathbf{p}$ is a "bubble function" and therefore $p(a_\kappa) = 0$ implies that $p = 0$ on κ. □

Remark 2.7. Except for $\int_{F_i} \mathbf{p} \cdot \mathbf{n}_i \, ds$ all degrees of freedom enumerated by (2.39) are preserved by affine transformations. Note that the space $\mathscr{P}_2(\kappa)$ itself is not invariant under affine transformations because of the normal vectors in the cubics $\mathbf{p}_i$.

From the above considerations, we choose the following finite element spaces for velocity and pressure:

$$(2.40) \qquad \begin{cases} W_h = \{\mathbf{w} \in \mathscr{C}^0(\bar{\Omega})^3;\ \mathbf{w}|_\kappa \in \mathscr{P}_2(\kappa) \quad \forall \kappa \in \mathscr{T}_h\}, \\ X_h = W_h \cap H_0^1(\Omega)^3, \end{cases}$$

$$(2.41) \qquad \begin{cases} Q_h = \{q \in L^2(\Omega);\ q|_\kappa \in P_1 \quad \forall \kappa \in \mathscr{T}_h\}, \\ M_h = Q_h \cap L_0^2(\Omega). \end{cases}$$

Lemma 2.9 provides an adequate interpolation operator r_h such that:

$$r_h \mathbf{v}|_\kappa = r_\kappa \mathbf{v} \qquad \forall \kappa \in \mathscr{T}_h \qquad \forall \mathbf{v} \in \mathscr{C}^0(\bar{\Omega})^3$$

where $r_\kappa \mathbf{v}$ is the unique polynomial of $\mathscr{P}_2(\kappa)$ that has the same degrees of freedom (2.39) as $\mathbf{v}$ on κ. Clearly we have:

$$r_h \in \mathscr{L}(\mathscr{C}^0(\bar{\Omega})^3;\ W_h) \cap \mathscr{L}([\mathscr{C}^0(\bar{\Omega}) \cap H_0^1(\Omega)]^3;\ X_h)$$

and the following lemma checks Hypothesis H1.

Lemma 2.10. *If $\mathscr{T}_h$ is a regular triangulation of $\bar{\Omega}$ then r_h satisfies the approximation property*:

$$(2.42) \qquad |\mathbf{v} - r_h \mathbf{v}|_{m,\Omega} \leqslant C h^{k-m} |\mathbf{v}|_{k,\Omega} \qquad \forall \mathbf{v} \in H^k(\Omega)^3,$$

$m = 0, 1$, $k = 2$ *or* 3, *with a positive constant* C *independent of* h *and* $\mathbf{v}$.

Proof. If r_κ were preserved by affine transformation the proof would be trivial; nevertheless, the proof is quite short because only the moments on the faces are not invariant by affine transformation.

Let Iv denote the polynomial of $P_2 \oplus (\lambda_1 \lambda_2 \lambda_3 \lambda_4 P_0)$ defined by:

$$Iv(a_i) = v(a_i) \qquad 1 \leqslant i \leqslant 4,$$

$$Iv(a_{ij}) = v(a_{ij}) \qquad 1 \leqslant i < j \leqslant 4, \qquad Iv(a_\kappa) = v(a_\kappa).$$

On the one hand this set of conditions determine Iv uniquely, and on the other hand the operator I is invariant under affine transformation. Moreover, as I preserves the polynomials of P_2, the standard argument of Corollary A.2 shows that

$$(2.43) \qquad |v - Iv|_{m,\kappa} \leqslant C_1 \|B_\kappa\|^k \|B_\kappa^{-1}\|^m |v|_{k,\kappa} \qquad m = 0, 1, \qquad k = 2, 3.$$

Like in Remark 2.1 we find that $r_\kappa \mathbf{v}$ can be expressed as:

$$r_\kappa \mathbf{v} = I\mathbf{v} + \sum_{i=1}^{4} \alpha_i \mathbf{p}_i + \boldsymbol{\beta} \lambda_1 \lambda_2 \lambda_3 \lambda_4$$

with

$$\alpha_i = \left[\int_{F_i} (\mathbf{v} - I\mathbf{v})\cdot \mathbf{n}_i \, ds\right] \Big/ \int_{F_i} \lambda_j \lambda_k \lambda_l \, ds$$

and

$$\boldsymbol{\beta} = -4 \sum_{i=1}^{4} \alpha_i \mathbf{n}_i.$$

Like in Lemma 2.2 we have:

$$|\mathbf{p}_i|_{m,\kappa} \leqslant C_2 |\det(B_\kappa)|^{1/2} \| B_\kappa^{-1} \|^m$$

with a similar expression for $|\lambda_1 \lambda_2 \lambda_3 \lambda_4|_{m,\kappa}$. Likewise, we infer that

$$|\alpha_i| \leqslant C_3 |\det(B_\kappa)|^{-1/2} \| B_\kappa \|^k |\mathbf{v}|_{k,\kappa}, \qquad k = 2, 3.$$

Therefore

$$\left| \sum_{i=1}^{4} \alpha_i \mathbf{p}_i + \boldsymbol{\beta} \lambda_1 \lambda_2 \lambda_3 \lambda_4 \right|_{m,\kappa} \leqslant C_4 \| B_\kappa \|^k \| B_\kappa^{-1} \|^m |\mathbf{v}|_{k,\kappa}.$$

Hence (2.43) and the regularity of $\mathscr{T}_h$ lead to

$$|r_\kappa \mathbf{v} - \mathbf{v}|_{m,\kappa} \leqslant C_5 \sigma^m h^{k-m} |\mathbf{v}|_{k,\kappa}. \qquad \square$$

As mentioned previously there is no need to establish directly the inf-sup condition because the space $\bar{X}_h$ is contained in X_h and the pair $(\bar{X}_h, \bar{M}_h)$ satisfies a uniform inf-sup condition. Instead it suffices to show that (X_h, M_h) satisfies an adequate local condition. In fact, it is easy to verify that the statement and proof of Theorem 2.2 are valid without change for the pair of local spaces:

$$X_h(\kappa) = \{\mathbf{v} \in \mathscr{P}_2(\kappa); \, \mathbf{v}|_{\partial\kappa} = \mathbf{0}\},$$
$$M_h(\kappa) = P_1 \cap L_0^2(\kappa).$$

(Observe that the crucial point in the proof is that the function $\lambda_1 \lambda_2 \lambda_3 \lambda_4 \, \mathbf{grad} \, q$ belongs to $X_h(\kappa)$ for all q in P_1). According to Theorem 1.12 this implies the global inf-sup condition (1.12):

Lemma 2.11. *Let $\mathscr{T}_h$ be a regular triangulation of $\bar{\Omega}$ and let the spaces X_h and M_h be defined by* (2.40) (2.41). *There exists a constant $\beta^* > 0$ independent of h such that:*

$$\sup_{\mathbf{v}_h \in X_h} \left\{ \left[\int_\Omega q_h \operatorname{div} \mathbf{v}_h \, dx\right] \Big/ |\mathbf{v}_h|_{1,\Omega} \right\} \geqslant \beta^* \| q_h \|_{0,\Omega} \qquad \forall q_h \in M_h.$$

Finally, Hypothesis H2 reduces to the standard approximation property of the L^2-projection operator ρ_h onto Q_h (cf. Lemma A.5):

$$\| q - \rho_h q \|_{0,\Omega} \leqslant C h^k |q|_{k,\Omega} \qquad \forall q \in H^k(\Omega), \qquad k = 1, 2.$$

Thus we have established the desired estimate for our second-order scheme.

Theorem 2.5. *Let Ω and $\mathscr{T}_h$ be like in Theorem 2.4 and let the solution* $(\mathbf{u}, p)$ *of the Stokes problem satisfy:*

$$\mathbf{u} \in [H^{k+1}(\Omega) \cap H_0^1(\Omega)]^3, \quad p \in H^k(\Omega) \cap L_0^2(\Omega), \quad k = 1 \text{ or } 2.$$

Then the solution $(\mathbf{u}_h, p_h)$ *of the scheme* (1.39) *with the spaces X_h and M_h defined respectively by* (2.40) *and* (2.41) *satisfies:*

$$(2.44) \quad |\mathbf{u} - \mathbf{u}_h|_{1,\Omega} + \|p - p_h\|_{0,\Omega} \leqslant C_1 h^k(|\mathbf{u}|_{k+1,\Omega} + |p|_{k,\Omega}), \quad k = 1, 2.$$

In addition, if the Stokes problem (1.48) *is regular, we have the L^2-estimate:*

$$(2.45) \quad \|\mathbf{u} - \mathbf{u}_h\|_{0,\Omega} \leqslant C_2 h^{k+1}(|\mathbf{u}|_{k+1,\Omega} + |p|_{k,\Omega}), \quad k = 1, 2.$$

Higher-order schemes of degree $l \geqslant 3$ are an easy generalization of the second-order scheme. If we want a velocity space that will match a polynomial pressure q of degree $l - 1$, we can try a velocity $\mathbf{w}$ in the subspace of P_{l+2}^3:

$$(2.46) \quad \mathscr{P}_l(\kappa) = [P_l \oplus \{\lambda_1 \lambda_2 \lambda_3 \lambda_4 (\tilde{P}_{l-2} \oplus \tilde{P}_{l-3})\}]^3.$$

Then each component of $\mathbf{w}$ reduces to an l-degree polynomial on the faces of κ and moreover $\lambda_1 \lambda_2 \lambda_3 \lambda_4 \,\mathbf{grad}\, q$ belongs to $\mathscr{P}_l(\kappa)$ whenever q belongs to P_{l-1} (cf. Remark 2.5). Thus we choose the following spaces:

$$(2.47) \quad \begin{cases} W_h = \{\mathbf{w} \in \mathscr{C}^0(\bar{\Omega})^3;\ \mathbf{w}|_\kappa \in \mathscr{P}_l(\kappa) \quad \forall \kappa \in \mathscr{T}_h\}, \\ X_h = W_h \cap H_0^1(\Omega)^3, \end{cases}$$

$$(2.48) \quad \begin{cases} Q_h = \{q \in L^2(\Omega);\ q|_\kappa \in P_{l-1} \quad \forall \kappa \in \mathscr{T}_h\}, \\ M_h = Q_h \cap L_0^2(\Omega). \end{cases}$$

Like in two dimensions, we can take for degrees of freedom of the velocity $\mathbf{w}$ its values associated with P_l on each face F_i of κ and its derivatives corresponding to P_{l-2} at the center of κ:

$$(2.49) \quad \begin{cases} \text{(a)} \quad w(a) \quad \forall a \in \Sigma_\kappa \cap F_i, \quad 1 \leqslant i \leqslant 4, \\ \text{(b)} \quad \partial^k w(a_\kappa) \quad \text{for all partial derivatives of order } k \text{ with } 0 \leqslant k \leqslant l - 2, \end{cases}$$

w denoting an arbitrary component of $\mathbf{w}$. If necessary, the derivatives can also be replaced by the interior moments:

$$(2.49) \quad \text{(c)} \qquad \int_\kappa wq\, dx \quad \forall q \in P_{l-2}.$$

On the one hand, either formula (2.49) defines a total of:

$$\begin{aligned} &4 + 6\dim(P_{l-2}(\mathbb{R})) + 4\dim(P_{l-3}(\mathbb{R}^2)) + \dim(P_{l-2}(\mathbb{R}^3)) \\ &= 4 + 6(l-1) + 2(l-2)(l-1) + (1/6)(l-1)l(l+1) \\ &= (1/6)\{(l^2 - 1)(l + 12) + 24\} \end{aligned}$$

degrees of freedom. On the other hand, each component p of $\mathscr{P}_l(\kappa)$ belongs to a space of dimension:

$$\begin{aligned}&\dim(P_l) + \dim(\tilde{P}_{l-2}) + \dim(\tilde{P}_{l-3}), \quad \text{all in } \mathbb{R}^3\\ &= (1/6)(l+1)(l+2)(l+3) + (1/2)(l-1)l + (1/2)(l-2)(l-1)\\ &= (1/6)\{(l^2-1)(l+12)+24\}.\end{aligned}$$

Recalling the argument of Lemma 2.3, it is an easy matter to prove that these degrees of freedom are $\mathscr{P}_l(\kappa)$-unisolvent.

Lemma 2.12. *Each component p of a polynomial of $\mathscr{P}_l(\kappa)$ is uniquely determined by the degrees of freedom* (2.49a) (2.49b) *or* (2.49a) (2.49c). *Moreover, on a given face of κ, p depends exclusively upon the degrees of freedom defined on that face.*

This lemma yields the following interpolation operator:

$$r_h \mathbf{v}|_\kappa = r_\kappa \mathbf{v} \quad \forall \kappa \in \mathscr{T}_h \quad \forall \mathbf{v} \in \mathscr{C}^0(\bar{\Omega})^3,$$

where $r_\kappa \mathbf{v}$ is the only polynomial of $\mathscr{P}_l(\kappa)$ that has the same degrees of freedom (2.49a) (2.49c) as the function $\mathbf{v}$. Obviously, the operator r_κ is invariant under affine transformations and as a consequence we have the following result:

Lemma 2.13. *The operator r_h belongs to $\mathscr{L}(\mathscr{C}^0(\bar{\Omega})^3; W_h) \cap \mathscr{L}([\mathscr{C}^0(\bar{\Omega}) \cap H_0^1(\Omega)]^3; X_h)$. In addition, if the triangulation $\mathscr{T}_h$ is regular, r_h has the following interpolation error*:

(2.50) $$|\mathbf{v} - r_h \mathbf{v}|_{m,\Omega} \leqslant C h^{k+1-m} |\mathbf{v}|_{k+1,\Omega} \quad \forall \mathbf{v} \in H^{k+1}(\Omega)^3, \quad m = 0 \ \text{or} \ 1,$$

$1 \leqslant k \leqslant l$, *with a positive constant C independent of h and $\mathbf{v}$.*

Like for the second-order scheme, the proof of the inf-sup condition here is trivially similar to that of Theorem 2.2. The local spaces involved are

$$\begin{aligned}X_h(\kappa) &= \{\mathbf{v} \in \mathscr{P}_l(\kappa);\ \mathbf{v}|_{\partial\kappa} = \mathbf{0}\},\\ M_h(\kappa) &= P_{l-1} \cap L_0^2(\Omega)\end{aligned}$$

and the salient property that links them is:

$$\lambda_1 \lambda_2 \lambda_3 \lambda_4 \, \mathbf{grad}\, q \in X_h(\kappa) \quad \forall q \in M_h(\kappa).$$

Hence the statement of Lemma 2.11 carries over to the pair (X_h, M_h) defined by (2.47) (2.48).

Again Hypothesis H2 follows from the approximation property of the L^2-projection operator ρ_h onto Q_h:

$$\|q - \rho_h q\|_{0,\Omega} \leqslant C h^k |q|_{k,\Omega} \quad \forall q \in H^k(\Omega), \quad 1 \leqslant k \leqslant l.$$

Collecting these results, we obtain the desired estimate for this last scheme.

Theorem 2.6. *Let Ω and $\mathscr{T}_h$ be like in Theorem 2.4 and let the solution $(\mathbf{u}, p)$ of the Stokes Problem satisfy*:

$$\mathbf{u} \in [H^{k+1}(\Omega) \cap H_0^1(\Omega)]^3, \qquad p \in H^k(\Omega) \cap L_0^2(\Omega)$$

for some integer $k \in [1, l]$. Then the solution $(\mathbf{u}_h, p_h)$ of (1.39) *with the spaces X_h and M_h defined by* (2.47) (2.48) *satisfies the error bound*:

$$(2.51) \qquad |\mathbf{u} - \mathbf{u}_h|_{1,\Omega} + \|p - p_h\|_{0,\Omega} \leqslant C_1 h^k (|\mathbf{u}|_{k+1,\Omega} + |p|_{k,\Omega}).$$

In addition, if the Stokes Problem (1.48) *is regular, we have the L^2-estimate*

$$(2.52) \qquad \|\mathbf{u} - \mathbf{u}_h\|_{0,\Omega} \leqslant C_2 h^{k+1} (|\mathbf{u}|_{k+1,\Omega} + |p|_{k,\Omega}).$$

Remark 2.8. It is easy to show that the statement of Remark 2.6 concerning the convergence under weak regularity assumptions on $(\mathbf{u}, p)$ is still valid in the three-dimensional case.

§ 3. Quadrilateral Finite Element Methods Using Discontinuous Pressures

There are two reasons for treating separately quadrilateral finite elements. On the one hand, isoparametric finite element methods are less transparent than simplicial ones and must be handled with some more care. On the other hand, quadrilateral elements (more precisely, rectangular elements) provide excellent examples of schemes which do not satisfy the inf-sup condition and yet can be proved to converge with optimal accuracy. Some of these schemes, being particularly simple, are preferred by a number of users.

For the sake of conciseness, we have only treated the two-dimensional case. The three-dimensional case is a straightforward adaptation of this material and that of Section 2.3.

3.1. A Quadrilateral Finite Element of Order One

The element discussed in this section is the analogue of the first-order element defined in Section 2.1. It has been introduced by Fortin [30].

Let Ω be a *bounded, plane polygon* and let $\mathscr{T}_h$ be a "triangulation" of $\bar{\Omega}$ made of **convex quadrilaterals** with diameters bounded by h. Consider one of these quadrilaterals κ with vertices a_1, a_2, a_3, a_4 (also numbered a_0); we denote by f_i the segment $[a_{i-1}, a_i]$ (cf. Figure 10) and by $\mathbf{n}_i$ its unit outward normal. To draw the parallel with Section 2.1 we replace the barycentric coordinates by the reference variables

$$\hat{x}_1, \hat{x}_2, \hat{x}_3 = 1 - \hat{x}_1 \quad \text{and} \quad \hat{x}_4 = 1 - \hat{x}_2.$$

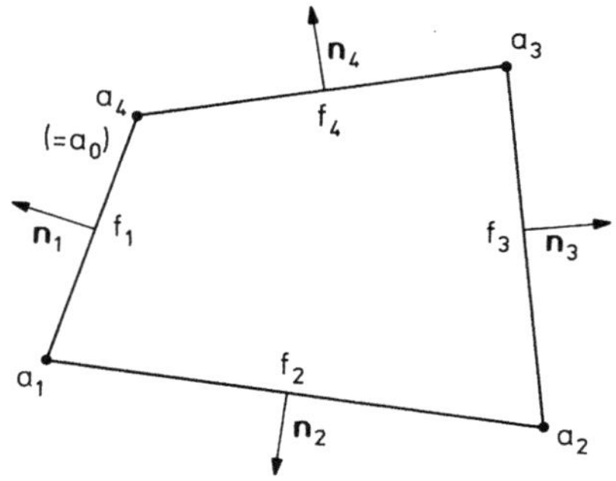

Figure 10

Now, we are looking for a velocity vector $\mathbf{w}$ that is compatible with a constant pressure in κ. Keeping in mind the material of Section 2.1 it is likely that $\mathbf{w}$ will belong to a space larger than $Q_1(\kappa)$, but that its tangential components on each side of κ will be affine. (The pair $(Q_1(\kappa), P_0)$ will in fact be the object of Section 3.3). As an example, the polynomial

$$\hat{q}_1 = \hat{x}_2 \hat{x}_3 \hat{x}_4$$

vanishes on the sides $\hat{f}_2$, $\hat{f}_3$ and $\hat{f}_4$ of the reference square $\hat{\kappa}$. Therefore the function

$$\mathbf{p}_1 = \mathbf{n}_1(\hat{q}_1 \circ F_\kappa^{-1})$$

has zero tangential components on the sides of κ. Generalizing this remark, we set

(3.1)
$$\hat{q}_2 = \hat{x}_3 \hat{x}_4 \hat{x}_1, \quad \hat{q}_3 = \hat{x}_4 \hat{x}_1 \hat{x}_2, \quad \hat{q}_4 = \hat{x}_1 \hat{x}_2 \hat{x}_3,$$
$$\mathbf{p}_i = \mathbf{n}_i(\hat{q}_i \circ F_\kappa^{-1})$$

and we take the velocities $\mathbf{w}$ in the space (of dimension 12):

(3.2)
$$\mathcal{Q}_1(\kappa) = Q_1(\kappa)^2 \oplus \operatorname{span}\{\mathbf{p}_1, \mathbf{p}_2, \mathbf{p}_3, \mathbf{p}_4\} \subset Q_2(\kappa)^2.$$

As will be seen in the next lemma, the degrees of freedom naturally attached to this space are the values of $\mathbf{w}$ at the vertices a_i and the flux of $\mathbf{w}$ through each side f_i of κ.

Lemma 3.1. *A polynomial* $\mathbf{p}$ *of* $\mathcal{Q}_1(\kappa)$ *is uniquely determined by the* 12 *quantities*:

(3.3)
$$\begin{cases} \mathbf{p}(a_i) & 1 \leqslant i \leqslant 4, \\ \displaystyle\int_{f_i} \mathbf{p} \cdot \mathbf{n}_i \, ds & 1 \leqslant i \leqslant 4. \end{cases}$$

Furthermore the restriction of $\mathbf{p}$ *to any side* f_i *of* κ *depends only upon the degrees of freedom defined on that side.*

Proof. Since the functions $\mathbf{p}_i$ vanish on all vertices of κ, we can write

$$\mathbf{p} = I_\kappa \mathbf{p} + \sum_{i=1}^{4} \alpha_i \mathbf{p}_i, \qquad \alpha_i \in \mathbb{R}, \tag{3.4}$$

where I_κ denotes the standard interpolation operator on $Q_1(\kappa)^2$. Furthermore,

$$(\mathbf{p} - I_\kappa \mathbf{p}) \cdot \mathbf{n}_i|_{f_i} = \alpha_i(\hat{q}_i \circ F_\kappa^{-1})|_{f_i}. \tag{3.5}$$

From these two expressions we can easily derive that zero moments yield only the zero polynomial. □

Thus, we choose the following velocity and pressure spaces:

$$\begin{cases} W_h = \{\mathbf{w} \in \mathscr{C}^0(\bar{\Omega})^2;\ \mathbf{w}|_\kappa \in \mathscr{Q}_1(\kappa) \quad \forall \kappa \in \mathscr{T}_h\}, \\ X_h = W_h \cap H_0^1(\Omega)^2, \end{cases} \tag{3.6}$$

$$\begin{cases} Q_h = \{q \in L^2(\Omega);\ q|_\kappa \in P_0 \quad \forall \kappa \in \mathscr{T}_h\}, \\ M_h = Q_h \cap L_0^2(\Omega). \end{cases} \tag{3.7}$$

Lemma 3.1 suggests the interpolation operator r_κ on $\mathscr{Q}_1(\kappa)$ defined by:

$$r_\kappa \mathbf{v}(a_i) = \mathbf{v}(a_i), \qquad \int_{f_j} (r_\kappa \mathbf{v} - \mathbf{v}) \cdot \mathbf{n}_j \, ds = 0, \qquad 1 \leqslant i < j \leqslant 4.$$

But once again, this operator does not satisfy Hypothesis H3 because it is not defined on $H^1(\Omega)^2$. Therefore, like in the simplicial case we replace the above values of $\mathbf{v}$ by those of the local regularization operator R_h, similar to that of Section A.3:

$$R_h \in \mathscr{L}(H_0^1(\Omega); \Phi_h)$$

with

$$\Phi_h = \{\phi \in \mathscr{C}^0(\bar{\Omega});\ \phi|_\kappa \in Q_1(\kappa) \quad \forall \kappa \in \mathscr{T}_h\} \cap H_0^1(\Omega).$$

Then we define the operator $\pi_h \in \mathscr{L}(H_0^1(\Omega)^2; X_h)$ by:

$$\begin{cases} \pi_h \mathbf{v}(a) = R_h \mathbf{v}(a) \qquad \forall \text{node } a \text{ of } \mathscr{T}_h, \\ \displaystyle\int_f (\pi_h \mathbf{v} - \mathbf{v}) \cdot \mathbf{n} \, ds = 0 \qquad \forall \text{side } f \text{ of } \mathscr{T}_h. \end{cases} \tag{3.8}$$

In order to establish the approximating properties of π_h we must assume that the triangulation $\mathscr{T}_h$ is regular in the sense of Definition A.2 with the parameters:

$$h_\kappa = \text{diameter of } \kappa, \qquad \rho_\kappa = 2 \operatorname*{Min}_{1 \leqslant i \leqslant 4} \{\text{diameter of circle inscribed in } S_i\}$$

where S_i denotes the triangle with vertices a_{i-1}, a_i, a_{i+1}.

Lemma 3.2. *The operator π_h defined by* (3.8) *satisfies*:

$$\int_\Omega \operatorname{div}(\mathbf{v} - \pi_h \mathbf{v}) q \, dx = 0 \qquad \forall q \in \bar{Q}_h.$$

Furthermore if the triangulation $\mathscr{T}_h$ is regular, π_h has the error bound:

(3.9) $$|\mathbf{v} - \pi_h \mathbf{v}|_{m,\Omega} \leqslant C h^{k-m} |\mathbf{v}|_{k,\Omega} \qquad \forall \mathbf{v} \in H^k(\Omega)^2$$

with $m = 0$ or 1 and $k = 1$ or 2.

Proof. From (3.4) and (3.5) we derive:

$$\pi_h \mathbf{v} = R_h \mathbf{v} + \sum_{i=1}^{4} \alpha_i \mathbf{p}_i$$

where

$$\alpha_i = \left[\int_{f_i} (\mathbf{v} - R_h \mathbf{v}) \cdot \mathbf{n}_i \, ds \right] \Big/ \int_{f_i} \hat{q}_i \circ F_\kappa^{-1} \, ds.$$

On the one hand, the operator R_h has the local interpolation error for $v \in H^k(\Omega)^2$:

(3.10) $$\|\mathbf{v} - R_h \mathbf{v}\|_{0,\kappa} + h_\kappa |\mathbf{v} - R_h \mathbf{v}|_{1,\kappa} \leqslant C_1 h_\kappa^k |\mathbf{v}|_{k,\Delta_\kappa} \qquad k = 1 \text{ or } 2$$

where again Δ_κ denotes the union of quadrilaterals which share at least a vertex with κ.

On the other hand, Lemma A.9 implies:

$$|\mathbf{p}_i|_{m,\kappa} \leqslant C_2 \sigma_\kappa^{2m} h_\kappa^{1-m} |\hat{q}_i|_{m,\hat{\kappa}} \leqslant C_3 \sigma_\kappa^{2m} h_\kappa^{1-m}, \qquad m = 0 \text{ or } 1.$$

Besides that

$$\int_{f_i} \hat{q}_i \circ F_\kappa^{-1} \, ds = \operatorname{meas}(f_i) \int_{\hat{f}_i} \hat{q}_i \, d\hat{s}$$

and

$$\left| \int_{f_i} (\mathbf{v} - R_h \mathbf{v}) \cdot \mathbf{n}_i \, ds \right| \leqslant \operatorname{meas}(f_i) \int_{\hat{f}_i} \|\hat{\mathbf{v}} - \widehat{R_h \mathbf{v}}\| \, d\hat{s}$$

because the restriction of F_κ to the sides of κ is affine. Then using the trace Theorem I.1.5 and Lemma A.9 we obtain:

$$\int_{\hat{f}_i} \|\hat{\mathbf{v}} - \widehat{R_h \mathbf{v}}\| \, d\hat{s} \leqslant (C_4/\rho_\kappa) \{ \|\mathbf{v} - R_h \mathbf{v}\|_{0,\kappa}^2 + h_\kappa^2 |\mathbf{v} - R_h \mathbf{v}|_{1,\kappa}^2 \}^{1/2}.$$

Therefore (3.10) yields:

$$\left| \int_{f_i} (\mathbf{v} - R_h \mathbf{v}) \cdot \mathbf{n}_i \, ds \right| \leqslant C_5 \sigma_\kappa \operatorname{meas}(f_i) h_\kappa^{k-1} |\mathbf{v}|_{k,\Delta_\kappa}.$$

Hence

$$|\alpha_i| \leqslant C_6 \sigma_\kappa h_\kappa^{k-1} |\mathbf{v}|_{k,\Delta_\kappa}$$

and

$$|\alpha_i| \, |\mathbf{p}_i|_{m,\kappa} \leqslant C_7 \sigma_\kappa^{2m+1} h_\kappa^{k-m} |\mathbf{v}|_{k,\Delta_\kappa}.$$

Then the proof ends like that of Lemma 2.2. □

Finally, the bound (2.13) is still valid for the above space Q_h. Therefore, we have established that our scheme is of order one:

Theorem 3.1. *Let Ω be a bounded polygon and assume that the solution* $(\mathbf{u}, p)$ *of the Stokes equations satisfies:*

$$\mathbf{u} \in [H^2(\Omega) \cap H_0^1(\Omega)]^2, \qquad p \in H^1(\Omega) \cap L_0^2(\Omega).$$

Then if the triangulation $\mathcal{T}_h$ is regular, the solution $(\mathbf{u}_h, p_h)$ *of* (1.39) *with the spaces $\bar{X}_h$ and $\bar{M}_h$ defined by* (3.6) (3.7) *satisfies the conclusion of Theorem* 2.1.

3.2. Higher-Order Quadrilateral Elements

We propose to discuss and generalize the widely used "Q_2–P_1" finite element scheme. In short, this method uses continuous velocities with components that are piecewise $Q_2(\kappa)$ and discontinuous pressures that are piecewise P_1 on each element κ. Its analysis is pretty straightforward and easily extended to arbitrary order l, so we can start directly with the general case.

Again, let κ be any quadrilateral of $\mathcal{T}_h$ and let us choose:

$$\text{(3.11)} \qquad \begin{cases} W_h = \{\mathbf{w} \in \mathscr{C}^0(\bar{\Omega})^2;\ \mathbf{w}|_\kappa \in Q_l(\kappa)^2 \quad \forall \kappa \in \mathcal{T}_h\}, \\ \qquad X_h = W_h \cap H_0^1(\Omega)^2, \end{cases}$$

$$\text{(3.12)} \qquad \begin{cases} Q_h = \{q \in L^2(\Omega);\ q|_\kappa \in P_{l-1} \quad \forall \kappa \in \mathcal{T}_h\}, \\ \qquad M_h = Q_h \cap L_0^2(\Omega), \end{cases}$$

with $l \geqslant 2$.

Right away, observe that $\bar{X}_h \subset X_h$ so that the inf-sup condition need only be checked locally. As a consequence, we can take the simplest degrees of freedom available such as:

the values of each component of $\mathbf{w}$ on the principal lattice Σ_κ of order l;
the moments corresponding to P_{l-1} for q:

$$\int_\kappa qf\,dx \qquad \forall f \in P_{l-1}.$$

To begin with, let us establish the local inf-sup condition. Here again, we take the triangulation as our partition:

$$\text{(3.13)} \qquad \begin{cases} X_h(\kappa) = \{\mathbf{v} \in Q_l(\kappa)^2;\ \mathbf{v}|_{\partial\kappa} = \mathbf{0}\}, \\ \quad M_h(\kappa) = P_{l-1} \cap L_0^2(\kappa). \end{cases}$$

Theorem 3.2. *Let the triangulation $\mathcal{T}_h$ be regular. Then the pair of spaces* $(X_h(\kappa), M_h(\kappa))$ *defined by* (3.13) *satisfies Hypothesis* H4.

Proof. The proof is much like that of Theorem 2.2 so we shall only dwell on the details inherent to quadrilaterals. We have:

$$\int_K q \operatorname{div} \mathbf{v}\, dx = -\int_K \mathbf{v} \cdot \mathbf{grad}\, q\, dx$$
$$= -\int_{\hat{K}} J_F \hat{\mathbf{v}} \cdot [(\mathbf{grad}\, q) \circ F_K]\, d\hat{x}.$$

Since $q \in P_{l-1}$, we have $\mathbf{grad}\, q \in P_{l-2}^2$ and therefore

$$(\mathbf{grad}\, q) \circ F_K \in Q_{l-2}^2 \quad \text{on } \hat{K}.$$

Let

$$b(\hat{x}) = \hat{x}_1 \hat{x}_2 \hat{x}_3 \hat{x}_4$$

denote the "bubble" function on $\hat{K}$ and let us choose

$$\hat{\mathbf{v}} = -b(\hat{x})[(\mathbf{grad}\, q) \circ F_K].$$

Then $\mathbf{v} \in X_h(K)$ and with this choice

$$\int_K q \operatorname{div} \mathbf{v}\, dx = \int_{\hat{K}} J_F b(\hat{x}) \, \|(\mathbf{grad}\, q) \circ F_K\|^2 \, d\hat{x}.$$

Of course the mapping

$$\phi \to \int_{\hat{K}} b(\hat{x}) |\phi| \, d\hat{x}$$

is a norm on any finite-dimensional space; thus

$$\int_K q \operatorname{div} \mathbf{v}\, dx \geqslant C_1 \int_{\hat{K}} J_F \|(\mathbf{grad}\, q) \circ F_K\|^2 \, d\hat{x}$$
$$\geqslant C_1 |q|_{1,K}^2.$$

Besides that

$$\|\mathbf{v}\|_{0,K} \leqslant |q|_{1,K}$$

and

$$|\mathbf{v}|_{1,K} \leqslant C_2(\sigma_K^2/\rho_K) \, \|\mathbf{v}\|_{0,K}$$

by applying to quadrilaterals the easy argument of Lemma A.6. Hence

$$\int_K q \operatorname{div} \mathbf{v}\, dx \geqslant C_3(\rho_K/\sigma_K^2) |q|_{1,K}.$$

Therefore the result follows from the next lemma and the regularity of $\mathcal{T}_h$. □

Lemma 3.3. *Let κ be a plane, convex quadrilateral. There exists a constant $C > 0$, independent of h and κ, such that*

$$\|q\|_{0,\kappa} \leqslant C\sigma_\kappa h_\kappa |q|_{1,\kappa} \quad \forall q \in H^1(\kappa) \cap L_0^2(\kappa). \tag{3.14}$$

We skip the proof because it is entirely similar to that of Lemma 2.5.

It remains to examine the approximation properties of W_h and Q_h. While the approximation error in W_h is completely standard since it stems directly from (A.49):

$$|\mathbf{w} - I_h\mathbf{w}|_{m,\Omega} \leqslant Ch^{k+1-m}|\mathbf{w}|_{k+1,\Omega} \quad \forall \mathbf{w} \in H^{k+1}(\Omega)^2, \quad 1 \leqslant k \leqslant l, \quad m = 0 \text{ or } 1, \tag{3.15}$$

the approximation error in Q_h is not so immediate because we are dealing with polynomials of P_{l-1} on quadrilaterals (instead of triangles). In particular, Theorem A.3 cannot be applied because $\{p \circ F_\kappa; p \in P_{l-1}\}$ is a proper subspace of Q_{l-1}. The following lemma is due to Bernardi (private communication).

Lemma 3.4. *If the triangulation $\mathcal{T}_h$ is regular, the operator ρ_h of orthogonal L^2-projection on Q_h satisfies the bound:*

$$\|f - \rho_h f\|_{0,\Omega} \leqslant Ch^k|f|_{k,\Omega} \quad \forall f \in H^k(\Omega) \quad \text{for } 0 \leqslant k \leqslant l. \tag{3.16}$$

Proof. Let κ be a quadrilateral of $\mathcal{T}_h$. Notice that (3.16) would be trivial if the mapping F_κ were affine instead of bilinear. So we propose to introduce another

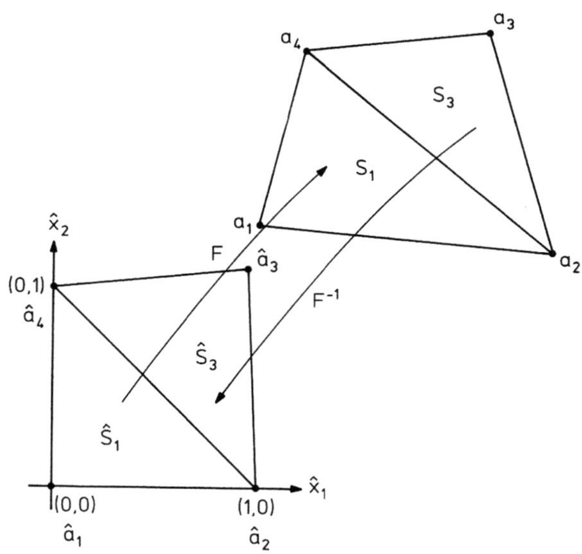

Figure 11

"reference" set $\hat{\kappa}$—not necessarily the unit square—that is related to κ by an *affine mapping* F. More precisely, let us split κ into the two subtriangles (cf. Figure 11):

$$\kappa = S_1 \cup S_3,$$

let $\hat{S}_1$ be the *reference unit triangle*, F the *affine mapping* such that

$$S_1 = F(\hat{S}_1), \qquad F(\hat{x}) = B\hat{x} + b$$

and set

$$\hat{S}_3 = F^{-1}(S_3).$$

In other words,

$$\hat{\kappa} = \hat{S}_1 \cup \hat{S}_3, \qquad \kappa = F(\hat{\kappa}).$$

As F is affine and κ is convex, the reference set $\hat{\kappa}$ is also a *convex quadrilateral*. In addition, we readily derive that on the one hand

$$\operatorname{meas}(\hat{S}_i) = (1/2)[\operatorname{meas}(S_i)/\operatorname{meas}(S_1)], \qquad 1 \leqslant i \leqslant 4$$

where $\hat{S}_i$ (resp. S_i) denotes any of the four subtriangles of $\hat{\kappa}$ (resp. κ). On the other hand any two vertices of $\hat{\kappa}$ satisfy:

$$\hat{a}_i - \hat{a}_j = B^{-1}(a_i - a_j).$$

Furthermore formulas (A.2) and (A.3) give here

$$\|B\| \leqslant (3/2)h_{s_1}, \qquad \|B^{-1}\| \leqslant \sqrt{2}/\rho_{s_1}, \qquad |\det(B)| = 2\operatorname{meas}(S_1).$$

Now, let $f \in H^k(\kappa)$ with $0 \leqslant k \leqslant l$. The L^2-projection ρ_h on P_{l-1} is invariant under affine transformations. Thus

$$\|f - \rho_h f\|_{0,\kappa} = |\det(B)|^{1/2} \|\hat{f} - \hat{\rho}\hat{f}\|_{0,\hat{\kappa}}.$$

But

$$\|\hat{f} - \hat{\rho}\hat{f}\|_{0,\hat{\kappa}} \leqslant \inf_{q \in P_j} \|\hat{f} + q\|_{0,\hat{\kappa}} = \|\hat{f}\|_{L^2(\hat{\kappa})/P_j} \qquad 0 \leqslant j \leqslant l-1.$$

Since $\hat{\kappa}$ *is a variable quadrilateral*, we must explicit the constant $C(\hat{\kappa})$ such that

$$\|\hat{f}\|_{L^2(\hat{\kappa})/P_j} \leqslant C(\hat{\kappa})|\hat{f}|_{j+1,\hat{\kappa}}.$$

This is done by induction on the degree j. When $j = 0$, which is the only case where we can use Theorem A.3, we get

$$\|\hat{f}\|_{L^2(\hat{\kappa})/P_0} \leqslant C_1 h_{\hat{\kappa}} \left[\frac{\operatorname{Max}_i \operatorname{meas}(\hat{S}_i)}{\operatorname{Min}_i \operatorname{meas}(\hat{S}_i)}\right]^{1/2} |\hat{f}|_{1,\hat{\kappa}},$$

with a constant C_1 independent of h, $\hat{\kappa}$ and $\hat{f}$. The above remarks concerning the geometry of $\hat{\kappa}$ imply that:

$$\left[\frac{\operatorname{Max}_i \operatorname{meas}(\hat{S}_i)}{\operatorname{Min}_i \operatorname{meas}(\hat{S}_i)}\right]^{1/2} \leqslant C_2 \sigma_\kappa, \qquad h_{\hat{\kappa}} \leqslant C_3 \sigma_\kappa.$$

Hence

$$\|\hat{f}\|_{L^2(\hat{K})/P_0} \leqslant C_4 \sigma_K^2 |\hat{f}|_{1,\hat{K}}. \tag{3.17}$$

Next assume that

$$\|\hat{f}\|_{L^2(\hat{K})/P_{j-1}} \leqslant (C_4 \sigma_K^2)^j |\hat{f}|_{j,\hat{K}}. \tag{3.18}$$

We can write

$$\begin{aligned}
\|\hat{f}\|_{L^2(\hat{K})/P_j} &= \inf_{(q,\tilde{q}) \in P_{j-1} \times \tilde{P}_j} \|\hat{f} + q + \tilde{q}\|_{0,\hat{K}} \\
&= \inf_{\tilde{q} \in \tilde{P}_j} \inf_{q \in P_{j-1}} \|(\hat{f} + \tilde{q}) + q\|_{0,\hat{K}} \\
&\leqslant (C_4 \sigma_K^2)^j \inf_{\tilde{q} \in \tilde{P}_j} |\hat{f} + \tilde{q}|_{j,\hat{K}}
\end{aligned}$$

by the induction hypothesis (3.18). Then (3.17) yields:

$$\|\hat{f}\|_{L^2(\hat{K})/P_j} \leqslant (C_4 \sigma_K^2)^{j+1} \left\{ \sum_{|\alpha|=j} |\partial^\alpha \hat{f}|_{1,\hat{K}}^2 \right\}^{1/2}.$$

Since the expression in brackets is $|\hat{f}|_{j+1,\hat{K}}$, this proves (3.18) for all j.

As a consequence, we have:

$$\|f - \rho_h f\|_{0,K} \leqslant (C_4 \sigma_K^2)^k |\det(B)|^{1/2} |\hat{f}|_{k,\hat{K}}$$

and (3.16) follows from (A.7) and the regularity of $\mathcal{T}_h$. □

From Lemma 3.4, (3.15) and Theorem 3.2, we derive the expected estimate for this scheme.

Theorem 3.3. *Let Ω be a bounded plane polygon and suppose the solution* $(\mathbf{u}, p)$ *of the Stokes system satisfies*:

$$\mathbf{u} \in [H^{k+1}(\Omega) \cap H_0^1(\Omega)]^2, \qquad p \in H^k(\Omega) \cap L_0^2(\Omega)$$

for some integer $k \in [1, l]$. If the triangulation $\mathcal{T}_h$ is regular, the solution $(\mathbf{u}_h, p_h)$ *of* (1.39) *with the spaces X_h and M_h defined by* (3.11) (3.12) *has the estimate*:

$$|\mathbf{u} - \mathbf{u}_h|_{1,\Omega} + \|p - p_h\|_{0,\Omega} \leqslant C_1 h^k \{|\mathbf{u}|_{k+1,\Omega} + |p|_{k,\Omega}\}.$$

In addition, if Ω is convex we have the L^2-estimate:

$$\|\mathbf{u} - \mathbf{u}_h\|_{0,\Omega} \leqslant C_2 h^{k+1} (|\mathbf{u}|_{k+1,\Omega} + |p|_{k,\Omega}).$$

3.3. An Example of Checkerboard Instability: the Q_1–P_0 Element

The most famous example of spaces failing to satisfy the inf-sup condition is that in which the velocity is made of piecewise polynomials of Q_1^2 and the pressure is piecewise constant on a rectangular grid. This combination is more familiarly called the "Q_1–P_0" element. More precisely, let us assume that Ω is a bounded polygon with *sides parallel to the axes* and, to simplify suppose that $\mathscr{T}_h$ is a *square grid*. We take:

$$\text{(3.19)}\qquad \begin{cases} X_h = \{\mathbf{v}_h \in \mathscr{C}^0(\bar{\Omega})^2;\ \mathbf{v}_h|_\kappa \in Q_1^2 \quad \forall\kappa\in\mathscr{T}_h,\ \mathbf{v}_h|_\Gamma = \mathbf{0}\}. \\ M_h = \{q_h \in L_0^2(\Omega);\ q_h|_\kappa \in P_0 \quad \forall\kappa\in\mathscr{T}_h\}. \end{cases}$$

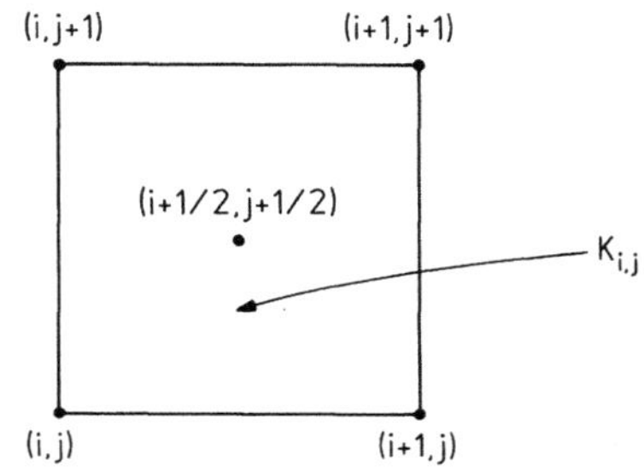

Figure 12

This pair of spaces was introduced a long time ago and because of its simplicity was used (and is still used) by many numerical analysts and engineers in connection with the Stokes Problem. But it was soon found out, through numerical instabilities in the approximate pressure, that there was something amiss with this choice of spaces.

The most conspicuous anomaly of (3.19) is that $\operatorname{Ker}(B_h')$ *is not reduced to* $\{0\}$. Indeed, let (i,j) be a cartesian enumeration of the nodes of $\mathscr{T}_h$ like in Figure 12, let $\kappa_{i,j}$ denote the square with bottom left vertex (i,j) and let $(i+1/2, j+1/2)$ be the index of the center of $\kappa_{i,j}$. To alleviate the notations, let $\mathbf{v} = (u, v)$ denote a function of X_h and let $u_{i,j}$ or $v_{i,j}$ denote the value of u or v at the node (i,j); similarly, we denote by $q_{i+1/2,j+1/2}$ the value of q at the center of $\kappa_{i,j}$. As $q \in M_h$ is constant on $\kappa_{i,j}$ we find immediately:

$$\begin{aligned}\int_{\kappa_{i,j}} q \operatorname{div}\mathbf{v}\,dx &= h^2 q_{i+1/2,j+1/2}(\operatorname{div}\mathbf{v})_{i+1/2,j+1/2} \\ &= h^2 q_{i+1/2,j+1/2}[1/(2h)]\{(u_{i+1,j+1} + u_{i+1,j} - u_{i,j+1} - u_{i,j}) \\ &\quad + (v_{i+1,j+1} + v_{i,j+1} - v_{i+1,j} - v_{i,j})\}.\end{aligned}$$

Thus a summation by parts yields:

$$\text{(3.20)}\qquad \int_\Omega q \operatorname{div}\mathbf{v}\,dx = -h^2 \sum_{i,j} \{u_{i,j}(\nabla_1 q)_{i,j} + v_{i,j}(\nabla_2 q)_{i,j}\}$$

where

$$(3.21)\quad \begin{aligned}(\nabla_1 q)_{i,j} &= [1/(2h)](q_{i+1/2,j+1/2} + q_{i+1/2,j-1/2} - q_{i-1/2,j+1/2} - q_{i-1/2,j-1/2}),\\ (\nabla_2 q)_{i,j} &= [1/(2h)](q_{i+1/2,j+1/2} + q_{i-1/2,j+1/2} - q_{i-1/2,j+1/2} - q_{i-1/2,j-1/2})\end{aligned}$$

and the summation runs over all interior nodes (i,j) of $\mathscr{T}_h$ (since $\mathbf{v}$ vanishes on Γ). Therefore, if q belongs to $\mathrm{Ker}(B_h')$, i.e. if $q \in M_h$ and

$$\int_\Omega q \operatorname{div} \mathbf{v}\, dx = 0 \qquad \forall \mathbf{v} \in X_h,$$

we must have

$$q_{i+1/2,j+1/2} = q_{i-1/2,j-1/2}, \qquad q_{i-1/2,j+1/2} = q_{i+1/2,j-1/2}.$$

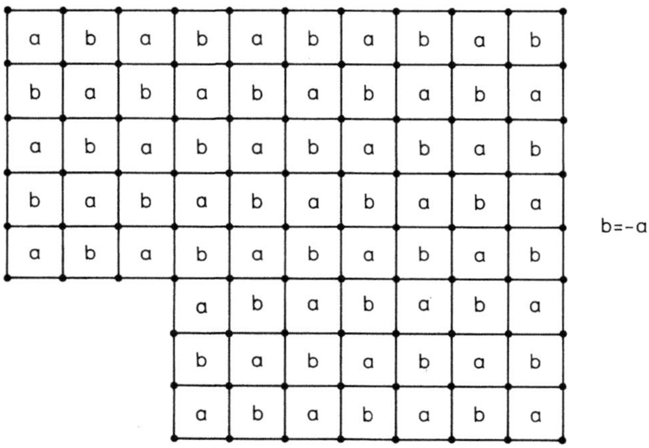

Figure 13

These equalities do not necessarily imply that q is a constant in Ω. Rather, the values of q can alternate between two constants on adjacent elements like in Figure 13. That these constants should be opposite numbers follows from the fact that $\int_\Omega q\, dx = 0$.

Let us characterize more precisely $\mathrm{Ker}(B_h')$. To simplify the discussion, it is convenient to suppose that Ω is the *square* $(-1,1) \times (-1,1)$ and that $\mathscr{T}_h$ is the even square grid with mesh size

$$h = 1/(2n)$$

and nodes

$$x_{i,j} = (ih, jh) \quad \text{with } -2n \leqslant i, j \leqslant 2n$$

(cf. Figure 14). Let $\mu \in M_h$ be defined by

$$(3.22)\qquad \mu|_{\kappa_{i,j}} = (-1)^{i+j} \qquad \forall \kappa_{i,j} \subset \mathscr{T}_h.$$

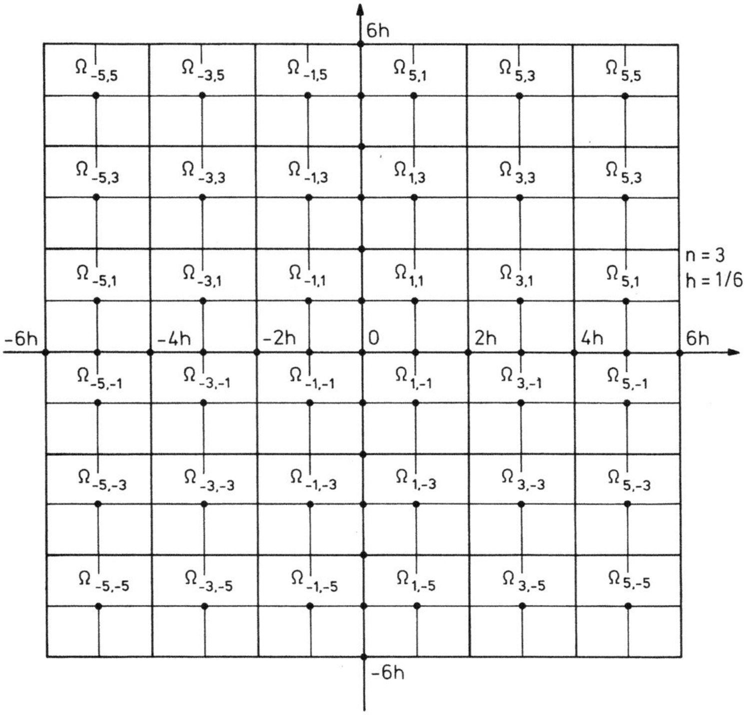

Figure 14

It stems from the above considerations that

(3.23) $$\operatorname{Ker}(B'_h) = \operatorname{span}(\mu).$$

Because of its alternate "plus and minus" pattern, the function μ is called a checkerboard function. Its connection with $\operatorname{Ker}(B'_h)$ was first reported by Fortin [28] and then by Sani et al. [70].

In view of (3.23), the pair of spaces (X_h, M_h) has no chance of satisfying the inf-sup condition. To save the situation, the first step we can take is to replace M_h by $[\operatorname{Ker}(B'_h)]^{\perp}$. Let us characterize this space. Let $I = 2i + 1$ and $J = 2j + 1$ for $-n \leqslant i, j \leqslant n - 1$ and let the macro-element $\Omega_{I,J}$ be the union of the four squares κ with common vertex (I, J). Following Johnson & Pitkäranta [46], for each (I, J) we introduce the four functions $(v_k)_{I,J}$ $1 \leqslant k \leqslant 4$ which take the value ± 1 on the subsquares of $\Omega_{I,J}$ according to the pattern of Figure 15.

Note that

(3.24) $$\int_{\Omega_{I,J}} (v_k)_{I,J}\, dx = 0 \quad \text{when } k \neq 1$$

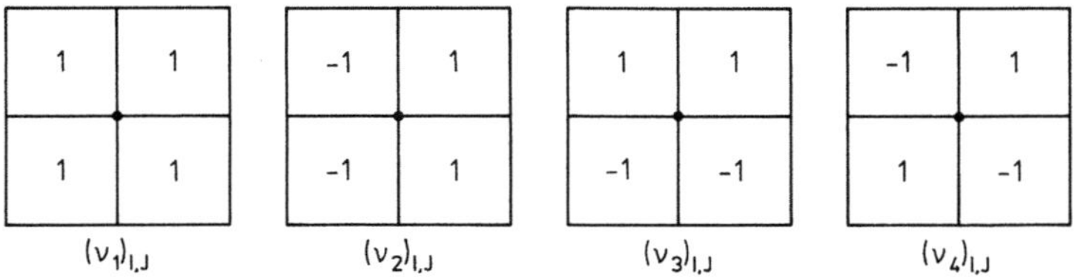

Figure 15

and

(3.25) $$\int_{\Omega_{I,J}} (v_k)_{I,J}(v_l)_{I,J}\,dx = 0 \quad \text{if } k \neq l.$$

Taking into account (3.24), it is easy to see that (3.19) defines M_h as follows:

$$M_h = \left\{q = \sum_{I,J}\sum_{k=1}^{4} (\alpha_k)_{I,J}(v_k)_{I,J};\quad \sum_{I,J}(\alpha_1)_{I,J} = 0\right\}.$$

Furthermore since the spurious function μ can only arise from the "local alternating" function v_4, we have in view of (3.25):

$$[\operatorname{Ker}(B_h')]^{\perp} = \left\{q = \sum_{I,J}\sum_{k=1}^{4} (\alpha_k)_{I,J}(v_k)_{I,J};\quad \sum_{I,J}(\alpha_1)_{I,J} = \sum_{I,J}(\alpha_4)_{I,J} = 0\right\}.$$

To simplify we use the notation:

(3.26) $$\mathscr{M}_h = [\operatorname{Ker}(B_h')]^{\perp}.$$

Since we are working with finite dimensional spaces the pair $(X_h, \mathscr{M}_h)$ satisfies the inf-sup condition (1.12). Unfortunately this is not the end of trouble for, as we are going to see below, the condition is *not uniformly satisfied with respect to h*,

Lemma 3.5. *Let Ω be like above and let the spaces X_h and $\mathscr{M}_h$ be defined by* (3.19) *and* (3.26) *respectively. There exists a constant $C > 0$, independent of h, such that*:

(3.27) $$\sup_{\mathbf{v}\in X_h}\left[\left(\int_\Omega q \operatorname{div} \mathbf{v}\,dx\right)\Big/|\mathbf{v}|_{1,\Omega}\right] \geqslant Ch\|q\|_{0,\Omega} \qquad \forall q \in \mathscr{M}_h.$$

Proof. Let q be an arbitrary function of $\mathscr{M}_h$; we introduce the discrete seminorm:

(3.28) $$|q|_{1,h} = \left(\sum_{i,j} h^2\{(\nabla_1 q)_{i,j}^2 + (\nabla_2 q)_{i,j}^2\}\right)^{1/2},$$

where the summation runs over all interior nodes (i,j) of $\mathscr{T}_h$. In view of (3.20), we define the function $\mathbf{v} = (u, v)$ of X_h by:

$$u_{i,j} = -(\nabla_1 q)_{i,j}, \qquad v_{i,j} = -(\nabla_2 q)_{i,j}$$

on all interior nodes (i,j) of $\mathcal{T}_h$. With this choice we have:

$$\int_\Omega q \operatorname{div} \mathbf{v}\, dx = |q|_{1,h}^2$$

and by virtue of Lemma A.6, an easy calculation gives:

$$|\mathbf{v}|_{1,\Omega} \leqslant (C_1/h)\|\mathbf{v}\|_{0,\Omega} \leqslant (C_2/h)|q|_{1,h}.$$

Therefore

$$\left(\int_\Omega q \operatorname{div} \mathbf{v}\, dx\right)\Big/ |\mathbf{v}|_{1,\Omega} \geqslant (h/C_2)|q|_{1,h}$$

and (3.27) will be established if we show the following analogue of Theorem I.1.9:

(3.29) $$\|q\|_{0,\Omega} \leqslant C_3 |q|_{1,h} \quad \forall q \in \mathcal{M}_h.$$

Let us prove (3.29). First a straightforward, constructive argument shows that (3.29) holds for every function q of Q_h that vanishes on two elements $\kappa_{i,j}$: one with $i+j$ even and one with $i+j$ odd. And of course the constant C_3 is independent of h and q. Next, if q belongs to $\mathcal{M}_h$ it is easy to find $\bar{q} \in \operatorname{Ker}(B_h') \oplus \mathbb{R}$ such that $q - \bar{q}$ is like above. Then the orthogonality of q and $\bar{q}$ implies that

$$\|q\|_{0,\Omega}^2 = \|q-\bar{q}\|_{0,\Omega}^2 - \|\bar{q}\|_{0,\Omega}^2 \leqslant C_3^2 |q-\bar{q}|_{1,h}^2 = C_3^2 |q|_{1,h}^2. \qquad \square$$

For a long time it was conjectured that (3.27) could not be improved; but it is only recently that Boland & Nicolaides [12] established it with the following counter-example. Roughly speaking, the idea is to find a function q in $\mathcal{M}_h$ such that

$$\left(\int_\Omega q \operatorname{div} \mathbf{v}\, dx\right)\Big/ |\mathbf{v}|_{1,\Omega}$$

is small while $\|q\|_{0,\Omega}$ is large.

More precisely let

(3.30) $$q = \sum_{I,J} \{I(v_4)_{I,J}\}.$$

On the one hand, q is indeed in $\mathcal{M}_h$ because I runs over integers of opposite signs. On the other hand, a simple calculation shows that:

$$\|q\|_{0,\Omega}^2 = 4h^2(2n)\sum_I I^2 = 4h(2n/3)(4n^2-1).$$

Thus

(3.31) $$\|q\|_{0,\Omega} = [2/(\sqrt{3}h)](1-h^2)^{1/2}.$$

Next, let us evaluate $\int_\Omega q \operatorname{div} \mathbf{v}\, dx$. According to (3.21) we have:

$$(\nabla_1 q)_{i,j} = 0 \quad \forall i,j, \quad (\nabla_2 q)_{i,j} = \begin{cases} 0 & \text{if } i \text{ is odd}, \\ (-1)^j(2/h) & \text{if } i \text{ is even}. \end{cases}$$

Therefore (3.20) yields:

$$\int_\Omega q \operatorname{div} \mathbf{v}\, dx = 2h \sum_{i=-(n-1)}^{n-1} \sum_{j=-n}^{n-1} \{v(2ih,(2j+1)h) - v(2ih, 2jh)\}$$

$$= h \sum_{i=-(n-1)}^{n-1} \left\{ \sum_{j=-n}^{n-1} \int_{2jh}^{(2j+1)h} \partial v(2ih, x_2)/\partial x_2\, dx_2 \right.$$

$$\left. - \sum_{j=-n+1}^{n} \int_{(2j-1)h}^{2jh} \partial v(2ih, x_2)/\partial x_2\, dx_2 \right\}.$$

Hence

$$\left| \int_\Omega q \operatorname{div} \mathbf{v}\, dx \right| \leqslant h \sum_{i=-(n-1)}^{n-1} \int_{-1}^{1} |\partial v(2ih, x_2)/\partial x_2|\, dx_2$$

$$\leqslant \sqrt{2h}(2n-1)^{1/2} \left\{ \int_{-1}^{1} \sum_{i=-(n-1)}^{n-1} |\partial v(2ih, x_2)/\partial x_2|^2\, dx_2 \right\}^{1/2}.$$

Now observe that for every affine function f, the following quadrature formula holds:

$$\int_0^1 f^2(x)\, dx = (1/3)\{f^2(0) + f(0)f(1) + f^2(1)\}.$$

$$\geqslant (1/4) \operatorname{Max}(f^2(0), f^2(1)),$$

in view of the inequality

$$ab \leqslant (1/4)a^2 + b^2 \qquad \forall a, b \geqslant 0.$$

As a consequence

$$\sum_{i=-(n-1)}^{n-1} |\partial v(2ih, x_2)/\partial x_2|^2 \leqslant (2/h) \int_{-1}^{1} |\partial v(x_1, x_2)/\partial x_2|^2\, dx_1.$$

(Here we use the fact that $\partial v/\partial x_2$ is a continuous and piecewise affine function of x_1). Therefore,

$$\left| \int_\Omega q \operatorname{div} \mathbf{v}\, dx \right| \leqslant 2(1-h)^{1/2} |\mathbf{v}|_{1,\Omega}.$$

Combined with (3.31), this becomes:

$$\left| \int_\Omega q \operatorname{div} \mathbf{v}\, dx \right| \bigg/ |\mathbf{v}|_{1,\Omega} \leqslant \sqrt{3h} \| q \|_{0,\Omega}.$$

Thus we have proved the following result:

Lemma 3.6. *Under the hypotheses of Lemma* 3.5, *the function q defined by* (3.30) *belongs to* $\mathscr{M}_h$ *and satisfies:*

$$\sup_{\mathbf{v}\in X_h}\left[\left(\int_\Omega q \operatorname{div} \mathbf{v}\, dx\right)\Big/ |\mathbf{v}|_{1,\Omega}\right] \leqslant \sqrt{3}h\|q\|_{0,\Omega}.$$

Together with Lemma 3.5, this means that the constant β^* is really $O(h)$.

In fact, it can be proved that this undesirable factor h arises exclusively from the local alternating component v_4 in the functions of $\mathscr{M}_h$. Again let us write $q \in \mathscr{M}_h$ in terms of the basis functions v_k:

$$q = \sum_{k=1}^{4} q^k,$$

where

$$q^k = \sum_{I,J} (\alpha_k v_k)_{I,J}, \qquad (\alpha_k)_{I,J} \in \mathbb{R}, \qquad \sum_{I,J} (\alpha_1)_{I,J} = \sum_{I,J} (\alpha_4)_{I,J} = 0.$$

Following Boland & Nicolaides [12] we split $\mathscr{M}_h$ as follows:

$$\mathscr{M}_h = A_h + \tilde{M}_h,$$

where

$$\text{(3.32)} \qquad \begin{cases} A_h = \{q \in \mathscr{M}_h;\ q|_{\Omega_{I,J}} = (\alpha_4 v_4)_{I,J}\}, \\ \tilde{M}_h = A_h^{\perp} = \{q \in \mathscr{M}_h;\ q|_{\Omega_{I,J}} = (\alpha_1 v_1 + \alpha_2 v_2 + \alpha_3 v_3)_{I,J}\} \end{cases}$$

and we associate with these spaces the following subspace of X_h:

$$\text{(3.33)} \qquad \tilde{V}_h = \{\mathbf{v}_h \in X_h;\ (q_h, \operatorname{div} \mathbf{v}_h) = 0 \quad \forall q_h \in A_h\}.$$

We propose to establish that the pair $(\tilde{V}_h, \tilde{M}_h)$ satisfies a uniform inf-sup condition.

To this end, let us start with a local condition.

Lemma 3.7. *With the above notations and hypotheses of Lemma* 3.5, *the pair* $(\tilde{V}_h, \tilde{M}_h)$ *satisfies uniformly a local inf-sup condition with respect to the partition* $\{\Omega_{I,J}\}$ *of* $\bar{\Omega}$.

Proof. Let

$$X_h(\Omega_{I,J}) = \{\mathbf{v} \in \tilde{V}_h;\ \mathbf{v}|_{\partial\Omega_{I,J}} = \mathbf{0}\},$$
$$M_h(\Omega_{I,J}) = \{q|_{\Omega_{I,J}};\ q \in \tilde{M}_h\} \cap L_0^2(\Omega_{I,J}).$$

We must show that all $q \in M_h(\Omega_{I,J})$ satisfy:

$$\text{(3.34)} \qquad \sup_{\mathbf{v}\in X_h(\Omega_{I,J})}\left[\left(\int_{\Omega_{I,J}} q \operatorname{div} \mathbf{v}\, dx\right)\Big/ |\mathbf{v}|_{1,\Omega_{I,J}}\right] \geqslant C\|q\|_{0,\Omega_{I,J}}.$$

First, observe that

$$X_h(\Omega_{I,J}) = \{\mathbf{v} = \mathbf{v}_{I,J}\phi_{I,J};\ \forall \mathbf{v}_{I,J} = (u_{I,J}, v_{I,J}) \in \mathbb{R}^2\},$$

where $\phi_{I,J}$ denotes the basis function of X_h that takes the value 1 at the node (I, J) and 0 at all other nodes of $\mathscr{T}_h$. Similarly, we have

$$M_h(\Omega_{I,J}) = \{q = \alpha_2(\nu_2)_{I,J} + \alpha_3(\nu_3)_{I,J};\ \forall \alpha_2, \alpha_3 \in \mathbb{R}\}.$$

Then formula (3.20) yields for all $\mathbf{v} \in X_h(\Omega_{I,J})$ and $q \in M_h(\Omega_{I,J})$:

$$\int_{\Omega_{I,J}} q \operatorname{div} \mathbf{v} \, dx = -2h(\alpha_2 u_{I,J} + \alpha_3 v_{I,J}),$$

where

$$\| q \|_{0,\Omega_{I,J}} = 2h(\alpha_2^2 + \alpha_3^2)^{1/2}.$$

By choosing

$$u_{I,J} = -2h\alpha_2, \qquad v_{I,J} = -2h\alpha_3$$

we immediately obtain (3.34) with $C = (3/8)^{1/2}$. □

Thus setting,

$$\overline{M}_h = \{q \in L_0^2(\Omega);\ q|_{\Omega_{I,J}} \in \mathbb{R} \quad \forall I, J\} = \{q \in \mathscr{M}_h;\ q|_{\Omega_{I,J}} = (\alpha_1 \nu_1)_{I,J}\}$$

it follows from Theorem 1.12 that $(\tilde{V}_h, \tilde{M}_h)$ satisfies a uniform inf-sup condition provided the same is true for the pair $(\tilde{V}_h, \overline{M}_h)$. This last property is less obvious. In order to prove it, it is convenient to group the macro-elements $\Omega_{I,J}$ four by four like in Figure 16; and of course we must assume that these super macro-elements $\mathcal{O}_{\alpha,\beta}$ form again a partition of $\overline{\Omega}$. Then we proceed in two steps. First we introduce the subspace of $\overline{M}_h$:

$$\overline{M}_{2h} = \{q \in L_0^2(\Omega);\ q|_{\mathcal{O}_{\alpha,\beta}} \in \mathbb{R} \quad \forall \alpha, \beta\}$$

and we prove that the pair $(\tilde{V}_h, \overline{M}_{2h})$ satisfies a uniform inf-sup condition. Next we show that the pair $(\tilde{V}_h, \overline{M}_h)$ satisfies a local inf sup-condition on each set $\mathcal{O}_{\alpha,\beta}$. This is achieved in the next two lemmas.

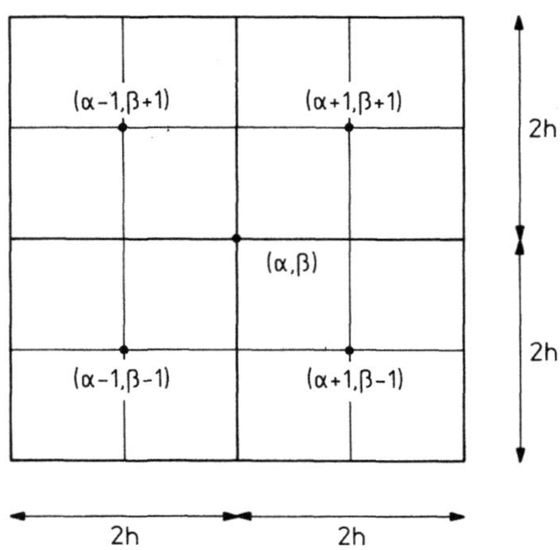

Figure 16. Super macro-element $\mathcal{O}_{\alpha,\beta}$

Lemma 3.8. *Assume that $\bar{\Omega}$ can be partitioned into groups $\mathcal{O}_{\alpha,\beta}$ of four macro-elements $\Omega_{I,J}$ like in Figure 16. Then the pair $(\tilde{V}_h, \overline{M_{2h}})$ satisfies a global uniform inf-sup condition.*

Proof. First, observe that the functions of X_{2h} belong necessarily to $\tilde{V}_h$ because their divergence reduces to polynomials of P_1 in each macro-element $\Omega_{I,J}$ and

$$\int_{\Omega_{I,J}} (v_4)_{I,J} p \, dx = 0 \qquad \forall p \in P_1.$$

Therefore let us prove the inf-sup condition for the pair $(X_{2h}, \overline{M_{2h}})$.

For this, we exhibit an adequate operator π_h very similar to that of Lemma 2.2. Let R_h be the local regularization operator of Section A.3 and let us fix one of the super macro-elements $\mathcal{O}_{\alpha,\beta}$. For v in $H^1(\mathcal{O}_{\alpha,\beta})$ we define $\pi v \in Q_1$ on each $\Omega_{I,J} \subset \mathcal{O}_{\alpha,\beta}$ by:

$$\pi v(x_{i,j}) = R_h v(x_{i,j}) \quad \text{at the four corners and center of } \mathcal{O}_{\alpha,\beta},$$

$$\int_T (\pi v - v) \, ds = 0 \quad \text{on each side } T \text{ of } \partial\mathcal{O}_{\alpha,\beta}.$$

Then, we take $\pi_h v = \pi v$ on each $\mathcal{O}_{\alpha,\beta}$.

By inspection, it easy to verify that $\pi_h \in \mathscr{L}(H_0^1(\Omega)^2; X_{2h})$ and

$$\int_\Omega \operatorname{div}(\pi_h \mathbf{v} - \mathbf{v}) q \, dx = 0 \qquad \forall q \in \overline{M_{2h}}.$$

Furthermore a simple argument shows that

$$|\pi_h \mathbf{v}|_{1,\Omega} \leqslant C |\mathbf{v}|_{1,\Omega} \quad \text{with a constant } C > 0 \text{ independent of } h.$$

This yields the desired inf-sup condition. □

Lemma 3.9. *On each $\mathcal{O}_{\alpha,\beta}$, the pair of spaces $(\tilde{V}_h, \overline{M_h})$ satisfies a local inf-sup condition.*

Proof. Let q belong to the space:

$$\overline{M_h}(\mathcal{O}_{\alpha,\beta}) = \{q^1|_{\mathcal{O}_{\alpha,\beta}}\} \cap L_0^2(\mathcal{O}_{\alpha,\beta}).$$

We must construct $\mathbf{v}$ in $\tilde{V}_h$ with $\mathbf{v}|_{\partial\mathcal{O}_{\alpha,\beta}} = \mathbf{0}$ such that

$$\left(\int_{\mathcal{O}_{\alpha,\beta}} q \operatorname{div} \mathbf{v} \, dx\right) \Big/ |\mathbf{v}|_{1,\mathcal{O}_{\alpha,\beta}} \geqslant C \|q\|_{0,\mathcal{O}_{\alpha,\beta}}. \tag{3.35}$$

Let us fix $\mathbf{v} = \mathbf{0}$ on the boundary and central nodes of $\mathcal{O}_{\alpha,\beta}$ and also at the central node of each macro-element $\Omega_{I,J}$ contained in $\mathcal{O}_{\alpha,\beta}$. Then, in view of the formula

$$\int_{\mathcal{O}_{\alpha,\beta}} q \operatorname{div} \mathbf{v} \, dx = -h^2 \sum_{i,j} \{u_{i,j}(\nabla_1 q)_{i,j} + v_{i,j}(\nabla_2 q)_{i,j}\}$$

we set

$$u_{i,j} = -h(\nabla_1 q)_{i,j}, \qquad v_{i,j} = -h(\nabla_2 q)_{i,j}$$

on all remaining nodes (i,j) of $\mathcal{O}_{\alpha,\beta}$ (i.e. $(\alpha \pm 1, \beta), (\alpha, \beta \pm 1)$). We can easily see that

$$(\nabla_k q^1)_{i,j}(\nabla_k q^4)_{i,j} = 0 \quad \text{for } k = 1, 2, \text{ on all such nodes } (i,j).$$

Hence the resulting function $\mathbf{v}$ belongs to $\tilde{V}_h$ and satisfies

$$\int_{\mathcal{O}_{\alpha,\beta}} q \operatorname{div} \mathbf{v} \, dx = h \sum_{i,j} \{|u_{i,j}|^2 + |v_{i,j}|^2\}.$$

Then (3.35) follows from the inequality:

$$|\mathbf{v}_h|_{1,\mathcal{O}_{\alpha,\beta}} \leqslant C_1 \left(\sum_{i,j} \{|u_{i,j}|^2 + |v_{i,j}|^2\} \right)^{1/2},$$

with a constant C_1 independent of h, α and β, and

$$\sum_{i,j} \{|u_{i,j}|^2 + |v_{i,j}|^2\} \geqslant 2 \sum_{I,J} |q_{I,J}|^2$$

considering that $\sum_{I,J} q_{I,J} = 0$ since $q \in L_0^2(\mathcal{O}_{\alpha,\beta})$. □

Lemmas 3.7, 3.8 and 3.9 yield immediately the next result.

Theorem 3.4. *Assume that $\bar{\Omega}$ can be partitioned into groups $\mathcal{O}_{\alpha,\beta}$ of four macro-elements like in Figure* 16. *Then the pair $(\tilde{V}_h, \tilde{M}_h)$ defined by* (3.33) *and* (3.32) *satisfies a uniform inf-sup condition.*

Remark 3.1. The argument of Lemma 3.8 can be used directly to show that the pair $(X_h, \bar{M}_h)$ satisfies a uniform inf-sup condition but this does not imply that $(\tilde{V}_h, \bar{M}_h)$ satisfies it as well.

Remark 3.2. The above analysis does not apply directly to arbitrary quadrilaterals. Usually, the "checkerboard" spurious pressure disappears from M_h but the inf-sup condition is not satisfied (cf. Sani et al [70]). However, it is possible to derive similar results for special quadrilateral meshes (cf. Pitkäranta & Stenberg [65]).

3.4. Error Estimates for the Q_1–P_0 Element

The object of this section is to show that, although it does not satisfy the inf-sup condition, the pair of spaces (X_h, M_h) can still be used to compute successfully the velocity $\mathbf{u}$ and (with some precautions) the pressure p. For this purpose, the statement of Theorem 3.4 will play a crucial role.

Let $(\mathbf{u}_h, p_h) \in X_h \times M_h$ be a solution of:

$$(3.26)\qquad \begin{cases} \nu(\mathbf{grad}\,\mathbf{u}_h, \mathbf{grad}\,\mathbf{v}_h) - (p_h, \operatorname{div}\mathbf{v}_h) = \langle \mathbf{f}, \mathbf{v}_h\rangle & \forall \mathbf{v}_h \in X_h, \\ (q_h, \operatorname{div}\mathbf{u}_h) = 0 & \forall q_h \in M_h, \end{cases}$$

with X_h and M_h defined by (3.19). We know that $\mathbf{u}_h$ is unique but that each p_h is of the form:

$$p_h = \tilde{p}_h + p_h^4 + C\mu$$

with μ defined by (3.22), p_h^4 and $\tilde{p}_h$ uniquely determined in $\tilde{A}_h$ and $\tilde{M}_h$ respectively and C arbitrary. Furthermore, $\mathbf{u}_h \in \tilde{V}_h$ with $\tilde{V}_h$ defined by (3.33) and the pair $(\mathbf{u}_h, \tilde{p}_h) \in \tilde{V}_h \times \tilde{M}_h$ is the unique solution of:

$$(3.37)\qquad \begin{cases} \nu(\mathbf{grad}\,\mathbf{u}_h, \mathbf{grad}\,\tilde{\mathbf{v}}_h) - (\tilde{p}_h, \operatorname{div}\tilde{\mathbf{v}}_h) = \langle \mathbf{f}, \tilde{\mathbf{v}}_h\rangle & \forall \tilde{\mathbf{v}}_h \in \tilde{V}_h, \\ (\tilde{q}_h, \operatorname{div}\mathbf{u}_h) = 0 & \forall \tilde{q}_h \in \tilde{M}_h. \end{cases}$$

Therefore, owing to Theorem 3.4, we can apply straight away Theorem 1.1 2°) with $\tilde{V}_h$ and $\tilde{M}_h$ instead of X_h and M_h respectively:

$$(3.38)\qquad \begin{cases} |\mathbf{u} - \mathbf{u}_h|_{1,\Omega} + \|p - \tilde{p}_h\|_{0,\Omega} \\ \leqslant C_1 \left\{ \inf\limits_{\tilde{\mathbf{v}}_h \in \tilde{V}_h} |\mathbf{u} - \tilde{\mathbf{v}}_h|_{1,\Omega} + \inf\limits_{\tilde{q}_h \in \tilde{M}_h} \|p - \tilde{q}_h\|_{0,\Omega} \right\} \end{cases}$$

with a constant $C_1 > 0$ independent of h.

Hence it remains to investigate the approximation properties of the spaces $\tilde{V}_h$ and $\tilde{M}_h$. As far as $\tilde{V}_h$ is concerned, recall that (cf. Lemma 3.8):

$$X_{2h} \subset \tilde{V}_h.$$

Thus, formula (A.49) yields:

$$(3.39)\qquad \inf_{\tilde{\mathbf{v}}_h \in \tilde{V}_h} |\mathbf{u} - \tilde{\mathbf{v}}_h|_{1,\Omega} \leqslant |\mathbf{u} - I_{2h}\mathbf{u}|_{1,\Omega} \leqslant C_2 h |\mathbf{u}|_{2,\Omega} \qquad \forall \mathbf{u} \in H^2(\Omega)^2.$$

Likewise, since $\overline{M}_h \subset \tilde{M}_h$, formula (A.51) gives:

$$(3.40)\qquad \inf_{\tilde{q}_h \in \tilde{M}_h} \|p - \tilde{q}_h\|_{0,\Omega} \leqslant \|p - \rho_{2h} p\|_{0,\Omega} \leqslant C_3 h |p|_{1,\Omega} \qquad \forall p \in H^1(\Omega).$$

These three inequalities are combined in the following theorem.

Theorem 3.5. *Assume that Ω is like in Theorem* 3.4 *and suppose the solution* $(\mathbf{u}, p)$ *of the Stokes system satisfies*:

$$\mathbf{u} \in [H^2(\Omega) \cap H_0^1(\Omega)]^2, \qquad p \in H^1(\Omega) \cap L_0^2(\Omega).$$

Then the solution $(\mathbf{u}_h, p_h)$ *of the scheme* (3.36) *has the error estimate*:

$$(3.41)\qquad |\mathbf{u} - \mathbf{u}_h|_{1,\Omega} + \|p - \tilde{p}_h\|_{0,\Omega} \leqslant C h \{|\mathbf{u}|_{2,\Omega} + |p|_{1,\Omega}\},$$

where $\tilde{p}_h$ is the component of p_h in $\tilde{M}_h$.

Here the component $\tilde{p}_h$ *acts as a filter* for the pressure p_h since it discards entirely the alternating functions v_4. It is clear, from Lemma 3.6, that no satisfying estimate can be expected for the whole of p_h. However, we are going to see that p_h^4, the supplementary component of p_h in A_h is bounded. Indeed, it stems from (3.36) that

$$(p_h^4, \operatorname{div} \mathbf{v}_h) = \nu(\mathbf{grad}(\mathbf{u}_h - \mathbf{u}), \mathbf{grad}\, \mathbf{v}_h) + (p - \tilde{p}_h, \operatorname{div} \mathbf{v}_h) \qquad \forall \mathbf{v}_h \in X_h.$$

Therefore,

$$\sup_{\mathbf{v}_h \in X_h} \left[\frac{(p_h^4, \operatorname{div} \mathbf{v}_h)}{|\mathbf{v}_h|_{1,\Omega}} \right] \leqslant C_4 h \{ |\mathbf{u}|_{2,\Omega} + |p|_{1,\Omega} \}.$$

It may happen, in the best of cases, that the left-hand side of this inequality is bounded below by $C_5 \| p_h^4 \|_{0,\Omega}$ with a constant C_5 that does not depend upon h and thus $\| p_h^4 \|_{0,\Omega}$ is $O(h)$. However, in the general case, all we can do is apply Lemma 3.5; it yields the next result.

Corollary 3.1. *Under the hypotheses of Theorem* 3.5, *the component* p_h^4 *of* p_h *in* A_h *is bounded as follows*:

(3.42) $$\| p_h^4 \|_{0,\Omega} \leqslant C \{ |\mathbf{u}|_{2,\Omega} + |p|_{1,\Omega} \}.$$

Finally, we can also apply Theorem 1.2 with the pair of spaces $(\tilde{V}_h, \tilde{M}_h)$ and derive an optimal error estimate for $\| \mathbf{u} - \mathbf{u}_h \|_{0,\Omega}$.

Corollary 3.2. *Under the hypotheses of Theorem* 3.5, *we have*:

(3.43) $$\| \mathbf{u} - \mathbf{u}_h \|_{0,\Omega} \leqslant C h^2 \{ |\mathbf{u}|_{2,\Omega} + |p|_{1,\Omega} \}.$$

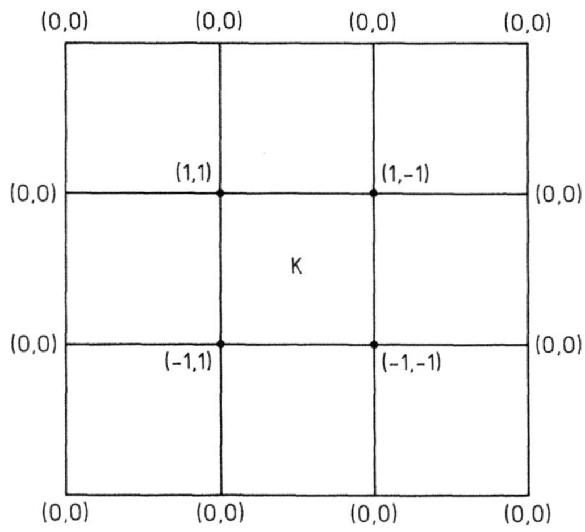

Figure 17

Numerical results confirm this theoretical analysis but it appears most often that it is the *entire pressure* p_h *that converges towards* p—not the component $\tilde{p}_h$ alone—. This renders the filtering of the pressure component p_h^4 seldom necessary. There are cases, though, where p_h^4 does diverge; the reader can refer to Boland & Nicolaides [12] for a specific example. Besides that, the reader will find in Malkus & Olsen [55] other examples of currently used finite element spaces which do not satisfy a uniform inf-sup condition.

From a practical point of view, Problem (3.36) can be solved by the penalty method of Section 1.3 (cf. (1.52)). Numerical results can be found for example in Carey & Krishnan [16]. But it is also possible to decouple directly the velocity from the pressure by using a basis of "divergence-free" velocities, i.e. a basis of V_h. Following Stephens *et al* [76] we introduce the vector field $\mathbf{v}_\kappa \in X_h$ that takes the values $(1, -1)$, $(1, 1)$, $(-1, 1)$, $(-1, -1)$ at the four vertices of κ like in Figure 17 and $(0, 0)$ at all other nodes of $\mathscr{T}_h$; then we define the set

$$S = \{\mathbf{v}_\kappa;\ \text{for all interior elements } \kappa \text{ of } \mathscr{T}_h\}.$$

By inspection, we can easily ascertain that each $\mathbf{v}_\kappa \in S$ belongs to V_h and that all these functions are linearly independent. In addition, a simple dimension argument yields:

$$\operatorname{card}(S) = \dim(X_h) - \dim(\mathscr{M}_h) = \dim(V_h).$$

Hence S is a convenient basis of V_h. The reader will find numerical results with this basis in the above reference.

§4. Continuous Approximation of the Pressure

So far, we have used approximate pressures that were (generally) discontinuous across interelement boundaries. But from the engineering point of view, continuous pressures are more natural because the pressures encountered in practice are usually continuous functions. The fact is that numerical analysts found Stokes solvers with $\mathscr{C}^0$ pressures more difficult to analyze than those with L^2 pressures. This difficulty accounts for the relatively meager literature on the theory of important schemes of the "Hood-Taylor" type.

The first rigorous error analysis of the very popular Hood & Taylor [44] scheme is due to Bercovier & Pironneau [7]. Later on this analysis was cleverly simplified by Verfürth [83]; but now the approach of Section 1.4 permits to insert directly the Hood-Taylor method into the framework of § 1.

Apart from the Hood-Taylor and closely related schemes, this paragraph studies an interesting variant introduced by Glowinski & Pironneau [38] which approximates the Stokes problem by a sequence of discrete Dirichlet problems for $-\Delta$.

4.1. A First Order Method: the "Mini" Finite Element

With minor modifications, the setting of the problem is that of Section 1.3. We take for Ω a bounded, plane polygon and we assume that the Stokes system

$$\text{(4.1)}\qquad \begin{cases} \left.\begin{aligned} -\nu\Delta\mathbf{u} + \mathbf{grad}\, p = \mathbf{f} \\ \operatorname{div}\mathbf{u} = 0 \end{aligned}\right\} \text{ in } \Omega, \\ \mathbf{u} = \mathbf{0} \quad \text{on } \Gamma \end{cases}$$

is such that:

$$p \text{ belongs to } H^1(\Omega).$$

Following Arnold *et al* [2], we construct a triangulation $\mathscr{T}_h$ of $\bar{\Omega}$ and we approximate the velocity on each element κ by a polynomial of

$$\text{(4.2)}\qquad \mathscr{P}_1(\kappa) = [P_1 \oplus \operatorname{span}\{\lambda_1\lambda_2\lambda_3\}]^2$$

and the pressure by a polynomial of P_1. Thus we choose the following finite element spaces:

$$\text{(4.3)}\qquad X_h = \{\mathbf{v}\in\mathscr{C}^0(\bar{\Omega})^2;\ \mathbf{v}|_\kappa\in\mathscr{P}_1(\kappa)\quad \forall\kappa\in\mathscr{T}_h,\ \mathbf{v}|_\Gamma = \mathbf{0}\},$$

$$\text{(4.4)}\qquad Q_h = \{q\in\mathscr{C}^0(\bar{\Omega});\ q|_\kappa\in P_1\quad \forall\kappa\in\mathscr{T}_h\},\ M_h = Q_h\cap L_0^2(\Omega).$$

The degrees of freedom are the simplest ones, namely the values of the *velocity* at the *vertices* and *center* of κ and the values of the *pressure* at the *vertices* of κ. As usual, the space V_h is defined by:

$$V_h = \{\mathbf{v}_h\in X_h;\ (\operatorname{div}\mathbf{v}_h, q_h) = 0\quad \forall q_h\in Q_h\}$$

and the approximate problem, called *Problem* (Q_h) reads:

Find a pair $(\mathbf{u}_h, p_h)$ *in* $X_h\times M_h$ *satisfying*:

$$\text{(4.5)}\qquad \begin{cases} a(\mathbf{u}_h,\mathbf{v}_h) - (p_h, \operatorname{div}\mathbf{v}_h) = \langle\mathbf{f},\mathbf{v}_h\rangle \quad \forall\mathbf{v}_h\in X_h, \\ (\operatorname{div}\mathbf{u}_h, q_h) = 0 \quad \forall q_h\in Q_h, \end{cases}$$

where the bilinear form $a(.,.)$ is unchanged:

$$a(\mathbf{u},\mathbf{v}) = 2\nu\sum_{i,j}^{2}(D_{ij}(\mathbf{u}), D_{ij}(\mathbf{v}))$$

or

$$a(\mathbf{u},\mathbf{v}) = \nu(\mathbf{grad}\,\mathbf{u}, \mathbf{grad}\,\mathbf{v}).$$

Note that because the space Q_h is contained in $H^1(\Omega)$, the bilinear form $b(.,.)$ can be written equivalently as:

$$b(\mathbf{v}_h, q_h) = -(\operatorname{div}\mathbf{v}_h, q_h) = (\mathbf{grad}\, q_h, \mathbf{v}_h)\quad \forall\mathbf{v}_h\in X_h,\quad \forall q_h\in Q_h.$$

The approximation properties of X_h and Q_h are well known. For instance, if the triangulation $\mathscr{T}_h$ is regular, the interpolation operator r_h defined by:

$$(4.6)\qquad \begin{cases} r_h\mathbf{v}(a) = \mathbf{v}(a) \quad \text{on all nodes } a \text{ of } \mathscr{T}_h, \\ r_h\mathbf{v}(a_\kappa) = \mathbf{v}(a_\kappa) \quad \text{on the center } a_\kappa \text{ of } \kappa, \quad \forall\kappa\in\mathscr{T}_h, \\ r_h\mathbf{v}\in\mathscr{P}_1(\kappa) \quad \text{on each } \kappa, \end{cases}$$

satisfies

$$r_h\in\mathscr{L}([H^2(\Omega)\cap H_0^1(\Omega)]^2;\, X_h)$$

and

$$(4.7)\qquad \|\mathbf{v} - r_h\mathbf{v}\|_{m,\Omega} \leqslant Ch^{2-m}|\mathbf{v}|_{2,\Omega} \quad \forall\mathbf{v}\in H^2(\Omega)^2, \quad m = 0 \text{ or } 1.$$

Indeed, r_h is preserved by affine transformations on each κ and leaves invariant the polynomials of P_1^2.

Likewise, the local regularization operator R_h on P_1 defined by (A.53), (A.54) satisfies $R_h\in\mathscr{L}(L^2(\Omega); Q_h)$ and

$$(4.8)\qquad \|q - R_hq\|_{m,\Omega} \leqslant Ch^{1-m}|q|_{1,\Omega} \quad \forall q\in H^1(\Omega), \quad m = 0 \text{ or } 1,$$

provided of course that $\mathscr{T}_h$ is regular.

Therefore the Hypotheses H1 and H2 are fulfilled and it remains to verify H3, namely the inf-sup condition:

$$(4.9)\qquad \sup_{\mathbf{v}_h\in X_h}\frac{(q_h, \operatorname{div}\mathbf{v}_h)}{|\mathbf{v}_h|_{1,\Omega}} \geqslant \beta^*\|q_h\|_{0,\Omega} \quad \forall q_h\in M_h.$$

This is achieved by the next lemma.

Lemma 4.1. *If the triangulation $\mathscr{T}_h$ is regular, the pair of spaces (X_h, M_h) defined by* (4.3) (4.4) *satisfies* (4.9) *with a constant $\beta^* > 0$ independent of h.*

Proof. Let us exhibit the operator π_h of Lemma 1.1. Take an arbitrary q_h in M_h. Since $q_h\in L_0^2(\Omega)$ there exists $\mathbf{v}$ in $H_0^1(\Omega)^2$ such that

$$(4.10)\qquad \operatorname{div}\mathbf{v} = q_h, \quad |\mathbf{v}|_{1,\Omega} \leqslant C_1\|q_h\|_{0,\Omega}.$$

Therefore, since $M_h \subset H^1(\Omega)$ we want to construct a function $\pi_h\mathbf{v}$ in X_h that satisfies

$$\sum_{\kappa\in\mathscr{T}_h}\int_\kappa \pi_h\mathbf{v}\cdot\mathbf{grad}\,\mu_h\,dx = \sum_{\kappa\in\mathscr{T}_h}\int_\kappa \mathbf{v}\cdot\mathbf{grad}\,\mu_h\,dx \quad \forall\mu_h\in M_h.$$

As $\mathbf{grad}\,\mu_h\in P_0^2$ on each κ, this equality induces us to define $\pi_h\mathbf{v}$ in X_h such that:

$$(4.11)\qquad \pi_h\mathbf{v}(a) = (R_h\mathbf{v})(a) \quad \forall\text{node } a \text{ of } \mathscr{T}_h,$$

and

(4.12)
$$\int_K \pi_h \mathbf{v}\,dx = \int_K \mathbf{v}\,dx \quad \forall \kappa \in \mathscr{T}_h,$$

where R_h denotes the now familiar local regularization operator on P_1^2.

Clearly, (4.11) and (4.12) determine uniquely $\pi_h \mathbf{v}$ in X_h and $\pi_h \in \mathscr{L}(H_0^1(\Omega)^2; X_h)$. Moreover

$$\int_\Omega \operatorname{div}(\pi_h \mathbf{v} - \mathbf{v})\mu_h\,dx = 0 \quad \forall \mu_h \in M_h.$$

Finally an argument similar to that of Lemma 2.2 shows that

$$|\pi_h \mathbf{v}|_{1,\Omega} \leqslant C_2 |\mathbf{v}|_{1,\Omega}$$

provided $\mathscr{T}_h$ is regular. This proves the lemma. □

These results are summarized in the following theorem.

Theorem 4.1. *Let Ω be a bounded plane polygon and let the solution* $(\mathbf{u}, p)$ *of the Stokes system* (4.1) *satisfy*:

$$\mathbf{u} \in [H^2(\Omega) \cap H_0^1(\Omega)]^2, \quad p \in H^1(\Omega) \cap L_0^2(\Omega).$$

If the triangulation $\mathscr{T}_h$ is regular, the solution $(\mathbf{u}_h, p_h)$ *of Problem* (4.5) *with the spaces X_h and M_h defined by* (4.3) *and* (4.4) *respectively satisfies the error bound*:

(4.13)
$$|\mathbf{u} - \mathbf{u}_h|_{1,\Omega} + \|p - p_h\|_{0,\Omega} \leqslant C_1 h\{|\mathbf{u}|_{2,\Omega} + |p|_{1,\Omega}\}.$$

In addition, when Ω is convex, we have the L^2-estimate:

(4.14)
$$\|\mathbf{u} - \mathbf{u}_h\|_{0,\Omega} \leqslant C_2 h^2\{|\mathbf{u}|_{2,\Omega} + |p|_{1,\Omega}\}.$$

This "mini" finite element method can easily be generalized to schemes of arbitrary order. The details can be found in Arnold *et al* [2].

4.2. The "Hood-Taylor" Finite Element Method

The results of the preceding section can be improved by taking a more accurate approximation of the velocity. Following Hood & Taylor [44], we keep the same space Q_h and we replace X_h by:

(4.15)
$$X_h = \{\mathbf{v} \in \mathscr{C}^0(\bar{\Omega})^2;\ \mathbf{v}|_\kappa \in P_2^2 \quad \forall \kappa \in \mathscr{T}_h,\ \mathbf{v}|_\Gamma = \mathbf{0}\},$$

with the function values at the principal lattice of order 2 (cf. (A.19)) as degrees of freedom. Therefore the standard interpolation operator I_h satisfies:

$$I_h \in \mathscr{L}([H^k(\Omega) \cap H_0^1(\Omega)]^2; X_h),$$

(4.16)
$$\|\mathbf{v} - I_h \mathbf{v}\|_{m,\Omega} \leqslant C h^{k-m} |\mathbf{v}|_{k,\Omega} \quad \forall \mathbf{v} \in H^k(\Omega)^2, \quad k = 2 \text{ or } 3, \quad m = 0 \text{ or } 1.$$

The remainder of this section is devoted to the proof of the inf-sup condition. We propose to establish it first locally and then extend it by Theorem 1.12. The reader can also refer to Verfürth [83] for a direct global proof.

Here, the difficulty in a local argument lies in the choice of an adequate partition of Ω. We propose to group together all the elements which share a common vertex like in Figure 18. More precisely, we make the following assumption on the triangulation $\mathscr{T}_h$:

$$\text{(4.17)}\quad \begin{cases} \mathscr{T}_h \text{ has a set of interior nodes } \{a_r\}_{r=1}^{R} \text{ such that } \{\Omega_r\}_{r=1}^{R} \text{ with} \\ \qquad \Omega_r = \overbrace{\bigcup_{\kappa \text{ has vertex } a_r} \kappa}^{\circ} \\ \text{is a partition of } \bar{\Omega}. \end{cases}$$

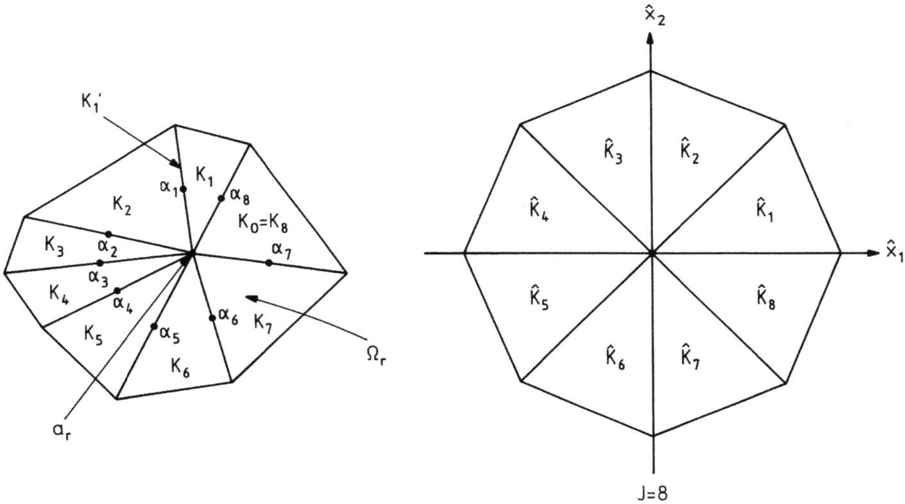

Figure 18

If this assumption holds, each element κ of $\mathscr{T}_h$ belongs to exactly one macro-element Ω_r. In addition, the fact that all the nodes a_r are *inside* Ω implies that each element κ has *exactly one side* on the boundary of its macro-element Ω_r and at *most one side* on the boundary Γ of Ω.

In practice, it is not difficult to construct a triangulation that satisfies (4.17). The usual procedure is to start with a coarse grid and then progressively refine it by adding interior nodes.

Recall the spaces defined by (1.54):

$$
(4.18) \qquad \begin{cases} X_h(\Omega_r) = \{\mathbf{v}_h \in X_h;\ \operatorname{supp}(\mathbf{v}_h) \subset \Omega_r\}, \\ Q_h(\Omega_r) = \{q|_{\Omega_r};\ q \in Q_h\}, \\ M_h(\Omega_r) = Q_h(\Omega_r) \cap L_0^2(\Omega_r). \end{cases}
$$

Let us prove that the pair $(X_h(\Omega_r), M_h(\Omega_r))$ satisfies a local inf-sup condition.

Theorem 4.2. *Suppose that $\mathcal{T}_h$ is a regular triangulation of $\bar{\Omega}$ and that $\mathcal{T}_h$ satisfies* (4.17). *Then there exists a constant $\lambda^* > 0$, independent of h and r, such that:*

$$
(4.19) \qquad \sup_{\mathbf{v} \in X_h(\Omega_r)} \left[\left(\int_{\Omega_r} q \operatorname{div} \mathbf{v}\, dx \right) \Big/ |\mathbf{v}|_{1,\Omega_r} \right] \geq \lambda^* \|q\|_{0,\Omega_r} \quad \forall q \in M_h(\Omega_r).
$$

Proof. Let J be the number of elements κ in Ω_r and let us number them with an index i ranging from 0 to J such that κ_i is adjacent to κ_{i-1} and κ_{i+1} and $\kappa_0 = \kappa_J$ (like in Figure 18):

$$
\Omega_r = \overbrace{\bigcup_{i=1}^{J} \kappa_i}^{\circ}.
$$

We denote by κ_i' the side shared by κ_i and κ_{i+1}, by α_i the midpoint of κ_i' and by a_r the vertex common to all the κ_i in Ω_r.

Like in Section A.3, we associate with Ω_r the reference set:

$$
\hat{\Omega} = \bigcup_{i=1}^{J} \hat{\kappa}_i \quad \text{(cf. Figure 18)}
$$

through the continuous, piecewise affine function F_r defined by:

$$
F_r(\hat{\kappa}_i) = \kappa_i, \qquad F_r(\hat{x}) = B_i\hat{x} + b_i \qquad \forall \hat{x} \in \hat{\kappa}_i.
$$

Since the triangulation is *regular*, the number J is bounded above by a fixed constant I independent of r and as a consequence there are at most I different reference sets $\hat{\Omega}$. This means that all geometrical constants related to $\hat{\Omega}$ and $\hat{\kappa}_i$ can be bounded independently of h and r.

Now let q_h be an arbitrary element of $Q_h(\Omega_r)$ and let $\mathbf{v}_h$ be a function in $X_h(\Omega_r)$ that satisfies $\mathbf{v}_h(a_r) = \mathbf{0}$. Since $\mathbf{v}_h$ vanishes on $\partial\Omega_r$ and q_h belongs to $H^1(\Omega_r)$ we have:

$$
\int_{\Omega_r} \operatorname{div} \mathbf{v}_h q_h\, dx = -\sum_{i=1}^{J} \int_{\kappa_i} \mathbf{v}_h \cdot \mathbf{grad}\, q_h\, dx.
$$

Observe that each component v of $\mathbf{v}_h$ is a polynomial of P_2 on κ_i that vanishes at the vertices of κ_i. Hence the following quadrature formula holds:

$$
\int_{\kappa_i} v\, dx = \operatorname{meas}(\kappa_i)(1/3)\{v(\alpha_i) + v(\alpha_{i-1})\}.
$$

As $\mathbf{grad}\, q_h$ is constant on each κ_i (say $\mathbf{grad}\, q_h|_{\kappa_i} = \mathbf{g}_i$), this formula yields:

$$
(4.20) \qquad \int_{\Omega_r} \operatorname{div} \mathbf{v}_h q_h\, dx = -(1/3) \sum_{i=1}^{J} \operatorname{meas}(\kappa_i)\{\mathbf{v}(\alpha_i) + \mathbf{v}(\alpha_{i-1})\} \cdot \mathbf{g}_i.
$$

Next, remark that $\partial q_h/\partial\tau$ is continuous at interelements boundaries. This suggests to choose

$$\mathbf{v}(\alpha_i) = -(\mathbf{g}_i \cdot \mathbf{t}_i)\mathbf{t}_i = -(\mathbf{g}_{i+1} \cdot \mathbf{t}_i)\mathbf{t}_i$$

where $\mathbf{t}_i$ is the tangent vector to κ_i' with length $\|\kappa_i'\|$ and (say) pointing outside Ω_r. With this choice we obtain

$$\int_{\Omega_r} \operatorname{div} \mathbf{v}_h q_h \, dx = (1/3) \sum_{i=1}^{J} \operatorname{meas}(\kappa_i)\{(\mathbf{g}_i \cdot \mathbf{t}_i)^2 + (\mathbf{g}_i \cdot \mathbf{t}_{i-1})^2\}.$$

But according to (A.9), $\mathbf{g}\cdot\mathbf{t}$ is preserved by affine transformations. Thus, with obvious notations we can write:

$$\int_{\Omega_r} \operatorname{div} \mathbf{v}_h q_h \, dx = (1/3) \sum_{i=1}^{J} \operatorname{meas}(\kappa_i)\{(\hat{\mathbf{g}}_i \cdot \hat{\mathbf{t}}_i)^2 + (\hat{\mathbf{g}}_i \cdot \hat{\mathbf{t}}_{i-1})^2\}.$$

Clearly, each set of vectors $\{\hat{\mathbf{t}}_{i-1}, \hat{\mathbf{t}}_i\}$ is a basis on the reference space. Therefore the mapping $\mathbf{g} \to \{(\mathbf{g}\cdot\hat{\mathbf{t}}_{i-1})^2 + (\mathbf{g}\cdot\hat{\mathbf{t}}_i)^2\}^{1/2}$ is equivalent to the Euclidean norm on the reference space. Hence, in view of

$$|\hat{q}|_{1,\hat{\kappa}_i}^2 = \operatorname{meas}(\hat{\kappa}_i)\,\|\hat{\mathbf{g}}_i\|^2,$$

there exists a constant $\hat{C}_1 > 0$ such that:

$$\int_{\Omega_r} \operatorname{div} \mathbf{v}_h q_h \, dx \geqslant \hat{C}_1 \sum_{i=1}^{J} \operatorname{meas}(\kappa_i)|\hat{q}|_{1,\hat{\kappa}_i}^2. \tag{4.21}$$

Next, on the one hand the definition of $\mathbf{v}_h$ yields:

$$\begin{aligned} \|\mathbf{v}_h\|_{0,\kappa_i}^2 &\leqslant \hat{C}_2 \operatorname{meas}(\kappa_i)\{\|\mathbf{v}(\alpha_{i-1})\|^2 + \|\mathbf{v}(\alpha_i)\|^2\} \\ &\leqslant \hat{C}_3 \operatorname{meas}(\kappa_i)[h_{\kappa_i}|\hat{q}|_{1,\hat{\kappa}_i}]^2. \end{aligned} \tag{4.22}$$

And on the other hand, the argument of Lemma A.6 gives:

$$\begin{aligned} |\mathbf{v}_h|_{1,\kappa_i}^2 &\leqslant \hat{C}_4[\|B_i^{-1}\|\,\|\mathbf{v}_h\|_{0,\kappa_i}]^2 \\ &\leqslant \hat{C}_5 \operatorname{meas}(\kappa_i)[\sigma_{\kappa_i}|\hat{q}|_{1,\hat{\kappa}_i}]^2 \end{aligned} \tag{4.23}$$

by virtue of (A.2) and (4.22). Hence it stems from (4.21), (4.23) and the regularity of $\mathscr{T}_h$ that

$$\int_{\Omega_r} \operatorname{div} \mathbf{v}_h q_h \, dx \geqslant \hat{C}_6 (1/\sigma)|\mathbf{v}_h|_{1,\Omega_r}\left\{\sum_{i=1}^{J} \operatorname{meas}(\kappa_i)|\hat{q}|_{1,\hat{\kappa}_i}^2\right\}^{1/2}. \tag{4.24}$$

It remains to show that, on a regular triangulation,

$$\left\{\sum_{i=1}^{J} \operatorname{meas}(\kappa_i)|\hat{q}|_{1,\hat{\kappa}_i}^2\right\}^{1/2} \geqslant \hat{C}\|q\|_{0,\Omega_r} \qquad \forall q \in H^1(\Omega_r) \cap L_0^2(\Omega_r).$$

The corresponding proof is a simple variant of that of Lemma 2.5. To begin with, observe that the composition with F_r maps $H^1(\Omega_r)$ into $H^1(\hat{\Omega})$ but does not preserve the zero mean value. This small difficulty can be handled by replacing

$q \in H^1(\Omega_r) \cap L_0^2(\Omega_r)$ by $\bar{q}$ where

$$\hat{\bar{q}} = \hat{q} - (1/\text{meas}(\hat{\Omega})) \int_{\hat{\Omega}} \hat{q}\, d\hat{x}.$$

Then q and $\bar{q}$ differ by a constant and we have:

$$\|q\|_{0,\Omega_r} = \inf_{c \in \mathbb{R}} \|q + c\|_{0,\Omega_r} \leqslant \|\bar{q}\|_{0,\Omega_r}.$$

But

$$\begin{aligned}
\|\bar{q}\|_{0,\Omega_r}^2 &= \sum_{i=1}^{J} \text{meas}(\kappa_i) \|\hat{\bar{q}}\|_{0,\hat{\kappa}_i}^2 \\
&\leqslant \hat{C}_7 \sup_{1 \leqslant i \leqslant J} (h_{\kappa_i}^2) |\hat{\bar{q}}|_{1,\hat{\Omega}}^2 \quad \text{since } \hat{\bar{q}} \in H^1(\hat{\Omega}) \cap L_0^2(\hat{\Omega}). \\
&\leqslant \hat{C}_8 \left\{ \sup_{1 \leqslant i \leqslant J} (h_{\kappa_i}^2) \Big/ \inf_{1 \leqslant i \leqslant J} (\rho_{\kappa_i}^2) \right\} \sum_{i=1}^{J} \text{meas}(\kappa_i) |\hat{q}|_{1,\hat{\kappa}_i}^2.
\end{aligned}$$

Hence

$$\left\{ \sum_{i=1}^{J} \text{meas}(\kappa_i) |\hat{q}|_{1,\hat{\kappa}_i}^2 \right\}^{1/2} \geqslant (\hat{C}_9/\sigma_r) \|q\|_{0,\Omega_r}, \tag{4.25}$$

where

$$\sigma_r = \sup_{1 \leqslant i \leqslant J} (h_{\kappa_i}) \Big/ \inf_{1 \leqslant i \leqslant J} (\rho_{\kappa_i}).$$

This finishes the proof because the regularity of $\mathcal{T}_h$ implies that (cf. Bernardi [9]):

$$\sigma_r \leqslant \hat{C}\sigma. \tag{4.26}$$

□

Owing to Theorem 1.12, the local inf-sup condition (4.19) yields readily the required global condition.

Corollary 4.1. *Under the assumptions of Theorem* 4.2 *the pair of spaces* (X_h, M_h) *defined by* (4.15) (4.4) *satisfies the inf-sup condition* (4.9) *with a constant* $\beta^* > 0$ *independent of* h.

Proof. Let $\bar{X}_h$ be the finite element space defined by (2.3) and let

$$\bar{M}_h = \{q \in L_0^2(\Omega);\ q|_{\Omega_r} \text{ is constant } \forall r\}.$$

Then on the one hand, $\bar{X}_h \subset X_h$ because the functions of $\bar{X}_h$ are piecewise incomplete polynomials of P_2^2. On the other hand, it follows from Lemma 2.2 that the pair $(\bar{X}_h, \bar{M}_h)$ satisfies a uniform inf-sup condition since the functions of $\bar{M}_h$ are a particular case of piecewise constants. Hence the result follows from Theorems 1.12 and 4.2. □

Remark 4.1. Most technical details in the proof of Theorem 4.2 have already been used by Bercovier & Pironneau [7] in establishing a weak form of the inf-sup condition (4.9). Albeit simple, the above proof is long because it deals with a reference region composed of several reference triangles instead of a single one. The crucial point in the proof is the particular choice of $\mathbf{v}$ at the midpoint of the interior segments. Apart from that, the major steps are essentially the same as those used in proving Theorem 2.2.

With Theorem 4.2 and its corollary we readily derive the major result of this section.

Theorem 4.3. *Let Ω be a bounded, plane polygon and let the solution $(\mathbf{u}, p)$ of the Stokes system* (4.1) *satisfy*

$$\mathbf{u} \in [H^{k+1}(\Omega) \cap H_0^1(\Omega)]^2, \qquad p \in H^k(\Omega) \cap L_0^2(\Omega), \qquad k = 1 \text{ or } 2.$$

If the triangulation $\mathscr{T}_h$ is regular and like in (4.17), *the solution $(\mathbf{u}_h, p_h)$ of Problem* (4.5) *with spaces X_h and M_h defined by* (4.15) (4.4) *satisfies the estimate*:

$$(4.27)\qquad |\mathbf{u} - \mathbf{u}_h|_{1,\Omega} + \|p - p_h\|_{0,\Omega} \leqslant C_1 h^k \{|\mathbf{u}|_{k+1,\Omega} + |p|_{k,\Omega}\}, \qquad k = 1 \text{ or } 2.$$

When Ω is convex, this can be refined:

$$(4.28)\qquad \|\mathbf{u} - \mathbf{u}_h\|_{0,\Omega} \leqslant C_2 h^{k+1} \{|\mathbf{u}|_{k+1,\Omega} + |p|_{k,\Omega}\}.$$

Furthermore, if $\mathscr{T}_h$ is uniformly regular (but Ω not necessarily convex) we also have:

$$(4.29)\qquad |p - p_h|_{1,\Omega} \leqslant C_3 h^{k-1} \{|\mathbf{u}|_{k+1,\Omega} + |p|_{k,\Omega}\}.$$

Remark 4.2. In order to establish (4.29) we can apply Corollary 4.1 and switch from $\|p_h - q_h\|_{0,\Omega}$ to $|p_h - q_h|_{1,\Omega}$ by Corollary A.3 (this is where the uniformity of $\mathscr{T}_h$ steps in).

Remark 4.3. Obviously, Remark 2.6 is also valid here.

We finish this section with a quick study of a popular variant of the Hood-Taylor method. Again, let $\mathscr{T}_h$ be a regular triangulation of $\bar{\Omega}$ that satisfies (4.17) and let us divide each of its triangles κ into four equal triangles $\tilde{\kappa}$ by joining the midpoints of the sides (cf. Figure 19). For the pressure, we retain the spaces Q_h and M_h defined by (4.4) and we replace the velocity space X_h by:

$$(4.30)\qquad \begin{aligned} X_h = \{&\mathbf{v} \in \mathscr{C}^0(\bar{\Omega})^2; \mathbf{v} \in P_1^2 \text{ on each subtriangle } \tilde{\kappa} \text{ of } \kappa \quad \forall \kappa \in \mathscr{T}_h, \\ &\mathbf{v}|_\Gamma = \mathbf{0}\}. \end{aligned}$$

The approximation properties of Q_h and M_h are unchanged while the approximation properties of X_h correspond to polynomials of P_1, i.e. we have:

$$(4.31)\qquad \|\mathbf{v} - I_h \mathbf{v}\|_{m,\Omega} \leqslant C h^{2-m} |\mathbf{v}|_{2,\Omega} \qquad \forall v \in H^2(\Omega)^2, \qquad m = 0 \text{ or } 1$$

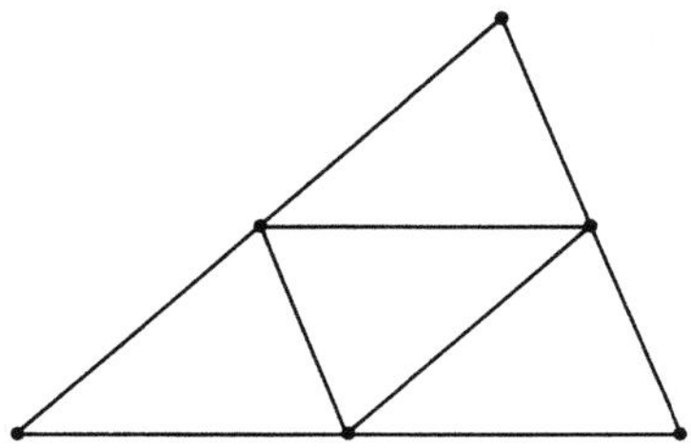

Figure 19. Triangle K divided into four subtriangles

where $I_h \in \mathscr{L}([H^2(\Omega) \cap H_0^1(\Omega)]^2; X_h)$ is the standard interpolator at the vertices of each subtriangle $\tilde{\kappa}$ of $\mathscr{T}_h$.

The proof of the inf-sup condition is almost exactly like above. We take the *same* partition (4.17) and observe that the space $X_h(\Omega_r)$ with X_h defined by (4.30) involves exactly the *same degrees of freedom* as if the space X_h were defined by (4.15). Just the degree of the polynomials varies and this affects only the factor 1/3 in the quadrature formula (4.20) and the reference constants $\hat{C}$ in subsequent formulas. Hence the *statement of Theorem* 4.2 *carries over here without modification.*

To switch from the local inf-sup condition to the global inf-sup condition, we choose the same space $\overline{M}_h$ but we take

$$\overline{X}_h = X_h.$$

It is easy to prove that the pair of spaces $(X_h, \overline{M}_h)$ satisfies a uniform inf-sup condition. This is in fact part of a more general result.

Lemma 4.2. *If the triangulation $\mathscr{T}_h$ is regular, the spaces X_h and*

$$\{q \in L_0^2(\Omega);\ q \text{ is constant on } \kappa \quad \forall \kappa \in \mathscr{T}_h\}$$

satisfy a uniform inf-sup condition.

We skip the proof since it is included in the argument of Lemma 3.8: it suffices to adapt the definition of the restriction operator π to the present case.

Summarizing, we have the analogue of Theorem 4.3.

Theorem 4.4. *Let Ω and $\mathscr{T}_h$ satisfy the hypotheses of Theorem* 4.3 *and suppose the solution* $(\mathbf{u}, p)$ *of* (4.1) *belongs to* $H^2(\Omega)^2 \times [H^1(\Omega) \cap L_0^2(\Omega)]$. *Then the solution* $(\mathbf{u}_h, p_h)$ *of Problem* (4.5) *with X_h and M_h defined by* (4.30) (4.4) *has the error bound:*

$$(4.32) \qquad |\mathbf{u} - \mathbf{u}_h|_{1,\Omega} + \|p - p_h\|_{0,\Omega} \leqslant C_1 h\{|\mathbf{u}|_{2,\Omega} + |p|_{1,\Omega}\}.$$

Moreover, when Ω is convex we have the L^2-estimate:

$$(4.33) \qquad \|\mathbf{u} - \mathbf{u}_h\|_{0,\Omega} \leqslant C_2 h^2\{|\mathbf{u}|_{2,\Omega} + |p|_{1,\Omega}\}.$$

Remark 4.4. Although it is less accurate than the preceding scheme while involving the same number of unknowns, this last method is often preferred because it leads to better conditioned linear systems.

4.3. The "Glowinski-Pironneau" Finite Element Method

To begin with, let the dimension N be two or three. The numerical scheme discussed in this section, introduced by Glowinski & Pironneau [38], is based on a Poisson equation for the pressure. By taking the divergence of both sides of the equation:

$$-\nu\Delta\mathbf{u} + \mathbf{grad}\, p = \mathbf{f}$$

and taking into account the condition $\operatorname{div}\mathbf{u} = 0$, we obtain:

$$\Delta p = \operatorname{div}\mathbf{f} \quad \text{in } \Omega.$$

Hence, if we know the trace $\rho = p|_\Gamma$ of p on Γ, the Stokes equations reduce to $N + 1$ Dirichlet problems for the Laplace operator:

$$\Delta p = \operatorname{div}\mathbf{f} \quad \text{in } \Omega, \qquad p = \rho \quad \text{on } \Gamma, \tag{4.34}$$

$$\nu\Delta\mathbf{u} = \mathbf{grad}\, p - \mathbf{f} \quad \text{in } \Omega, \qquad \mathbf{u} = \mathbf{0} \quad \text{on } \Gamma. \tag{4.35}$$

In fact, ρ is the major unknown of the problem. We shall show that ρ can be in turn determined by the constraint $\operatorname{div}\mathbf{u} = 0$.

More precisely, observe that $\mathbf{u}$ and p can be split into two components:

$$\mathbf{u} = \mathbf{u}^0 + \mathbf{u}(\rho), \qquad p = p^0 + p(\rho), \tag{4.36}$$

where p^0 and $\mathbf{u}^0$ are the solutions of the Dirichlet problems:

$$\Delta p^0 = \operatorname{div}\mathbf{f} \quad \text{in } \Omega, \qquad p^0 = 0 \quad \text{on } \Gamma, \tag{4.37a}$$

$$\nu\Delta\mathbf{u}^0 = \mathbf{grad}\, p^0 - \mathbf{f} \quad \text{in } \Omega, \qquad \mathbf{u}^0 = \mathbf{0} \quad \text{on } \Gamma, \tag{4.37b}$$

and, for each boundary value g, $p(g)$ and $\mathbf{u}(g)$ are the solutions of:

$$\Delta p(g) = 0 \quad \text{in } \Omega, \qquad p(g) = g \quad \text{on } \Gamma, \tag{4.38a}$$

$$\nu\Delta\mathbf{u}(g) = \mathbf{grad}\, p(g) \quad \text{in } \Omega, \qquad \mathbf{u}(g) = \mathbf{0} \quad \text{on } \Gamma. \tag{4.38b}$$

The space G of the boundary functions g is chosen so that the mapping $g \to p(g)$ defined by (4.38a) is an isomorphism from G onto the space

$$\{q \in L^2(\Omega);\ \Delta q = 0\}.$$

Then, since the solution $\mathbf{u}$ of (4.34) (4.35) must satisfy

$$\Delta(\operatorname{div}\mathbf{u}) = 0 \quad \text{in } \Omega,$$

the constraint $\operatorname{div}\mathbf{u} = 0$ can be expressed by:

$$(\operatorname{div} \mathbf{u}, p(g)) = 0 \quad \forall g \in G$$

or equivalently by

$$(4.39) \qquad (\operatorname{div} \mathbf{u}(\rho), p(g)) = -(\operatorname{div} \mathbf{u}^0, p(g)) \quad \forall g \in G.$$

We are going to see below that, for a proper choice of the space G, the equation (4.39) defines a unique boundary function ρ.

In order to choose G, let us put Problem (4.38a) in variational form. Assuming for the moment that the function $p(g)$ is smooth enough, Green's formula yields:

$$(4.40) \qquad \int_\Omega p(g) \Delta\mu \, dx = \int_\Gamma g \partial\mu/\partial n \, ds \quad \forall \mu \in H^2(\Omega) \cap H_0^1(\Omega).$$

When the boundary Γ is $\mathscr{C}^{1,1}$, we know from Theorem I.1.6 that the mapping $\mu \to \partial\mu/\partial n$ is continuous from $H^2(\Omega) \cap H_0^1(\Omega)$ onto $H^{1/2}(\Gamma)$. When Γ is a plane polygon, made of segments Γ_j for $1 \leqslant j \leqslant J$, Remark I.1.1 asserts that the mapping $\mu \to (\partial\mu/\partial n_j; 1 \leqslant j \leqslant J)$ is continuous from $H^2(\Omega) \cap H_0^1(\Omega)$ onto $\prod H^{1/2}(\Gamma_j)$. These considerations suggest the following choice for G:

$$(4.41) \qquad G = \begin{cases} [H^{1/2}(\Gamma)]' = H^{-1/2}(\Gamma) & \text{if } \Gamma \text{ is } \mathscr{C}^{1,1}, \\ \left[\prod_{1 \leqslant j \leqslant J} H^{1/2}(\Gamma_j)\right]' & \text{if } \Gamma \text{ is a two-dimensional polygon} \end{cases}$$

equipped with the usual dual norm which, for the sake of simplicity, we denote in both cases by $\|.\|_{-1/2,\Gamma}$. Then, in either case, when g and $p(g)$ are related by (4.40) we have:

$$\begin{aligned} \|g\|_{-1/2,\Gamma} &\leqslant C_1 \sup_{\mu \in H^2(\Omega) \cap H_0^1(\Omega)} \left(\frac{1}{\|\mu\|_{2,\Omega}} \int_\Gamma g \partial\mu/\partial n \, ds \right) \\ &\leqslant \sqrt{N} C_1 \|p(g)\|_{0,\Omega}. \end{aligned}$$

Conversely, if either Γ is $\mathscr{C}^{1,1}$ or if Ω is a convex polygon, it stems from Remark I.1.2 that for all q in $L^2(\Omega)$, the problem:

$$\Delta\mu = q \quad \text{in } \Omega, \qquad \mu = 0 \quad \text{on } \Gamma$$

has a unique solution μ in $H^2(\Omega) \cap H_0^1(\Omega)$ and $\|\mu\|_{2,\Omega} \leqslant C_2 \|q\|_{0,\Omega}$. Hence Problem (4.40) has a unique solution $p(g)$ and

$$\begin{aligned} \|p(g)\|_{0,\Omega} &\leqslant C_2 \sup_{\mu \in H^2(\Omega) \cap H_0^1(\Omega)} \left(\frac{1}{\|\mu\|_{2,\Omega}} \int_\Omega p(g) \Delta\mu \, dx \right) \\ &\leqslant C_3 \|g\|_{-1/2,\Gamma}. \end{aligned}$$

Moreover, using the fact that $\mathscr{D}(\bar{\Omega})$ is dense in the space

$$L(\Delta; \Omega) = \{q \in L^2(\Omega); \Delta q \in L^2(\Omega)\}$$

we can readily define the trace mappping γ: $L(\Delta; \Omega) \to H^{-1/2}(\Gamma)$ and establish the following Green's formula for all $p \in L(\Delta; \Omega)$:

$$(\Delta p, \mu) = (p, \Delta \mu) - \int_\Gamma (\gamma p) \partial \mu / \partial n \, ds \qquad \forall \mu \in H^2(\Omega) \cap H_0^1(\Omega).$$

Collecting these results, we find that Problems (4.38a) and (4.40) are equivalent, have a unique solution $p(g)$ for each g in G and

$$(4.42) \qquad [1/(\sqrt{N} C_1)] \|g\|_{-1/2,\Gamma} \leqslant \|p(g)\|_{0,\Omega} \leqslant C_3 \|g\|_{-1/2,\Gamma} \qquad \forall g \in G.$$

Next, we write Problem (4.38b) in variational form:

$$(4.43) \qquad \nu(\mathbf{grad}\, \mathbf{u}(g), \mathbf{grad}\, \mathbf{v}) = (p(g), \operatorname{div} \mathbf{v}) \qquad \forall \mathbf{v} \in H_0^1(\Omega)^N.$$

From this and Corollary I.2.4 we derive immediately that

$$(4.44) \qquad (C_1/\nu) \|p(g)\|_{L^2(\Omega)/\mathbb{R}} \leqslant |\mathbf{u}(g)|_{1,\Omega} \leqslant (\sqrt{N}/\nu) \|p(g)\|_{L^2(\Omega)/\mathbb{R}}.$$

Thus, combining (4.42) and (4.44) and using the fact that

$$p(g + c) = p(g) + c$$

we obtain

$$(4.45) \qquad \frac{C_4}{\nu \sqrt{N} C_1} \inf_{c \in \mathbb{R}} \|g + c\|_{-1/2,\Gamma} \leqslant |\mathbf{u}(g)|_{1,\Omega} \leqslant (\sqrt{N}/\nu) C_3 \inf_{c \in \mathbb{R}} \|g + c\|_{-1/2,\Gamma}$$

Finally, by substituting (4.43) into (4.39) we derive:

$$\nu(\mathbf{grad}\, \mathbf{u}(\rho), \mathbf{grad}\, \mathbf{u}(l)) = -(\operatorname{div} \mathbf{u}^0, p(l)) \qquad \forall l \in G.$$

In other words, with the notation

$$(4.46) \qquad a(\rho, l) = \nu(\mathbf{grad}\, \mathbf{u}(\rho), \mathbf{grad}\, \mathbf{u}(l)),$$

Problem (4.39) reads:

Find ρ in $G/\mathbb{R}$ such that:

$$(4.47) \qquad a(\rho, l) = -(\operatorname{div} \mathbf{u}^0, p(l)) \qquad \forall l \in G/\mathbb{R}.$$

Clearly, in view of (4.44) and (4.45) this problem has a unique solution. Note also that the bilinear form $a(.,.)$ is symmetric.

The above results are summarized in the following theorem.

Theorem 4.5. *Let $N = 2$ or 3. Assume that Ω is bounded with either a $\mathscr{C}^{1,1}$ boundary or a polygonal boundary with no reentrant corners. For $\mathbf{f}$ given in $L^2(\Omega)^N$, the solution $(\mathbf{u}, p)$ of the Stokes system* (4.1) *can be split into:*

$$\mathbf{u} = \mathbf{u}^0 + \mathbf{u}(\rho), \qquad p = p^0 + p(\rho),$$

where ρ is the unique solution of Problem (4.47) *and the pairs $(\mathbf{u}^0, p^0)$, $(\mathbf{u}(\rho), p(\rho))$ are the only solutions of* (4.37) *and* (4.38) *respectively.*

For the sake of simplicity, let us restrict now the discussion to the case $N = 2$. The Glowinski-Pironneau scheme is a very straightforward approximation of Problems (4.37) and (4.38), on a polygonal domain Ω, with the Hood-Taylor finite element spaces for the velocity and pressure:

$$X_h \text{ defined by (4.15)}, \qquad Q_h \text{ defined by (4.4)}.$$

The space $G/\mathbb{R}$ is represented by:

$$G_h = \left\{ q_h \in Q_h;\ q_h(a) = 0 \quad \forall \text{node } a \text{ of } \mathscr{T}_h \cap \Omega,\ \int_\Gamma q_h \, ds = 0 \right\}. \tag{4.48}$$

Observe that, on the one hand the support of the functions of G_h is a neighborhood of Γ. On the other hand, the additive constant of these functions is fixed by the condition $\int_\Gamma q_h \, ds = 0$. In addition, we introduce the space

$$\Phi_h = Q_h \cap H_0^1(\Omega). \tag{4.49}$$

Note that we have the decomposition:

$$\left\{ q_h \in Q_h;\ \int_\Gamma q_h \, ds = 0 \right\} = \Phi_h \oplus G_h.$$

With these spaces, Problems (4.37) and (4.38) are discretized as follows:

Find $p_h^0 \in \Phi_h$ *such that*:

$$(\mathbf{grad}\, p_h^0, \mathbf{grad}\, q_h) = (\mathbf{f}, \mathbf{grad}\, q_h) \quad \forall q_h \in \Phi_h; \tag{4.50}$$

Find $\mathbf{u}_h^0 \in X_h$ *such that*:

$$\nu(\mathbf{grad}\, \mathbf{u}_h^0, \mathbf{grad}\, \mathbf{v}_h) = -(\mathbf{grad}\, p_h^0, \mathbf{v}_h) + (\mathbf{f}, \mathbf{v}_h) \quad \forall \mathbf{v}_h \in X_h;$$

For g_h *given in* G_h, *find* $p_h(g_h) \in Q_h$ *such that*:

$$(\mathbf{grad}\, p_h(g_h), \mathbf{grad}\, q_h) = 0 \quad \forall q_h \in \Phi_h,$$

$$p_h(g_h) - g_h = 0 \quad \text{on } \Gamma; \tag{4.51}$$

Find $\mathbf{u}_h(g_h) \in X_h$ *such that*:

$$\nu(\mathbf{grad}\, \mathbf{u}_h(g_h), \mathbf{grad}\, \mathbf{v}_h) = -(\mathbf{grad}\, p_h(g_h), \mathbf{v}_h) \quad \forall \mathbf{v}_h \in X_h.$$

Finally, the boundary function ρ is discretized by the analogue of (4.47):

Find $\rho_h \in G_h$ *satisfying*:

$$\nu(\mathbf{grad}\, \mathbf{u}_h(\rho_h), \mathbf{grad}\, \mathbf{u}_h(l_h)) = (\mathbf{u}_h^0, \mathbf{grad}\, p_h(l_h)) \quad \forall l_h \in G_h. \tag{4.52}$$

Then the approximate velocity and pressure calculated by the Glowinski-Pironneau scheme are:

$$\mathbf{u}_h = \mathbf{u}_h^0 + \mathbf{u}_h(\rho_h), \qquad p_h = p_h^0 + p_h(\rho_h), \tag{4.53}$$

where $(\mathbf{u}_h^0, p_h^0)$ is the solution of (4.50), ρ_h is the solution of (4.52) and $(\mathbf{u}_h(\rho_h), p_h(\rho_h))$ is the solution of (4.51) for this ρ_h.

To stress the parallel with the continuous case, we set

$$a_h(g_h, l_h) = \nu(\mathbf{grad}\,\mathbf{u}_h(g_h), \mathbf{grad}\,\mathbf{u}_h(l_h)) \qquad \forall g_h, l_h \in G_h.$$

Clearly, Problem (4.50) has a unique solution, and so does Problem (4.51) for a given g_h. Moreover, it is easy to check that Problem (4.52) has also a unique solution ρ_h. Indeed, if $\mathbf{u}_h(\rho_h) = \mathbf{0}$ then

$$(\mathbf{grad}\, p_h(\rho_h), \mathbf{v}_h) = 0 \qquad \forall \mathbf{v}_h \in X_h.$$

But the inf-sup condition established by Corollary 4.1 implies that

$$p_h(\rho_h) = (1/\text{meas}(\Omega)) \int_\Omega p_h(\rho_h)\, dx,$$

i.e. p_h is constant in Ω. As $p_h(\rho_h)$ also satisfies $\int_\Gamma p_h(\rho_h)\, ds = 0$ we conclude that $p_h(\rho_h) = 0$ and in particular, $\rho_h = 0$. Therefore we have the following result:

Lemma 4.3. *Let the right-hand side* $\mathbf{f}$ *belong to* $L^2(\Omega)^2$ *where* Ω *is a plane, bounded polygon and assume that the triangulation* $\mathscr{T}_h$ *is like in* (4.17). *Then the Glowinski-Pironneau scheme* (4.50) ... (4.53) *determines a unique pair* $(\mathbf{u}_h, p_h)$ *with* $\mathbf{u}_h$ *in* X_h *and* p_h *in* Q_h, $\int_\Gamma p_h\, ds = 0$.

Moreover, the pair $(\mathbf{u}_h, p_h)$ *satisfies:*

$$\begin{aligned} &\nu(\mathbf{grad}\,\mathbf{u}_h, \mathbf{grad}\,\mathbf{v}_h) + (\mathbf{grad}\, p_h, \mathbf{v}_h - \mathbf{grad}\, q_h) \\ &\quad = (\mathbf{f}, \mathbf{v}_h - \mathbf{grad}\, q_h) \qquad \forall (\mathbf{v}_h, q_h) \in X_h \times \Phi_h. \end{aligned} \tag{4.54}$$

Note that (4.54) amounts to two independent equations: one for $\mathbf{u}_h$ and one for p_h. They are obtained by combining the last equations (resp. the first equations) of (4.50) and (4.51).

It is important to point out that, although the finite element spaces coincide with those of the Hood-Taylor method and (4.54) with $q_h = 0$ is satisfied in both cases, the above pair $(\mathbf{u}_h, p_h)$ is not, in general, the solution of the Hood-Taylor algorithm because it does not satisfy the discrete divergence-free constraint:

$$(\mathbf{u}_h, \mathbf{grad}\, q_h) = 0 \qquad \forall q_h \in Q_h.$$

Indeed, it follows from (4.51) that we have:

$$\begin{aligned} (\mathbf{grad}\, p_h(g_h), \mathbf{u}_h) &= -\nu(\mathbf{grad}\,\mathbf{u}_h(g_h), \mathbf{grad}\,\mathbf{u}_h) \\ &= -\nu(\mathbf{grad}\,\mathbf{u}_h(g_h), \mathbf{grad}(\mathbf{u}_h^0 + \mathbf{u}_h(\rho_h))) \\ &= -(\mathbf{u}_h^0, \mathbf{grad}\, p_h(g_h)) - \nu(\mathbf{grad}\,\mathbf{u}_h(g_h), \mathbf{grad}\,\mathbf{u}_h^0) \end{aligned}$$

by virtue of (4.52). Hence, another application of (4.51) shows that we have:

$$(\mathbf{grad}\, p_h(g_h), \mathbf{u}_h) = 0 \qquad \forall g_h \in G_h.$$

Unlike the continuous case, this equality does not necessarily carry over to all q_h in Q_h. In this sense, the *Glowinski-Pironneau algorithm relaxes the divergence-free constraint*. In fact, we must add something to $\mathbf{u}_h$ in order to verify this constraint. To be specific, if we define λ_h in Φ_h by

$$(\mathbf{grad}\,\lambda_h, \mathbf{grad}\,q_h) = (\mathbf{u}_h, \mathbf{grad}\,q_h) \quad \forall q_h \in \Phi_h \tag{4.55}$$

then the sum $\mathbf{u}_h - \mathbf{grad}\,\lambda_h$ does satisfy:

$$(\mathbf{u}_h - \mathbf{grad}\,\lambda_h, \mathbf{grad}\,q_h) = 0 \quad \forall q_h \in Q_h.$$

Indeed, let $q_h \in Q_h$ with $\int_\Gamma q_h\,ds = 0$ and let $g_h \in G_h$ denote the boundary value of q_h:

$$q_h|_\Gamma = g_h|_\Gamma.$$

Then q_h has the orthogonal decomposition (with respect to $|.|_{1,\Omega}$):

$$q_h = q_h(g_h) + q_h^0,$$

where $q_h(g_h)$ is the solution of (4.51) and $q_h^0 \in \Phi_h$. We have:

$$\begin{aligned}(\mathbf{u}_h, \mathbf{grad}\,q_h) &= (\mathbf{u}_h, \mathbf{grad}\,q_h^0)\\ &= (\mathbf{grad}\,\lambda_h, \mathbf{grad}\,q_h^0) \quad \text{by (4.55)}\\ &= (\mathbf{grad}\,\lambda_h, \mathbf{grad}\,q_h) \quad \text{in view of (4.51).}\end{aligned}$$

We shall see below that λ_h is indeed small, so that $\mathbf{u}_h$ is nearly "divergence-free". In addition, although λ_h has only been introduced here for a theoretical purpose, it will prove to be useful in practice for solving efficiently (4.52).

Remark 4.5. The triple $(\mathbf{u}_h, p_h, \lambda_h)$ can also be introduced directly as the unique solution in $X_h \times (Q_h/\mathbb{R}) \times \Phi_h$ of:

$$\nu(\mathbf{grad}\,\mathbf{u}_h, \mathbf{grad}\,\mathbf{v}_h) + (\mathbf{grad}\,p_h, \mathbf{v}_h - \mathbf{grad}\,q_h) = (\mathbf{f}, \mathbf{v}_h - \mathbf{grad}\,q_h)$$
$$\forall (\mathbf{v}_h, q_h) \in X_h \times \Phi_h,$$
$$(\mathbf{u}_h - \mathbf{grad}\,\lambda_h, \mathbf{grad}\,q_h) = 0 \quad \forall q_h \in Q_h.$$

But the advantage of the formulations (4.50)...(4.53) is that it appears as the solution of a sequence of decoupled Dirichlet problems for the Laplace operator.

The error analysis closely resembles that of the corresponding Problem (4.5). In particular, its inf-sup condition is precisely (4.9) and therefore, Theorem 4.2 and its corollary are valid.

Theorem 4.6. *Let Ω be a bounded, convex, plane polygon and suppose the right-hand side* $\mathbf{f}$ *of the Stokes Problem* (4.1) *belongs to* $L^2(\Omega)^2$. *If the solution* $(\mathbf{u}, p)$ *of* (4.1) *has the regularity*:

$$\mathbf{u} \in H^{k+1}(\Omega)^2, \quad p \in H^k(\Omega) \cap L_0^2(\Omega) \quad \textit{for } k = 1, 2$$

and if the triangulation $\mathcal{T}_h$ is uniformly regular and like in (4.17), *we have the error estimates*:

$$|\mathbf{u} - \mathbf{u}_h|_{1,\Omega} + |\lambda_h|_{1,\Omega} + \|p - \bar{p}_h\|_{0,\Omega} \leqslant C_1 h^k \{|\mathbf{u}|_{k+1,\Omega} + |p|_{k,\Omega}\}, \tag{4.56}$$

(4.57) $$|p - p_h|_{1,\Omega} \leqslant C_2 h^{k-1}\{|\mathbf{u}|_{k+1,\Omega} + |p|_{k,\Omega}\},$$

(4.58) $$\|\mathbf{u} - \mathbf{u}_h + \mathbf{grad}\,\lambda_h\|_{0,\Omega} \leqslant C_3 h^{k+1}\{|\mathbf{u}|_{k+1,\Omega} + |p|_{k,\Omega}\},$$

where $(\mathbf{u}_h, p_h)$ *is the solution of* (4.50)...(4.53), λ_h *is given by* (4.55) *and* $\bar{p}_h$ *is the representative of* p_h *in* $L_0^2(\Omega)$.

Proof. We have:

(4.59) $$\begin{cases} \nu(\mathbf{grad}(\mathbf{u} - \mathbf{u}_h), \mathbf{grad}\,\mathbf{v}_h) + (\mathbf{grad}(p - p_h), \mathbf{v}_h - \mathbf{grad}\,q_h) = 0 \\ \qquad\qquad \forall (\mathbf{v}_h, q_h) \in X_h \times \Phi_h, \\ (\mathbf{u}_h - \mathbf{grad}\,\lambda_h, \mathbf{grad}\,q_h) = 0 \quad \forall q_h \in Q_h. \end{cases}$$

Let us restrict the pair $(\mathbf{v}_h, q_h)$ to the space:

$$B_h = \{(\mathbf{v}_h, q_h) \in X_h \times \Phi_h;\ (\mathbf{v}_h - \mathbf{grad}\,q_h, \mathbf{grad}\,\mu_h) = 0 \quad \forall \mu_h \in Q_h\}.$$

Note that $(\mathbf{u}_h, \lambda_h) \in B_h$. Then (4.59) reads:

$$\nu(\mathbf{grad}(\mathbf{u} - \mathbf{u}_h), \mathbf{grad}\,\mathbf{v}_h) + (\mathbf{grad}(p - \mu_h), \mathbf{v}_h - \mathbf{grad}\,q_h) = 0$$
$$\forall (\mathbf{v}_h, q_h) \in B_h, \quad \forall \mu_h \in Q_h.$$

To get rid of $\mathbf{grad}\,q_h$, we choose $\mu_h = P_h p$, the H^1-projection of p on M_h defined by (A.25). Hence

$$\nu(\mathbf{grad}(\mathbf{u} - \mathbf{u}_h), \mathbf{grad}\,\mathbf{v}_h) = (\operatorname{div} \mathbf{v}_h, p - P_h p) \quad \forall (\mathbf{v}_h, q_h) \in B_h.$$

As $(\mathbf{u}_h, \lambda_h) \in B_h$, this equation readily implies:

$$|\mathbf{u} - \mathbf{u}_h|_{1,\Omega} \leqslant 2|\mathbf{u} - \mathbf{w}_h|_{1,\Omega} + (\sqrt{2}/\nu)\,\|p - P_h p\|_{0,\Omega} \qquad \forall (\mathbf{w}_h, q_h) \in B_h.$$

Clearly, we may choose here the pair $(\mathbf{w}_h, 0)$ with $\mathbf{w}_h$ in V_h and since the spaces (X_h, M_h) satisfy the inf-sup condition (4.9) we can apply (1.16):

$$\inf_{\mathbf{w}_h \in V_h} |\mathbf{u} - \mathbf{w}_h|_{1,\Omega} \leqslant (1 + \sqrt{2}/\beta^*) \inf_{\mathbf{v}_h \in X_h} |\mathbf{u} - \mathbf{v}_h|_{1,\Omega}.$$

As a consequence,

$$|\mathbf{u} - \mathbf{u}_h|_{1,\Omega} \leqslant 2(1 + \sqrt{2}/\beta^*) \inf_{\mathbf{v}_h \in X_h} |\mathbf{u} - \mathbf{v}_h|_{1,\Omega} + (\sqrt{2}/\nu)\,\|p - P_h p\|_{0,\Omega},$$

and the velocity bound in (4.56) follows from (4.16) and (A.26). Notice that it is (A.26) alone which requires the convexity of Ω and the uniform regularity of $\mathcal{T}_h$.

The bound for λ_h is obtained from the above inequality and the fact that $\operatorname{div} \mathbf{u} = 0$:

(4.60) $$(\mathbf{grad}\,\lambda_h, \mathbf{grad}\,q_h) = -(\mathbf{u} - \mathbf{u}_h, \mathbf{grad}\,q_h) \quad \forall q_h \in Q_h.$$

Therefore

$$|\lambda_h|_{1,\Omega} \leqslant \|\mathbf{u} - \mathbf{u}_h\|_{0,\Omega}.$$

As far as the pressure is concerned, we revert to the standard situation by taking $q_h = 0$ in (4.59). Thus, by virtue of the inf-sup condition, we can apply (1.17):

$$\|p - \bar{p}_h\|_{0,\Omega} \leqslant (1 + \sqrt{2}/\beta^*) \inf_{q_h \in M_h} \|p - q_h\|_{0,\Omega} + (\nu/\beta^*)|\mathbf{u} - \mathbf{u}_h|_{1,\Omega}.$$

This yields (4.56); in turn a familiar argument gives (4.57).

Finally we establish an L^2-estimate for the velocity. The proof is an easy variant of that of Theorem 1.2. As Ω is convex, there exists a unique pair $(\boldsymbol{\phi}, \mu)$ in $[V \cap H^2(\Omega)^2] \times [H^1(\Omega) \cap L_0^2(\Omega)]$ such that

$$(4.61) \qquad \begin{aligned} (\mathbf{g}, \mathbf{u} - \mathbf{u}_h) &= \nu(\mathbf{grad}\, \boldsymbol{\phi}, \mathbf{grad}(\mathbf{u} - \mathbf{u}_h)) + (\mathbf{grad}\, \mu, \mathbf{u} - \mathbf{u}_h), \\ \|\boldsymbol{\phi}\|_{2,\Omega} &+ |\mu|_{1,\Omega} \leqslant C \|\mathbf{g}\|_{0,\Omega}. \end{aligned}$$

Combining this equality with (4.59) we readily derive:

$$\begin{aligned} (\mathbf{g}, \mathbf{u} - \mathbf{u}_h) = {} & \nu(\mathbf{grad}(\boldsymbol{\phi} - \boldsymbol{\phi}_h), \mathbf{grad}(\mathbf{u} - \mathbf{u}_h)) + (p - p_h, \operatorname{div}(\boldsymbol{\phi}_h - \boldsymbol{\phi})) \\ & - (\mu - q_h, \operatorname{div}(\mathbf{u} - \mathbf{u}_h)) + (\mathbf{grad}\, \lambda_h, \mathbf{grad}(\mu - q_h)) \\ & - (\mathbf{g}, \mathbf{grad}\, \lambda_h) \qquad \forall \boldsymbol{\phi}_h \in X_h, \quad \forall q_h \in Q_h. \end{aligned}$$

(Here we use the fact that $\Delta\mu = \operatorname{div} \mathbf{g}$ in Ω). Therefore choosing $q_h = P_h \mu$ we obtain:

$$\begin{aligned} (\mathbf{g}, \mathbf{u} - \mathbf{u}_h + \mathbf{grad}\, \lambda_h) = {} & \nu(\mathbf{grad}(\boldsymbol{\phi} - \boldsymbol{\phi}_h), \mathbf{grad}(\mathbf{u} - \mathbf{u}_h)) + (p - p_h, \operatorname{div}(\boldsymbol{\phi}_h - \boldsymbol{\phi})) \\ & - (\mu - P_h \mu, \operatorname{div}(\mathbf{u} - \mathbf{u}_h)) \qquad \forall \boldsymbol{\phi}_h \in X_h. \end{aligned}$$

In view of (4.56) and (4.61) this yields (4.58). □

Remark 4.6. It does not appear possible to find an L^2 bound like (4.58) for $\mathbf{u} - \mathbf{u}_h$ alone. On the contrary, (4.60) implies that

$$\|\mathbf{u} - \mathbf{u}_h + \mathbf{grad}\, \lambda_h\|_{0,\Omega}^2 = \|\mathbf{u} - \mathbf{u}_h\|_{0,\Omega}^2 - |\lambda_h|_{1,\Omega}^2$$

so that neither $\mathbf{u} - \mathbf{u}_h$ nor λ_h can be isolated from (4.58). But this is not surprising since λ_h acts as a correction on the velocity $\mathbf{u}_h$.

The same analysis can be applied when X_h is defined by (4.30) (and M_h is unchanged). Because of (4.31), the statement of Theorem 4.6 holds *only with* $k = 1$.

4.4. Implementation of the Glowinski-Pironneau Scheme

It is not absolutely straightforward to compute the solution of (4.50) (4.51) (4.52) because the test function l_h does not appear explicitly in the left-hand side of (4.52). It is easier to split the computation by calculating the auxiliary function λ_h, defined by (4.55), that compensates for the fact that $\mathbf{u}_h$ does not satisfy the discrete divergence-free constraint.

To be specific, following the pattern of the equations (4.50) and (4.51), let us define $\lambda_h(g_h)$ and λ_h^0 by:

$$(4.62)\quad \begin{cases} \lambda_h^0 \in \Phi_h, & (\mathbf{grad}\, \lambda_h^0, \mathbf{grad}\, q_h) = (\mathbf{u}_h^0, \mathbf{grad}\, q_h) \quad \forall q_h \in \Phi_h, \\ \lambda_h(g_h) \in \Phi_h, & (\mathbf{grad}\, \lambda_h(g_h), \mathbf{grad}\, q_h) = (\mathbf{u}_h(g_h), \mathbf{grad}\, q_h) \quad \forall q_h \in \Phi_h. \end{cases}$$

Then

$$\lambda_h = \lambda_h^0 + \lambda_h(\rho_h)$$

and we have an alternate expression for the bilinear form $a_h(.,.)$.

Lemma 4.4. *We have*:

$$(4.63)\qquad a_h(g_h, l_h) = (\mathbf{grad}\, \lambda_h(g_h) - \mathbf{u}_h(g_h), \mathbf{grad}\, l_h) \quad \forall g_h, l_h \in G_h.$$

Likewise, the right-hand side of (4.52) *reads*:

$$(4.64)\qquad (\mathbf{u}_h^0, \mathbf{grad}\, p_h(l_h)) = -(\mathbf{grad}\, \lambda_h^0 - \mathbf{u}_h^0, \mathbf{grad}\, l_h) \quad \forall l_h \in G_h.$$

Proof. By definition and (4.51) we have:

$$\begin{aligned} a_h(g_h, l_h) &= \nu(\mathbf{grad}\, \mathbf{u}_h(g_h), \mathbf{grad}\, \mathbf{u}_h(l_h)) = -(\mathbf{grad}\, p_h(l_h), \mathbf{u}_h(g_h)) \\ &= -(\mathbf{grad}(p_h(l_h) - l_h), \mathbf{u}_h(g_h)) - (\mathbf{grad}\, l_h, \mathbf{u}_h(g_h)) \\ &= -(\mathbf{grad}\, \lambda_h(g_h), \mathbf{grad}(p_h(l_h) - l_h)) - (\mathbf{grad}\, l_h, \mathbf{u}_h(g_h)) \end{aligned}$$

by (4.62). Then (4.51) implies that

$$a_h(g_h, l_h) = (\mathbf{grad}\, \lambda_h(g_h) - \mathbf{u}_h(g_h), \mathbf{grad}\, l_h),$$

thus proving (4.63). The proof of (4.64) is similar. □

Hence the Problem (4.52) takes the more manageable form:

Find ρ_h in G_h such that:

$$(4.65)\quad (\mathbf{grad}\, \lambda_h(\rho_h) - \mathbf{u}_h(\rho_h), \mathbf{grad}\, l_h) = (\mathbf{grad}\, \lambda_h^0 - \mathbf{u}_h^0, \mathbf{grad}\, l_h) \quad \forall l_h \in G_h.$$

From the preceding lemma and the definition of the bilinear form $a_h(.,.)$ we know that the left-hand side of (4.65) is a bilinear, symmetric and positive definite form on $G_h \times G_h$. Let us show briefly how (4.65) is solved in practice; the reader will find more details in Glowinski *et al.*, Chapter 13 [37].

Assume that the nodes of $\mathscr{T}_h \cap \Gamma$ are numbered from 1 to N_h and let $\{\mu\}_{1 \leqslant i \leqslant N_h}$ be the set of basis functions of Q_h defined by:

$$\begin{aligned} \mu_i(a_i) &= 1 \quad \text{for each node } a_i \text{ of } \mathscr{T}_h \cap \Gamma, \\ \mu_i(b_k) &= 0 \quad \text{for all other nodes of } \mathscr{T}_h. \end{aligned}$$

Then (dropping for the moment the condition $\int_\Gamma g_h\, ds = 0$) the functions g_h of G_h have the expression:

$$g_h = \sum_{i=1}^{N_h} g_i \mu_i \quad \text{with } g_i = g_h(a_i).$$

With this notation, (4.65) is equivalent to:

$$\sum_{j=1}^{N_h} \rho_j a_h(\mu_j, \mu_i) = -(\mathbf{grad}\, \lambda_h^0 - \mathbf{u}_h^0, \mathbf{grad}\, \mu_i) \qquad 1 \leqslant i \leqslant N_h.$$

In other words we have to solve the system of linear equations:

$$A_h \boldsymbol{\rho} = \mathbf{b} \tag{4.66}$$

where

$$(A_h)_{i,j} = a_h(\mu_j, \mu_i) = (\mathbf{grad}\, \lambda_h(\mu_j) - \mathbf{u}_h(\mu_j), \mathbf{grad}\, \mu_i),$$
$$b_i = -(\mathbf{grad}\, \lambda_h^0 - \mathbf{u}_h^0, \mathbf{grad}\, \mu_i).$$

To compute **b**, we have to solve the three Dirichlet problems (4.50) to find $\mathbf{u}_h^0$ plus the first Dirichlet problem (4.62) to obtain λ_h^0—a total of four Dirichlet problems. To compute the jth column of the matrix A_h we must solve the three Dirichlet problems (4.51) with $g_h = \mu_j$ to find $\mathbf{u}_h(\mu_j)$ plus the second problem (4.62) to get $\lambda_h(\mu_j)$—again a total of four Dirichlet problems.

From the above considerations, it follows that the matrix A_h is symmetric and semi-positive definite with zero as a simple eigenvalue. Furthermore, when the nodes of $\mathscr{T}_h$ are properly numbered, it can be shown that $\text{Ker}(A_h)$ is the constant vector and that the principal block $\tilde{A}_h = (a_h(\mu_j, \mu_i))_{1 \leqslant i,j \leqslant N_h - 1}$ is positive definite. Therefore, setting $\tilde{\rho}_{N_h} = 0$, we can solve (4.66) by the Cholesky factorisation for the first $N_h - 1$ components of a representative $\tilde{\boldsymbol{\rho}}$ of $\boldsymbol{\rho}$. Then the solution ρ_h of (4.65) that satisfies $\int_\Gamma \rho_h \, ds = 0$ is given by:

$$\tilde{\rho}_h - (1/\text{meas}(\Gamma)) \int_\Gamma \tilde{\rho}_h \, ds.$$

Once ρ_h is known, the pressure $p_h(\rho_h)$ and velocity $\mathbf{u}_h(\rho_h)$ are computed by solving the three Dirichlet problems (4.51).

The problem (4.65) can also be solved by the conjugate-gradient algorithm (cf. Glowinski *et al. loc. cit.*).

Chapter III. Incompressible Mixed Finite Element Methods for Solving the Stokes Problem

§ 1. Mixed Approximation of an Abstract Problem

In this paragraph, we concentrate again upon the abstract problem studied in Chapter I § 4, but we put it into a weaker setting leading to a (generally) different mixed formulation. The mixed approximation derived from this formulation will give rise to the important class of exactly incompressible methods to solve the Stokes and Navier-Stokes equations.

1.1. A Mixed Variational Problem

We put ourselves into the situation of Section I.4.1. Recall that X and M are two Hilbert spaces and that $a(.,.)$ and $b(.,.)$ are two *continuous bilinear* forms on $X \times X$ and $X \times M$ respectively. Recall that *Problem* (Q) is:

For (l, χ) given in $X' \times M'$, find a pair (u, λ) in $X \times M$ with:

(1.1) $$a(u, v) + b(v, \lambda) = \langle l, v \rangle \quad \forall v \in X,$$

(1.2) $$b(u, \mu) = \langle \chi, \mu \rangle \quad \forall \mu \in M.$$

As usual, we set

$$V(\chi) = \{v \in X;\, b(v, \mu) = \langle \chi, \mu \rangle \quad \forall \mu \in M\},$$
$$V = V(0).$$

Again, we assume that $a(.,.)$ and $b(.,.)$ satisfy the two hypotheses:
there exists a constant $\alpha > 0$ such that

(1.3) $$a(v, v) \geqslant \alpha \| v \|_X^2 \quad \forall v \in V;$$

there exists a constant $\beta > 0$ such that

(1.4) $$\sup_{v \in X} \frac{b(v, \mu)}{\| v \|_X} \geqslant \beta \| \mu \|_M \quad \forall \mu \in M.$$

These two assumptions guarantee that the Problem (Q) and its corresponding

Problem (P):

Find u in $V(\chi)$ *such that*

$$a(u, v) = \langle l, v \rangle \quad \forall v \in V, \tag{1.5}$$

are well posed.

Let us elaborate a weaker formulation of Problem (Q) better suited to the approximation we have in mind. We introduce two reflexive Banach spaces $\tilde{X}$ and $\tilde{M}$ normed respectively by $\|.\|_{\tilde{X}}$ and $\|.\|_{\tilde{M}}$ such that:

$$X \underset{d}{\subset} \tilde{X}, \qquad \tilde{M} \underset{d}{\subset} M,$$

where the sign $\underset{d}{\subset}$ means that the imbedding is dense and continuous.

Next, we consider two continuous bilinear forms:

$$\tilde{a}(.,.): \tilde{X} \times \tilde{X} \to \mathbb{R}, \qquad \tilde{b}(.,.): \tilde{X} \times \tilde{M} \to \mathbb{R},$$

with the norms:

$$\|\tilde{a}\| = \sup_{u, v \in \tilde{X}} \frac{\tilde{a}(u, v)}{\|u\|_{\tilde{X}} \|v\|_{\tilde{X}}}, \qquad \|\tilde{b}\| = \sup_{v \in \tilde{X}, \mu \in \tilde{M}} \frac{\tilde{b}(v, \mu)}{\|v\|_{\tilde{X}} \|\mu\|_{\tilde{M}}}.$$

These two bilinear forms are extensions of a and b in the sense that

$$\tilde{a}(u, v) = a(u, v) \quad \forall u, v \in X, \tag{1.6}$$

$$\tilde{b}(v, \mu) = b(v, \mu) \quad \forall v \in X, \quad \forall \mu \in \tilde{M}. \tag{1.7}$$

In addition, we assume that the right-hand side l of (1.1) belongs to the dual space $\tilde{X}'$ of $\tilde{X}$ and we denote by $\langle .,. \rangle$ the duality pairing between $\tilde{X}$ and $\tilde{X}'$. Then we introduce the following *Problem* ($\tilde{\mathrm{Q}}$):

Find a pair $(\tilde{u}, \tilde{\lambda}) \in \tilde{X} \times \tilde{M}$ *such that*:

$$\tilde{a}(\tilde{u}, v) + \tilde{b}(v, \tilde{\lambda}) = \langle l, v \rangle \quad \forall v \in \tilde{X}, \tag{1.8}$$

$$\tilde{b}(\tilde{u}, \mu) = \langle \chi, \mu \rangle \quad \forall \mu \in \tilde{M}. \tag{1.9}$$

For each $\chi \in M'$ we define the affine variety:

$$\tilde{V}(\chi) = \{v \in \tilde{X}; \tilde{b}(v, \mu) = \langle \chi, \mu \rangle \quad \forall \mu \in \tilde{M}\}, \tag{1.10}$$

and the following closed subspace of $\tilde{X}$:

$$\tilde{V} = \tilde{V}(0) = \{v \in \tilde{X}; \tilde{b}(v, \mu) = 0 \quad \forall \mu \in \tilde{M}\}. \tag{1.11}$$

Equality (1.7) implies that:

$$V \subset \tilde{V}, \qquad V(\chi) \subset \tilde{V}(\chi). \tag{1.12}$$

With the Problem ($\tilde{\mathrm{Q}}$) we associate the following *Problem* ($\tilde{\mathrm{P}}$):

Find $\tilde{u}$ *in* $\tilde{V}(\chi)$ *such that*

$$\tilde{a}(\tilde{u}, v) = \langle l, v \rangle \quad \forall v \in \tilde{V}. \tag{1.13}$$

In order to analyze conveniently Problems ($\tilde{\mathrm{P}}$) and ($\tilde{Q}$) we make the following assumption on $\tilde{a}(.,.)$:

the form $\tilde{a}(.,.)$ is $\tilde{V}$-elliptic, i.e. there exists a constant $\tilde{\alpha} > 0$ such that:

$$(1.14) \qquad \tilde{a}(v,v) \geqslant \tilde{\alpha}\,\|v\|_{\tilde{X}}^2 \qquad \forall v \in \tilde{V}.$$

Note that on the one hand (1.14) does not stem from (1.3) because $\tilde{V}$ is (usually) a larger space than V. On the other hand there is no inf-sup condition on $\tilde{b}(.,.)$ except the one which follows from (1.4) and (1.7):

$$(1.15) \qquad \sup_{v \in \tilde{X}} \frac{\tilde{b}(v,\mu)}{\|v\|_{\tilde{X}}} \geqslant (1/C) \sup_{v \in X} \frac{\tilde{b}(v,\mu)}{\|v\|_X} \geqslant (\beta/C)\,\|\mu\|_M \qquad \forall \mu \in \tilde{M},$$

where C is the continuity constant of the imbedding $X \subset \tilde{X}$. Strictly speaking, this is not sufficient to ensure that Problem ($\tilde{\mathrm{Q}}$) is well posed. The next theorem tackles this difficulty.

Theorem 1.1. *Let* (u,λ) *be the solution of Problem* (Q) *and let* $\tilde{a}$ *satisfy* (1.14).

1°) *Problem* ($\tilde{\mathrm{P}}$) *has exactly one solution* $\tilde{u}$ *in* $\tilde{V}(\chi)$. *Moreover, if* $\tilde{u}$ *also belongs to* $V(\chi)$ *or if* V *is dense in* $\tilde{V}$ *then* $\tilde{u} = u$.

2°) *In addition, if* λ *belongs to* $\tilde{M}$ *then the pair* (u,λ) *is the only solution of Problem* ($\tilde{\mathrm{Q}}$).

Proof. To begin with, recall that the theory of Section I.4.1 applies also to reflexive Banach spaces (cf. Remark I.4.2).

1°) The inf-sup condition (1.4) on $b(.,.)$ implies that $V(\chi)$ is not empty; hence $\tilde{V}(\chi)$ is not empty. Then the ellipticity of $\tilde{a}$ implies that Problem ($\tilde{\mathrm{P}}$) has one and only one solution $\tilde{u}$ in $\tilde{V}(\chi)$. If $\tilde{u} \in V(\chi)$, we see from (1.6) that $\tilde{u}$ is a solution of Problem (P); therefore $\tilde{u} = u$ since (P) has exactly one solution. Otherwise, we assume that V is dense in $\tilde{V}$; then (1.5) and (1.6) imply that u is a solution of Problem ($\tilde{\mathrm{P}}$). Thus $u = \tilde{u}$.

2°) In addition, suppose that $\lambda \in \tilde{M}$. Then by virtue of (1.6) and (1.7), (1.1) becomes:

$$\tilde{a}(u,v) + \tilde{b}(v,\lambda) = \langle l, v\rangle \qquad \forall v \in X.$$

As X is dense in $\tilde{X}$, this shows that the pair (u,λ) is a solution of Problem ($\tilde{\mathrm{Q}}$). Finally, we must prove that it is the only solution of ($\tilde{\mathrm{Q}}$). Obviously, the first component u is unique. Then assume that

$$\tilde{b}(v,\lambda) = 0 \qquad \forall v \in \tilde{X}.$$

With (1.15) this implies that $\lambda = 0$. □

Remark 1.1. The assumption $\lambda \in \tilde{M}$ is in fact a regularity condition.

In subsequent paragraphs, we shall study applications where the approximation of Problem ($\tilde{\mathrm{Q}}$) is simpler than that of Problem (Q).

1.2. Abstract Mixed Approximation

Throughout this section we assume that the hypotheses of Theorem 1.1 hold.

For each h let X_h and M_h be two *finite-dimensional* subspaces of $\tilde{X}$ and $\tilde{M}$ respectively. We approximate Problem ($\tilde{\mathrm{Q}}$) by *Problem* (Q_h):

Find a pair $(u_h, \lambda_h) \in X_h \times M_h$ *such that*

$$\tilde{a}(u_h, v_h) + \tilde{b}(v_h, \lambda_h) = \langle l, v_h \rangle \quad \forall v_h \in X_h, \tag{1.16}$$

$$\tilde{b}(u_h, \mu_h) = \langle \chi, \mu_h \rangle \quad \forall \mu_h \in M_h. \tag{1.17}$$

Again we define

$$\begin{aligned} V_h(\chi) &= \{v_h \in X_h;\ \tilde{b}(v_h, \mu_h) = \langle \chi, \mu_h \rangle \quad \forall \mu_h \in M_h\}, \\ V_h &= V_h(0). \end{aligned} \tag{1.18}$$

Next we associate with Problem (Q_h) the following *Problem* (P_h):

Find $u_h \in V_h(\chi)$ *such that*

$$\tilde{a}(u_h, v_h) = \langle l, v_h \rangle \quad \forall v_h \in V_h. \tag{1.19}$$

Here also V_h is generally not included in $\tilde{V}$ and therefore Problem (P_h) is an external approximation of Problem ($\tilde{\mathrm{P}}$).

In order to derive error estimates for u_h and λ_h we make the following assumptions, analogous to (1.14) and (1.15):

i) there exists a constant $\alpha^* > 0$ such that

$$\tilde{a}(v_h, v_h) \geqslant \alpha^* \| v_h \|_{\tilde{X}}^2 \quad \forall v_h \in V_h; \tag{1.20}$$

ii) there exists a constant $\beta^* > 0$ such that

$$\sup_{v_h \in X_h} \frac{\tilde{b}(v_h, \mu_h)}{\| v_h \|_{\tilde{X}}} \geqslant \beta^* \| \mu_h \|_M \quad \forall \mu_h \in M_h. \tag{1.21}$$

The next theorem is a natural extension of Theorem II.1.1.

Theorem 1.2. 1°) *Suppose* $V_h(\chi)$ *is not empty and* $\tilde{a}(.,.)$ *satisfies* (1.20). *Then Problem* (P_h) *has a unique solution* $u_h \in V_h(\chi)$ *and the following error bound holds:*

$$\| u - u_h \|_{\tilde{X}} \leqslant (1 + \| \tilde{a} \| / \alpha^*) \inf_{v_h \in V_h(\chi)} \| u - v_h \|_{\tilde{X}} + (1/\alpha^*) \inf_{\mu_h \in M_h} \left[\sup_{v_h \in V_h} \frac{\tilde{b}(v_h, \lambda - \mu_h)}{\| v_h \|_{\tilde{X}}} \right]. \tag{1.22}$$

2°) *Suppose moreover that* $\tilde{b}(.,.)$ *satisfies* (1.21). *Then* $V_h(\chi)$ *is not empty and Problem* (Q_h) *has exactly one solution* (u_h, λ_h) *where* u_h *is the solution of Problem* (P_h). *Furthermore* λ_h *satisfies the error estimate:*

$$\| \lambda - \lambda_h \|_M \leqslant \left(\frac{\| \tilde{a} \|}{\beta^*} \right) \| u - u_h \|_{\tilde{X}} + \inf_{\mu_h \in M_h} \left\{ \left(\frac{\| \tilde{b} \|}{\beta^*} \right) \| \lambda - \mu_h \|_{\tilde{M}} + \| \lambda - \mu_h \|_M \right\}. \tag{1.23}$$

Proof. 1°) The idea of the proof is very similar to that of Theorem II.1.1. The existence and uniqueness of the solution u_h of Problem (P_h) follow from (1.20) and Lax & Milgram's Theorem I.1.7, provided $V_h(\chi)$ is not empty.

Now, let w_h be any element of $V_h(\chi)$ and let $v_h = u_h - w_h \in V_h$. Then formula (II.1.15) holds:

$$\tilde{a}(v_h, v_h) = \tilde{a}(u - w_h, v_h) + \tilde{b}(v_h, \lambda - \mu_h) \qquad \forall \mu_h \in M_h, \qquad \forall w_h \in V_h(\chi).$$

Thus (1.20) implies:

$$\alpha^* \|v_h\|_{\tilde{X}} \leqslant \|\tilde{a}\| \, \|u - w_h\|_{\tilde{X}} + \sup_{v_h \in V_h} \left\{ \frac{\tilde{b}(v_h, \lambda - \mu_h)}{\|v_h\|_{\tilde{X}}} \right\} \qquad \forall \mu_h \in M_h, \qquad \forall w_h \in V_h(\chi)$$

and this gives immediately (1.22).

2°) Since the dimension of M_h is finite, the condition (1.21) implies the classical inf-sup condition on M_h, eventually with a constant that depends upon h. Therefore $V_h(\chi)$ is not empty and Problem (Q_h) has exactly one solution (u_h, λ_h) where u_h satisfies Problem (P_h). Moreover, the following equality holds for any v_h in X_h and μ_h in M_h:

$$\tilde{b}(v_h, \lambda_h - \mu_h) = \tilde{a}(u - u_h, v_h) + \tilde{b}(v_h, \lambda - \mu_h). \tag{1.24}$$

Then it stems from (1.21) that:

$$\beta^* \|\lambda_h - \mu_h\|_M \leqslant \|\tilde{a}\| \, \|u - u_h\|_{\tilde{X}} + \|\tilde{b}\| \, \|\lambda - \mu_h\|_{\tilde{M}}$$

and (1.23) is established. □

Note that the estimate (1.23) is not optimal inasmuch as it gives an upper bound for $\|\lambda - \lambda_h\|_M$ in terms of $\|\lambda - \mu_h\|_{\tilde{M}}$ while the two norms are usually not meant to be equivalent. The examples of § 3 will show how to overcome this defect.

Observe also that it is often difficult to evaluate directly an expression like

$$\inf_{v_h \in V_h(\chi)} \|u - v_h\|_{\tilde{X}}.$$

Like in Chapter II, it is possible to reduce this term to the approximation error in X_h although here the process is not always optimal. As M_h is finite-dimensional, there exists a constant $K(h) > 0$ such that

$$\|\mu\|_{\tilde{M}} \leqslant K(h) \|\mu\|_M \qquad \forall \mu \in M_h. \tag{1.25}$$

With this we readily prove the following result.

Corollary 1.1. *With the assumptions* (1.20) *and* (1.21) *each v in $V(\chi)$ satisfies:*

$$\inf_{v_h \in V_h(\chi)} \|v - v_h\|_{\tilde{X}} \leqslant [1 + K(h)(\|\tilde{b}\|/\beta^*)] \inf_{v_h \in X_h} \|v - v_h\|_{\tilde{X}}. \tag{1.26}$$

Furthermore, there exists a constant $C > 0$, depending solely upon α^*, β^*, $\|\tilde{a}\|$ and $\|\tilde{b}\|$ such that:

$$\|u - u_h\|_{\tilde{X}} + \|\lambda - \lambda_h\|_M \leqslant C\left\{(1 + K(h)) \inf_{v_h \in X_h} \|u - v_h\|_{\tilde{X}} + \inf_{\mu_h \in M_h} \|\lambda - \mu_h\|_{\tilde{M}}\right\}.$$

(1.27)

Proof. Clearly (1.27) is an immediate consequence of (1.26) and Theorem 1.2.

Next, the proof of (1.26) follows very closely the lines of Theorem II.1.1. Indeed, in view of (1.25) the inf-sup condition (1.21) becomes:

(1.28) $$\sup_{v_h \in X_h} \frac{\tilde{b}(v_h, \mu_h)}{\|v_h\|_{\tilde{X}}} \geqslant (\beta^*/K(h)) \|\mu_h\|_{\tilde{M}} \qquad \forall \mu_h \in M_h.$$

Therefore, applying the argument of Theorem II.1.1 with the constant $\beta^*/K(h)$ instead of β^*, we obtain directly (1.26). □

We shall see in the examples of the next paragraph that $K(h)$ usually becomes infinite when the dimension of M_h tends to infinity. But again the techniques developed in § 3 will show in some cases how to estimate sharply the approximation error in $V_h(\chi)$ without introducing the factor $K(h)$.

§ 2. The "Stream Function-Vorticity-Pressure" Method for the Stokes Problem in Two Dimensions

The widespread "stream function-vorticity-pressure" formulation of the two-dimensional Stokes system of equations is based on the two identities:

$$\mathbf{curl}(\operatorname{curl} \mathbf{u}) = -\Delta \mathbf{u} + \mathbf{grad}(\operatorname{div} \mathbf{u}), \qquad \operatorname{curl}(\mathbf{curl}\, \phi) = -\Delta \phi.$$

Thus the Stokes problem

$$\left.\begin{aligned} -\nu \Delta \mathbf{u} + \mathbf{grad}\, p &= \mathbf{f} \\ \operatorname{div} \mathbf{v} &= 0 \end{aligned}\right\} \text{ in } \Omega, \qquad \mathbf{u}|_\Gamma = \mathbf{0}$$

can be expressed either as:

$$\left.\begin{aligned} \operatorname{curl} \mathbf{u} = \omega, \qquad \operatorname{div} \mathbf{u} &= 0 \\ \nu\, \mathbf{curl}\, \omega + \mathbf{grad}\, p &= \mathbf{f} \end{aligned}\right\} \text{ in } \Omega, \qquad \mathbf{u}|_\Gamma = \mathbf{0},$$

or as

$$\left.\begin{aligned} \mathbf{u} = \mathbf{curl}\, \phi, \qquad \omega &= -\Delta \phi \\ \nu\, \mathbf{curl}\, \omega + \mathbf{grad}\, p &= \mathbf{f} \end{aligned}\right\} \text{ in } \Omega, \qquad \mathbf{curl}\, \phi|_\Gamma = \mathbf{0}.$$

The mixed method discussed here and in the next paragraph is derived from these two formulations.

2.1. A Mixed Formulation

Let Ω be a bounded domain of $\mathbb{R}^2$ with a Lipschitz-continuous boundary Γ whose components are denoted by Γ_i $(0 \leqslant i \leqslant p)$ like in Figure 2. For $\mathbf{f}$ given in $L^r(\Omega)^2$ with $r > 1$, recall the classical formulation of the Stokes equations in Ω:

Find a pair $(\mathbf{u}, p)$ *in* $H_0^1(\Omega)^2 \times L_0^2(\Omega)$ *such that*

$$(2.1)\qquad \begin{cases} \nu(\mathbf{grad}\,\mathbf{u}, \mathbf{grad}\,\mathbf{v}) - (p, \operatorname{div} \mathbf{v}) = (\mathbf{f}, \mathbf{v}) & \forall \mathbf{v} \in H_0^1(\Omega)^2, \\ (q, \operatorname{div} \mathbf{u}) = 0 & \forall q \in L_0^2(\Omega). \end{cases}$$

Since $\operatorname{div} \mathbf{u} = 0$ we have proved in Theorem I.5.5 that

$$(2.2)\qquad (\mathbf{grad}\,\mathbf{u}, \mathbf{grad}\,\mathbf{v}) = (\operatorname{curl} \mathbf{u}, \operatorname{curl} \mathbf{v}) \qquad \forall \mathbf{v} \in H_0^1(\Omega)^2.$$

Thus introducing the vorticity ω:

$$\omega = \operatorname{curl} \mathbf{u},$$

the equations (2.1) are equivalent to:

Find a triple $(\mathbf{u}, \omega, p) \in H_0^1(\Omega)^2 \times L^2(\Omega) \times L_0^2(\Omega)$ *solution of*

$$(2.3)\qquad \begin{cases} \nu(\omega, \operatorname{curl} \mathbf{v}) - (p, \operatorname{div} \mathbf{v}) = (\mathbf{f}, \mathbf{v}) \qquad \forall \mathbf{v} \in H_0^1(\Omega)^2, \\ (\operatorname{curl} \mathbf{u} - \omega, \mu) - (q, \operatorname{div} \mathbf{u}) = 0 \qquad \forall \mu \in L^2(\Omega), \qquad \forall q \in L_0^2(\Omega). \end{cases}$$

Let us put Problem (2.3) into the framework of Section I.4.1. Set

$$X = H_0^1(\Omega)^2 \times L^2(\Omega),$$

equipped with the norm

$$\|(\mathbf{v}, \theta)\|_X = (|\mathbf{v}|_{1,\Omega}^2 + \|\theta\|_{0,\Omega}^2)^{1/2} \qquad \forall (\mathbf{v}, \theta) \in X$$

and

$$M = L^2(\Omega) \times L_0^2(\Omega),$$

normed by:

$$\|(\mu, q)\|_M = (\|\mu\|_{0,\Omega}^2 + \|q\|_{0,\Omega}^2)^{1/2} \qquad \forall (\mu, q) \in M.$$

Next, we introduce the bilinear forms:

$$(2.4)\qquad \begin{cases} a((\mathbf{u}, \omega), (\mathbf{v}, \theta)) = \nu(\omega, \theta), \\ b((\mathbf{v}, \theta), (\mu, q)) = (\operatorname{curl} \mathbf{v} - \theta, \mu) - (q, \operatorname{div} \mathbf{v}) \end{cases}$$

and the right-hand sides l and χ defined by:

$$(2.5)\qquad \langle l, (\mathbf{v}, \theta) \rangle = (\mathbf{f}, \mathbf{v}), \qquad \chi = 0.$$

With these data the corresponding *Problem* (Q) reads:

Find $(\mathbf{u}, \omega) \in X$ *and* $(\lambda, p) \in M$ *solution of*

$$(2.6)\quad \begin{cases} \nu(\omega, \theta) + (\operatorname{curl} \mathbf{v} - \theta, \lambda) - (p, \operatorname{div} \mathbf{v}) = (\mathbf{f}, \mathbf{v}) & \forall (\mathbf{v}, \theta) \in X, \\ (\operatorname{curl} \mathbf{u} - \omega, \mu) - (q, \operatorname{div} \mathbf{u}) = 0 & \forall (\mu, q) \in M; \end{cases}$$

and its associated space V is:

$$V = \{(\mathbf{v}, \theta) \in X;\ \theta = \operatorname{curl} \mathbf{v},\ \operatorname{div} \mathbf{v} = 0\}.$$

Remark I.2.7 asserts that $a(.,.)$ is V-elliptic. As far as $b(.,.)$ is concerned, Corollary I.2.4 easily yields the inf-sup condition (1.4). Thus we have the following equivalence theorem.

Theorem 2.1. *Problem* (2.6) *has a unique solution* $((\mathbf{u}, \omega), (\lambda, p)) \in X \times M$; *in addition,* $(\mathbf{u}, p)$, *or equivalently* $(\mathbf{u}, \omega, p)$, *is the solution of the Stokes Problem* (2.1), *or* (2.3), *with* $\omega = \operatorname{curl} \mathbf{u}$ *and* $\lambda = \nu\omega$.

Proof. The above considerations imply that Problem (2.6) has a unique solution. (The uniqueness can also be established by a plain, direct argument.) Next, a quick inspection shows that if $(\mathbf{u}, \omega = \operatorname{curl} \mathbf{u}, p)$ satisfies (2.3) then the pair $((\mathbf{u}, \omega), (\nu\omega, p))$ also satisfies (2.6). □

Problems (2.3) and (2.6) can also be expressed in terms of stream functions. Recall the space Ψ defined in Section I.3.1:

$$(2.7)\quad \Psi = \{\chi \in H^2(\Omega);\ \chi|_{\Gamma_0} = 0,\ \chi|_{\Gamma_i} = \text{a constant } c_i,\ 1 \leqslant i \leqslant p,\ \partial\chi/\partial n|_{\Gamma} = 0\}.$$

We know from Corollary I.3.2 that the mapping **curl** is an isomorphism from Ψ onto $\{\mathbf{v} \in H_0^1(\Omega)^2; \operatorname{div} \mathbf{v} = 0\}$. Thus we have

$$\mathbf{u} = \mathbf{curl}\, \psi \quad \text{for a unique } \psi \quad \text{in } \Psi.$$

And by choosing $\mathbf{v} = \mathbf{curl}\, \phi$ for all $\phi \in \Psi$ in (2.3) or (2.6) we find that the two components $\mathbf{u}$, ω of the solution of (2.3) or the three components $\mathbf{u}$, ω, λ of the solution of (2.6) can be determined by solving respectively:

Find a pair $(\psi, \omega) \in \Psi \times L^2(\Omega)$ *such that*

$$(2.8)\quad \begin{cases} \nu(\omega, -\Delta\phi) = (\mathbf{f}, \mathbf{curl}\, \phi) & \forall \phi \in \Psi, \\ (\Delta\psi + \omega, \mu) = 0 & \forall \mu \in L^2(\Omega), \end{cases}$$

or

Find a triple $(\psi, \omega, \lambda) \in \Psi \times L^2(\Omega) \times L^2(\Omega)$ *such that*

$$(2.9)\quad \begin{cases} \nu(\omega, \theta) - (\Delta\phi + \theta, \lambda) = (\mathbf{f}, \mathbf{curl}\, \phi) & \forall (\phi, \theta) \in \Psi \times L^2(\Omega), \\ (\Delta\psi + \omega, \mu) = 0 & \forall \mu \in L^2(\Omega). \end{cases}$$

Observe that this problem fits into the framework of Section I.4.1 with the following interpretation:

$$(2.10)\quad \begin{cases} X = \Psi \times L^2(\Omega), \qquad M = L^2(\Omega), \\ a((\psi,\omega),(\phi,\theta)) = \nu(\omega,\theta), \qquad \forall(\psi,\omega),(\phi,\theta)\in X, \\ b((\phi,\theta),\mu) = -(\Delta\phi + \theta,\mu) \qquad \forall(\phi,\theta)\in X, \qquad \forall\mu\in M, \\ \langle l,(\phi,\theta)\rangle = (\mathbf{f},\mathbf{curl}\,\phi) \qquad \forall(\phi,\theta)\in X. \end{cases}$$

Then Theorem I.5.5 immediately implies the following result.

Theorem 2.2. *Problem* (2.8) *has a unique solution* $(\psi, \omega = -\Delta\psi) \in \Psi \times L^2(\Omega)$ *and there exists a unique* p *in* $L_0^2(\Omega)$ *such that* $(\mathbf{u} = \mathbf{curl}\,\psi, \omega, p)$ *is the solution of Problem* (2.3).

However, none of these problems are entirely satisfactory for our purpose because their internal approximation requires the construction of finite-dimensional subspaces of either divergence-free velocities with H^1-regularity or equivalently stream functions with H^2-regularity. In practice, this is far from desirable. To resolve this difficulty, we propose to relax the regularity of the test velocities $\mathbf{v}$ or stream functions ϕ.

Let us start with Problem (2.6). The bilinear form $a(.,.)$ can be left as such whereas the form $b(.,.)$ which involves $\operatorname{curl}\mathbf{v}$ and $\operatorname{div}\mathbf{v}$ must be modified. As $\mathbf{v}\in H_0^1(\Omega)^2$ we have:

$$\left.\begin{aligned} (\operatorname{curl}\mathbf{v},\mu) &= (\mathbf{v},\mathbf{curl}\,\mu) \\ (q,\operatorname{div}\mathbf{v}) &= -(\mathbf{grad}\,q,\mathbf{v}) \end{aligned}\right\} \quad \forall\mu, q\in H^1(\Omega).$$

Thus the regularity of $\mathbf{v}$ can be decreased if that of μ and q is increased. The most straightforward choice is: $\mathbf{v}\in L^2(\Omega)^2$ and μ and $q\in H^1(\Omega)$; as far as the Stokes problem is concerned this is pretty adequate. But in view of subsequent applications to the Navier-Stokes equations, it is preferable to allow μ and q to cover a wider range of spaces.

To this end, let us *fix* a real $s \geqslant 2$ and associate $r\in\mathbb{R}$ by

$$1/r + 1/s = 1.$$

Then take the spaces:

$$(2.11)\quad \tilde{X} = L^s(\Omega)^2 \times L^2(\Omega), \qquad \tilde{M} = W^{1,r}(\Omega) \times [W^{1,r}(\Omega)\cap L_0^2(\Omega)],$$

with the respective norms:

$$\|(\mathbf{v},\theta)\|_{\tilde{X}} = \|\mathbf{v}\|_{0,s,\Omega} + \|\theta\|_{0,\Omega}, \qquad \|(\mu,q)\|_{\tilde{M}} = \|\mu\|_{1,r,\Omega} + |q|_{1,r,\Omega},$$

and the bilinear forms:

$$(2.12)\quad \begin{cases} \tilde{a}((\mathbf{u},\omega),(\mathbf{v},\theta)) = \nu(\omega,\theta), \\ \tilde{b}((\mathbf{v},\theta),(\mu,q)) = (\mathbf{v},\mathbf{curl}\,\mu) - (\theta,\mu) + (\mathbf{grad}\,q,\mathbf{v}), \\ \qquad\qquad \text{for all } (\mathbf{u},\omega),(\mathbf{v},\theta)\in\tilde{X} \quad\text{and}\quad (\mu,q)\in\tilde{M}. \end{cases}$$

Now assume that the right-hand side $\mathbf{f}$ of (2.1) belongs to $L^r(\Omega)^2$ and consider the following *Problem* $(\tilde{\mathrm{Q}})$.

Find $(\mathbf{u}, \omega) \in \tilde{X}$ *and* $(\lambda, p) \in \tilde{M}$ *such that*:

$$(2.13)\qquad \begin{cases} \nu(\omega, \theta) + (\mathbf{v}, \mathbf{curl}\,\lambda) - (\theta, \lambda) + (\mathbf{grad}\,p, \mathbf{v}) = (\mathbf{f}, \mathbf{v}) & \forall (\mathbf{v}, \theta) \in \tilde{X}, \\ (\mathbf{u}, \mathbf{curl}\,\mu) - (\omega, \mu) + (\mathbf{grad}\,q, \mathbf{u}) = 0 & \forall (\mu, q) \in \tilde{M}. \end{cases}$$

Before going any further, let us examine the space $\tilde{V}$ associated with Problem $(\tilde{\mathrm{Q}})$:

$$(2.14)\qquad \tilde{V} = \{(\mathbf{v}, \theta) \in \tilde{X};\ (\mathbf{v}, \mathbf{curl}\,\mu) - (\theta, \mu) + (\mathbf{grad}\,q, \mathbf{v}) = 0 \quad \forall (\mu, q) \in \tilde{M}\}.$$

We introduce the following seminorm on $\tilde{X}$:

$$(2.15)\qquad |(\mathbf{v}, \theta)| = \|\theta\|_{0,\Omega} \quad \forall (\mathbf{v}, \theta) \in \tilde{X}.$$

Lemma 2.1. *The spaces V and $\tilde{V}$ coincide algebraically and topologically. Moreover the seminorm $|.|$ is a norm equivalent to $\|.\|_{\tilde{X}}$ on V*:

$$|(\mathbf{v}, \theta)| \cong \|(\mathbf{v}, \theta)\|_{\tilde{X}} \quad \forall (\mathbf{v}, \theta) \in V.$$

Proof. Let $(\mathbf{v}, \theta) \in \tilde{V}$. Straightforward applications of Green's formulas (I.2.17) and (I.2.22) yield in succession:

$$\operatorname{div} \mathbf{v} = 0 \ \text{ in } \Omega, \quad \mathbf{v}\cdot\mathbf{n} = 0 \ \text{ on } \Gamma;$$
$$\theta = \operatorname{curl} \mathbf{v} \ \text{ in } \Omega, \quad \mathbf{v}\cdot\boldsymbol{\tau} = 0 \ \text{ on } \Gamma.$$

Hence $\mathbf{v} \in H_0(\operatorname{div}; \Omega) \cap H_0(\operatorname{curl}; \Omega)$ and Lemma I.2.5 implies that

$$\mathbf{v} \in H_0^1(\Omega)^2$$

with (cf. Remark I.2.7)

$$(2.16)\qquad |\mathbf{v}|_{1,\Omega} \leqslant C\|\theta\|_{0,\Omega}.$$

Therefore $\tilde{V} \subset V$ and the converse is obvious.

The equivalence of norms is an immediate consequence of the stronger result (2.16) and Sobolev's Imbedding Theorem I.1.3. □

Now we can prove that problems (Q) and $(\tilde{\mathrm{Q}})$ are equivalent.

Theorem 2.3. *Assume that the solution* $(\mathbf{u}, p)$ *of the Stokes problem* (2.1) *satisfies*:

$$(2.17)\qquad \operatorname{curl} \mathbf{u} \in W^{1,r}(\Omega), \quad p \in W^{1,r}(\Omega) \cap L_0^2(\Omega);$$

then Problem $(\tilde{\mathrm{Q}})$ *has the unique solution*: $\mathbf{u}$, $\omega = \operatorname{curl} \mathbf{u}$, $\lambda = \nu\omega$, p.

Proof. The proof is an easy application of Theorem 1.1 and Lemma 2.1.

The inclusions $X \underset{d}{\subset} \tilde{X}$ and $\tilde{M} \underset{d}{\subset} M$ follow from Theorem I.1.3 and Lemma

I.1.1. The equalities (1.6) and (1.7) are obvious. The ellipticity condition (1.14) stems from (2.16) and we have already checked (1.3) and (1.4). In view of (2.17) and Lemma 2.1, this means that all the assumptions of Theorem 1.1 are satisfied. □

Remark 2.1. If Γ is of class $\mathscr{C}^2$ (resp. if Ω is a convex polygon) Theorem I.5.4 (resp. Remark I.5.6) says that $\mathbf{u} \in W^{2,r}(\Omega)^2$ and $p \in W^{1,r}(\Omega)$ as long as $\mathbf{f}$ belongs to $L^r(\Omega)^2$. Therefore, this mild regularity assumption on $\mathbf{f}$ and Ω is sufficient to ensure that Problem $(\tilde{\mathrm{Q}})$ has a unique solution.

As far as stream functions are concerned, recall Corollary I.3.1 which states that the operator **curl** is an isomorphism from

$$\Phi = \{\chi \in H^1(\Omega);\ \chi|_{\Gamma_0} = 0,\ \chi|_{\Gamma_i} = \text{a constant } c_i,\ 1 \leqslant i \leqslant p\}$$

onto

$$H = \{\mathbf{v} \in L^2(\Omega)^2;\ \operatorname{div} \mathbf{v} = 0,\ \mathbf{v} \cdot \mathbf{n}|_\Gamma = 0\}.$$

An immediate consequence of this result (cf. Section I.3.1) is that **curl** is also an *isomorphism* from

$$\Phi_s = \Phi \cap W^{1,s}(\Omega) \tag{2.18}$$

onto

$$H \cap L^s(\Omega)^2 = \{\mathbf{v} \in L^s(\Omega)^2;\ \operatorname{div} \mathbf{v} = 0,\ \mathbf{v} \cdot \mathbf{n}|_\Gamma = 0\}.$$

Therefore, if $((\mathbf{u}, \omega), (\lambda, p))$ is the solution of (2.13) then on the one hand $(\mathbf{u}, \omega) \in V$ (owing to Lemma 2.1); on the other hand, by choosing $\mathbf{v} = \mathbf{curl}\, \phi$ with $\phi \in \Phi_s$ we find: $\mathbf{u} = \mathbf{curl}\, \psi$ where $((\psi, \omega), \lambda)$ is the solution of

Find $(\psi, \omega) \in \Phi_s \times L^2(\Omega)$ *and* $\lambda \in W^{1,r}(\Omega)$ *satisfying*:

$$\begin{cases} \nu(\omega, \theta) - (\lambda, \theta) + (\mathbf{curl}\, \phi, \mathbf{curl}\, \lambda) = (\mathbf{f}, \mathbf{curl}\, \phi) & \forall (\phi, \theta) \in \Phi_s \times L^2(\Omega), \\ (\mathbf{curl}\, \psi, \mathbf{curl}\, \mu) - (\omega, \mu) = 0 & \forall \mu \in W^{1,r}(\Omega). \end{cases} \tag{2.19}$$

Conversely, it is easy to check that (2.19) has at most one solution. Hence Theorem 2.3 has the following important consequences.

Corollary 2.1. *Under the hypotheses of Theorem* 2.3, *Problem* (2.19) *has the unique solution*: ψ, $\omega = -\Delta\psi$, $\lambda = \nu\omega$ *where* $\mathbf{curl}\, \psi = \mathbf{u}$ *is the solution of* (2.1).

Theorem 2.4. *Under the hypotheses of Theorem* 2.3, *Problem* (2.19) *is equivalent to*:

Find $\psi \in \Phi_s$, $\omega \in W^{1,r}(\Omega)$ *such that*:

$$\begin{cases} \nu(\mathbf{curl}\, \omega, \mathbf{curl}\, \phi) = (\mathbf{f}, \mathbf{curl}\, \phi) & \forall \phi \in \Phi_s, \\ (\mathbf{curl}\, \psi, \mathbf{curl}\, \mu) = (\omega, \mu) & \forall \mu \in W^{1,r}(\Omega). \end{cases} \tag{2.20}$$

In addition, the pressure p is the unique solution of:

Find $p \in W^{1,r}(\Omega) \cap L_0^2(\Omega)$ *such that*

$$(\mathbf{grad}\, p, \mathbf{grad}\, q) = (\mathbf{f} - \nu\, \mathbf{curl}\, \omega, \mathbf{grad}\, q) \quad \forall q \in W^{1,s}(\Omega). \tag{2.21}$$

Proof. Equation (2.21) is obtained by taking $\mathbf{v} = \mathbf{grad}\, q$ for all $q \in W^{1,s}(\Omega)$ in (2.13) and using the fact that $\lambda = \nu\omega$. □

The approximation discussed in the next sections is based essentially on the formulations (2.20) and (2.21). It is important to point out that (2.20) can be interpreted as two Dirichlet problems for the Laplace operator. Indeed, for the sake of simplicity, consider the case where Ω is simply-connected ($p = 0$). Let ω_0 denote the trace of ω on Γ; then Problem (2.20) has the following interpretation:

$$-\nu\Delta\omega = \operatorname{curl} \mathbf{f} \quad \text{in } \Omega, \qquad -\Delta\psi = \omega \quad \text{in } \Omega,$$

$$\omega|_\Gamma = \omega_0 \qquad\qquad \psi|_\Gamma = 0.$$

Likewise, Problem (2.21) is a Neumann's problem for the Laplace operator.

Although the two problems of (2.20) are coupled (because ω_0 is unknown), problem (2.20) is entirely decoupled from (2.21) in the sense that one solves first (2.20) and afterward (2.21). Besides that, these two formulations are well-adapted for approximations with:

1°) exactly divergence-free velocities;

2°) continuous pressures.

2.2. Mixed Approximation and Application to Finite Elements of Degree l

Throughout this section we retain the hypotheses of Theorem 2.3:

$$\operatorname{curl} \mathbf{u} \in W^{1,r}(\Omega), \qquad p \in W^{1,r}(\Omega) \cap L_0^2(\Omega), \tag{2.22}$$

but to simplify the discussion, we shall postpone the approximation of the pressure and discretize for the time being the stream function and vorticity alone.

We propose to adapt to Problem (2.19) the general approximation developed in Section 1.2. The *Problem* (Q) that we want to solve is Problem (2.9):

Find $(u = (\mathbf{curl}\, \psi, \omega), \lambda) \in X \times M$ *such that*

$$\begin{cases} \nu(\omega, \theta) - (\Delta\phi + \theta, \lambda) = (\mathbf{f}, \mathbf{curl}\, \phi) & \forall v = (\mathbf{curl}\, \phi, \theta) \in X, \\ (\Delta\psi + \omega, \mu) = 0 \quad \forall \mu \in M, \end{cases} \tag{2.9}$$

with the spaces

$$X = \{v = (\mathbf{curl}\, \phi, \theta);\ \phi \in \Psi,\ \theta \in L^2(\Omega)\}, \qquad M = L^2(\Omega).$$

Then Problem (2.19) is in fact a *Problem* ($\tilde{\mathrm{Q}}$) associated with (2.9). Here the spaces are:

$$\tilde{X} = \{v = (\mathbf{curl}\,\phi, \theta);\ \phi \in \Phi_s,\ \theta \in L^2(\Omega)\}, \qquad \tilde{M} = W^{1,r}(\Omega),$$

equipped with the norms:

$$\|(\mathbf{curl}\,\phi, \theta)\|_{\tilde{X}} = |\phi|_{1,s,\Omega} + \|\theta\|_{0,\Omega}, \qquad \|\mu\|_{\tilde{M}} = \|\mu\|_{1,r,\Omega}.$$

The bilinear forms are:

$$\tilde{a}(u, v) = \nu(\omega, \theta) \qquad \forall u = (\mathbf{curl}\,\psi, \omega), \qquad v = (\mathbf{curl}\,\phi, \theta) \in \tilde{X},$$

$$\tilde{b}(v, \mu) = (\mathbf{curl}\,\phi, \mathbf{curl}\,\mu) - (\theta, \mu) \qquad \forall v = (\mathbf{curl}\,\phi, \theta) \in \tilde{X}, \qquad \forall \mu \in \tilde{M}$$

and the corresponding space $\tilde{V}$ is:

$$\tilde{V} = \{v \in \tilde{X};\ \tilde{b}(v, \mu) = 0 \quad \forall \mu \in \tilde{M}\}.$$

In view of the approximation, we introduce three finite-dimensional spaces:

$$\Phi_h \subset \Phi_s, \qquad \Theta_h \subset L^2(\Omega), \qquad M_h \subset W^{1,r}(\Omega) \tag{2.23}$$

and we discretize (2.19) by the following *Problem* (Q_h):

Find $(\psi_h, \omega_h) \in \Phi_h \times \Theta_h$ *and* $\lambda_h \in M_h$ *such that*

$$\begin{cases} \nu(\omega_h, \theta_h) - (\lambda_h, \theta_h) + (\mathbf{curl}\,\phi_h, \mathbf{curl}\,\lambda_h) = (\mathbf{f}, \mathbf{curl}\,\phi_h) & \forall (\phi_h, \theta_h) \in \Phi_h \times \Theta_h, \\ (\mathbf{curl}\,\psi_h, \mathbf{curl}\,\mu_h) = (\omega_h, \mu_h) & \forall \mu_h \in M_h. \end{cases} \tag{2.24}$$

Here, the spaces X_h and V_h are:

$$X_h = \{v_h = (\mathbf{curl}\,\phi_h, \theta_h);\ \phi_h \in \Phi_h,\ \theta_h \in \Theta_h\},$$

$$V_h = \{v_h = (\mathbf{curl}\,\phi_h, \theta_h) \in X_h;\ (\mathbf{curl}\,\phi_h, \mathbf{curl}\,\mu_h) = (\theta_h, \mu_h) \quad \forall \mu_h \in M_h\}.$$

The next lemma checks the inf-sup condition (1.21).

Lemma 2.2. *If* $M_h \subset \Theta_h$ *then*

$$\sup_{v_h \in X_h} \frac{\tilde{b}(v_h, \mu_h)}{\|v_h\|_{\tilde{X}}} \geqslant \|\mu_h\|_{0,\Omega} \qquad \forall \mu_h \in M_h. \tag{2.25}$$

Indeed, (2.25) stems from the choice $v_h = (\mathbf{0}, -\mu_h)$.

When $s = 2$, the uniform ellipticity of $\tilde{a}(.,.)$ follows from the next lemma, but when $s > 2$ there seems to be no obvious way of establishing this property and it has to be introduced by means of a hypothesis.

Lemma 2.3. *If* $\Phi_h \subset M_h$ *there exists a constant* $C > 0$ *such that:*

$$|\phi_h|_{1,\Omega} \leqslant C\|\theta_h\|_{0,\Omega} \qquad \forall (\mathbf{curl}\,\phi_h, \theta_h) \in V_h. \tag{2.26}$$

Proof. By definition a function $(\mathbf{curl}\,\phi_h, \theta_h)$ in V_h verifies:

$$(\mathbf{curl}\,\phi_h, \mathbf{curl}\,\mu_h) = (\theta_h, \mu_h) \qquad \forall \mu_h \in M_h.$$

As $\Phi_h \subset M_h$ we can take $\mu_h = \phi_h$; with Lemma I.3.1 this implies:

$$|\phi_h|_{1,\Omega}^2 \leqslant \|\theta_h\|_{0,\Omega}\|\phi_h\|_{0,\Omega} \leqslant C\|\theta_h\|_{0,\Omega}|\phi_h|_{1,\Omega},$$

thus proving (2.26). □

When $s = 2$, it follows from this lemma that there exists $\alpha^* > 0$:

$$\text{(2.27)} \qquad \nu\|\theta_h\|_{0,\Omega}^2 \geqslant \alpha^*\|v_h\|_{\tilde{X}}^2 \qquad \forall v_h = (\mathbf{curl}\,\phi_h, \theta_h) \in V_h,$$

provided $\Phi_h \subset M_h$. In other words, the *seminorm* $|.|$ *defined by* (2.15) *and* $\|.\|_{\tilde{X}}$ *are two uniformly equivalent norms on* V_h. When $s > 2$, we take it as an assumption which will be verified in the forthcoming examples.

Hypothesis H1. *For* $s > 2$, $|.|$ *and* $\|.\|_{\tilde{X}}$ *are two uniformly equivalent norms (with respect to h) on* V_h.

Lemma 2.4. 1°) *Suppose that*

$$\text{(2.28)} \qquad \Phi_h \subset M_h \subset \Theta_h$$

and when $s > 2$ *assume that Hypothesis* H1 *holds. Then Problem* (2.24) *has exactly one solution* $u_h = (\mathbf{curl}\,\psi_h, \omega_h) \in X_h$ *and* $\lambda_h \in M_h$.

2°) *If in addition* $M_h = \Theta_h$ *then* $\lambda_h = \nu\omega_h$.

Proof. 1°) Part 1 follows directly from Lemmas 2.2 and 2.3, Hypothesis H1 and Theorem 1.2.

2°) Let $M_h = \Theta_h$ and let us show that $(u_h, \nu\omega_h)$ satisfies (2.24), i.e. that

$$\nu(\mathbf{curl}\,\omega_h, \mathbf{curl}\,\phi_h) = (\mathbf{f}, \mathbf{curl}\,\phi_h) \qquad \forall\phi_h \in \Phi_h.$$

If $(\mathbf{curl}\,\phi_h, \theta_h)$ belongs to V_h, (2.24) becomes:

$$\nu(\omega_h, \theta_h) = (\mathbf{f}, \mathbf{curl}\,\phi_h).$$

But since $M_h = \Theta_h$ the functions ϕ_h and θ_h satisfy in particular:

$$(\mathbf{curl}\,\phi_h, \mathbf{curl}\,\omega_h) = (\theta_h, \omega_h).$$

Hence

$$\nu(\mathbf{curl}\,\phi_h, \mathbf{curl}\,\omega_h) = (\mathbf{f}, \mathbf{curl}\,\phi_h) \qquad \forall(\mathbf{curl}\,\phi_h, \theta_h) \in V_h.$$

This is the desired result because when $M_h = \Theta_h$, each function ϕ_h in Φ_h has (exactly) one function θ_h in Θ_h such that the pair $(\mathbf{curl}\,\phi_h, \theta_h)$ belongs to V_h. □

Note that the above proof does not require that $\Phi_h \subset M_h$ when $s > 2$; but this inclusion always holds in practice.

Lemma 2.4 has the following vital consequence.

Theorem 2.5. *Under the assumptions of Lemma* 2.4, *Problem* (2.24) is equivalent to:

Find $\psi_h \in \Phi_h$ *and* $\omega_h \in \Theta_h$ *such that*

$$(2.29)\qquad \begin{cases} \nu(\mathbf{curl}\,\omega_h, \mathbf{curl}\,\phi_h) = (\mathbf{f}, \mathbf{curl}\,\phi_h) & \forall \phi_h \in \Phi_h, \\ (\mathbf{curl}\,\psi_h, \mathbf{curl}\,\mu_h) = (\omega_h, \mu_h) & \forall \mu_h \in \Theta_h. \end{cases}$$

Problem (2.29) is the formulation that is used as a basis for practical computation. Like in the continuous case, it enjoys the following attractive features:

1°) Problem (2.29) is formulated in terms of stream functions (i.e. exactly divergence-free velocities) with H^1-regularity;

2°) it consists of two Dirichlet problems for the Laplace operator;

3°) the computation of the stream function and vorticity is independent of any future computation of the pressure.

Here again, the two problems in (2.29) are coupled. They can either be solved together or dissociated by a technique very similar to that discussed in Chapter II, Sections 4.3 and 4.4 (cf. Glowinski & Pironneau [39]).

In view of Theorem 1.2, the hypotheses (2.28) and H1 imply the error estimates (1.22) and (1.23) with constants that do not depend on h. Moreover, if $M_h = \Theta_h$, the bound (1.23) can be slightly improved by introducing the projection operator P_h from $W^{1,r}(\Omega)$ onto Θ_h defined by (A.25):

$$(\text{A.25})\qquad \begin{cases} (\mathbf{grad}(P_h\mu - \mu), \mathbf{grad}\,\theta_h) = 0 & \forall \theta_h \in \Theta_h, \\ (P_h\mu - \mu, 1) = 0. \end{cases}$$

Then the resulting estimates are given by the following theorem.

Theorem 2.6. *Let* $u = (\mathbf{curl}\,\psi, \omega = -\Delta\psi)$ *and* $u_h = (\mathbf{curl}\,\psi_h, \omega_h)$ *be the respective solutions of Problems* (2.20) *and* (2.29). *Under the assumptions of Theorem* 2.3 *and Lemma* 2.4, *we have the error bounds*:

$$(2.30)\qquad \|u - u_h\|_{\tilde{X}} \leqslant C_1 \left\{ \inf_{v_h \in V_h} \|u - v_h\|_{\tilde{X}} + \|\omega - P_h\omega\|_{0,\Omega} \right\},$$

$$(2.31)\qquad \begin{aligned} \|u - u_h\|_{\tilde{X}} \leqslant C_2 \Big[(1 + K_r(h)) \Big\{ &\inf_{\phi_h \in \Phi_h} |\psi - \phi_h|_{1,s,\Omega} \\ &+ \inf_{\theta_h \in \Theta_h} \|\omega - \theta_h\|_{0,\Omega} \Big\} + \|\omega - P_h\omega\|_{0,\Omega} \Big], \end{aligned}$$

where P_h *is defined by* (A.25) *and for* $r \leqslant 2$, $K_r(h)$ *is given here by*:

$$K_r(h) = \sup_{\theta_h \in \Theta_h} (\|\theta_h\|_{1,r,\Omega} / \|\theta_h\|_{0,\Omega}).$$

Proof. Let us establish (2.30). From (2.20) and (2.29) we infer that:

$$(\mathbf{curl}(\mu_h - \omega_h), \mathbf{curl}\,\phi_h) = (\mathbf{curl}(\mu_h - \omega), \mathbf{curl}\,\phi_h) \quad \forall \phi_h \in \Phi_h, \quad \forall \mu_h \in \Theta_h.$$

In particular, by choosing $\mu_h = P_h\omega$ we obtain:

$$(2.32)\qquad (\mathbf{curl}(P_h\omega - \omega_h), \mathbf{curl}\,\phi_h) = 0 \quad \forall \phi_h \in \Phi_h.$$

Therefore the definition of V_h implies that

(2.33) $$(P_h\omega - \omega_h, \theta_h) = 0 \qquad \forall v_h = (\mathbf{curl}\,\phi_h, \theta_h) \in V_h.$$

As a consequence,

(2.34) $$\|\omega_h - \theta_h\|_{0,\Omega} \leqslant \|P_h\omega - \theta_h\|_{0,\Omega} \qquad \forall v_h = (\mathbf{curl}\,\phi_h, \theta_h) \in V_h$$

and the equivalence of the norms $|.|$ and $\|.\|_{\tilde{X}}$ on V_h yields:

$$\|u_h - v_h\|_{\tilde{X}} \leqslant C\{\|P_h\omega - \omega\|_{0,\Omega} + |u - v_h|\} \qquad \forall v_h \in V_h.$$

This proves (2.30).

The bound (2.31) follows immediately from (2.30) and (1.26). □

Remark 2.2. Formula (2.34) provides a slightly sharper estimate for $|u - u_h|$:

$$|u - u_h| \leqslant 2 \inf_{v_h \in V_h} |u - v_h| + \|\omega - P_h\omega\|_{0,\Omega}.$$

Now we turn to the pressure. To solve for the pressure, we introduce a fourth finite-dimensional space:

(2.35) $$Q_h \subset W^{1,s}(\Omega) \cap L_0^2(\Omega)$$

with which we propose a straightforward discretization of Problem (2.21):

Find $p_h \in Q_h$ solution of

(2.36) $$(\mathbf{grad}\,p_h, \mathbf{grad}\,q_h) = (\mathbf{f} - \nu\,\mathbf{curl}\,\omega_h, \mathbf{grad}\,q_h) \qquad \forall q_h \in Q_h.$$

For a fixed ω_h, this square system of linear equations has a unique solution because the fact that $\mathbf{grad}\,p = 0$ and $p \in L_0^2(\Omega)$ implies that $p = 0$.

Remark 2.3. Problems (2.29) and (2.36) can be combined in a single one:

Find $\psi_h \in \Phi_h$, $\omega_h \in \Theta_h$ and $p_h \in Q_h$ such that

$$\nu(\mathbf{v}_h, \mathbf{curl}\,\omega_h) + (\mathbf{grad}\,p_h, \mathbf{v}_h) = (\mathbf{f}, \mathbf{v}_h) \qquad \forall \mathbf{v}_h = \mathbf{curl}\,\phi_h + \mathbf{grad}\,q_h$$

$$\text{with } \phi_h \in \Phi_h \quad \text{and} \quad q_h \in Q_h,$$

$$(\mathbf{curl}\,\psi_h, \mathbf{curl}\,\mu_h) = (\omega_h, \mu_h) \qquad \forall \mu_h \in \Theta_h.$$

Note the analogy with Problem (2.13).

Remark 2.4. Like in the continuous case, Problem (2.36) uses pressures with H^1-regularity and solves a discrete Neumann's problem for the Laplace operator.

To estimate the error attached to Problem (2.36) we shall make use of a projection operator like the one defined by (A.25) (still denoted by P_h) from $W^{1,r}(\Omega) \cap L_0^2(\Omega)$ onto Q_h. In addition, we shall require the following hypothesis concerning the regularity of the decomposition of vector fields.

Hypothesis H2. *Every vector field* $\mathbf{v} \in H_0^1(\Omega)^2$ *has the following decomposition:*

$$\mathbf{v} = \mathbf{grad}\, q + \mathbf{curl}\, \phi \quad \text{with } \phi \in \Phi_s, \qquad q \in W^{1,s}(\Omega) \cap L_0^2(\Omega)$$

and

$$(2.37) \qquad |P_h q - q|_{1,\beta,\Omega} + \inf_{\phi_h \in \Phi_h} |\phi_h - \phi|_{1,\beta,\Omega} \leqslant C h^{2/\beta} |\mathbf{v}|_{1,\Omega} \qquad \forall \beta \in [2, s],$$

with a constant C independent of h, ϕ, *q and* $\mathbf{v}$.

Theorem 2.7. *Let p and* p_h *denote the solutions of Problems* (2.21) *and* (2.36) *respectively. In addition to the assumptions of Theorem* 2.6, *suppose that Hypothesis* H2 *holds and* $p \in W^{1,\alpha}(\Omega)$, $\omega \in W^{1,\alpha}(\Omega)$ *for some real* $\alpha \in [r, 2]$. *Then the error on p is bounded as follows:*

$$(2.38) \qquad \begin{aligned} \|p - p_h\|_{0,\Omega} \leqslant C \Big\{ & h^{2/\beta} \inf_{q_h \in Q_h} |p - q_h|_{1,\alpha,\Omega} + \nu \|\omega - \omega_h\|_{0,\Omega} \\ & + \nu \inf_{\theta_h \in \Theta_h} [h K_2(h) \|\omega_h - \theta_h\|_{0,\Omega} + h^{2/\beta} |\theta_h - \omega|_{1,\alpha,\Omega}] \Big\}, \end{aligned}$$

with $1/\alpha + 1/\beta = 1$.

Proof. As $p - p_h \in L_0^2(\Omega)$, there exists $\mathbf{v} \in H_0^1(\Omega)^2$ with

$$\operatorname{div} \mathbf{v} = p - p_h, \qquad |\mathbf{v}|_{1,\Omega} \leqslant C \|p - p_h\|_{0,\Omega}.$$

Hence by virtue of Hypothesis H2 we have:

$$\mathbf{v} = \mathbf{grad}\, q + \mathbf{curl}\, \phi,$$

$$\|p - p_h\|_{0,\Omega}^2 = -(\mathbf{grad}(p - p_h), \mathbf{v}) = (\mathbf{grad}(p_h - p), \mathbf{grad}\, q).$$

Introducing $P_h q$ we can write:

$$\|p - p_h\|_{0,\Omega}^2 = (\mathbf{grad}(q_h - p), \mathbf{grad}(q - P_h q)) + (\mathbf{grad}(p_h - p), \mathbf{grad}(P_h q)) \qquad \forall q_h \in Q_h.$$

Now (2.21) and (2.36) yield:

$$(\mathbf{grad}(p - p_h), \mathbf{grad}(P_h q)) = \nu(\mathbf{curl}(\omega_h - \omega), \mathbf{grad}(P_h q)).$$

Thus taking any ϕ_h in Φ_h and setting

$$\mathbf{v}_h = \mathbf{grad}(P_h q) + \mathbf{curl}\, \phi_h,$$

we obtain in view of (2.20) and (2.29):

$$\begin{aligned} (\mathbf{grad}(p - p_h), \mathbf{grad}(P_h q)) &= \nu(\mathbf{curl}(\omega_h - \omega), \mathbf{v}_h - \mathbf{curl}\, \phi_h) \\ &= \nu(\mathbf{curl}(\omega_h - \omega), \mathbf{v}_h - \mathbf{v}) + \nu(\omega_h - \omega, \operatorname{curl} \mathbf{v}) \\ &= \nu(\mathbf{curl}(\omega_h - \theta_h), \mathbf{v}_h - \mathbf{v}) + \nu(\mathbf{curl}(\theta_h - \omega), \mathbf{v}_h - \mathbf{v}) \\ &\quad + \nu(\omega_h - \omega, \operatorname{curl} \mathbf{v}) \qquad \forall \theta_h \in \Theta_h, \quad \phi_h \in \Phi_h. \end{aligned}$$

Collecting these equalities we get:

$$\|p - p_h\|_{0,\Omega}^2 = (\mathbf{grad}(q_h - p), \mathbf{grad}(q - P_h q)) - \nu(\omega_h - \omega, \operatorname{curl} \mathbf{v})$$
$$- \nu(\mathbf{curl}(\omega_h - \theta_h), \mathbf{v}_h - \mathbf{v}) - \nu(\mathbf{curl}(\theta_h - \omega), \mathbf{v}_h - \mathbf{v})$$
$$\forall q_h \in Q_h, \quad \theta_h \in \Theta_h, \quad \phi_h \in \Phi_h.$$

Then (2.37) implies (2.38). □

We end this section with a specific example of spaces Φ_h, Θ_h, M_h and Q_h. To simplify the discussion, we assume that Ω is a *polygonal domain* of $\mathbb{R}^2$. Let $\mathscr{T}_h$ be a family of triangulations of $\bar{\Omega}$ consisting of triangles and/or convex quadrilaterals whose diameters are bounded by h. We suppose that the family $\mathscr{T}_h$ is *uniformly regular* as h tends to zero (cf. Definition A.2):

(A.16) $$\tau h \leqslant h_\kappa \leqslant \sigma \rho_\kappa \quad \forall \kappa \in \mathscr{T}_h, \quad \tau > 0, \quad \sigma > 0.$$

For the sake of conciseness, the proofs are derived for a triangulation which is made exclusively of triangles but they can be extended, with very few modifications, to the case where $\mathscr{T}_h$ also includes quadrilaterals.

For a given integer $l \geqslant 1$, we choose the following standard finite element spaces:

(2.39) $$\begin{cases} \Theta_h = M_h = \{\theta_h \in \mathscr{C}^0(\bar{\Omega});\ \theta_h|_\kappa \in P_l \quad \forall \kappa \in \mathscr{T}_h\} \subset W^{1,\infty}(\Omega), \\ \Phi_h = \Theta_h \cap \Phi = \{\phi_h \in \Theta_h;\ \phi_h|_{\Gamma_0} = 0,\ \phi_h|_{\Gamma_i} = \text{an arbitrary constant } c_i \\ \qquad 1 \leqslant i \leqslant p\}, \\ Q_h = \{q_h \in \mathscr{C}^0(\bar{\Omega}) \cap L_0^2(\Omega);\ q_h|_\kappa \in P_k \quad \forall \kappa \in \mathscr{T}_h\}, \quad k = \min(1, l-1). \end{cases}$$

The reason for taking polynomials of degree $l - 1$ for the pressure appears in the bound (2.38). Indeed, the order of convergence of $p - p_h$ cannot be higher than that of $\omega - \omega_h$, whatever the degree of p_h.

Here, we have $\Phi_h \subset \Theta_h = M_h$. Now the approximation properties of the operator P_h are stated in Theorem A.2 and the constants $K(h)$ are estimated by Corollary A.3 (both results require the uniformity of $\mathscr{T}_h$):

(2.40) $$\|\theta_h\|_{1,r,\Omega} \leqslant C(r)(1/h)\|\theta_h\|_{0,\Omega} \quad \forall \theta_h \in \Theta_h, \quad \forall r \leqslant 2,$$

where the constant $C(r)$ depends upon r and Ω. As a consequence, we have the following estimate for the approximation error in V_h:

(2.40′) $$\inf_{v_h \in V_h} \|v - v_h\|_{\tilde{X}} \leqslant C h^{k-1}(|\phi|_{k+1,s,\Omega} + \|\theta\|_{k,\Omega}) \quad 0 \leqslant k \leqslant l,$$
$$\forall v = (\mathbf{curl}\,\phi, \theta) \in V \cap (W^{k+1,s}(\Omega) \times H^k(\Omega)).$$

Thus it remains to check Hypotheses H1 (when $s > 2$) and H2. This is the object of the next two lemmas. Both include the assumption that Ω is a convex polygon because their proofs require extra regularity for the solution of the Dirichlet problem for the Laplace operator.

Lemma 2.5. *In addition to the above assumptions, suppose that Ω is a convex polygon. Then for each real $s > 2$ there exists a positive constant C_s, independent of h, such that:*

$$(2.41) \qquad |\phi_h|_{1,s,\Omega} \leqslant C_s \|\theta_h\|_{0,\Omega} \qquad \forall v_h = (\mathbf{curl}\,\phi_h, \theta_h) \in V_h.$$

Proof. First observe that because Ω is convex, its boundary Γ has only one component (i.e. $\Gamma = \Gamma_0$) and $\Phi_h \subset W_0^{1,s}(\Omega)$.

Next, in view of (2.16) and Sobolev's Theorem I.1.3, every function $v = (\mathbf{curl}\,\phi, \theta)$ in V satisfies (2.41). Moreover we have

$$(2.42) \qquad -\Delta\phi = \theta \quad \text{in } \Omega.$$

Of course, this equality does not hold for the functions of V_h but nevertheless they satisfy

$$(\mathbf{curl}\,\phi_h, \mathbf{curl}\,\mu_h) = (\theta_h, \mu_h) \qquad \forall \mu_h \in \Theta_h$$

which can be interpreted as a discrete version of (2.42). This suggests to compare ϕ_h with the auxiliary function $\phi(h)$, solution of the homogeneous Dirichlet problem:

$$\begin{cases} -\Delta\phi(h) = \theta_h & \text{in } \Omega, \\ \phi(h) = 0 & \text{on } \Gamma. \end{cases}$$

Since Ω is convex, Theorem I.1.8 claims that $\phi(h) \in H^2(\Omega)$ with

$$(2.43) \qquad \|\phi(h)\|_{2,\Omega} \leqslant C_1 \|\theta_h\|_{0,\Omega}.$$

Moreover, we can readily see that $\phi_h = \mathring{P}_h\phi(h)$, the H^1-projection of $\phi(h)$ onto Φ_h (cf. (A.24)). Therefore Theorem A.2 yields:

$$|\phi(h) - \phi_h|_{1,s,\Omega} \leqslant C_2 \|\phi(h)\|_{1,s,\Omega}.$$

Hence applying (2.43) and Sobolev's Theorem I.1.3, we get:

$$|\phi_h|_{1,s,\Omega} \leqslant (1 + C_2) \|\phi(h)\|_{1,s,\Omega} \leqslant C_3 \|\theta_h\|_{0,\Omega}. \qquad \square$$

Remark 2.5. Lemma 2.5 has a more elementary proof that does not call for Theorem A.2 (which is a difficult result). On the one hand, the fact that $\phi_h = \mathring{P}_h\phi(h)$ implies that

$$|\phi_h - \chi_h|_{1,\Omega} \leqslant |\phi(h) - \chi_h|_{1,\Omega} \qquad \forall \chi_h \in \Phi_h,$$

and Lemma A.7 gives for all $s > 2$:

$$|\phi_h - \chi_h|_{1,s,\Omega} \leqslant C(s) h^{2/s-1} |\phi_h - \chi_h|_{1,\Omega}.$$

On the other hand, we can write:

$$|\phi_h|_{1,s,\Omega} \leqslant |\phi_h - \chi_h|_{1,s,\Omega} + |\chi_h|_{1,s,\Omega} \qquad \forall \chi_h \in \Phi_h.$$

Collecting these three inequalities, we get:

$$|\phi_h|_{1,s,\Omega} \leqslant \inf_{\chi_h \in \Phi_h} \{C(s)h^{2/s-1}|\phi(h) - \chi_h|_{1,\Omega} + |\chi_h|_{1,s,\Omega}\}.$$

But as $s > 2$, Sobolev's Theorem I.1.3 states that $W^{1,s}(\Omega) \hookrightarrow \mathscr{C}^0(\bar{\Omega})$ and therefore we can take $\chi_h = I_h\phi(h)$, where I_h is the standard finite element interpolation operator defined by Lemma A.2. This lemma gives the estimates:

$$|I_h\phi(h)|_{1,s,\Omega} \leqslant C_1|\phi(h)|_{1,s,\Omega},$$

$$|\phi(h) - I_h\phi(h)|_{1,\Omega} \leqslant C_2 h|\phi(h)|_{2,\Omega}.$$

In view of (2.43) this proves (2.41).

Remark 2.6. When s is not too large, it can be shown that the convexity hypothesis on Ω is not necessary for (2.41). Notice that the alternate proof given above only uses this hypothesis to deduce that $\phi(h) \in H^2(\Omega)$, but in fact it does not require so much regularity. For example, if $s = 4$ (and this is the exponent most often used for Navier-Stokes equations) it suffices that $\phi(h) \in H^{3/2}(\Omega)$ because $H^{3/2}(\Omega) \hookrightarrow W^{1,4}(\Omega)$. But an extension of Theorem I.1.8 (cf. Grisvard [42]) proves that $\phi(h)$ belongs to $H^{3/2}(\Omega)$ whenever the angles of Γ are bounded by $3\pi/2$ and this allows a much wider range of domains. The reader will find more details in Section IV.4.3.

Lemma 2.6. *Hypothesis* H2 *is satisfied for all real* $s \geqslant 2$ *when* Ω *is a convex polygon and* $\mathscr{T}_h$ *is uniformly regular.*

Proof. Let $\mathbf{v} \in H_0^1(\Omega)^2$. We use the orthogonal decomposition of $\mathbf{v}$ established by Theorem I.3.2:

$$\mathbf{v} = \mathbf{grad}\, q + \mathbf{curl}\, \phi,$$

where $\phi \in \Phi$ and, in view of Remark I.3.2 $q \in H^1(\Omega) \cap L_0^2(\Omega)$ is the solution of:

$$\begin{cases} \Delta q = \operatorname{div} \mathbf{v} & \text{in } \Omega, \\ \partial q/\partial n = 0 & \text{on } \Gamma. \end{cases}$$

Since Ω is convex, Theorem I.1.10 asserts that $q \in H^2(\Omega) \cap L_0^2(\Omega)$ with:

$$\|q\|_{2,\Omega} \leqslant C_1 \|\operatorname{div} \mathbf{v}\|_{0,\Omega} \leqslant C_1\sqrt{2}|\mathbf{v}|_{1,\Omega}.$$

As a consequence $\phi \in H^2(\Omega) \cap H_0^1(\Omega)$ and

$$\|\phi\|_{2,\Omega} \leqslant C_2 \|\mathbf{v}\|_{1,\Omega} \leqslant C_3|\mathbf{v}|_{1,\Omega}.$$

By virtue of Sobolev's Theorem I.1.3, we have

$$H^2(\Omega) \hookrightarrow W^{1+2/s,s}(\Omega).$$

Then, since the polynomials of Q_h are at least of degree one, (A.26) implies:

$$|q - P_h q|_{1,s,\Omega} \leqslant C_4 h^{2/s}|\mathbf{v}|_{1,\Omega}.$$

Likewise, (A.17) and Jensen's inequality (A.34) yield:

$$|\phi - I_h\phi|_{1,s,\Omega} \leqslant C_5 h^{2/s} |\mathbf{v}|_{1,\Omega}.$$

Hence Hypothesis H2 holds for all $s \geqslant 2$. □

From Theorems 2.6, 2.7 and these two lemmas we immediately derive the main result of this section.

Theorem 2.8. *Let Ω be an open, bounded convex polygon and let $\mathscr{T}_h$ be a uniformly regular family of triangulations of $\bar{\Omega}$. Let $(u = (\mathbf{curl}\,\psi, \omega = -\Delta\psi), p)$ be the solution of* (2.20) *and* (2.21) *and assume that*

(2.44) $\psi \in W^{k+1,s}(\Omega)$, $\Delta\psi \in H^k(\Omega)$, $p \in H^m(\Omega) \cap L_0^2(\Omega)$, $m = \max(1, k-1)$

for some integer k such that $1 \leqslant k \leqslant l$. Then we have the error bounds:

(2.45) $\|u - u_h\|_{\tilde{X}} \leqslant C_1[h^{k-1}(|\psi|_{k+1,s,\Omega} + |\Delta\psi|_{k,\Omega}) + h^k \|\Delta\psi\|_{k,\Omega}]$,

(2.46) $\|p - p_h\|_{0,\Omega} \leqslant C_2[h^{k-1}(|p|_{m,\Omega} + |\psi|_{k+1,s,\Omega} + |\Delta\psi|_{k,\Omega}) + h^k \|\Delta\psi\|_{k,\Omega}]$,

with constants $C_1 > 0$ and $C_2 > 0$ independent of h, ψ and p.

Remark 2.7. Theorem 2.8 uses the convexity of Ω in three instances: to verify Hypotheses H1 and H2 and to evaluate $\|\omega - P_h\omega\|_{0,\Omega}$. In many cases, Hypothesis H1 does not demand that Ω be convex (cf. Remark 2.6). But the L^2-estimate for $P_h\omega$, which stems from a duality argument, and Hypothesis H2 both require the convexity of Ω (because Γ has corners). As far as (2.45) is concerned, we can always replace $\|\omega - P_h\omega\|_{0,\Omega}$ by $\inf_{\theta_h \in \Theta_h} \|\omega - \theta_h\|_{1,r,\Omega}$ (cf. (1.27)) and, if ω is sufficiently smooth, derive an estimate of the same order for $\|u - u_h\|_{\tilde{X}}$ without Ω being convex.

But unfortunately, our proof of (2.46) does require the convexity of Ω and we see no possibility of discarding this assumption.

Because of the factor $K(h)$ the estimates of Theorem 2.8 are not optimal; in particular (2.45) and (2.46) do not imply convergence when $l = 1$ (i.e. when piecewise linear elements are used). As mentioned previously, Section 3.1 will take care of this difficulty.

2.3. The Technique of Mesh-Dependent Norms

In this section we propose to improve to some extent the estimate (2.45) by the use of mesh-dependent norms, introduced by Babuška, Osborn & Pitkäranta [5], which are better adapted here than $\|\,.\,\|_{\tilde{X}}$. The underlying idea for introducing these norms is an integration by parts on each element κ of $\mathscr{T}_h$. To be specific, let $\phi \in H^2(\kappa)$ and $\theta \in H^1(\kappa)$; Green's formula gives:

$$\int_\kappa \mathbf{curl}\,\phi\cdot\mathbf{curl}\,\theta\,dx = -\int_\kappa \Delta\phi\theta\,dx + \int_{\partial\kappa}(\partial\phi/\partial n)\theta\,ds,$$

where $\mathbf{n}$ denotes the exterior unit normal to κ. In order to sum this identity over all elements κ of $\mathscr{T}_h$, it is convenient to define the set:

$$\Gamma_h = \{\bigcup \kappa';\text{ for all sides }\kappa'\text{ of }\kappa\text{ and all elements }\kappa\text{ of }\mathscr{T}_h\}$$

and the jump $S(\partial\phi/\partial n)$ of $\partial\phi/\partial n$ across the side κ' of κ. More precisely, if κ' is a side shared by κ_1 and κ_2 and $\mathbf{n}_i$ denotes the exterior normal to κ_i we set

$$S(\partial\phi/\partial n)|_{\kappa'} = \partial\phi/\partial n_1 + \partial\phi/\partial n_2$$

and if $\kappa' \subset \Gamma$ we set

$$S(\partial\phi/\partial n)|_{\kappa'} = \partial\phi/\partial n.$$

With these notations we have

$$\text{(2.47)}\quad \begin{cases} (\mathbf{curl}\,\phi, \mathbf{curl}\,\theta) = -\sum_{\kappa\in\mathscr{T}_h}\int_\kappa \Delta\phi\theta\,dx + \int_{\Gamma_h} S(\partial\phi/\partial n)\theta\,ds \\ \qquad \forall\phi\in H^1(\Omega)\text{ with }\phi\in H^2(\kappa)\text{ on each }\kappa\text{ of }\mathscr{T}_h, \quad \forall\theta\in H^1(\Omega). \end{cases}$$

From now on, we are going to work with L^2 and H^m-spaces. The identity (2.47) is the basis for another Problem (Q) that is equivalent to Problem (2.20). Indeed, if the solution $\mathbf{u}$ and right-hand side $\mathbf{f}$ of the Stokes Problem (2.1) satisfy: $\mathbf{f}\in L^2(\Omega)^2$, $\omega = \operatorname{curl}\mathbf{u}\in H^1(\Omega)$ then an easy argument shows that the pair (ψ, ω) satisfying (2.20) is also the only solution of *Problem* (Q):

$$\begin{cases} -\nu\sum_{\kappa\in\mathscr{T}_h}\int_\kappa \Delta\phi\omega\,dx + \nu\int_{\Gamma_h} S(\partial\phi/\partial n)\omega\,ds = (\mathbf{f}, \mathbf{curl}\,\phi), \\ (\omega,\mu) + \sum_{\kappa\in\mathscr{T}_h}\int_\kappa \Delta\psi\mu\,dx - \int_{\Gamma_h} S(\partial\psi/\partial n)\mu\,ds = 0, \\ \qquad\qquad \forall\mu\in H^1(\Omega), \quad \forall\phi\in\tilde{\Psi}\text{ with} \\ \tilde{\Psi} = \{\phi\in\Phi;\ \phi|_\kappa\in H^2(\kappa)\quad \forall\kappa\in\mathscr{T}_h\}. \end{cases}$$

As far as the approximation is concerned, we notice that the functions of Φ_h (cf. (2.39)) belong to Φ globally and to $H^2(\kappa)$ (in particular) on each κ. Thus the identity (2.47) is valid for ϕ_h in Φ_h and θ_h in Θ_h and consequently, Problem (2.29) is equivalent to the following *Problem* (Q_h):

Find $\psi_h\in\Phi_h$ and $\omega_h\in\Theta_h$ satisfying:

$$\text{(2.48)}\quad \begin{cases} -\nu\sum_{\kappa\in\mathscr{T}_h}\int_\kappa \Delta\phi_h\omega_h\,dx + \nu\int_{\Gamma_h} S(\partial\phi_h/\partial n)\omega_h\,ds = (\mathbf{f}, \mathbf{curl}\,\phi_h) \quad \forall\phi_h\in\Phi_h, \\ (\omega_h,\mu_h) + \sum_{\kappa\in\mathscr{T}_h}\int_\kappa \Delta\psi_h\mu_h\,dx - \int_{\Gamma_h} S(\partial\psi_h/\partial n)\mu_h\,ds = 0 \quad \forall\mu_h\in\Theta_h. \end{cases}$$

A look at (2.47) and (2.48) suggests to introduce the following bilinear forms:

$$a_h(\omega, \mu) = (\omega, \mu) \qquad \forall \omega, \mu \in L^2(\Omega), \tag{2.49}$$

$$b_h(\theta, \phi) = \sum_{\kappa \in \mathscr{T}_h} \int_\kappa \Delta\phi\theta \, dx - \int_{\Gamma_h} S(\partial\phi/\partial n)\theta \, ds \tag{2.50}$$
$$\forall \theta \in H^1(\Omega), \quad \forall \phi \in H^1(\Omega) \text{ with } \phi \in H^2(\kappa) \text{ on all } \kappa \text{ of } \mathscr{T}_h.$$

(The subscript h is introduced to emphasize the fact that some of the spaces involved depend upon the triangulation.) With these bilinear forms, Problem (2.48) appears indeed as a *Problem* (Q_h):

Find a pair $(\omega_h, \psi_h) \in \Theta_h \times \Phi_h$ *such that*

$$\begin{cases} a_h(\omega_h, \theta_h) + b_h(\theta_h, \psi_h) = 0 & \forall \theta_h \in \Theta_h, \\ b_h(\omega_h, \phi_h) = -(1/\nu)(\mathbf{f}, \mathbf{curl}\, \phi_h) & \forall \phi_h \in \Phi_h. \end{cases} \tag{2.48'}$$

Obviously,

$$b_h(\theta, \phi) = -(\mathbf{curl}\, \phi, \mathbf{curl}\, \theta) \tag{2.51a}$$
$$\forall \phi \in H^1(\Omega) \text{ with } \phi|_\kappa \in H^2(\kappa), \qquad \forall \theta \in H^1(\Omega),$$

$$b_h(\theta, \phi) = \int_\Omega \Delta\phi\theta \, dx \qquad \forall \phi \in H^2(\Omega) \text{ with } (\partial\phi/\partial n)|_\Gamma = 0, \qquad \forall \theta \in H^1(\Omega). \tag{2.51b}$$

In addition, observe that $b_h(.,.)$ is well defined if θ is only in $L^2(\Omega)$ with $\theta|_{\Gamma_h} \in L^2(\Gamma_h)$. These considerations induce us to associate with these bilinear forms the following mesh-dependent seminorms:

$$\|\theta\|_{0,h} = (\|\theta\|_{0,\Omega}^2 + h\|\theta\|_{0,\Gamma_h}^2)^{1/2} \qquad \forall \theta \in H^1(\Omega), \tag{2.52}$$

$$\|\phi\|_{2,h} = \left(\sum_{\kappa \in \mathscr{T}_h} |\phi|_{2,\kappa}^2 + (1/h)\|S(\partial\phi/\partial n)\|_{0,\Gamma_h}^2\right)^{1/2} \tag{2.53}$$
$$\forall \phi \in H^1(\Omega) \text{ with } \phi|_\kappa \in H^2(\kappa).$$

Clearly,

$$|a_h(\omega, \theta)| \leqslant \|\omega\|_{0,\Omega}\|\theta\|_{0,\Omega} \qquad \forall \omega, \theta \in L^2(\Omega),$$

$$|b_h(\theta, \phi)| \leqslant \sqrt{2}\|\theta\|_{0,h}\|\phi\|_{2,h} \qquad \forall \theta \in H^1(\Omega), \qquad \forall \phi \in H^1(\Omega) \text{ with } \phi|_\kappa \in H^2(\kappa).$$

Nearly all the remainder of this section is devoted to the study of these bilinear forms and seminorms. We are going to see, in particular, that these seminorms are norms on Θ_h and Φ_h respectively and that $b_h(.,.)$ satisfies a uniform inf-sup condition with respect to them.

The following lemma shows that $\|.\|_{0,h}$ and $\|.\|_{0,\Omega}$ are *two uniformly equivalent norms* on Θ_h.

Lemma 2.7. *If $\mathscr{T}_h$ is a uniformly regular family of triangulations of $\bar{\Omega}$, there exists a constant $C > 0$, independent of h, such that:*

$$\|\theta_h\|_{0,h} \leqslant C\|\theta_h\|_{0,\Omega} \qquad \forall \theta_h \in \Theta_h. \tag{2.54}$$

The proof is left as an exercise to the reader. (Hint: show that

$$\|\theta_h\|^2_{0,\Gamma_h} \leqslant (C/h)\,\|\theta_h\|^2_{0,\Omega} \qquad \forall\theta_h \in \Theta_h.$$

The next lemma derives an analogue of Corollary A.3.

Lemma 2.8. *If $\mathscr{T}_h$ is a uniformly regular family of triangulations of $\bar{\Omega}$, there exists a constant $C > 0$, independent of h, such that:*

$$(2.55) \qquad \|\theta_h\|_{2,h} \leqslant (C/h)\,|\theta_h|_{1,\Omega} \qquad \forall\theta_h \in \Theta_h.$$

Proof. If we apply formula (A.31) with $r = p = 2$ to $\mathbf{grad}\,\theta_h$ in each κ we obtain:

$$|\theta_h|_{2,\kappa} \leqslant (C_1/h)\,|\theta_h|_{1,\kappa} \qquad \forall\kappa \in \mathscr{T}_h.$$

Thus it remains to establish that

$$(2.56) \qquad \|S(\partial\theta_h/\partial n)\|^2_{0,\Gamma_h} \leqslant (C_2/h)\,|\theta_h|^2_{1,\Omega} \qquad \forall\theta_h \in \Theta_h.$$

Let κ be an arbitrary element of $\mathscr{T}_h$. In view of (A.9) and (A.2) we get:

$$\int_{\partial\kappa} |\partial\theta_h/\partial n|^2\,ds \leqslant \int_{\partial\kappa} \|\mathbf{grad}\,\theta_h\|^2\,ds$$
$$\leqslant C_3(h_\kappa/\rho_\kappa^2)\,\|\mathbf{grad}\,\hat{\theta}_h\|^2_{0,\partial\hat{\kappa}}.$$

But $\mathbf{grad}\,\hat{\theta}_h$ belongs to the finite-dimensional space P^2_{l-1} and therefore

$$\int_{\partial\kappa} |\partial\theta_h/\partial n|^2\,ds \leqslant C_4(\sigma_\kappa/\rho_\kappa)\,|\hat{\theta}_h|^2_{1,\hat{\kappa}}.$$

Then (A.7), (A.2) and (A.4) give (2.56). □

Of course, it is not possible to extend (2.54) to $H^1(\Omega)$ but we can readily prove the next relation:

Lemma 2.9. *If $\mathscr{T}_h$ is uniformly regular there exists a constant $C > 0$, independent of h, such that:*

$$(2.57) \qquad \|\theta\|_{0,h} \leqslant C(\|\theta\|^2_{0,\Omega} + h^2|\theta|^2_{1,\Omega})^{1/2} \qquad \forall\theta \in H^1(\Omega).$$

Proof. Owing to the trace Theorem I.1.5 there exists a constant C_1, independent of the geometry of κ such that

$$\|\theta\|^2_{0,\kappa'} \leqslant C_1 h_\kappa \|\hat{\theta}\|^2_{1,\hat{\kappa}} \qquad \forall\theta \in H^1(\kappa).$$

Then (A.7), (A.2), (A.4) and (A.16) yield:

$$\|\theta\|^2_{0,\kappa'} \leqslant C_2\sigma^2[1/(\tau h)\,\|\theta\|^2_{0,\kappa} + h|\theta|^2_{1,\kappa}],$$

thus proving (2.57). □

This lemma has the following interesting consequence.

Corollary 2.2. *If $\mathscr{T}_h$ is uniformly regular, there exists a constant $C > 0$, independent of h, such that*

$$\|\phi\|_{2,h} \geqslant C\left\{\|\phi\|_{1,\Omega}^2 + \sum_{\kappa\in\mathscr{T}_h} |\phi|_{2,\kappa}^2\right\}^{1/2}, \tag{2.58}$$

for all $\phi \in \tilde{\Psi}$.

Proof. By virtue of Lemma I.3.1 (a generalized Poincaré inequality), it suffices to show that:

$$\|\phi\|_{2,h} \geqslant C_1 |\phi|_{1,\Omega}.$$

But it follows from (2.47) and (2.50) that

$$|\phi|_{1,\Omega}^2 = -b_h(\phi,\phi) \leqslant \sqrt{2}\,\|\phi\|_{2,h}\|\phi\|_{0,h}$$

and (2.57) combined with Lemma I.3.1 yields (2.58). □

Corollary 2.2 means that the space $\tilde{\Psi}$ is a Hilbert space for the norm $\|\,.\,\|_{2,h}$.
The next two lemmas derive approximation properties of the space Θ_h.

Lemma 2.10. *Let $\mathscr{T}_h$ be a regular triangulation of $\bar{\Omega}$. For each real $k \in [1, l+1]$ there exists a constant $C > 0$, independent of h, such that:*

$$\inf_{\theta_h\in\Theta_h} \|v - \theta_h\|_{0,h} \leqslant Ch^k|v|_{k,\Omega} \quad \forall v \in H^k(\Omega). \tag{2.59}$$

Proof. If $k \geqslant 2$ we take $\theta_h = I_h v$, the interpolation defined by (A.20); otherwise, we take $\theta_h = R_h v$, the local regularization defined by (A.53), (A.54). Then either (A.21a) or (A.55) gives:

$$\|v - \theta_h\|_{0,\Omega} \leqslant C_1 h^k |v|_{k,\Omega} \quad \forall v \in H^k(\Omega).$$

Thus it remains to evaluate $\|v - \theta_h\|_{0,\Gamma_h}$.
Like in Lemma 2.9, we have

$$\|v - \theta_h\|_{0,\kappa'}^2 \leqslant C_1 h_\kappa \|\hat{v} - \hat{\theta}_h\|_{1,\hat{\kappa}}^2.$$

Then if $\theta_h = I_h v$, this gives

$$\|v - I_h v\|_{0,\kappa'}^2 \leqslant C_2 h_\kappa |\hat{v}|_{k,\hat{\kappa}}^2;$$

hence we infer from (A.7), (A.2) and (A.4) that:

$$\|v - I_h v\|_{0,\kappa'}^2 \leqslant C_3 \sigma^2 h^{2k-1} |v|_{k,\kappa}^2.$$

This proves (2.59).
When $\theta_h = R_h v$, we use the inequality arising in the proof of Lemma 2.9:

$$\|v - \theta_h\|_{0,\kappa'}^2 \leqslant C_4 \sigma_\kappa [(1/\rho_\kappa)\|v - \theta_h\|_{0,\kappa}^2 + \sigma_\kappa h_\kappa |v - \theta_h|_{1,\kappa}^2].$$

Then, applying the local error estimate (A.55) of R_h:

(2.60) $$\|v - R_h v\|_{0,\kappa} + h_\kappa |v - R_h v|_{1,\kappa} \leqslant C_5 h_\kappa^k |v|_{k,\Delta_\kappa} \quad \forall v \in H^k(\Delta_\kappa),$$

we obtain

$$\|v - R_h v\|_{0,\kappa'}^2 \leqslant C_6 \sigma^2 h^{2k-1} |v|_{k,\Delta_\kappa}^2.$$

Recall that Δ_κ denotes the union of all elements of $\mathscr{T}_h$ that share a vertex or a side with κ. In addition, it stems from the regularity of $\mathscr{T}_h$ that the maximum number of occurrences of a given element κ in the sets Δ_κ is bounded by a fixed constant M independent of h and κ. Therefore, summing the last inequality over all segments κ' of Γ_h yields again (2.59). □

Lemma 2.11. *Let $\mathscr{T}_h$ be a regular triangulation of $\bar{\Omega}$. For each real $k \in [2, l+1]$ there exists a constant $C > 0$, independent of h, such that:*

(2.61) $$\inf_{\theta_h \in \Theta_h} \|v - \theta_h\|_{2,h} \leqslant C h^{k-2} |v|_{k,\Omega} \quad \forall v \in H^k(\Omega).$$

Proof. Since $k \geqslant 2$ we can take $\theta_h = I_h v$. Then according to (A.21a), it suffices to show that:

$$\|S(\partial(v - I_h v)/\partial n)\|_{0,\Gamma_h}^2 \leqslant C_1 h^{2k-3} |v|_{k,\Omega}^2 \quad \forall v \in H^k(\Omega).$$

Like in Lemma 2.8 we obtain

$$\begin{aligned}\|\partial(v - \theta_h)/\partial n\|_{0,\partial\kappa}^2 &\leqslant C_2(\sigma_\kappa/\rho_\kappa) \|\mathbf{grad}(\hat{v} - \hat{\theta}_h)\|_{0,\partial\hat{\kappa}}^2, \\ &\leqslant C_3(\sigma_\kappa/\rho_\kappa)(|\hat{v} - \hat{\theta}_h|_{1,\hat{\kappa}}^2 + |\hat{v} - \hat{\theta}_h|_{2,\hat{\kappa}}^2) \\ &\leqslant C_4(\sigma_\kappa/\rho_\kappa) |\hat{v}|_{k,\hat{\kappa}}^2.\end{aligned}$$

Hence

$$\|\partial(v - \theta_h)/\partial n\|_{0,\partial\kappa}^2 \leqslant C_5 \sigma^4 h^{2k-3} |v|_{k,\kappa}^2,$$

which implies (2.61). □

As an application of Lemmas 2.10 and 2.11, the best estimates are:

$$\left.\begin{aligned}\inf_{\theta_h \in \Theta_h} \|v - \theta_h\|_{0,h} &\leqslant C h^{l+1} |v|_{l+1,\Omega} \\ \inf_{\theta_h \in \Theta_h} \|v - \theta_h\|_{2,h} &\leqslant C h^{l-1} |v|_{l+1,\Omega}.\end{aligned}\right\} \quad \forall v \in H^{l+1}(\Omega).$$

Next, let us examine the ellipticity and inf-sup condition that will enable us to apply the standard approximation theory of Section II.1.1. It follows from (2.54) that the bilinear form a_h is *uniformly elliptic* on Θ_h equipped with the norm $\|\cdot\|_{0,h}$:

(2.62) $$a_h(\theta_h, \theta_h) \geqslant \alpha^* \|\theta_h\|_{0,h}^2 \quad \forall \theta_h \in \Theta_h, \quad \alpha^* > 0 \ \text{ independent of } h.$$

But more important, the next lemma establishes that the bilinear form b_h satisfies a uniform inf-sup condition on Θ_h and Φ_h equipped respectively with the norms $\|\cdot\|_{0,h}$ and $\|\cdot\|_{2,h}$.

Lemma 2.12. *Assume that Ω is a bounded, convex polygon and that the triangulation $\mathscr{T}_h$ is uniformly regular. There exists a constant $\beta^* > 0$, independent of h, such that*:

(2.63)
$$\sup_{\theta_h \in \Theta_h} \left\{ \frac{b_h(\theta_h, \phi_h)}{\|\theta_h\|_{0,h}} \right\} \geqslant \beta^* \|\phi_h\|_{2,h} \quad \forall \phi_h \in \Phi_h.$$

Proof. First recall that

$$b_h(\theta_h, \phi_h) = -(\mathbf{curl}\, \phi_h, \mathbf{curl}\, \theta_h) \quad \forall \phi_h \in \Phi_h, \quad \forall \theta_h \in \Theta_h.$$

Next for a given ϕ_h in Φ_h, define θ_h in Θ_h by:

(2.64)
$$(\theta_h, \mu_h) = (\mathbf{curl}\, \phi_h, \mathbf{curl}\, \mu_h) \quad \forall \mu_h \in \Theta_h$$

(note the analogy with the second equation in (2.24)). Hence, by virtue of (2.62) we have found θ_h in Θ_h such that

$$-b_h(\theta_h, \phi_h) \geqslant \alpha^* \|\theta_h\|_{0,h}^2.$$

Therefore, in order to derive (2.63), we are going to show that:

(2.65)
$$\|\phi_h\|_{2,h} \leqslant C \|\theta_h\|_{0,\Omega}.$$

Now the proof of (2.65) is very similar to that of (2.41) for ϕ_h and θ_h bear here the same relation. But instead, it is convenient to interpret (2.64) as the discretization in Θ_h of a homogeneous Neumann's problem. This leads us to introduce the solution $\phi(h)$ (unique up to an additive constant) of

$$-\Delta \phi(h) = \theta_h \quad \text{in } \Omega, \qquad \partial \phi(h)/\partial n = 0 \quad \text{on } \Gamma.$$

Note that by choosing $\mu_h = 1$ ($\in \Theta_h$) in (2.64), we find that

$$\int_\Omega \theta_h \, dx = 0,$$

which is precisely the compatibility condition of this Neumann's problem. In addition, the convexity of Ω implies that $\phi(h) \in H^2(\Omega)/\mathbb{R}$ with:

$$|\phi(h)|_{2,\Omega} \leqslant C_1 \|\theta_h\|_{0,\Omega}.$$

Like in Remark 2.5 we derive that

$$|\phi_h - \chi_h|_{1,\Omega} \leqslant |\phi(h) - \chi_h|_{1,\Omega} \quad \forall \chi_h \in \Theta_h.$$

Hence in view of (2.55) we obtain

$$\|\phi_h - \chi_h\|_{2,h} \leqslant C_2(1/h) |\phi(h) - \chi_h|_{1,\Omega}.$$

Therefore

$$\|\phi_h\|_{2,h} \leqslant \inf_{\chi_h \in \Theta_h} \{C_2(1/h)|\phi(h) - \chi_h|_{1,\Omega} + \|\chi_h - \phi(h)\|_{2,h} + \|\phi(h)\|_{2,h}\}.$$

The choice $\chi_h = I_h \phi(h)$ gives

$$\|\chi_h - \phi(h)\|_{2,h} \leqslant C_3 |\phi(h)|_{2,\Omega}$$

by virtue of Lemma 2.11 and

$$|\chi_h - \phi(h)|_{1,\Omega} \leqslant C_4 h |\phi(h)|_{2,\Omega} \quad \text{by (A.21a).}$$

In addition, since $\phi(h) \in H^2(\Omega)$ and $\partial\phi(h)/\partial n = 0$ on Γ it follows that $S(\partial\phi(h)/\partial n) = 0$ and so

$$\|\phi(h)\|_{2,h} = |\phi(h)|_{2,\Omega}.$$

By collecting these inequalities we derive (2.65). □

Remark 2.8. In the above proof, since θ_h and ϕ_h are related by (2.64), it follows from Lemma 2.5 that we have:

$$|\phi_h|_{1,s,\Omega} \leqslant C_s \|\theta_h\|_{0,\Omega} \qquad \forall \text{real } s \geqslant 2.$$

Therefore, under the assumptions of Lemma 2.12, in addition to (2.63), the following inf-sup condition holds:

$$\sup_{\theta_h \in \Theta_h} \left\{ \frac{b_h(\theta_h, \phi_h)}{\|\theta_h\|_{0,h}} \right\} \geqslant \gamma(s) |\phi_h|_{1,s,\Omega} \qquad \forall \text{real } s \geqslant 2,$$

with a constant $\gamma(s) > 0$ independent of h.

Now, Problem (2.48) can be placed into the abstract framework of Section 1.2 but it is in fact quicker to elaborate directly its error analysis. Let us introduce the space V_h naturally attached to the formulation (2.48′):

$$V_h = \{(\theta_h, \phi_h) \in \Theta_h \times \Phi_h;\ a_h(\theta_h, \mu_h) + b_h(\mu_h, \phi_h) = 0 \quad \forall \mu_h \in \Theta_h\}.$$

Then by applying the arguments of Section II.1.1, it is an easy exercise to derive the following approximation results:

$$\|\omega - \omega_h\|_{0,h} \leqslant (1 + \sqrt{2}C/\alpha^*) \inf_{(\theta_h, \phi_h) \in V_h} \|\omega - \theta_h\|_{0,h},$$

$$\text{(2.66)} \quad \|\omega - \omega_h\|_{0,h} \leqslant (1 + \sqrt{2}C/\alpha^*) \left\{ (1 + 1/\alpha^*) \inf_{\theta_h \in \Theta_h} \|\omega - \theta_h\|_{0,h} + (\sqrt{2}/\alpha^*) \inf_{\phi_h \in \Phi_h} \|\psi - \phi_h\|_{2,h} \right\},$$

$$\|\psi - \psi_h\|_{2,h} \leqslant (1 + \sqrt{2}/\beta^*) \inf_{\theta_h \in \Theta_h} \|\psi - \phi_h\|_{2,h} + (1/\beta^*) \|\omega - \omega_h\|_{0,\Omega},$$

where α^*, β^* and C are the constants of (2.62), (2.63) and (2.65) respectively. With Lemmas 2.10 and 2.11 we derive immediately the next theorem.

Theorem 2.9. *Let Ω be a bounded, convex polygon and $\mathcal{T}_h$ a uniformly regular family of triangulations of $\bar{\Omega}$. If the solution $\mathbf{u} = \mathbf{curl}\,\psi$ and right-hand side $\mathbf{f}$ of the*

Stokes problem (2.1) *satisfy*

$$\mathbf{f} \in L^2(\Omega)^2, \qquad \psi \in H^{k+1}(\Omega), \qquad \Delta\psi \in H^k(\Omega)$$

for some real $k \in [1, l]$, *then the solution* (ω_h, ψ_h) *of Problem* (2.48) *satisfies the error estimate*:

$$\|\omega - \omega_h\|_{0,h} + \|\psi - \psi_h\|_{2,h} \leqslant C(h^k|\Delta\psi|_{k,\Omega} + h^{k-1}|\psi|_{k+1,\Omega}). \tag{2.67}$$

The best estimate is obtained when $\psi \in H^{l+1}(\Omega)$:

$$\|\omega - \omega_h\|_{0,h} + \|\psi - \psi_h\|_{2,h} \leqslant Ch^{l-1}|\psi|_{l+1,\Omega}. \tag{2.68}$$

Finally, a simple duality argument permits to sharpen the error bound for ψ in the H^1-norm.

Theorem 2.10. *Under the assumptions of Theorem* 2.9 *we have*:

$$|\psi - \psi_h|_{1,\Omega} \leqslant Ch^k|\psi|_{k+1,\Omega} \tag{2.69}$$

provided the stream function $\psi \in H^{k+1}(\Omega)$ *for a real* $k \in [2, l]$. *When* $\psi \in H^{l+1}(\Omega)$, *the best error bound is*:

$$|\psi - \psi_h|_{1,\Omega} \leqslant Ch^l|\psi|_{l+1,\Omega}, \qquad l \geqslant 2. \tag{2.70}$$

Proof. For $\mathbf{g} \in L^2(\Omega)^2$, consider the auxiliary Stokes problem:

$$a_h(\lambda_g, \mu) + b_h(\mu, \phi_g) = 0 \qquad \forall \mu \in H^1(\Omega),$$
$$b_h(\lambda_g, \chi) = -(\mathbf{g}, \mathbf{curl}\,\chi) \qquad \forall \chi \in \tilde{\Psi}.$$

Because of the convexity of Ω, the unique solution (λ_g, ϕ_g) of this problem belongs to $H^1(\Omega) \times H^3(\Omega)$ and (cf. Remark I.5.6):

$$\|\phi_g\|_{3,\Omega} + \|\lambda_g\|_{1,\Omega} \leqslant C_1\|\mathbf{g}\|_{0,\Omega}.$$

From this and Problem (2.48), we easily derive the following equalities:

$$\begin{aligned}(\mathbf{g}, \mathbf{curl}(\psi - \psi_h)) &= -b_h(\lambda_g - \lambda_h, \psi - \psi_h) + a_h(\omega - \omega_h, \lambda_h)\\ &= b_h(\lambda_h - \lambda_g, \psi - \psi_h) + a_h(\omega - \omega_h, \lambda_h - \lambda_g)\\ &\quad - b_h(\omega - \omega_h, \phi_g - \phi_h) \qquad \forall \lambda_h \in \Theta_h, \quad \forall \phi_h \in \Phi_h.\end{aligned}$$

Then (2.69) follows from Theorem 2.9 and Lemmas 2.10 and 2.11. □

As an interesting consequence, we have the following estimate for ψ in the $W^{1,s}$-norm.

Corollary 2.3. *Suppose the hypotheses of Theorem* 2.9 *hold. If the stream function* ψ *belongs to* $H^{k+1}(\Omega)$ *for a real* $k \in [2, l]$, *then for each real* $s \geqslant 2$ *there exists a constant* C_s, *independent of* h, *such that*:

(2.71) $$|\psi - \psi_h|_{1,s,\Omega} \leqslant C_s h^{k-1+2/s} |\psi|_{k+1,\Omega}.$$

This estimate is a routine consequence of (2.69), Lemma A.7 and (A.21a).

The last three results call for a number of comments. The first obvious remark is that the approach of this section provides sharper estimates than the preceding section. Indeed, as long as $l \geqslant 2$ and the stream function ψ belongs to $H^{l+1}(\Omega)$, (2.70) provides an optimal error bound for $|\psi - \psi_h|_{1,\Omega}$. Unfortunately, this section brings no improvement on the error estimate for the vorticity ω. And furthermore, it fails just as much to establish the convergence when piecewise linear elements are used (i.e. $l = 1$). In this case, the loss of accuracy arises from the approximation error in the norm $\|\,.\,\|_{2,h}$: clearly, this norm is not adapted to piecewise linear elements.

Finally, it is worth pointing out that, in this section, there is little hope of discarding the convexity hypothesis on Ω because Lemma 2.12 requires that $\phi(h)$ belong necessarily to $H^2(\Omega)$.

Remark 2.9. Owing to the trace terms in $\|\,.\,\|_{0,h}$ and $\|\,.\,\|_{2,h}$, (2.67) yields the additional estimates:

$$\|\omega - \omega_h\|_{0,\Gamma_h} \leqslant C h^{k-3/2} |\psi|_{k+1,\Omega},$$

$$\|S(\partial\psi_h/\partial n)\|_{0,\Gamma_h} \leqslant C h^{k-1/2} |\psi|_{k+1,\Omega},$$

for $1 \leqslant k \leqslant l$, provided of course that $\psi \in H^{k+1}(\Omega)$. This last inequality shows that the jump of $\partial\psi_h/\partial n$ across interelement boundaries (as well as $\partial\psi_h/\partial n$ on Γ) tends to zero like $h^{k-1/2}$.

§ 3. Further Topics on the "Stream Function-Vorticity-Pressure" Scheme

This paragraph gives further developments on the "stream function-vorticity-pressure" method for the two-dimensional Stokes problem. In particular it derives better error estimates for ψ_h, ω_h and p_h in the general case. These can also be improved by the use of special meshes.

3.1. Refinement of the Error Analysis

We place ourselves in the situation of Section 2.2 and, unless otherwise specified, we use the same notations and concrete spaces $\Theta_h = M_h$, Φ_h and Q_h defined by (2.39). As mentioned at the section's end, the error bounds given in Theorem 2.8 are quite poor because our estimate for the approximation error in V_h is too coarse. Now, we are going to concentrate on this estimate. The following lemma establishes an analogue of (II.1.16) in abstract situations.

Lemma 3.1. *If $M_h \subset \Theta_h$, we have for all $v = (\mathbf{curl}\,\phi, \theta) \in V$:*

$$\inf_{w_h \in V_h} |v - w_h| \leqslant \inf_{v_h = (\mathbf{curl}\,\phi_h, \theta_h) \in X_h} \left[2|v - v_h| + \sup_{\mu_h \in M_h} \frac{|(\mathbf{curl}(\phi - \phi_h), \mathbf{curl}\,\mu_h)|}{\|\mu_h\|_{0,\Omega}} \right]. \tag{3.1}$$

The same bound is valid for $\inf_{w_h \in V_h} \|v - w_h\|_{\tilde{X}}$ with the norm $|.|$ in the right-hand side of (3.1) *replaced by $\|.\|_{\tilde{X}}$.*

Proof. Let $v_h = (\mathbf{curl}\,\phi_h, \theta_h)$ be an arbitrary element of X_h and let $\tau_h \in M_h$ be defined by:

$$(\tau_h, \mu_h) = (\theta - \theta_h, \mu_h) - (\mathbf{curl}(\phi - \phi_h), \mathbf{curl}\,\mu_h) \qquad \forall \mu_h \in M_h.$$

Then

$$\|\tau_h\|_{0,\Omega} \leqslant \|\theta - \theta_h\|_{0,\Omega} + \sup_{\mu_h \in M_h} \frac{|(\mathbf{curl}(\phi - \phi_h), \mathbf{curl}\,\mu_h)|}{\|\mu_h\|_{0,\Omega}}. \tag{3.2}$$

Furthermore, since $M_h \subset \Theta_h$ the pair $z_h = (\mathbf{0}, \tau_h)$ belongs to X_h. Let us set

$$w_h = v_h + z_h$$

and note that $w_h \in V_h$ because $v \in \tilde{V}$. Thus, we have

$$|v - w_h| \leqslant |v - v_h| + \|\tau_h\|_{0,\Omega},$$

$$\|v - w_h\|_{\tilde{X}} \leqslant \|v - v_h\|_{\tilde{X}} + \|\tau_h\|_{0,\Omega}$$

and the bound (3.1) follows from (3.2) and the fact that v_h is arbitrary. □

By inspecting (3.1), we see that the difficulties lie in evaluating properly the term

$$\sup_{\mu_h \in M_h} \frac{|(\mathbf{curl}(\phi - \phi_h), \mathbf{curl}\,\mu_h)|}{\|\mu_h\|_{0,\Omega}},$$

because the numerator involves the first derivatives of μ_h while the denominator does not. Of course, here we cannot resort to the simple trick of § 2 which consists in writing that

$$|\mu_h|_{1,\Omega} \leqslant (C/h) \|\mu_h\|_{0,\Omega} \qquad \forall \mu_h \in M_h,$$

for this process is too crude. Instead, we propose to use a clever argument due to Scholz [72]. Roughly speaking, this author remarks that by choosing

$$\phi_h = \mathring{P}_h \phi,$$

the projection of ϕ onto Φ_h then the expression $(\mathbf{curl}(\phi - \phi_h), \mathbf{curl}\,\mu_h)$ reduces to a sum of integrals taken only on boundary elements. Since these elements are few compared to the total number of elements in Ω, a much sharper bound can be derived for this expression. The details are given in the next lemma.

Lemma 3.2. *Suppose that Ω is a bounded, convex polygon and $\mathcal{T}_h$ a uniformly regular family of triangulations of $\bar{\Omega}$. Let k be an integer and p a real number such that $1 \leqslant k \leqslant l$ and $2 \leqslant p \leqslant \infty$. There exists a constant $C > 0$, independent of h, such that for all functions $\phi \in W^{k+1,p}(\Omega) \cap H_0^1(\Omega)$, we have*:

$$(3.3) \qquad \sup_{\mu_h \in \Theta_h} \frac{|(\mathbf{curl}(\phi - \mathring{P}_h \phi), \mathbf{curl}\, \mu_h)|}{\| \mu_h \|_{0,\Omega}} \leqslant C h^{k-1/2-1/p} \| \phi \|_{k+1,p,\Omega}.$$

Proof. First, note that $\Phi_h \subset H_0^1(\Omega)$ because Ω is convex. Next, let μ_h be an arbitrary element of Θ_h and let λ_h denote the function of Φ_h which coincides with μ_h at all interior finite element nodes. In particular, if Σ_h denotes the union of the boundary elements of $\mathcal{T}_h$ then $\operatorname{supp}(\mu_h - \lambda_h) \subset \Sigma_h$. Therefore since

$$(\mathbf{curl}(\phi - \mathring{P}_h \phi), \mathbf{curl}\, \phi_h) = 0 \qquad \forall \phi_h \in \Phi_h,$$

this implies:

$$(\mathbf{curl}(\phi - \mathring{P}_h \phi), \mathbf{curl}\, \mu_h) = \sum_{\kappa \subset \Sigma_h} \int_\kappa \mathbf{curl}(\phi - \mathring{P}_h \phi) \cdot \mathbf{curl}(\mu_h - \lambda_h)\, dx.$$

Hence by Hölder's inequality, we have:

$$(3.4) \qquad |(\mathbf{curl}(\phi - \mathring{P}_h \phi), \mathbf{curl}\, \mu_h)| \leqslant |\phi - \mathring{P}_h \phi|_{1,p,\Sigma_h} |\mu_h - \lambda_h|_{1,q,\Sigma_h},$$

where

$$1/p + 1/q = 1, \quad \text{with } p \geqslant 2.$$

Now, (A.32) and the argument of Corollary A.3 yield:

$$|\mu_h - \lambda_h|_{1,q,\Sigma_h} \leqslant C_1 h^{-1} (\operatorname{meas}(\Sigma_h))^{1/q-1/2} \| \mu_h - \lambda_h \|_{0,\Sigma_h}.$$

On the one hand

$$\| \mu_h - \lambda_h \|_{0,\Sigma_h} \leqslant C_2 \| \mu_h \|_{0,\Sigma_h}$$

because $\mu_h - \lambda_h$ vanishes on all interior nodes of $\mathcal{T}_h$ and reduces to μ_h on all boundary nodes. On the other hand,

$$\operatorname{meas}(\Sigma_h) \leqslant C_3 h \operatorname{meas}(\Gamma).$$

Therefore,

$$|\mu_h - \lambda_h|_{1,q,\Sigma_h} \leqslant C_4 h^{-1/2-1/p} \| \mu_h \|_{0,\Sigma_h}.$$

When substituted into (3.4), this inequality gives:

$$|(\mathbf{curl}(\phi - \mathring{P}_h \phi), \mathbf{curl}\, \mu_h)| \leqslant C_4 h^{-1/2-1/p} |\phi - \mathring{P}_h \phi|_{1,p,\Omega} \| \mu_h \|_{0,\Omega}$$

and the desired result follows from the estimate (A.26), whatever the value of k since $p \geqslant 2$. □

Remark 3.1. Let us denote

$$c(\phi, \mu_h) = (\mathbf{curl}(\phi - \mathring{P}_h \phi), \mathbf{curl}\, \mu_h) \qquad \forall \phi \in W^{1,p}(\Omega), \qquad \forall \mu_h \in \Theta_h,$$

and let $\langle .,. \rangle$ denote the duality between Θ_h and its dual Θ_h' for the L^2-norm. Then

$$c(\phi, \mu_h) = \langle C_h(\phi), \mu_h \rangle$$

where $C_h \in \mathscr{L}(W^{k+1,p}(\Omega) \cap H_0^1(\Omega); \Theta_h')$ for all integers $k \in [1, l]$ and

$$\| C_h \|_{\mathscr{L}(W^{k+1,p}(\Omega) \cap H_0^1(\Omega); \Theta_h')} \leqslant C h^{k-1/2-1/p}.$$

Therefore by interpolating between two consecutive integral values of k, we derive from Theorem I.1.4 that $C_h \in \mathscr{L}(W^{k+1,p}(\Omega) \cap H_0^1(\Omega); \Theta_h')$ for all *real* $k \in [1, l]$ and formula (I.1.10) permits to extend (3.3) to non integral k:

$$\text{(3.5)} \quad \sup_{\mu_h \in \Theta_h} \frac{|(\mathbf{curl}(\phi - \mathring{P}_h \phi), \mathbf{curl}\, \mu_h)|}{\| \mu_h \|_{0,\Omega}} \leqslant C h^{k-1/2-1/p} \| \phi \|_{k+1,p,\Omega} \quad \forall \text{real } k \in [1, l].$$

With these two lemmas, we are able to improve the approximation result in V_h.

Lemma 3.3. *Assume that Ω and $\mathscr{T}_h$ are like in Lemma 3.2. For each real $k \in [1, l]$ and real $p \in [2, \infty]$, there exists a constant $C > 0$, independent of h, such that all functions $v = (\mathbf{curl}\, \phi, \theta = -\Delta\phi) \in \tilde{V}$ with $\phi \in W^{k+1,p}(\Omega) \cap H_0^1(\Omega)$ satisfy:*

$$\text{(3.6)} \quad \begin{cases} \inf\limits_{w_h \in V_h} |v - w_h| \leqslant 2 \inf\limits_{\theta_h \in \Theta_h} \| \theta - \theta_h \|_{0,\Omega} + C h^{k-1/2-1/p} \| \phi \|_{k+1,p,\Omega}, \\ \inf\limits_{w_h \in V_h} \| v - w_h \|_{\tilde{X}} \leqslant 2 \left\{ |\phi - \mathring{P}_h \phi|_{1,s,\Omega} + \inf\limits_{\theta_h \in \Theta_h} \| \theta - \theta_h \|_{0,\Omega} \right\} \\ \qquad + C h^{k-1/2-1/p} \| \phi \|_{k+1,p,\Omega}. \end{cases}$$

When $\phi \in H^{l+3/2}(\Omega) \cap W^{l+1,\infty}(\Omega)$ (and of course $\phi|_\Gamma = 0$), the best estimate is:

$$\text{(3.7)} \quad \left. \begin{array}{l} \inf\limits_{w_h \in V_h} |v - w_h| \\ \inf\limits_{w_h \in V_h} \| v - w_h \|_{\tilde{X}} \end{array} \right\} \leqslant C h^{l-1/2} (\| \phi \|_{l+3/2,\Omega} + \| \phi \|_{l+1,\infty,\Omega}).$$

Proof. Recall that

$$\| (\mathbf{curl}\, \phi, \theta) \|_{\tilde{X}} = |\phi|_{1,s,\Omega} + \| \theta \|_{0,\Omega}.$$

Then, the inequality (3.6) is a direct consequence of Lemmas 3.1 and 3.2 together with Remark 3.1.

Next, let $\phi \in H^{l+3/2}(\Omega) \cap W^{l+1,\infty}(\Omega)$. Lemma 3.2 with $k = l$ and $p = \infty$ gives:

$$\sup_{\mu_h \in \Theta_h} \frac{|(\mathbf{curl}(\phi - \mathring{P}_h \phi), \mathbf{curl}\, \mu_h)|}{\| \mu_h \|_{0,\Omega}} \leqslant C_1 h^{l-1/2} \| \phi \|_{l+1,\infty,\Omega}.$$

Moreover, since $H^{l+3/2}(\Omega) \subset W^{l+1/2,s}(\Omega)$ for all $s < \infty$, an application of (A.26) yields:

$$|\phi - \mathring{P}_h \phi|_{1,s,\Omega} \leqslant C_2 h^{l-1/2} \| \phi \|_{l+1/2,s,\Omega} \leqslant C_3 h^{l-1/2} \| \phi \|_{l+3/2,\Omega}.$$

Finally, $\theta = -\Delta\phi \in H^{l-1/2}(\Omega)$ and it follows from (A.26) (or (A.21a) when $l \geqslant 3$) that:

$$\inf_{\theta_h \in \Theta_h} \|\theta - \theta_h\|_{0,\Omega} \leqslant C_4 h^{l-1/2} \|\theta\|_{l-1/2,\Omega} \leqslant C_4 h^{l-1/2} \|\phi\|_{l+3/2,\Omega}.$$

Hence (3.7) stems from (3.6) and these three inequalities. □

As an immediate consequence, we derive the following corollary by using a simple and classical density argument.

Corollary 3.1. *Under the assumptions of Lemma 3.2, we have for all $v \in \tilde{V}$:*

$$\lim_{h\to 0} \inf_{w_h \in V_h} \|v - w_h\|_{\tilde{X}} = 0.$$

In addition, Lemma 3.3 implies a number of error estimates for the "stream function-vorticity" method. First, let us consider the simplest case where the solution of the Stokes problem has sufficient regularity.

Theorem 3.1. *Let Ω be a bounded, convex polygon and let $\mathscr{T}_h$ be a uniformly regular family of triangulations of $\bar{\Omega}$. Assume that the solution ($\mathbf{u} = \mathbf{curl}\,\psi, p$) of the Stokes problem (2.1) is such that $\psi \in H^{k+3/2}(\Omega) \cap W^{k+1,\infty}(\Omega)$, $p \in H^{k-1/2}(\Omega) \cap L_0^2(\Omega)$ for some real $k \in [3/2, l]$. Then the solution ($u_h = (\mathbf{curl}\,\psi_h, \omega_h), p_h$) of Problem (2.29) (2.36) satisfies the estimate:*

$$\|u - u_h\|_{\tilde{X}} \leqslant C_1 h^{k-1/2}(\|\psi\|_{k+3/2,\Omega} + \|\psi\|_{k+1,\infty,\Omega}), \tag{3.8}$$

$$\|p - p_h\|_{0,\Omega} \leqslant C_2 h^{k-1/2}(|p|_{k-1/2,\Omega} + \|\psi\|_{k+3/2,\Omega} + \|\psi\|_{k+1,\infty,\Omega}). \tag{3.9}$$

When $1 \leqslant k \leqslant \min(3/2, l)$ and $\psi \in W^{k+1,\infty}(\Omega)$, $\Delta\psi$ and $p \in H^1(\Omega)$ we have:

$$\|u - u_h\|_{\tilde{X}} \leqslant C_3(h|\Delta\psi|_{1,\Omega} + h^{k-1/2}\|\psi\|_{k+1,\infty,\Omega}), \tag{3.10}$$

$$\|p - p_h\|_{0,\Omega} \leqslant C_4(h|p|_{1,\Omega} + h|\Delta\psi|_{1,\Omega} + h^{k-1/2}\|\psi\|_{k+1,\infty,\Omega}). \tag{3.11}$$

Proof. Owing to (3.6), Remark 2.2 and Theorems 2.6 and 2.7 we get:

$$|u - u_h| \leqslant C_1\{\|\omega - P_h\omega\|_{0,\Omega} + h^{k-1/2}\|\psi\|_{k+1,\infty,\Omega}\},$$

$$\|u - u_h\|_{\tilde{X}} \leqslant C_2\{\|\omega - P_h\omega\|_{0,\Omega} + |\psi - \mathring{P}_h\psi|_{1,s,\Omega} + h^{k-1/2}\|\psi\|_{k+1,\infty,\Omega}\},$$

$$\|p - p_h\|_{0,\Omega} \leqslant C_3\left\{h \inf_{q_h \in Q_h} |p - q_h|_{1,\Omega} + |u - u_h| \right.$$
$$\left. + \inf_{\theta_h \in \Theta_h} (\|\omega - \theta_h\|_{0,\Omega} + h|\omega - \theta_h|_{1,\Omega})\right\}.$$

Applying (A.26), this gives (3.8) and (3.9) or (3.10) and (3.11) according to the value of k. □

When $l \geqslant 2$ and $\psi \in H^{l+3/2}(\Omega) \cap W^{l+1,\infty}(\Omega)$, $p \in H^{l-1/2}(\Omega)$ the best estimate is:

$$\|u - u_h\|_{\tilde{X}} \leqslant Ch^{l-1/2}(\|\psi\|_{l+3/2,\Omega} + \|\psi\|_{l+1,\infty,\Omega}),$$

$$\|p - p_h\|_{0,\Omega} \leqslant Ch^{l-1/2}(|p|_{l-1/2,\Omega} + \|\psi\|_{l+3/2,\Omega} + \|\psi\|_{l+1,\infty,\Omega}).$$

When $l = 1$, $\psi \in W^{2,\infty}(\Omega)$, $\Delta\psi$ and $p \in H^1(\Omega)$ the best estimate is:

$$\|u - u_h\|_{\tilde{X}} \leqslant Ch(|\Delta\psi|_{1,\Omega} + h^{-1/2}\|\psi\|_{2,\infty,\Omega}),$$

$$\|p - p_h\|_{0,\Omega} \leqslant Ch(|p|_{1,\Omega} + |\Delta\psi|_{1,\Omega} + h^{-1/2}\|\psi\|_{2,\infty,\Omega}).$$

Remark 3.2. When ψ belongs to $H^{k+2}(\Omega)$ and not to $W^{k+1,\infty}(\Omega)$, the estimates of Theorem 3.1 are nearly valid. More precisely, by applying Lemma 3.2 with this k and arbitrary $p \geqslant 2$ and using Sobolev's Imbedding Theorem I.1.3 we can replace (3.6) by:

$$\inf_{w_h \in V_h} \|v - w_h\|_{\tilde{X}} \leqslant C_p h^{k-1/2-1/p}\|\phi\|_{k+2,\Omega} \qquad \forall p \geqslant 2.$$

Hence, by letting p tend to infinity, we see that for each $\varepsilon > 0$ there exists a constant $C(\varepsilon) > 0$ such that

$$\inf_{w_h \in V_h} \|v - w_h\|_{\tilde{X}} \leqslant C(\varepsilon) h^{k-1/2-\varepsilon}\|\phi\|_{k+2,\Omega}.$$

In turn, this implies the estimate

$$\|u - u_h\|_{\tilde{X}} \leqslant C(\varepsilon) h^{k-1/2-\varepsilon}\|\psi\|_{k+2,\Omega},$$

with no restriction upon k or $\Delta\psi$ as in this case $\Delta\psi$ belongs to $H^k(\Omega)$ with $k \geqslant 1$.

Here again, a standard density argument establishes that the "stream function-vorticity" method is convergent.

Corollary 3.2. *Let* Ω *and* $\mathcal{T}_h$ *be like in Theorem* 3.1 *and let the solution* ($\mathbf{u} = \mathbf{curl}\,\psi$; p) *of the Stokes problem* (2.1) *belong to* $H_0^1(\Omega)^2 \times L_0^2(\Omega)$. *Then the solution* (u_h, p_h) *of Problem* (2.29) (2.36) *satisfies*:

$$\lim_{h \to 0}(\|u - u_h\|_{\tilde{X}} + \|p - p_h\|_{0,\Omega}) = 0.$$

When studying Navier-Stokes equations in Chapter IV, we shall encounter right-hand sides $\mathbf{f}$ with no better than L^r-regularity, $1 < r \leqslant 2$. Now since Ω is assumed to be convex, the solution (ψ, ω, p) of the Stokes problem belongs to $W^{3,r}(\Omega) \times W^{1,r}(\Omega) \times W^{1,r}(\Omega)$ whenever its right-hand side belongs to $L^r(\Omega)^2$. In this case, we cannot apply directly (A.26) to evalute $\|\omega - P_h\omega\|_{0,\Omega}$. Instead, we prove the following approximation result.

Lemma 3.4. *We retain the hypotheses of Theorem* 3.1 *on* Ω *and* $\mathcal{T}_h$. *Let* $1 < q \leqslant 2$ *and* $1/p + 1/q = 1$. *There exists a constant* $C > 0$ *independent of* h, *such that*:

(3.12) $$\|v - P_h v\|_{0,\Omega} \leqslant C h^{2/p} |\ln(h)|^{\beta} |v|_{1,q,\Omega} \qquad \forall v \in W^{1,q}(\Omega),$$

where $\beta = 0$ *if* $l \geqslant 2$ *and* $\beta = 1 - 2/p$ *if* $l = 1$.

Proof. Let $R_h \in \mathscr{L}(H^1(\Omega); \Theta_h)$ be the local regularization operator defined by (A.53) (A.54) and let us write:

$$\|v - P_h v\|_{0,\Omega} \leqslant \|v - R_h v\|_{0,\Omega} + \|R_h v - P_h v\|_{0,\Omega}.$$

As $R_h v - P_h v \in \Theta_h$, Lemma A.7 implies:

$$\|R_h v - P_h v\|_{0,\Omega} \leqslant C h^{1-2/q} \|R_h v - P_h v\|_{0,q,\Omega}.$$

Hence

$$\|v - P_h v\|_{0,\Omega} \leqslant \|v - R_h v\|_{0,\Omega} + C h^{1-2/q}(\|v - R_h v\|_{0,q,\Omega} + \|v - P_h v\|_{0,q,\Omega}).$$

Thus (3.12) is a straightforward consequence of Theorem A.4, Jensen's inequality (A.34) and (A.26) with a logarithmic factor if $q < 2$. □

This lemma enables us to extend Theorem 3.1.

Theorem 3.2. *Let* Ω *and* $\mathscr{T}_h$ *be like in Theorem* 3.1 *and assume that the stream function* ψ *belongs to* $W^{3,\alpha}(\Omega)$ *and the pressure* p *to* $W^{1,\alpha}(\Omega)$ *with* $\alpha \in [r, 2)$. *Let* β *satisfy* $1/\alpha + 1/\beta = 1$. *We have the following estimates:*

(3.13) $$\|u - u_h\|_{\tilde{X}} \leqslant C_1 \begin{cases} h^{2/\beta} |\psi|_{3,\alpha,\Omega} & \text{if } l \geqslant 2, \\ h^{1/\beta} \|\psi\|_{3,\alpha,\Omega} & \text{if } l = 1, \end{cases}$$

(3.14) $$\|p - p_h\|_{0,\Omega} \leqslant C_2(\|\psi\|_{3,\alpha,\Omega} + |p|_{1,\alpha,\Omega}) \begin{cases} h^{2/\beta} & \text{if } l \geqslant 2, \\ h^{1/\beta} & \text{if } l = 1, \end{cases}$$

with constants C_1 *and* C_2 *independent of* h, ψ *and* p.

Proof. When $l \geqslant 2$ we can apply the material of Section 2.3. Indeed, the arguments of Lemmas 2.10 and 2.11 can be readily extended to obtain:

$$\inf_{\theta_h \in \Theta_h} \|\omega - \theta_h\|_{0,h} \leqslant C_1 h^{2/\beta} |\omega|_{1,\alpha,\Omega} \quad \text{for } l \geqslant 1,$$

$$\inf_{\phi_h \in \Theta_h} \|\psi - \phi_h\|_{2,h} \leqslant C_2 h^{2/\beta} |\psi|_{3,\alpha,\Omega} \quad \text{for } l \geqslant 2.$$

(Note that ψ belongs to $H^2(\Omega)$ because $W^{3,\alpha}(\Omega) \subsetneq H^2(\Omega)$). When substituted into (2.66) these two bounds yield a somewhat sharper estimate than (3.13), namely:

(3.15) $$\|\omega - \omega_h\|_{0,h} + \|\psi - \psi_h\|_{2,h} \leqslant C_3 h^{2/\beta} |\psi|_{3,\alpha,\Omega} \quad \text{for } l \geqslant 2.$$

In turn, (2.38) and (3.13) give immediately (3.14).

When $l = 1$, we must resort to (2.30) and (3.6):

(3.16) $$\|u - u_h\|_{\tilde{X}} \leqslant C_4\{\|\omega - P_h\omega\|_{0,\Omega} + |\psi - \mathring{P}_h\psi|_{1,s,\Omega} + h^{k-1/2-1/p}\|\psi\|_{k+1,p,\Omega}\}.$$

On the one hand, Sobolev's Theorem I.1.3 implies that $W^{3,\alpha}(\Omega) \subsetneq W^{2,p}(\Omega)$ with $1/p = 1/\alpha - 1/2 > 0$. Therefore the last term in (3.16) with this p and $k = 1$ is bounded by:

$$h^{1/2-1/p}\|\psi\|_{2,p,\Omega} \leqslant C_5 h^{1/\beta}\|\psi\|_{3,\alpha,\Omega}.$$

On the other hand, another application of Theorem I.1.3 yields $W^{3,\alpha}(\Omega) \subsetneq W^{1+2/\beta,p}(\Omega)$ for all $p > 0$. Hence (A.26) gives in particular:

$$|\psi - \mathring{P}_h\psi|_{1,s,\Omega} \leqslant C_6 h^{2/\beta}\|\psi\|_{3,\alpha,\Omega}.$$

Finally, we infer from Lemma 3.4 that:

$$\|\omega - P_h\omega\|_{0,\Omega} \leqslant C_7 h^{2/\beta}|\ln(h)|^{1-2/\beta}|\omega|_{1,\alpha,\Omega}.$$

Thus the dominating power of h in (3.16) is $h^{1/\beta}$. This establishes (3.13). Then (3.14) follows again from (2.38). □

Remark 3.3. When $\alpha = 2$, the statement of Theorem 3.2 is still valid for $l = 2$ but not for $l = 1$ because of the last term in (3.16). Instead, like in Remark 3.2 we obtain

$$\left.\begin{array}{l}\|u - u_h\|_{\tilde{X}}\\ \|p - p_h\|_{0,\Omega}\end{array}\right\} \leqslant C(\varepsilon)h^{1/2-\varepsilon}\begin{cases}\|\psi\|_{3,\Omega}\\ |p|_{1,\Omega} + \|\psi\|_{3,\Omega}\end{cases} \quad \forall\varepsilon > 0.$$

Finally, by using a familiar duality argument, the following theorem completes the statement of Theorem 2.10 and establishes a nearly optimal upper bound for $|\psi - \psi_h|_{1,\Omega}$ when $l = 1$.

Theorem 3.3. *Let Ω and $\mathscr{T}_h$ be like in Theorem* 3.1. *Suppose the solution* $\mathbf{u} = \mathbf{curl}\,\psi$ *of the Stokes problem* (2.1) *satisfies*:

$$\psi \in H^{k+1}(\Omega) \quad \text{for } 2 \leqslant k \leqslant l \quad \text{or} \quad \psi \in H^3(\Omega) \quad \text{if } l = 1.$$

Then the solution $u_h = (\mathbf{curl}\,\psi_h, \omega_h)$ *of Problem* (2.29) *satisfies the error estimate*:

$$|\psi - \psi_h|_{1,\Omega} \leqslant \begin{cases} Ch^k|\psi|_{k+1,\Omega} & \text{if } 2 \leqslant k \leqslant l,\\ C(\varepsilon)h^{1-\varepsilon}\|\psi\|_{3,\Omega} & \text{if } l = 1,\end{cases} \tag{3.17}$$

where $\varepsilon > 0$ is arbitrary.

Proof. When $l \geqslant 2$, the bound (3.17) is established by Theorem 2.10. When $l = 1$, we use a very similar duality argument. Thus, for each $\mathbf{g} \in \mathbf{L}^2(\Omega)^2$, the auxiliary Stokes problem:

$$(\mathbf{curl}\,\lambda_g, \mathbf{curl}\,\chi) = (\mathbf{g}, \mathbf{curl}\,\chi) \quad \forall\chi \in H_0^1(\Omega),$$

$$(\mathbf{curl}\,\phi_g, \mathbf{curl}\,\mu) = (\lambda_g, \mu) \quad \forall\mu \in H^1(\Omega),$$

has a unique solution $\lambda_g \in H^1(\Omega)$, $\phi_g \in H^3(\Omega) \cap H_0^1(\Omega)$ and

$$\|\phi_g\|_{3,\Omega} + \|\lambda_g\|_{1,\Omega} \leqslant C_1\|\mathbf{g}\|_{0,\Omega}. \tag{3.18}$$

In addition, we have the identity:

$$(\mathbf{g}, \mathbf{curl}(\psi - \psi_h)) = (\mathbf{curl}(\lambda_g - \lambda_h), \mathbf{curl}(\psi - \psi_h)) + (\omega - \omega_h, \lambda_h - \lambda_g) + (\mathbf{curl}(\phi_g - \phi_h), \mathbf{curl}(\omega - \omega_h)) \quad \forall \lambda_h \in \Theta_h, \quad \forall \phi_h \in \Phi_h.$$

On the one hand, we can choose $\lambda_h = P_h \lambda_g$ in the first two terms:

$$(\mathbf{curl}(\lambda_g - P_h\lambda_g), \mathbf{curl}(\psi - \chi_h)) + (\omega - \omega_h, P_h\lambda_g - \lambda_g) \quad \forall \chi_h \in \Phi_h.$$

On the other hand, we can split the third term as follows:

$$(\mathbf{curl}(\phi_g - \phi_h), \mathbf{curl}(\omega - \omega_h)) = (\mathbf{curl}(\phi_g - \phi_h), \mathbf{curl}(\omega - P_h\omega)) + (\mathbf{curl}(\phi_g - \phi_h), \mathbf{curl}(P_h\omega - \omega_h)).$$

But

$$(\mathbf{curl}\, \phi_g, \mathbf{curl}(P_h\omega - \omega_h)) = (\lambda_g, P_h\omega - \omega_h)$$

and according to (2.33) and (2.32) we have:

$$(P_h\omega - \omega_h, \theta_h) = 0 \quad \forall (\mathbf{curl}\, \sigma_h, \theta_h) \in V_h,$$
$$(\mathbf{curl}(P_h\omega - \omega_h), \mathbf{curl}\, \phi_h) = 0 \quad \forall \phi_h \in \Phi_h.$$

Therefore, collecting these equalities we obtain:

$$(3.19) \quad \begin{aligned} (\mathbf{g}, \mathbf{curl}(\psi - \psi_h)) &= (\mathbf{curl}(\lambda_g - P_h\lambda_g), \mathbf{curl}(\psi - \chi_h)) + (\omega - \omega_h, P_h\lambda_g - \lambda_g) \\ &+ (\mathbf{curl}(\phi_g - \phi_h), \mathbf{curl}(\omega - P_h\omega)) \\ &+ (\lambda_g - \theta_h, P_h\omega - \omega_h) \\ &\forall \chi_h, \phi_h \in \Phi_h, \quad \forall (\mathbf{curl}\, \sigma_h, \theta_h) \in V_h. \end{aligned}$$

As ϕ_g and ψ belong at least to $H^2(\Omega)$ we take

$$\chi_h = I_h\psi, \quad \phi_h = I_h\phi_g.$$

Then

$$|(\mathbf{curl}(\lambda_g - P_h\lambda_g), \mathbf{curl}(\psi - I_h\psi))| \leqslant C_2 h |\lambda_g|_{1,\Omega} |\psi|_{2,\Omega},$$
$$|(\omega - \omega_h, P_h\lambda_g - \lambda_g)| \leqslant C_3 h |\omega|_{1,\Omega} |\lambda_g|_{1,\Omega},$$
$$|(\mathbf{curl}(\phi_g - I_h\phi_g), \mathbf{curl}(\omega - P_h\omega))| \leqslant C_4 h |\phi_g|_{2,\Omega} |\omega|_{1,\Omega},$$
$$\inf_{(\mathbf{curl}\, \sigma_h, \theta_h) \in V_h} \| \lambda_g - \theta_h \|_{0,\Omega} \leqslant C_5(\varepsilon) h^{1/2-\varepsilon} \| \phi_g \|_{3,\Omega}$$

according to Remark 3.2 and similarly Remark 3.3 gives:

$$\| P_h\omega - \omega_h \|_{0,\Omega} \leqslant C_6(\varepsilon) h^{1/2-\varepsilon} |\omega|_{1,\Omega} \quad \forall \varepsilon > 0.$$

By substituting these estimates into (3.19) and applying (3.18) we find (3.17). □

3.2. Super Convergence Using Quadrilateral Finite Elements of Degree l

In this section, we propose to study again the expression

$$\inf_{\phi_h \in \Phi_h} \sup_{\mu_h \in \Theta_h} \frac{|(\mathbf{curl}(\phi - \phi_h), \mathbf{curl}\, \mu_h)|}{\| \mu_h \|_{0,\Omega}}$$

and show that it is of the order of h^l, for ϕ in $H^{l+2}(\Omega)$, provided that the triangulation $\mathcal{T}_h$ has a favorable configuration. Essentially, this will be achieved by choosing the function ϕ_h as a suitable interpolate of ϕ.

To begin with, we assume that $\mathcal{T}_h$ is composed exclusively of convex quadrilaterals and we refer to Section A.2 for the notations and approximation results pertaining to quadrilaterals. Let us *fix* an integer $l \geqslant 1$ and let

$$0 = \alpha_0 < \alpha_1 < \cdots < \alpha_{l-1} < \alpha_l = 1$$

be $l + 1$ distinct points of the interval $[0, 1]$. Here, we denote by $\hat{I} \in \mathcal{L}(\mathscr{C}^0(\hat{K}); Q_l)$ the standard interpolation operator at the $(l + 1)^2$ points

$$\{(\alpha_i, \alpha_j);\ 0 \leqslant i, j \leqslant l\}$$

of the reference unit square $\hat{K}$. The following technical result gives a first approximation of the difference $\hat{\phi} - \hat{I}\hat{\phi}$.

Lemma 3.5. *Let $R \in \mathcal{L}(H^{l+1}(\hat{K}); P_{l+1})$ be defined by:*

$$R(\hat{\phi}) = [1/(l + 1)!] \sum_{i=1}^{2} \left(\int_{\hat{K}} (\partial^{l+1} \hat{\phi}/\partial \hat{x}_i^{l+1})\, d\hat{x} \right) \prod_{k=0}^{l} (\hat{x}_i - \alpha_k).$$

Then, for each integer m with $0 \leqslant m \leqslant l + 2$, there exists a positive constant $\hat{C}$ depending only upon $\hat{K}$, $\hat{I}$, R, m and l, such that

$$\| \hat{\phi} - \hat{I}\hat{\phi} - R(\hat{\phi}) \|_{m,\hat{K}} \leqslant \hat{C} |\hat{\phi}|_{l+2,\hat{K}} \quad \forall \hat{\phi} \in H^{l+2}(\hat{K}). \tag{3.20}$$

Proof. Suppose first that $\hat{\phi} \in \mathscr{C}^{l+1}(\hat{K})$. Recall that the interpolation operator $\hat{I}$ has the expression

$$\hat{I}\hat{\phi}(\hat{x}_1, \hat{x}_2) = \sum_{k=0}^{l} L_k(\hat{x}_1) \sum_{j=0}^{l} L_j(\hat{x}_2) \hat{\phi}(\alpha_k, \alpha_j)$$

where $L_k \in P_l$ is the polynomial of one variable such that:

$$L_i(\alpha_j) = \delta_{ij}, \qquad 0 \leqslant i, j \leqslant l.$$

The well-known remainder formula for the interpolation error gives on the one hand:

$$\hat{\phi}(\alpha_k, \hat{x}_2) = \sum_{j=0}^{l} L_j(\hat{x}_2) \hat{\phi}(\alpha_k, \alpha_j) + [1/(l + 1)!] [\partial^{l+1} \hat{\phi}(\alpha_k, \mu_k)/\partial \hat{x}_2^{l+1}] \prod_{p=0}^{l} (\hat{x}_2 - \alpha_p),$$

and on the other hand,

$$\hat{\phi}(\hat{x}_1, \hat{x}_2) = \sum_{k=0}^{l} L_k(\hat{x}_1) \hat{\phi}(\alpha_k, \hat{x}_2) + [1/(l + 1)!] [\partial^{l+1} \hat{\phi}(v, \hat{x}_2)/\partial \hat{x}_1^{l+1}] \prod_{p=0}^{l} (\hat{x}_1 - \alpha_p)$$

where the coordinates v and μ_k depend upon $\hat{\phi}$, $(\hat{x}_1, \hat{x}_2)$ and the interpolation points (α_i, α_j). Thus we can write:

$$(\hat{\phi} - \hat{I}\hat{\phi})(\hat{x}_1, \hat{x}_2) = [1/(l+1)!]\Bigg[\Bigg(\sum_{k=0}^{l} L_k(\hat{x}_1)\partial^{l+1}\hat{\phi}(\alpha_k, \mu_k)/\partial\hat{x}_2^{l+1}\Bigg)\prod_{p=0}^{l}(\hat{x}_2 - \alpha_p)$$
$$+ \partial^{l+1}\hat{\phi}(v, \hat{x}_2)/\partial\hat{x}_1^{l+1}\prod_{p=0}^{l}(\hat{x}_1 - \alpha_p)\Bigg].$$

By comparing the right-hand side of this expression with $R(\hat{\phi})$, and using the fact that

$$\sum_{k=0}^{l} L_k(\hat{x}) \equiv 1 \quad \text{on } [0, 1],$$

we readily derive that the mapping $\hat{\phi} \to \hat{\phi} - \hat{I}\hat{\phi} - R(\hat{\phi})$ vanishes on P_{l+1}. Since this mapping belongs to $\mathscr{L}(H^{l+2}(\hat{\kappa}); H^{l+2}(\hat{\kappa}))$, (3.20) follows from Corollary A.1. □

The next result is obtained by a simple rearrangement of terms.

Lemma 3.6. *The following identity holds for all ϕ and $\mu \in H^1(\kappa)$:*

$$\begin{aligned}\int_\kappa \mathbf{grad}\,\phi \cdot \mathbf{grad}\,\mu\, dx = &\int_{\hat{\kappa}} (1/J_F)[\|\partial F_\kappa/\partial\hat{x}_2\|^2(\partial\hat{\phi}/\partial\hat{x}_1)(\partial\hat{\mu}/\partial\hat{x}_1)\\ &+ \|\partial F_\kappa/\partial\hat{x}_1\|^2(\partial\hat{\phi}/\partial\hat{x}_2)(\partial\hat{\mu}/\partial\hat{x}_2)]\, d\hat{x}\\ &- \int_{\hat{\kappa}} (1/J_F)(\partial F_\kappa/\partial\hat{x}_1)\cdot(\partial F_\kappa/\partial\hat{x}_2)[(\partial\hat{\phi}/\partial\hat{x}_1)(\partial\hat{\mu}/\partial\hat{x}_2)\\ &+ (\partial\hat{\phi}/\partial\hat{x}_2)(\partial\hat{\mu}/\partial\hat{x}_1)]\, d\hat{x},\end{aligned} \tag{3.21}$$

where $\|.\|$ denotes the Euclidean norm of $\mathbb{R}^2$ and J_F the Jacobian of the bilinear mapping F_κ.

In order to evaluate properly the right-hand side of (3.21), it is convenient to shift out of the integrals the factors involving J_F and $\partial F_\kappa/\partial\hat{x}_i$. As these two quantities are not constant, we must therefore assume that their derivatives are "small". To be specific, we make the following hypothesis:

there exists a constant $C > 0$, independent of h and κ, such that

$$\|\partial^2 F_\kappa/\partial\hat{x}_1\partial\hat{x}_2\| \leqslant Ch_\kappa^2 \quad \forall\kappa\in\mathscr{T}_h. \tag{3.22}$$

Note that $\partial^2 F_\kappa/\partial\hat{x}_1\partial\hat{x}_2 = 0$ when κ is a parallelogram; in fact, (3.22) holds only if κ is almost a parallelogram. It easy to prove that when (3.22) holds, the full seminorm $|.|_{k,\hat{\kappa}}$ and $[.]_{k,\hat{\kappa}}$ (cf. formula (A.41)) have upper bounds of the same order. More precisely, we have:

Lemma 3.7. *Let κ be a convex quadrilateral that satisfies* (3.22). *Then for each integer $k \geqslant 1$ there exists a constant $C > 0$, independent of h and κ, such that:*

$$(3.23)\qquad |\hat{\phi}|_{k,\hat{\kappa}} \leqslant C\sigma_\kappa h_\kappa^{k-1}\left(\sum_{i=1}^{k} |\phi|_{i,\kappa}^2\right)^{1/2} \qquad \forall \phi \in H^k(\kappa).$$

Moreover there exists a constant $C > 0$, independent of h and κ, such that:

$$(3.24)\qquad \|\partial[(1/J_F)(\partial F_\kappa/\partial \hat{x}_i)\cdot(\partial F_\kappa/\partial \hat{x}_j)]/\partial \hat{x}_k\|_{0,\infty,\hat{\kappa}} \leqslant C\sigma_\kappa^2 h_\kappa(1+\sigma_\kappa^2) \qquad 1 \leqslant i,j,k \leqslant 2.$$

As a consequence, if the triangulation $\mathscr{T}_h$ is regular, it follows that (3.24) implies:

$$[(1/J_F)(\partial F_\kappa/\partial \hat{x}_i)\cdot(\partial F_\kappa/\partial \hat{x}_j)](\hat{x}_1,\hat{x}_2) = [(1/J_F)(\partial F_\kappa/\partial \hat{x}_i)\cdot(\partial F_\kappa/\partial \hat{x}_j)](1/2,1/2) + R_\kappa$$

where the remainder R_κ is bounded by:

$$|R_\kappa| \leqslant Ch.$$

Since we are specifically interested in recovering one power of h, we can neglect that remainder and it follows from (3.21) and (3.24) that the study of the expression

$$\int_\kappa \mathbf{curl}(\phi - \phi_h)\cdot \mathbf{curl}\,\mu_h\,dx$$

reduces for us to that of the four terms

$$(3.25)\qquad \int_{\hat{\kappa}} [\partial(\hat{\phi} - \hat{\phi}_h)/\partial \hat{x}_i][\partial \hat{\mu}/\partial \hat{x}_j]\,d\hat{x}, \qquad 1 \leqslant i,j \leqslant 2.$$

Furthermore, if we choose $\phi_h = I_h\phi$, where I_h is the interpolation operator on κ corresponding to $\hat{I}$, then according to Lemma 3.5, $\int_{\hat{\kappa}} [\partial(\hat{\phi} - \hat{I}\hat{\phi})/\partial \hat{x}_i][\partial \hat{\mu}/\partial \hat{x}_j]\,d\hat{x}$ involves in particular factors of the form

$$\int_{\hat{\kappa}} \partial(\omega_i)/\partial \hat{x}_i \partial \hat{\mu}/\partial \hat{x}_j\,d\hat{x}$$

where
$$\omega_i = \prod_{0 \leqslant p \leqslant l} (\hat{x}_i - \alpha_p).$$

Now, if the points α_k coincide with the nodes of a highly precise quadrature formula, this last integral will possibly vanish. This remark induces us to choose for the set $\{\alpha_k\}$ the $l+1$ *nodes of the Gauss-Lobatto quadrature formula on* $[0,1]$. Recall that this formula is exact for polynomials of degree $2l-1$. Then we have the following lemma.

Lemma 3.8. *Let κ be like in Lemma 3.7 and let $\{\alpha_k\}_{0 \leqslant k \leqslant l}$ be the $l+1$ nodes of the Gauss-Lobatto quadrature formula. Then there exists a constant $C > 0$, independent of h and κ, such that:*

$$(3.26)\qquad \left|\int_{\hat{\kappa}} [\partial(\hat{\phi} - \hat{I}\hat{\phi})/\partial \hat{x}_i][\partial \hat{\mu}/\partial \hat{x}_i]\,d\hat{x}\right| \leqslant C\sigma_\kappa^2 h_k^l \|\phi\|_{l+2,\kappa}\|\mu_h\|_{0,\kappa}$$
$$i = 1, 2, \qquad \forall \mu_h \in \Theta_h, \qquad \forall \phi \in H^{l+2}(\kappa).$$

Proof. As mentioned above, Lemma 3.5 implies that

$$\int_{\hat{K}} [\partial(\hat{\phi} - \hat{I}\hat{\phi})/\partial\hat{x}_i][\partial\hat{\mu}/\partial\hat{x}_i]\,d\hat{x}$$

$$= [1/(l+1)!]\left[\int_{\hat{K}} \partial^{l+1}\hat{\phi}/\partial\hat{x}_i^{l+1}\,d\hat{x}\right]\left[\int_{\hat{K}} (\partial\omega_i/\partial\hat{x}_i)(\partial\hat{\mu}/\partial\hat{x}_i)\,d\hat{x}\right] + E_i$$

where the remainder term E_i is bounded by

$$|E_i| \leqslant C_1 |\hat{\phi}|_{l+2,\hat{K}} |\hat{\mu}|_{1,\hat{K}} \leqslant C_2 |\hat{\phi}|_{l+2,\hat{K}} \|\hat{\mu}\|_{0,\hat{K}}$$
$$\leqslant C_3 \sigma_\kappa^2 h_\kappa^l \|\phi\|_{l+2,\kappa} \|\mu_h\|_{0,\kappa}$$

by virtue of (3.23) and (A.45).

Now, a simple integration by parts shows that

$$\int_{\hat{K}} (\partial\omega_i/\partial\hat{x}_i)(\partial\hat{\mu}/\partial\hat{x}_i)\,d\hat{x} = 0$$

because $\omega_i \hat{n}_i$ vanishes on the boundary of $\hat{K}$ ($\hat{n}_i$ is the ith-component of the normal $\hat{\mathbf{n}}$ to $\hat{K}$) and the integrand $\omega_i \partial^2\hat{\mu}/\partial\hat{x}_i^2$ is a polynomial of Q_{2l-1} which vanishes on the set $\{\alpha_k\}$. This proves (3.26). □

This lemma takes care of the first two terms in (3.25). But the third and fourth terms which involve

$$[\partial(\hat{\phi} - \hat{I}\hat{\phi})/\partial\hat{x}_i][\partial\hat{\mu}/\partial\hat{x}_j]$$

with $i \neq j$ are not so amenable, as can be seen by inspecting $R(\hat{\phi})$. Indeed, the argument of Lemma 3.8 leads to the integral of polynomials of Q_{2l}, whereas the Gauss-Lobatto formula is only exact for polynomials of degree $2l - 1$. So the particular choice of points $\{\alpha_k\}$ is of no avail here.

Let us examine closely $\int_{\hat{K}} [\partial(\hat{\phi} - \hat{I}\hat{\phi})/\partial\hat{x}_i][\partial\hat{\mu}/\partial\hat{x}_j]\,d\hat{x}$. To get rid of the derivative of $\hat{\mu}$ we integrate by parts with respect to $\hat{x}_j$. Applying again Lemma 3.5 and using the fact that

$$\partial^2 R(\hat{\phi})/\partial\hat{x}_i\partial\hat{x}_j = 0 \quad \text{for } i \neq j$$

we obtain:

$$\int_{\hat{K}} [\partial(\hat{\phi} - \hat{I}\hat{\phi})/\partial\hat{x}_i][\partial\hat{\mu}/\partial\hat{x}_j]\,d\hat{x} = \int_{\partial\hat{K}} [\partial(\hat{\phi} - \hat{I}\hat{\phi})/\partial\hat{x}_i]\hat{n}_j\hat{\mu}\,d\hat{s} + E_j$$

where

(3.27) $$|E_j| \leqslant C_1 \sigma_\kappa^2 h_\kappa^l \|\phi\|_{l+2,\kappa} \|\mu_h\|_{0,\kappa}.$$

But for $i \neq j$, the inverse image on κ of $[\partial(\hat{\phi} - \hat{I}\hat{\phi})/\partial\hat{x}_i]\hat{n}_j\hat{\mu}$ is continuous (but with opposite signs) across interelement boundaries and vanishes on Γ. In addition, Theorem A.3 and Lemma A.9 give only:

(3.28)
$$\left|\int_{\partial\hat{\kappa}} [\partial(\hat{\phi} - \hat{I}\hat{\phi})/\partial\hat{x}_i]\hat{n}_j\hat{\mu}\, d\hat{s}\right| \leqslant C_2[\hat{\phi}]_{l+1,\hat{\kappa}}\|\hat{\mu}\|_{0,\hat{\kappa}} \leqslant C_3\sigma_\kappa^2 h_\kappa^{l-1}|\phi|_{l+1,\kappa}\|\mu_h\|_{0,\kappa}.$$

Hence, in order to sharpen this estimate, we must first sum these terms over all elements κ of $\mathscr{T}_h$ and next evaluate their contribution on a given interelement boundary. Thus a better estimate can only be attained if the difference in the factor

$$X_\kappa = (1/J_F)(\partial F_\kappa/\partial\hat{x}_1)(\partial F_\kappa/\partial\hat{x}_2)$$

arising from two adjacent elements is "small".

To evaluate this difference, let κ_1 and κ_2 denote two adjacent elements of $\mathscr{T}_h$. First we remark that, in view of (3.24), it suffices to estimate the difference

$$X_{\kappa_1}(a_1, b_1) - X_{\kappa_2}(a_2, b_2)$$

where (a_i, b_i) is an arbitrary point of κ_i. The simplest choice is to pick a convenient point (a, b) on the interface κ' between κ_1 and κ_2 and to set

$$(a_1, b_1) = (a_2, b_2) = (a, b).$$

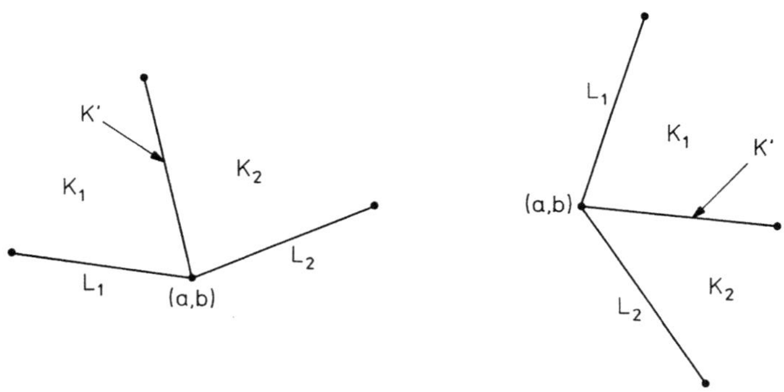

Figure 20

Let i be the index such that $\partial F_\kappa/\partial\hat{x}_i$ is continuous across the interface $\hat{\kappa}'$. Then a straightforward calculation shows that

$$|J_{F_1}J_{F_2}(X_{\kappa_1} - X_{\kappa_2})(a, b)| = \|\partial F_\kappa(a, b)/\partial\hat{x}_i\|^2 \|(\partial F_{\kappa_1}/\partial\hat{x}_j) \times (\partial F_{\kappa_2}/\partial\hat{x}_j)(a, b)\|$$

where $j \neq i$. Now, let us choose for (a, b) one of the end points of κ' and let L_1 and L_2 denote the sides of κ_1 and κ_2 (other than κ') that meet at (a, b) (cf. Figure 20). It easy to see that

(3.29) $$|J_{F_1}J_{F_2}(X_{\kappa_1} - X_{\kappa_2})(a, b)| = \text{meas}^2(\kappa')\,\text{meas}(L_1)\,\text{meas}(L_2)|\boldsymbol{\tau}_1 \cdot \mathbf{n}_2|$$

where $\boldsymbol{\tau}_1$ is the unit direction vector of L_1 and $\mathbf{n}_2$ the unit normal vector to L_2. Therefore, the difference $X_{\kappa_1} - X_{\kappa_2}$ is $O(h)$ if $\boldsymbol{\tau}_1 \cdot \mathbf{n}_2$ is also $O(h)$; in other words if L_1 and L_2 are nearly parallel. Thus with the notations of Figure 20, we make the following hypothesis:

there exists a constant $C > 0$, independent of h, such that

$$|\boldsymbol{\tau}_1 \cdot \mathbf{n}_2| \leqslant Ch \tag{3.30}$$

for all pairs of adjacent segments L_1 and L_2 of $\mathscr{T}_h$.

This hypothesis means that a mesh $\mathscr{T}_h$ that satisfies (3.22) as well as (3.30) is obtained by slightly distorting two pencils of parallel lines. Of course it is a very stringent condition on the mesh and only few domains Ω lend themselves readily to such a triangulation. Albeit so, with (3.30) we derive the result announced at the beginning of this section. First, we infer the next lemma from the above considerations.

Lemma 3.9. *Let Ω be a bounded polygon and let $\mathscr{T}_h$ be a regular triangulation of $\bar{\Omega}$ made of convex quadrilaterals satisfying* (3.22) *and* (3.30). *Then there exists a constant $C > 0$, independent of h, such that*:

$$\left| \sum_{\kappa \in \mathscr{T}_h} \int_{\hat{\kappa}} X_\kappa [\partial(\hat{\phi} - \hat{I}\hat{\phi})/\partial \hat{x}_i] [\partial \hat{\mu}/\partial \hat{x}_j] \, d\hat{x} \right| \leqslant Ch^l \|\phi\|_{l+2,\Omega} \|\mu_h\|_{0,\Omega} \tag{3.31}$$
$$i \neq j, \quad \forall \phi \in H^{l+2}(\Omega), \quad \forall \mu_h \in \Theta_h.$$

Then, using Lemmas 3.6 to 3.9 we can prove the following theorem.

Theorem 3.4. *Let Ω and $\mathscr{T}_h$ be like in Lemma* 3.9 *and let I_h be the interpolation operator on κ corresponding to $\hat{I}$*:

$$(I_h \phi) \circ F_\kappa = \hat{I}(\phi \circ F_\kappa) \quad \textit{on each } \kappa,$$

where $\hat{I} \in \mathscr{L}(\mathscr{C}^0(\hat{\kappa}); Q_l)$ is the standard interpolation operator at the $(l + 1)^2$ Gauss-Lobatto quadrature nodes. Then there exists a constant $C > 0$, independent of h, such that the following bound holds for all ϕ in $H^{l+2}(\Omega)$:

$$\sup_{\mu_h \in \Theta_h} \frac{|(\mathbf{curl}(\phi - I_h \phi), \mathbf{curl}\, \mu_h)|}{\|\mu_h\|_{0,\Omega}} \leqslant Ch^l \|\phi\|_{l+2,\Omega}. \tag{3.32}$$

With the material of the preceding section, we immediately derive the following consequence of Theorem 3.4.

Corollary 3.3. *Under the assumptions of Theorem* 3.4, *there exists a constant $C > 0$, independent of h, such that*:

$$\inf_{v_h \in V_h} \|v - v_h\|_{\tilde{X}} \leqslant Ch^l \|\phi\|_{l+2,\Omega} \tag{3.33}$$
$$\forall v = (\mathbf{curl}\, \phi, \theta) \in V \quad \textit{with } \phi \in H^{l+2}(\Omega).$$

Moreover, assuming that Ω is convex, the solution $u_h = (\mathbf{curl}\,\psi_h, \omega_h)$ *of Problem* (2.29) *satisfies the estimate* (*with another constant* $C > 0$):

$$\|u - u_h\|_{\tilde{X}} \leqslant Ch^l \|\psi\|_{l+2,\Omega}, \tag{3.34}$$

provided the solution of the Stokes Problem (2.1) *has its stream function* ψ *in* $H^{l+2}(\Omega)$. *When* Ω *is not convex, the same order of accuracy is attained if the vorticity* ω *belongs to* $H^{l+1}(\Omega)$.

Finally, if the solution $(u = (\mathbf{curl}\,\psi, \omega), p)$ *of Problem* (2.1) *has the regularity:* $\psi \in H^{l+2}(\Omega)$, $p \in H^l(\Omega)$ *and* Ω *is convex, we have the estimate for the pressure solution of Problem* (2.36):

$$\|p - p_h\|_{0,\Omega} \leqslant Ch^l\{|p|_{l,\Omega} + \|\psi\|_{l+2,\Omega}\}. \tag{3.35}$$

§4. A "Stream Function-Gradient of Velocity Tensor" Method in Two Dimensions

The method discussed in this paragraph is obtained by taking for dependent variables the stream function ψ and the gradient of the velocity:

$$\sigma_{ij} = \partial^2\psi/\partial x_i \partial x_j, \qquad 1 \leqslant i, j \leqslant 2.$$

In other words it solves for the stream function and all its second derivatives instead of the Laplacian, as was the case of the previous method. The reader will discover that this approach leads to a scheme—called the Hellan-Herrmann-Johnson scheme—which is economical and optimal. This, and related schemes, are thoroughly analyzed in Brezzi & Raviart [15]. The analysis presented here is a slight variant of this analysis, proposed by Babuška *et al* [5]. It is an elegant application of the use of mesh-dependent norms.

4.1. The Hellan-Herrmann-Johnson Formulation

Let $(\mathbf{u}, p)$ denote the solution of Problem (2.1), but instead of the vorticity ω, let us first introduce the non-symmetric gradient of velocity tensor $\lambda = (\lambda_{ij})$:

$$\lambda_{ij} = \partial u_i/\partial x_j, \qquad 1 \leqslant i, j \leqslant 2 \tag{4.1}$$

as a new dependent variable. Then the equations of Problem (2.1) read:

$$\left.\begin{cases} -\nu \displaystyle\sum_{j=1}^{2} \partial\lambda_{ij}/\partial x_j + \partial p/\partial x_i = f_i, & i = 1, 2 \\ \displaystyle\sum_{j=1}^{2} \lambda_{jj} = 0, & \end{cases}\right\} \text{ in } \Omega \tag{4.2}$$

whence a new formulation will be deduced by suitable integrations by parts.

More precisely, let κ be a bounded and open subset of $\mathbb{R}^2$ with a Lipschitz-continuous boundary $\partial\kappa$, unit exterior normal $\mathbf{n} = (n_1, n_2)$ and unit tangent vector $\mathbf{t} = (-n_2, n_1)$. For a vector $\mathbf{v}$ in $H^1(\kappa)^2$ and a tensor $\tau = (\tau_{ij})$ in $H^1(\kappa)^4$ Green's formula gives:

$$(4.3)\qquad \int_\kappa (\partial v_i/\partial x_j)\tau_{ij}\,dx = -\int_\kappa v_i \partial\tau_{ij}/\partial x_j\,dx + \int_{\partial\kappa} v_i n_j \tau_{ij}\,ds, \qquad 1 \leqslant i, j \leqslant 2.$$

Using the fact that $v_i = (\mathbf{v}\cdot\mathbf{n})n_i + (\mathbf{v}\cdot\mathbf{t})t_i$, $i = 1, 2$, the boundary integral in (4.3) can also be written as:

$$\int_{\partial\kappa} v_i n_j \tau_{ij}\,ds = \int_{\partial\kappa} (\mathbf{v}\cdot\mathbf{n})\tau_{ij} n_j n_i\,ds + \int_{\partial\kappa} (\mathbf{v}\cdot\mathbf{t})\tau_{ij} n_j t_i\,ds.$$

Thus summing over all i and j and defining the quantities:

$$(4.4)\qquad \begin{cases} M_n(\tau) = \sum_{i,j=1}^{2} \tau_{ij} n_j n_i, \\ M_{nt}(\tau) = \sum_{i,j=1}^{2} \tau_{ij} n_j t_i, \end{cases}$$

the identity (4.3) becomes:

$$\int_\kappa \sum_{i,j=1}^{2} (\partial v_i/\partial x_j)\tau_{ij}\,dx = -\int_\kappa \sum_{i,j=1}^{2} v_i(\partial\tau_{ij}/\partial x_j)\,dx + \int_{\partial\kappa} \mathbf{v}\cdot\mathbf{n} M_n(\tau)\,ds + \int_{\partial\kappa} \mathbf{v}\cdot\mathbf{t} M_{nt}(\tau)\,ds.$$

Or, with the well-known summation convention that a repeated index corresponds to a sum, we have the more compact expression:

$$(4.5)\qquad \int_\kappa (\partial v_i/\partial x_j)\tau_{ij}\,dx = -\int_\kappa v_i(\partial\tau_{ij}/\partial x_j)\,dx + \int_{\partial\kappa} \mathbf{v}\cdot\mathbf{n} M_n(\tau)\,ds + \int_{\partial\kappa} \mathbf{v}\cdot\mathbf{t} M_{nt}(\tau)\,ds, \quad \forall\mathbf{v}\in H^1(\kappa)^2, \quad \forall\tau\in H^1(\kappa)^4.$$

Together with (4.2), this identity is the foundation of several schemes, including the Hellan-Herrmann-Johnson scheme.

Here again, there are several ways of formulating the present method; we propose a mesh-dependent formulation closely related to that of Section 2.3. So, we assume that Ω is a bounded domain of $\mathbb{R}^2$ with a polygonal boundary Γ and we introduce a triangulation $\mathscr{T}_h$ of $\bar{\Omega}$ made of triangles and/or convex quadrilaterals with diameters bounded by h.

Like in Section 2.3, assume that the velocity $\mathbf{u}$ and right-hand side $\mathbf{f}$ satisfy: $\mathbf{u}\in H^2(\Omega)^2$, $\mathbf{f}\in L^2(\Omega)^2$. Let us examine the equations (4.2) in the light of (4.1) and (4.5). Observe that if the tensor τ belongs globally to $H^1(\Omega)^4$ and if the vector field $\mathbf{v}$ belongs to

$$H_0(\mathrm{div};\Omega) = \{\mathbf{v} \in L^2(\Omega)^2;\ \mathrm{div}\,\mathbf{v} \in L^2(\Omega),\ \mathbf{v}\cdot\mathbf{n}|_\Gamma = 0\},$$

then the sum of the surface integrals: $\sum_{\kappa\in\mathscr{T}_h}\int_{\partial\kappa}\mathbf{v}\cdot\mathbf{n}M_n(\tau)\,ds$ is zero. Therefore the assumption that $\mathbf{u}$ belongs to $H^2(\Omega)^2$ permits to apply (4.5) to the first equations (4.2), thus giving:

$$(4.6)\quad \nu\sum_{\kappa\in\mathscr{T}_h}\left\{\int_\kappa \lambda_{ij}(\partial v_i/\partial x_j)\,dx - \int_{\partial\kappa} M_{nt}(\lambda)\mathbf{v}\cdot\mathbf{t}\,ds\right\} - \int_\Omega p\,\mathrm{div}\,\mathbf{v}\,dx = \int_\Omega \mathbf{f}\cdot\mathbf{v}\,dx$$
$$\forall\mathbf{v}\in H_0(\mathrm{div};\Omega)\quad\text{with } \mathbf{v}|_\kappa \in H^1(\kappa)^2\quad \forall\kappa\in\mathscr{T}_h.$$

Likewise, to match (4.6) we can also express the relation (4.1) between $\mathbf{u}$ and λ by:

$$(4.7)\quad \int_\Omega \lambda_{ij}\tau_{ij}\,dx = \sum_{\kappa\in\mathscr{T}_h}\left\{\int_\kappa \tau_{ij}(\partial u_i/\partial x_j)\,dx - \int_{\partial\kappa} M_{nt}(\tau)\mathbf{u}\cdot\mathbf{t}\,ds\right\}\quad \forall\tau\in H^1(\Omega)^4$$

since the sum

$$\sum_{\kappa\in\mathscr{T}_h}\int_{\partial\kappa} M_{nt}(\tau)\mathbf{u}\cdot\mathbf{t}\,ds$$

is zero for $\mathbf{u}$ in $H_0^1(\Omega)^2$ and τ in $H^1(\Omega)^4$.

Finally a look at (4.6) and (4.7) shows that while the regularity of the test function $\mathbf{v}$ can hardly be decreased, the test tensor τ need not belong globally to $H^1(\Omega)^4$. In fact, (4.7) makes sense if τ belongs to $H^1(\kappa)^4$ on each κ. Besides that, it is convenient to assume that $M_{nt}(\tau)$ is continuous across interelement boundaries—i.e. across all segments of Γ_h.

Summing up, we see that if $(\mathbf{u}, p)$ is the solution of the Stokes problem (2.1) with $\mathbf{u}$ in $(H^2(\Omega)\cap H_0^1(\Omega))^2$ and p in $H^1(\Omega)\cap L_0^2(\Omega)$ then the triple $(\mathbf{u}, \lambda, p)$ with

$$\lambda_{ij} = \partial u_i/\partial x_j$$

is also a solution of:

$$(4.8)\quad \begin{cases} \mathbf{u}\in H_0(\mathrm{div};\Omega)\quad \text{with } \mathbf{u}|_\kappa\in H^1(\kappa)^2 \quad \forall\kappa\in\mathscr{T}_h,\\ \lambda\in H^1(\kappa)^4\quad \forall\kappa\in\mathscr{T}_h\quad \text{with } M_{nt}(\lambda) \text{ continuous on } \Gamma_h \text{ and}\\ \quad \lambda_{11}+\lambda_{22}=0,\quad p\in L_0^2(\Omega),\\ \nu\sum_{\kappa\in\mathscr{T}_h}\left\{\int_\kappa \lambda_{ij}(\partial v_i/\partial x_j)\,dx - \int_{\partial\kappa} M_{nt}(\lambda)\mathbf{v}\cdot\mathbf{t}\,ds\right\} - \int_\Omega p\,\mathrm{div}\,\mathbf{v}\,dx\\ \quad = \int_\Omega \mathbf{f}\cdot\mathbf{v}\,dx \quad \forall\mathbf{v}\in H_0(\mathrm{div};\Omega)\quad \text{with } \mathbf{v}|_\kappa\in H^1(\kappa)^2\quad \forall\kappa\in\mathscr{T}_h,\\ \int_\Omega \lambda_{ij}\tau_{ij}\,dx - \sum_{\kappa\in\mathscr{T}_h}\left\{\int_\kappa \tau_{ij}(\partial u_i/\partial x_j)\,dx - \int_{\partial\kappa} M_{nt}(\tau)\mathbf{u}\cdot\mathbf{t}\,ds\right\} = 0\\ \qquad \forall\tau\in H^1(\kappa)^4\quad \forall\kappa\in\mathscr{T}_h\quad \text{with } M_{nt}(\tau) \text{ continuous on } \Gamma_h. \end{cases}$$

Conversely, a routine argument shows that Problem (4.8) has at most one solution. Indeed, the second equations (4.8) imply:

$$\lambda_{ij} = \partial u_i/\partial x_j \quad \text{in each element } \kappa,$$

and in turn $\operatorname{div}\mathbf{u} = 0$ in Ω. Then, supposing that the right-hand side $\mathbf{f} = \mathbf{0}$ we easily derive that $\lambda_{ij} = 0$ for all i, j. Thus, $\mathbf{u}$ is constant in each κ and moreover

$$\sum_{\kappa} \int_{\partial\kappa} M_{nt}(\tau)\mathbf{u}\cdot\mathbf{t}\,ds = 0$$

$$\forall \tau \in H^1(\kappa)^4 \quad \text{with } M_{nt}(\tau) \text{ continuous across the segments of } \Gamma_h.$$

In view of the fact that $\mathbf{u}$ belongs to $H_0(\operatorname{div};\Omega)$, this last relation readily implies that $\mathbf{u} = \mathbf{0}$ in Ω. As a consequence, *Problem* (4.8) *is an equivalent formulation of the Stokes problem* (2.1) *whenever the solution of Problem* (2.1) *has sufficient regularity.*

To eliminate the pressure, Problem (4.8) can also be expressed in terms of stream functions. Indeed, recall that $\mathbf{v} \in H_0(\operatorname{div};\Omega)$ satisfies $\operatorname{div}\mathbf{v} = 0$ iff:

$$\mathbf{v} = \mathbf{curl}\,\phi \quad \text{with } \phi \in \Phi = \{\phi \in H^1(\Omega);\ \phi|_{\Gamma_0} = 0,\ \phi|_{\Gamma_i} \text{ is constant, } 1 \leqslant i \leqslant p\}.$$

If in addition $\mathbf{v}|_\kappa \in H^1(\kappa)^2$ then $\phi|_\kappa \in H^2(\kappa)$ and conversely. Then we can rewrite directly Problem (4.8) in terms of stream functions, but the notations are simplified if instead of working with the tensor λ we now introduce the symmetric tensor σ:

$$\sigma_{ij} = \partial^2\psi/\partial x_i \partial x_j \quad \text{with } \mathbf{curl}\,\psi = \mathbf{u}. \tag{4.9}$$

The correspondence between σ and λ is:

$$\begin{cases} \lambda_{11} = -\lambda_{22} = \sigma_{12}, \quad \lambda_{21} = -\sigma_{11}, \quad \lambda_{12} = \sigma_{22}, \\ M_{nt}(\lambda) = -M_n(\sigma), \\ M_n(\lambda) = M_{nt}(\sigma). \end{cases} \tag{4.10}$$

This induces us to define the following space of tensors:

$$\begin{aligned} \Sigma = \{\tau = (\tau_{ij}) \in L^2(\Omega)^4;\ \tau|_\kappa \in H^1(\kappa)^4 \quad \forall \kappa \in \mathscr{T}_h,\ \tau_{12} = \tau_{21}, \\ M_n(\tau) \text{ is continuous on each segment of } \Gamma_h\}, \end{aligned} \tag{4.11}$$

together with the space $\tilde{\Psi}$ of stream functions (already introduced in Section 2.3):

$$\tilde{\Psi} = \{\phi \in \Phi;\ \phi|_\kappa \in H^2(\kappa) \quad \forall \kappa \in \mathscr{T}_h\}. \tag{4.12}$$

With these spaces and the above correspondence, the sum of the surface integrals in (4.8) has also the following expression:

$$\sum_{\kappa \in \mathscr{T}_h} \int_{\partial\kappa} M_{nt}(\lambda)\mathbf{u}\cdot\mathbf{t}\,ds = \int_{\Gamma_h} M_n(\sigma) S(\partial\psi/\partial n)\,ds,$$

where $S(\chi)$ denotes the jump of χ over the segments of $\mathscr{T}_h$ (cf. Section 2.3). Therefore, Problem (4.8) has the following equivalent formulation, called the *Hellan-Herrmann-Johnson formulation*:

Find a pair $(\sigma, \psi) \in \Sigma \times \tilde{\Psi}$ *satisfying*:

$$(4.13)\quad \begin{cases} \nu \sum_{\kappa\in\mathscr{T}_h} \int_\kappa \sigma_{ij}\partial^2\phi/\partial x_i \partial x_j \, dx - \nu \int_{\Gamma_h} M_n(\sigma) S(\partial\phi/\partial n)\, ds \\ \quad = \int_\Omega \mathbf{f}\cdot \mathbf{curl}\, \phi\, dx \qquad \forall\phi\in\tilde{\Psi}, \\ \int_\Omega \sigma_{ij}\tau_{ij}\, dx - \left\{ \sum_{\kappa\in\mathscr{T}_h} \int_\kappa \tau_{ij}\partial^2\psi/\partial x_i\partial x_j\, dx - \int_{\Gamma_h} M_n(\tau) S(\partial\psi/\partial n)\, ds \right\} \\ \quad = 0 \qquad \forall\tau\in\Sigma. \end{cases}$$

Here again, it is possible to express (4.13) by means of two bilinear forms $a_h(.,.)$ and $b_h(.,.)$:

$$(4.14)\qquad a_h(\sigma,\tau) = \int_\Omega \sigma_{ij}\tau_{ij}\, dx \qquad \forall\text{tensors } \sigma \text{ and } \tau \text{ in } L^2(\Omega)^4,$$

$$(4.15)\qquad b_h(\tau,\phi) = -\sum_{\kappa\in\mathscr{T}_h} \int_\kappa \tau_{ij}\partial^2\phi/\partial x_i\partial x_j\, dx + \int_{\Gamma_h} M_n(\tau) S(\partial\phi/\partial n)\, ds$$

$$\forall\tau\in\Sigma, \quad \forall\phi\in H^1(\Omega) \quad \text{with } \phi|_\kappa \in H^2(\kappa).$$

Then Problem (4.13) reads:

Find a pair $(\sigma,\psi)\in\Sigma\times\tilde{\Psi}$ *such that:*

$$(4.13')\qquad \begin{cases} b_h(\sigma,\phi) = -(1/\nu)(\mathbf{f}, \mathbf{curl}\,\phi) \qquad \forall\phi\in\tilde{\Psi}, \\ a_h(\sigma,\tau) + b_h(\tau,\psi) = 0 \qquad \forall\tau\in\Sigma. \end{cases}$$

Note the analogy with Problems (2.48) and (2.48′) of Section 2.3. Observe also that when τ belongs to $\Sigma\cap H^1(\Omega)^4$ and ϕ belongs to $\tilde{\Psi}$, $b_h(\tau,\phi)$ reduces to:

$$(4.16)\qquad b_h(\tau,\phi) = \int_\Omega (\partial\tau_{ij}/\partial x_j)(\partial\phi/\partial x_i)\, dx,$$

a property which is similar to (2.51a).

The following theorem summarizes the results of this section.

Theorem 4.1. *Let Ω be a bounded plane polygon and $\mathscr{T}_h$ a triangulation of $\bar{\Omega}$. Suppose that the solution* ($\mathbf{u} = \mathbf{curl}\,\psi, p$) *of the Stokes problem* (2.1) *has the regularity*:

$$\mathbf{u}\in(H^2(\Omega)\cap H_0^1(\Omega))^2, \qquad p\in H^1(\Omega)\cap L_0^2(\Omega).$$

Then

1°) *the triple* ($\mathbf{u}$, ($\lambda_{ij} = \partial u_i/\partial x_j$), p) *is the unique solution of Problem* (4.8);

2°) *the pair* (($\sigma_{ij} = \partial^2\psi/\partial x_i\partial x_j$), ψ) *is the unique solution of Problem* (4.13).

In the next section, we shall assume that:

$$(4.17)\qquad \mathbf{u}\in H^2(\Omega)^2, \qquad p\in H^1(\Omega),$$

in order to be able to work with either Problem (4.8) or Problem (4.13).

4.2. Approximation with Triangular Finite Elements of Degree l

We propose to approximate Problem (4.13). To simplify the discussion, we assume that $\mathscr{T}_h$ consists of triangles, but the present method can easily be extended to the case where $\mathscr{T}_h$ also contains quadrilaterals. First, we choose the finite element spaces. As usual, we take

$$\Theta_h = \{\theta \in \mathscr{C}^0(\bar{\Omega});\ \theta|_\kappa \in P_l \quad \forall \kappa \in \mathscr{T}_h\}, \qquad \Phi_h = \Theta_h \cap \Phi, \tag{4.18}$$

for some *integer* $l \geqslant 1$. But as far as tensors are concerned, since the tensors of Σ need not be globally continuous, we can approximate it with a space Σ_h that involves less degrees of freedom:

$$\Sigma_h = \{\tau = (\tau_{ij}) \in \Sigma;\ \tau|_\kappa \in P_{l-1}^4 \quad \forall \kappa \in \mathscr{T}_h\}. \tag{4.19}$$

With these spaces, Problem (4.13) is discretized by the following *Problem* (Q_h):

Find a pair $(\sigma_h, \psi_h) \in \Sigma_h \times \Phi_h$ *satisfying*:

$$\begin{cases} \nu \displaystyle\sum_{\kappa \in \mathscr{T}_h} \int_\kappa (\partial^2 \phi/\partial x_i \partial x_j)(\sigma_h)_{ij}\, dx - \nu \int_{\Gamma_h} M_n(\sigma_h) S(\partial \phi/\partial n)\, ds \\ \quad = (\mathbf{f}, \mathbf{curl}\, \phi) \qquad \forall \phi \in \Phi_h, \\ \displaystyle\int_\Omega (\sigma_h)_{ij} \tau_{ij}\, dx - \sum_{\kappa \in \mathscr{T}_h} \int_\kappa (\partial^2 \psi_h/\partial x_i \partial x_j) \tau_{ij}\, dx - \int_{\Gamma_h} M_n(\tau) S(\partial \psi_h/\partial n)\, ds \\ \quad = 0 \qquad \forall \tau \in \Sigma_h. \end{cases} \tag{4.20}$$

When expressed in terms of the bilinear forms $a_h(.,.)$ and $b_h(.,.)$ defined by (4.14) and (4.15), the equations (4.20) can be written more compactly as:

$$\begin{cases} b_h(\sigma_h, \phi_h) = -(1/\nu)(\mathbf{f}, \mathbf{curl}\, \phi_h) \qquad \forall \phi_h \in \Phi_h, \\ a_h(\sigma_h, \tau_h) + b_h(\tau_h, \psi_h) = 0 \qquad \forall \tau_h \in \Sigma_h. \end{cases} \tag{4.20'}$$

In order to analyze Problem (Q_h), we must first study thoroughly the above finite element spaces and in particular equip them with appropriate norms and specify their degrees of freedom. As far as norms are concerned, the expression of the bilinear form $b_h(.,.)$ suggests to choose:

$$\|\tau\|_{0,h} = \left\{ \sum_{i,j=1}^{2} \|\tau_{ij}\|_{0,\Omega}^2 + h \|M_n(\tau)\|_{0,\Gamma_h}^2 \right\}^{1/2} \qquad \forall \tau \in \Sigma, \tag{4.21}$$

$$\|\phi\|_{2,h} = \left\{ \sum_{\kappa \in \mathscr{T}_h} |\phi|_{2,\kappa}^2 + (1/h) \|S(\partial \phi/\partial n)\|_{0,\Gamma_h}^2 \right\}^{1/2} \qquad \forall \phi \in H^1(\Omega) \quad \text{with } \phi|_\kappa \in H^2(\kappa). \tag{4.22}$$

Observe again the analogy with (2.52) and (2.53). Note also that

$$|b_h(\tau, \phi)| \leqslant \|\tau\|_{0,h} \|\phi\|_{2,h} \qquad \forall \tau \in \Sigma, \quad \forall \phi \in H^1(\Omega) \quad \text{with } \phi|_\kappa \in H^2(\kappa),$$

$$|a_h(\tau, \mu)| \leqslant \|\tau\|_{0,\Omega} \|\mu\|_{0,\Omega} \qquad \forall \tau, \mu \in L^2(\Omega)^4.$$

Now let us determine the degrees of freedom of the tensors of Σ_h. First we consider the special case where κ is the reference triangle $\hat{\kappa}$ (cf. Appendix A).

Lemma 4.1. *Let $\hat{\kappa}$ be the reference unit triangle. A symmetric tensor-valued function $\tau \in P_{l-1}^4$ is uniquely determined on $\hat{\kappa}$ by the following moments:*

$$(4.23)\qquad \begin{cases} \displaystyle\int_{\hat{\kappa}'} M_{\hat{n}}(\tau) q \, d\hat{s} & \forall q \in P_{l-1}(\hat{\kappa}') \quad \text{for all sides } \hat{\kappa}' \text{ of } \hat{\kappa}, \\ \displaystyle\int_{\hat{\kappa}} \tau_{ij} q \, d\hat{x} & \forall q \in P_{l-2}(\hat{\kappa}), \quad 1 \leqslant i, j \leqslant 2, \quad \text{if } l \geqslant 2. \end{cases}$$

Proof. To begin with, we observe that (4.23) consists of

$$3l + (3/2)l(l-1) = (3/2)l(l+1)$$

degrees of freedom and that

$$\dim\{\tau \in P_{l-1}^4;\, \tau_{12} = \tau_{21}\} = (3/2)l(l+1).$$

Thus it suffices to prove that the set of homogeneous equations:

$$(4.24)\qquad \begin{cases} \displaystyle\int_{\hat{\kappa}'} M_{\hat{n}}(\tau) q \, d\hat{s} = 0 & \forall q \in P_{l-1}(\hat{\kappa}'), \quad \text{for all sides } \hat{\kappa}' \text{ of } \hat{\kappa}, \\ \displaystyle\int_{\hat{\kappa}} \tau_{ij} q \, d\hat{x} = 0 & \forall q \in P_{l-2}(\hat{\kappa}), \quad 1 \leqslant i, j \leqslant 2, \end{cases}$$

has the unique solution $\tau = 0$.

Now the first equation of (4.24) is equivalent to $M_{\hat{n}}(\tau) = 0$ on $\partial\hat{\kappa}$. And, taking advantage of the position and shape of $\hat{\kappa}$, this amounts to:

$$\tau_{11} = 0 \quad \text{on } \hat{x}_1 = 0,$$

$$\tau_{22} = 0 \quad \text{on } \hat{x}_2 = 0,$$

$$\tau_{11} + 2\tau_{12} + \tau_{22} = 0 \quad \text{on } \hat{x}_1 + \hat{x}_2 = 1.$$

Next, by taking $q = \partial\tau_{ij}/\partial\hat{x}_k$ in the second equation of (4.24), we obtain:

$$\int_{\partial\hat{\kappa}} \tau_{ij}^2 \hat{n}_k \, d\hat{s} = 0 \quad \text{for } k = 1, 2, \quad 1 \leqslant i, j \leqslant 2.$$

Combining with the three equations above, this yields first that $\tau = 0$ on $\hat{x}_1 + \hat{x}_2 = 1$ and then that $\tau = 0$ on $\partial\hat{\kappa}$.

Then (4.24) immediately implies that $\tau = 0$ on the whole of $\hat{\kappa}$. □

In order to extend Lemma 4.1 to an arbitrary triangle κ, we must introduce a suitable transformation that maps a symmetric tensor on $\hat{\kappa}$ into a symmetric tensor on κ while preserving in some sense $M_n(.)$ on $\partial\kappa$. To this end, we recall on the one hand that the normal $\mathbf{n}$ to κ and the normal $\hat{\mathbf{n}}$ to $\hat{\kappa}$ are related by:

$$\hat{\mathbf{n}} = [(B_\kappa^T \mathbf{n})/\| B_\kappa^T \mathbf{n} \|] \circ F_\kappa$$

or

$$\mathbf{n} \circ F_\kappa = [(B_\kappa^{-1})^T \hat{\mathbf{n}}] / \| (B_\kappa^{-1})^T \hat{\mathbf{n}} \|$$

where $x = F_\kappa(\hat{x}) = B_\kappa \hat{x} + b_\kappa$ and B_κ is a nonsingular matrix with constant coefficients. On the other hand, we remark that

$$M_n(\tau)(x) = ((\tau(x)\mathbf{n}(x), \mathbf{n}(x)))$$

where $((.,.))$ denotes the Euclidean scalar product of $\mathbb{R}^2$ associated with $\| . \|$. Thus, we can write

$$(M_n(\tau)) \circ F_\kappa = \| (B_\kappa^{-1})^T \hat{\mathbf{n}} \|^{-2} ((B_\kappa^{-1}(\tau \circ F_\kappa)(B_\kappa^{-1})^T \hat{\mathbf{n}}, \hat{\mathbf{n}})).$$

This suggests to establish (in each κ) the correspondence between tensor-valued functions:

$$\tau = B_\kappa(\hat{\tau} \circ F_\kappa^{-1}) B_\kappa^T = \mathscr{G}_\kappa(\hat{\tau}), \tag{4.25a}$$

or equivalently

$$\hat{\tau} = B_\kappa^{-1}(\tau \circ F_\kappa)(B_\kappa^{-1})^T = \mathscr{G}_\kappa^{-1}(\tau). \tag{4.25b}$$

The first equation can be written explicitly as follows:

$$\tau_{ij} = \sum_{r,s=1}^{2} (\hat{\tau}_{rs} \circ F_\kappa^{-1})(\partial F_i / \partial \hat{x}_r)(\partial F_j / \partial \hat{x}_s).$$

Obviously, the transformation $\mathscr{G}_\kappa$ preserves the symmetry and regularity of tensors and furthermore:

$$\widehat{M_n(\tau)} = \| (B_\kappa^{-1})^T \hat{\mathbf{n}} \|^{-2} M_{\hat{n}}(\hat{\tau}), \tag{4.26}$$

or

$$\widehat{M_n(\tau)} = \| B_\kappa^T (\mathbf{n} \circ F_\kappa) \|^2 M_{\hat{n}}(\hat{\tau}), \tag{4.26'}$$

since

$$\| B_\kappa^T(\mathbf{n} \circ F_\kappa) \| = \| (B_\kappa^{-1})^T \hat{\mathbf{n}} \|^{-1}.$$

Hence it is easy to check that the statement of Lemma 4.1 carries over to an arbitrary triangle κ of $\mathscr{T}_h$. As a consequence, we can take the following values as degrees of freedom for the tensors τ of Σ_h:

$$\int_{\kappa'} M_n(\tau) q \, ds \quad \forall q \in P_{l-1}(\kappa'), \quad \forall \kappa' \text{ of } \Gamma_h,$$

$$\int_\kappa \tau_{ij} q \, dx \quad \forall q \in P_{l-2}(\kappa), \quad 1 \leqslant i, j \leqslant 2, \quad \forall \kappa \in \mathscr{T}_h, \quad \text{if } l \geqslant 2.$$

Remark 4.1. The simplest example of spaces Σ_h corresponds to $l = 1$:

$$\Sigma_h = \{ \tau \in L^2(\Omega)^4;\ \tau_{ij} = \tau_{ji},\ \tau_{ij}|_\kappa = \text{a constant } c_{ij} \quad \forall \kappa \text{ of } \mathscr{T}_h,$$
$$M_n(\tau) \text{ is continuous on each segment of } \Gamma_h \}.$$

Clearly, the tensors τ of Σ_h are uniquely determined by the data $M_n(\tau)$ on each segment κ' of Γ_h.

At the same time, we can easily define a convenient restriction operator π_h from Σ onto Σ_h with attractive properties. First, if τ is a symmetric tensor of $L^1(\kappa)^4$ with $M_n(\tau) \in L^1(\partial\kappa)$ we define the symmetric tensor $\pi_\kappa \tau$ of P_{l-1}^4 by:

$$(4.27) \quad \begin{cases} \displaystyle\int_{\kappa'} M_n(\pi_\kappa \tau - \tau) q \, ds = 0 \quad \forall q \in P_{l-1}(\kappa') \quad \forall \text{sides } \kappa' \text{ of } \kappa, \\ \displaystyle\int_\kappa (\pi_\kappa \tau - \tau)_{ij} q \, dx = 0 \quad \forall q \in P_{l-2}(\kappa), \quad 1 \leqslant i, j \leqslant 2, \quad \text{if } l \geqslant 2. \end{cases}$$

Then for $\tau \in \Sigma$, we define $\pi_h \tau \in \Sigma_h$ by:

$$(4.28) \qquad \pi_h \tau|_\kappa = \pi_\kappa(\tau|_\kappa) \quad \forall \kappa \in \mathscr{T}_h.$$

Clearly, $\pi_h \tau$ is a symmetric tensor and the continuity of $M_n(\tau)$ implies that $M_n(\pi_h \tau)$ is continuous across each interelement boundary κ'. Therefore $\pi_h \tau$ belongs indeed to Σ_h. In addition, we readily derive from (4.25a) and (4.26) that (4.27) holds iff

$$(4.29) \quad \begin{cases} \displaystyle\int_{\hat\kappa'} M_{\hat n}(\widehat{\pi_\kappa \tau} - \hat\tau) q \, d\hat s = 0 \quad \forall q \in P_{l-1}(\hat\kappa') \quad \forall \text{sides } \hat\kappa' \text{ of } \hat\kappa, \\ \displaystyle\int_{\hat\kappa} (\widehat{\pi_\kappa \tau} - \hat\tau)_{ij} q \, d\hat x = 0 \quad \forall q \in P_{l-2}(\hat\kappa), \quad 1 \leqslant i, j \leqslant 2, \quad \text{if } l \geqslant 2, \end{cases}$$

where, as usual, $\hat\kappa$ denotes the unit reference triangle. Hence, applying the definition (4.27) to $\hat\kappa$, we find that π_κ is preserved by affine transformations:

$$(4.30) \qquad \widehat{\pi_\kappa \tau} = \pi_{\hat\kappa} \hat\tau \quad \forall \kappa \in \mathscr{T}_h.$$

The remaining properties of π_h are stated in the next lemma.

Lemma 4.2. *Assume that the triangulation $\mathscr{T}_h$ is regular. The operator π_h defined by* (4.27) *and* (4.28) *is a linear mapping from Σ onto Σ_h and satisfies*:

$$(4.31) \qquad b_h(\tau - \pi_h \tau, \phi_h) = 0 \quad \forall \phi_h \in \Phi_h, \qquad \forall \tau \in \Sigma.$$

$$(4.32) \qquad \|\pi_h \tau\|_{0,h} \leqslant C_1 \|\tau\|_{0,h}$$

Moreover, if $\tau \in H^k(\Omega)^4 \cap \Sigma$ for some real $k \in [1, l]$, the following estimate holds:

$$(4.33) \qquad \|\pi_h \tau - \tau\|_{0,h} \leqslant C_2 h^k |\tau|_{k,\Omega}.$$

Proof. It is clear from (4.27) that the operator π_h is a linear mapping from Σ onto Σ_h. Besides that, (4.31) follows immediately from (4.27), the expression (4.15) of $b_h(.,.)$ and the definition of Θ_h.

Let us turn to (4.32). To begin with, we observe from (4.27) that

$$(4.34) \qquad \|M_n(\pi_h \tau)\|_{0,\Gamma_h} \leqslant \|M_n(\tau)\|_{0,\Gamma_h} \quad \forall \tau \in \Sigma.$$

Now, let us show that there exists a constant $C > 0$ independent of h, such that

(4.35) $$\|\pi_h \tau\|_{0,\Omega} \leqslant C\|\tau\|_{0,h} \quad \forall \tau \in \Sigma,$$

where, for the sake of simplicity, the norms of tensors and scalars are denoted alike. For each κ of $\mathscr{T}_h$, (4.30) and (4.25a) imply:

$$\|\pi_\kappa \tau\|_{0,\Omega}^2 \leqslant |\det(B_\kappa)|\,\|B_\kappa\|^4\,\|\pi_{\hat\kappa}\hat\tau\|_{0,\hat\kappa}^2.$$

But it follows easily from (4.29) that

$$\|\pi_{\hat\kappa}\hat\tau\|_{0,\hat\kappa}^2 \leqslant C_1(\|\hat\tau\|_{0,\hat\kappa}^2 + \|M_{\hat n}(\hat\tau)\|_{0,\partial\hat\kappa}^2),$$

where the constant C_1 is independent of h. Next, (4.26) implies:

$$\|M_{\hat n}(\hat\tau)\|_{0,\hat\kappa'}^2 \leqslant (C_2/\rho_\kappa)\,\|B_\kappa^{-1}\|^4\,\|M_n(\tau)\|_{0,\kappa'}^2 \quad \forall \text{sides } \kappa' \text{ of } \kappa.$$

Likewise, we derive from (4.25b) that

$$\|\hat\tau\|_{0,\hat\kappa}^2 \leqslant |\det(B_\kappa)|^{-1}\,\|B_\kappa^{-1}\|^4\,\|\tau\|_{0,\kappa}^2.$$

Collecting these four inequalities and applying (A.2) and (A.4) we obtain:

$$\|\pi_\kappa\tau\|_{0,\kappa}^2 \leqslant C_3\sigma_\kappa^4(\|\tau\|_{0,\kappa}^2 + \sigma_\kappa h_\kappa\|M_n(\tau)\|_{0,\partial\kappa}^2).$$

Since $\mathscr{T}_h$ is regular, this proves (4.35) and in turn (4.32).

Finally, let us establish (4.33). Like above, we have

$$\|\pi_\kappa\tau - \tau\|_{0,\kappa}^2 \leqslant |\det(B_\kappa)|\,\|B_\kappa\|^4\,\|\pi_{\hat\kappa}\hat\tau - \hat\tau\|_{0,\hat\kappa}^2.$$

As the mapping $\pi_{\hat\kappa}$ leaves invariant the symmetric tensors with coefficients in P_{l-1}, (A.12) gives:

$$\|\pi_{\hat\kappa}\hat\tau - \hat\tau\|_{0,\hat\kappa} \leqslant C_4|\hat\tau|_{k,\hat\kappa} \qquad 1 \leqslant k \leqslant l.$$

According to (4.25b), we have:

$$|\hat\tau|_{k,\hat\kappa} \leqslant \|B_\kappa^{-1}\|^2|\tau\circ F_\kappa|_{k,\hat\kappa};$$

and in view of (A.7) this becomes

(4.36) $$|\hat\tau|_{k,\hat\kappa} \leqslant C_5\|B_\kappa\|^k\|B_\kappa^{-1}\|^2|\det(B_\kappa)|^{-1/2}|\tau|_{k,\kappa}.$$

Since $\mathscr{T}_h$ is regular, the above inequalities yield:

$$\|\pi_h\tau - \tau\|_{0,\Omega} \leqslant C_6h^k|\tau|_{k,\Omega}.$$

Similarly, it stems from (4.26′) that:

$$\|M_n(\pi_\kappa\tau - \tau)\|_{0,\kappa'}^2 \leqslant C_7h_\kappa\|B_\kappa\|^4\|M_{\hat n}(\pi_{\hat\kappa}\hat\tau - \hat\tau)\|_{0,\hat\kappa'}^2,$$

and the trace Theorem I.1.5 implies:

$$\|M_{\hat n}(\pi_{\hat\kappa}\hat\tau - \hat\tau)\|_{0,\hat\kappa'}^2 \leqslant C_8\|\pi_{\hat\kappa}\hat\tau - \hat\tau\|_{1,\hat\kappa}^2 \leqslant C_9|\hat\tau|_{k,\hat\kappa}^2.$$

Therefore, owing to (4.36) and the regularity of $\mathscr{T}_h$, we get:

$$\|M_n(\pi_\kappa\tau - \tau)\|_{0,\kappa'}^2 \leqslant C_{10}h^{2k-1}|\tau|_{k,\kappa}^2,$$

thus establishing (4.33). □

Besides (4.31), the bilinear form $b_h(.,.)$ satisfies the following striking property.

Lemma 4.3. *For each τ_h in Σ_h the following equivalence holds*:

(4.37) $\{b_h(\tau_h, \phi_h) = 0 \quad \forall \phi_h \in \Phi_h\}$ *iff* $\{b_h(\tau_h, \phi) = 0 \quad \forall \phi \in \tilde{\Psi} \cap \mathscr{C}^0(\bar{\Omega})\}$.

Proof. Obviously, it is the "only if" part of (4.37) which must be established. To this end, let us take ϕ in $\mathscr{C}^0(\bar{\Omega})$ with $\phi|_\kappa \in H^2(\kappa)$ (in which case $\phi \in H^1(\Omega)$) and prove that

(4.38) $$b_h(\tau_h, \phi - \tilde{I}_h \phi) = 0 \quad \forall \tau_h \in \Sigma_h,$$

where $\tilde{I}_h$ is the interpolation operator defined by (A.22).

Indeed, two integrations by parts yield:

$$\int_\kappa f \partial^2 \phi / \partial x_i \partial x_j \, dx = \int_\kappa (\partial^2 f / \partial x_i \partial x_j) \phi \, dx - \int_{\partial\kappa} (\partial f / \partial x_j) \phi n_i \, ds + \int_{\partial\kappa} f (\partial \phi / \partial x_i) n_j \, ds$$
$$\forall f, \phi \in H^2(\kappa).$$

Thus, when $f \in P_{l-1}$, the formulas (A.22) give:

$$\int_\kappa f \partial^2 (\phi - \tilde{I}_h \phi) / \partial x_i \partial x_j \, dx = \int_{\partial\kappa} f n_j \partial (\phi - \tilde{I}_h \phi) / \partial x_i \, ds \quad \forall \phi \in H^2(\kappa), \quad \forall f \in P_{l-1}.$$

Substituting into the definition (4.15) of b_h, we obtain:

$$b_h(\tau_h, \phi - \tilde{I}_h \phi) = - \sum_{\kappa \in \mathscr{T}_h} \int_{\partial\kappa} M_{nt}(\tau_h) \partial (\phi - \tilde{I}_h \phi) / \partial t \, ds.$$

Then integrating by parts on each segment κ' of $\partial\kappa$ and applying again (A.22) we readily find that $b_h(\tau_h, \phi - \tilde{I}_h \phi) = 0$. This proves (4.37). □

From the definition (A.22) it is easy to derive that the statement of Lemma 2.11 holds with $\theta_h = \tilde{I}_h v$:

(4.39) $$\| v - \tilde{I}_h v \|_{2,h} \leqslant C h^{k-2} |v|_{k,\Omega} \quad \forall v \in H^k(\Omega),$$

provided the triangulation $\mathscr{T}_h$ is regular. Thus, combining (4.38) and (4.39) we have the next result.

Corollary 4.1. *The operator $\tilde{I}_h$ defined by* (A.22) *satisfies*:

$$b_h(\tau_h, \phi - \tilde{I}_h \phi) = 0 \quad \forall \phi \in \mathscr{C}^0(\bar{\Omega}) \ \textit{with} \ \phi|_\kappa \in H^2(\kappa), \quad \forall \tau_h \in \Sigma_h.$$

In addition, if $\mathscr{T}_h$ is a regular triangulation of $\bar{\Omega}$, there exists a constant $C > 0$, independent of h and ϕ, such that:

$$\| \phi - \tilde{I}_h \phi \|_{2,h} \leqslant C h^{k-2} |\phi|_{k,\Omega} \quad \forall \phi \in H^k(\Omega),$$

provided the real k belongs to $[2, l + 1]$.

Now we turn to the inf-sup condition. Let us first restrict ourselves to the space of tensors $\Sigma \cap \Theta_h^4$ and more specifically to tensors of the form

$$\tau_{ij} = 0 \quad \text{if } i \neq j, \qquad \tau_{ii} = \theta_h \qquad i = 1, 2.$$

All such tensors satisfy

$$b_h(\tau_h, \phi_h) = (\mathbf{curl}\,\theta_h, \mathbf{curl}\,\phi_h) \qquad \forall \phi_h \in \Phi_h$$

and

$$\|\tau_h\|_{0,h}^2 = 2\|\theta_h\|_{0,\Omega}^2 + h\|\theta_h\|_{0,\Gamma_h}^2.$$

Hence applying Lemma 2.12 we obtain the preliminary result:

$$\sup_{\tau_h \in \Sigma \cap \Theta_h^4} \frac{b_h(\tau_h, \phi_h)}{\|\tau_h\|_{0,h}} \geqslant (1/\sqrt{2})\beta^* \|\phi_h\|_{2,h} \qquad \forall \phi_h \in \Phi_h,$$

where β^* is the constant of Lemma 2.12, provided Ω is convex and the triangulation is uniformly regular. By virtue of (4.31) and (4.32) this condition implies the inf-sup condition on the space Σ_h.

Lemma 4.4. *Let Ω be a bounded, convex polygon and let $\mathscr{T}_h$ be a uniformly regular triangulation of $\bar{\Omega}$. Then we have:*

$$\sup_{\tau_h \in \Sigma_h} \frac{b_h(\tau_h, \phi_h)}{\|\tau_h\|_{0,h}} \geqslant [1/(\sqrt{2}C_1)]\beta^* \|\phi_h\|_{2,h} \qquad \forall \phi_h \in \Phi_h,$$

where β^* and C_1 are the constants of (2.63) *and* (4.32) *respectively.*

Remark 4.2. Owing to Remark 2.8, we also have

$$\sup_{\tau_h \in \Sigma \cap \Theta_h^4} \frac{b_h(\tau_h, \phi_h)}{\|\tau_h\|_{0,h}} \geqslant (\gamma(s)/\sqrt{2})|\phi_h|_{1,s,\Omega} \qquad \forall \text{real } s \geqslant 2, \qquad \forall \phi_h \in \Phi_h.$$

Thus, the assumptions of Lemma 4.4 imply the additional inf-sup condition:

$$\sup_{\tau_h \in \Sigma_h} \frac{b_h(\tau_h, \phi_h)}{\|\tau_h\|_{0,h}} \geqslant [1/(\sqrt{2}C_1)]\gamma(s)|\phi_h|_{1,s,\Omega} \qquad \forall \text{real } s \geqslant 2, \qquad \forall \phi_h \in \Phi_h.$$

Remark 4.3. The construction of Lemma 4.4 can also be applied to prove that Problem (4.20) has a unique solution *without restriction on Ω and $\mathscr{T}_h$*. Indeed, since we are working with finite-dimensional spaces, all we need to show is that the set $\{\phi_h \in \Phi_h; b_h(\tau_h, \phi_h) = 0\ \forall \tau_h \in \Sigma_h\}$ is reduced to the zero function. Now, by proceeding like above, we construct $\tau \in \Sigma \cap \Phi_h^4$ such that $b_h(\tau, \phi_h) = |\phi_h|_{1,\Omega}^2 = b_h(\pi_h \tau, \phi_h) = 0$. Hence $\phi_h = 0$.

Finally when $\mathscr{T}_h$ is uniformly regular we can show like in Lemma 2.7 that $\|\,.\,\|_{0,h}$ and $\|\,.\,\|_{0,\Omega}$ are *two uniformly equivalent norms on Σ_h*. The proof, which is left as an exercise, stems from the inequality:

$$\|M_n(\tau)\|_{0,\Gamma_h}^2 \leqslant (C/h)\|\tau\|_{0,h}^2 \qquad \forall \tau \in \Sigma_h.$$

We are now in a position to establish optimal error estimates for Problem (4.20).

Theorem 4.2. *Let Ω be a bounded, plane polygon and $\mathcal{T}_h$ a triangulation of $\bar{\Omega}$. Then Problem* (4.20) *has a unique solution $(\sigma_h, \psi_h) \in \Sigma_h \times \Phi_h$. Next, assume that the solution $(\mathbf{u} = \mathbf{curl}\,\psi, p)$ of the Stokes Problem* (2.1) *satisfies* (4.17); *let $\sigma = (\partial^2 \psi / \partial x_i \partial x_j)_{i,j}$.*

1°) *If the triangulation $\mathcal{T}_h$ is regular, we have the bound*

$$\|\sigma - \sigma_h\|_{0,\Omega} \leqslant C_1 h^k |\psi|_{k+2,\Omega} \quad \forall k \in [1, l], \quad l \geqslant 1, \tag{4.40}$$

if $\psi \in H^{k+2}(\Omega)$. If in addition Ω is convex, we have either

$$|\psi - \psi_h|_{1,\Omega} \leqslant C_2 h |\psi|_{3,\Omega} \quad \text{if } l = 1 \quad \text{and} \quad \psi \in H^3(\Omega), \tag{4.41}$$

or

$$|\psi - \psi_h|_{1,\Omega} \leqslant C_3 h^k |\psi|_{k+1,\Omega} \quad \forall k \in [2, l] \quad \text{if } l \geqslant 2 \quad \text{and} \quad \psi \in H^{k+1}(\Omega). \tag{4.42}$$

2°) *If $\mathcal{T}_h$ is uniformly regular, we have:*

$$\|\sigma - \sigma_h\|_{0,h} \leqslant C_4 h^k |\psi|_{k+2,\Omega} \quad \forall \kappa \in [1, l], \quad l \geqslant 1 \tag{4.43}$$

and if in addition Ω is convex, we have:

$$\|\psi - \psi_h\|_{2,h} \leqslant C_5 h^k |\psi|_{k+2,\Omega} \quad \forall k \in [1, l-1], \quad \text{if } l \geqslant 2 \quad \text{and} \quad \psi \in H^{k+2}(\Omega). \tag{4.44}$$

Proof. As usual, we have:

$$\begin{cases} b_h(\sigma - \sigma_h, \phi_h) = 0 & \forall \phi_h \in \Phi_h, \\ a_h(\sigma - \sigma_h, \tau_h) + b_h(\tau_h, \psi - \psi_h) = 0 & \forall \tau_h \in \Sigma_h. \end{cases} \tag{4.45}$$

Owing to Lemma 4.2 and Corollary 4.1, the relations (4.45) yield:

$$b_h(\pi_h \sigma - \sigma_h, \phi_h) = 0 \quad \forall \phi_h \in \Phi_h,$$

$$a_h(\sigma - \sigma_h, \pi_h \sigma - \sigma_h) = 0$$

and observe that these equalities hold without constraint on Ω and $\mathcal{T}_h$. Now, the last equation implies directly that

$$\|\sigma - \sigma_h\|_{0,\Omega} \leqslant \|\sigma - \pi_h \sigma\|_{0,\Omega}. \tag{4.46a}$$

In addition, when $\mathcal{T}_h$ is uniformly regular, the equivalence between the norms $\|\cdot\|_{0,h}$ and $\|\cdot\|_{0,\Omega}$ gives

$$\|\sigma - \sigma_h\|_{0,h} \leqslant C_1 \|\sigma - \pi_h \sigma\|_{0,h}. \tag{4.46b}$$

Therefore (4.40) and (4.43) follow from (4.33).

Next, the second equation (4.45) and Corollary 4.1 yield:

$$b_h(\tau_h, \tilde{I}_h \psi - \psi_h) = a_h(\sigma_h - \sigma, \tau_h) \quad \forall \tau_h \in \Sigma_h. \tag{4.47}$$

Therefore, when Ω is convex and $\mathcal{T}_h$ uniformly regular, it stems from Lemma 4.4 that

(4.48) $$\|\tilde{I}_h\psi - \psi_h\|_{2,h} \leqslant C_2 \|\sigma_h - \sigma\|_{0,\Omega}.$$

Hence (4.44) follows from (4.48), (4.40) and Corollary 4.1.

To establish (4.41) and (4.42) we use a familiar duality argument. For $\mathbf{g}$ in $L^2(\Omega)^2$ we introduce the auxiliary Stokes Problem:

(4.49) $$\begin{cases} b_h(\mu_g, \phi) = (\mathbf{g}, \mathbf{curl}\,\phi) & \forall \phi \in \tilde{\Psi}, \\ a_h(\mu_g, \tau) + b_h(\tau, \lambda_g) = 0 & \forall \tau \in \Sigma. \end{cases}$$

Since Ω is convex, the solution (μ_g, λ_g) belongs to $H^1(\Omega)^4 \times H^3(\Omega)$ with

(4.50) $$\|\mu_g\|_{1,\Omega} + \|\lambda_g\|_{3,\Omega} \leqslant C_3 \|\mathbf{g}\|_{0,\Omega}.$$

Then a straightforward combination of (4.45), (4.49), (4.31) and Corollary 4.1 leads to:

$$\begin{aligned}(\mathbf{g}, \mathbf{curl}(\psi - \psi_h)) = b_h(\mu_g - \pi_h\mu_g, \psi - \phi_h) + a_h(\sigma - \sigma_h, \mu_g - \pi_h\mu_g) \\ + b_h(\sigma - \tau_h, \lambda_g - \tilde{I}_h\lambda_g) \quad \forall \phi_h \in \Phi_h, \quad \forall \tau_h \in \Sigma_h.\end{aligned}$$

When $l \geqslant 2$, Corollary 4.1, (4.33) and (4.50) yield:

$$|\psi - \psi_h|_{1,\Omega} \leqslant C_4 h \left\{ \inf_{\phi_h \in \Phi_h} \|\psi - \phi_h\|_{2,h} + \|\sigma - \sigma_h\|_{0,\Omega} + \inf_{\tau_h \in \Sigma_h} \|\sigma - \tau_h\|_{0,h} \right\}.$$

(4.51)

When $l = 1$, we only have:

(4.52) $$\begin{aligned}|\psi - \psi_h|_{1,\Omega} \leqslant C_5 h \left\{ \inf_{\phi_h \in \Phi_h} \|\psi - \phi_h\|_{2,h} + \|\sigma - \sigma_h\|_{0,\Omega} \right\} \\ + C_6 \inf_{\tau_h \in \Sigma_h} \|\sigma - \tau_h\|_{0,h}.\end{aligned}$$

In view of (4.40), Corollary 4.1 and (4.33), this implies (4.42) and (4.41). □

Corollary 4.2. *We retain all the assumptions of Theorem* 4.2. *If* ψ *belongs to* $H^{k+2}(\Omega)$ *for some real* $k \in [1, l]$ *with* $l \geqslant 1$, *we have*:

(4.53) $$|\psi - \psi_h|_{1,s,\Omega} \leqslant C(s) h^k |\psi|_{k+2,\Omega} \quad \textit{for each } s \geqslant 2.$$

Proof. Formula (4.47) and the inf-sup condition proved in Remark 4.2 give:

(4.54) $$|\tilde{I}_h\psi - \psi_h|_{1,s,\Omega} \leqslant C(s) \|\sigma - \sigma_h\|_{0,\Omega}.$$

Hence (4.53) follows from (4.40) and (A.23). □

Remark 4.4. The above theorem calls for a number of comments. First of all, it is obvious that this approach yields very neatly optimal error estimates for polynomials of all degrees. In addition, the scheme considered is fairly inexpensive.

On the other hand, all results are stated for a right-hand side $\mathbf{f}$ in $L^2(\Omega)^2$ whereas one is often interested in solving the Stokes problem when the right-hand side is in $L^r(\Omega)^2$ with $r < 2$. The next section extends the error analysis to this case.

4.3. Additional Results for the Hellan-Herrmann-Johnson Scheme

In this short section, we propose to extend part of the preceding results to the case where the right-hand side $\mathbf{f}$ belongs to $L^r(\Omega)^2$ with $1 < r < 2$. Since the spaces of Problem (4.20) are finite-dimensional, and in particular Φ_h is included in $W^{1,\infty}(\Omega)$, it is clear that Problem (4.20) still has a unique solution when the right-hand side $\mathbf{f}$ is only in $L^r(\Omega)^2$, for $1 < r < 2$. Thus, we must focus our attention on the equations (4.13) of the continuous problem and see how to adapt them to such a right-hand side. This is achieved much like in Section 2.1: the regularity of the tensor-valued functions τ is decreased while that of the test stream functions ϕ is increased.

If the solution $(\mathbf{u} = \mathbf{curl}\,\psi, p)$ of the Stokes Problem (2.1) is such that $\psi \in W^{3,r}(\Omega)$, then σ belongs to $W^{1,r}(\Omega)^4$ and therefore, according to Sobolev's Imbedding Theorem I.1.3 and the trace Theorem I.1.5 we have:

$$\sigma \in L^2(\Omega)^4, \qquad M_n(\sigma) \in L^{r/(2-r)}(\Gamma_h).$$

Hence we replace the space of tensors Σ by:

$$\tilde{\Sigma} = \{\tau \in L^2(\Omega)^4;\ \tau|_\kappa \in W^{1,r}(\kappa)^4,\ \tau_{12} = \tau_{21},\ M_n(\tau) \text{ is continuous on each segment of } \Gamma_h\}.$$

Likewise, since $\mathbf{f}$ is only in $L^r(\Omega)^2$, we replace the space $\tilde{\Psi}$ by

$$\tilde{\Psi} \cap W^{1,s}(\Omega), \qquad 1/s + 1/r = 1.$$

Then, it is a matter of routine to verify that the pair $(\psi, \sigma = (\partial^2\psi/\partial x_i \partial x_j))$ is the unique solution of:

$$\text{(4.55)} \qquad \begin{cases} b_h(\sigma, \phi) = -(1/\nu)\displaystyle\int_\Omega \mathbf{f}\cdot\mathbf{curl}\,\phi\,dx & \forall \phi \in \tilde{\Psi} \cap W^{1,s}(\Omega), \\ a_h(\sigma, \tau) + b_h(\tau, \psi) = 0 & \forall \tau \in \tilde{\Sigma}. \end{cases}$$

Now, a glance at Theorem 4.2 and its corollary shows that, in order to derive adequate error estimates in this case, we must verify that (4.46a), (4.48) and (4.54) are still valid here. To begin with, (4.46a) is a consequence of (4.45) together with the equations:

$$b_h(\sigma - \pi_h\sigma, \phi_h) = 0 \qquad \forall \phi_h \in \Phi_h,$$

$$b_h(\tau_h, \psi - \tilde{I}_h\psi) = 0 \qquad \forall \tau_h \in \Sigma_h.$$

But for ψ in $W^{3,r}(\Omega)$ and σ in $W^{1,r}(\Omega)^4$, both $\tilde{I}_h\psi$ and $\pi_h\sigma$ are well-defined and satisfy the above equations. And of course the equations (4.45) hold here. Therefore (4.46a) is verified. Likewise, (4.48) and (4.54) stem from (4.47) and the inf-sup conditions of Lemma 4.4 and Remark 4.2. Since the finite element spaces are unchanged, the inf-sup conditions carry over without modification; and the above considerations show that (4.47) is still valid here. The next lemma summarizes these results.

Lemma 4.5. *Let Ω be a bounded, plane polygon and let the solution* ($\mathbf{u} = \mathbf{curl}\,\psi, p$) *of the Stokes Problem* (2.1) *satisfy*:

(4.56) $$\psi \in W^{3,r}(\Omega), \quad p \in W^{1,r}(\Omega) \quad \textit{for some } r \in (1, 2].$$

Then (4.46a) *is valid. If in addition Ω is convex and $\mathcal{T}_h$ is uniformly regular, then* (4.48) *and* (4.54) *also hold.*

Next, it is easy to extend the approximation property of Lemma 4.2 to the case where $\tau \in W^{1,r}(\Omega)^4$.

Lemma 4.6. *Let $\mathcal{T}_h$ be a regular family of triangulations of $\bar{\Omega}$. We have*:

(4.57) $$\|\pi_h\tau - \tau\|_{0,\Omega} \leqslant C h^{2(1-1/r)} |\tau|_{1,r,\Omega}$$

for all symmetric tensors τ in $W^{1,r}(\Omega)^4$ with $1 < r \leqslant 2$.

Finally, combining these two lemmas we easily obtain the desired extension of Theorem 4.2 and its corollary.

Theorem 4.3. *Suppose that the regularity conditions* (4.56) *hold. If the triangulation $\mathcal{T}_h$ is regular, the solution (σ_h, ψ_h) of Problem* (4.20) *satisfies the estimate*:

(4.58) $$\|\sigma - \sigma_h\|_{0,\Omega} \leqslant C_1 h^{2(1-1/r)} |\psi|_{3,r,\Omega},$$

where $\sigma = (\partial^2\psi/\partial x_i \partial x_j)$.

If in addition $\mathcal{T}_h$ is uniformly regular and Ω is convex, then for each real $\beta \geqslant 2$, there exists a constant $C_2(\beta)$ such that:

(4.59) $$|\psi - \psi_h|_{1,\beta,\Omega} \leqslant C_2(\beta) h^{2(1-1/r)} |\psi|_{3,r,\Omega}.$$

Furthermore, when the polynomials are of degree $l \geqslant 2$ we also have:

(4.60) $$\left(\sum_{\kappa \in \mathcal{T}_h} |\psi - \psi_h|_{2,\kappa}^2\right)^{1/2} \leqslant C_3 h^{2(1-1/r)} |\psi|_{3,r,\Omega}.$$

4.4. Discontinuous Approximation of the Pressure

This section is devoted to a brief analysis of a numerical method that recovers the pressure in the Hellan-Herrmann-Johnson scheme. The pressure is obtained by a suitable approximation of Problem (4.8)—suitable in the sense that it reduces to the equations (4.20) when divergence-free test functions are used. The reader will find that it corresponds to a discontinuous approximation of the pressure.

To be specific, we want to construct finite-dimensional subspaces D_{0h} of $H_0(\text{div}; \Omega)$ and Q_h of $L_0^2(\Omega)$ that satisfy:

$$\{\mathbf{v}_h \in D_{0h} \text{ and } (q_h, \operatorname{div} \mathbf{v}_h) = 0 \quad \forall q_h \in Q_h\} \Rightarrow \mathbf{v}_h = \mathbf{curl}\,\phi_h \quad \text{with } \phi_h \text{ in } \Phi_h,$$

together with an adequate inf-sup condition. Let us start with the reference element $\hat{\kappa}$. If $\hat{\kappa}$ is the unit triangle, we introduce the polynomial space of dimension $l(l+2)$, with $l \geqslant 1$:

$$\hat{D} = P_{l-1}^2 \oplus \{p(\hat{x})\hat{\mathbf{x}};\ p \in \tilde{P}_{l-1}\}, \tag{4.61a}$$

where $\tilde{P}_k$ denotes the space of homogeneous polynomials of degree k. If $\hat{\kappa}$ is the unit square, we simply take the space of dimension $2l(l+1)$:

$$\hat{D} = Q_{l,l-1} \times Q_{l-1,l}, \tag{4.61b}$$

where exceptionally $Q_{r,s}$ denotes the space of all polynomials of degree at most r in x_1 and s in x_2. Then, it is easy to check that:

$$\operatorname{div}(\hat{D}) = P_{l-1}, \quad \operatorname{Ker}(\operatorname{div}) = \mathbf{curl}(P_l) \quad \text{if } \hat{\kappa} \text{ is the unit triangle,}$$

$$\operatorname{div}(\hat{D}) = Q_{l-1}, \quad \operatorname{Ker}(\operatorname{div}) = \mathbf{curl}(Q_l) \quad \text{if } \hat{\kappa} \text{ is the unit square.}$$

In addition, owing to the geometry of $\hat{\kappa}$, $\hat{\mathbf{v}} \cdot \hat{\mathbf{n}}$ reduces to a polynomial of P_{l-1} on $\partial\hat{\kappa}$ for $\hat{\mathbf{v}}$ in $\hat{D}$.

As a consequence, we can choose the following degrees of freedom for the vectors $\hat{\mathbf{v}}$ of $\hat{D}$:

(i) the boundary moments of order $l-1$ for $\hat{\mathbf{v}} \cdot \hat{\mathbf{n}}$:

$$\int_{\hat{\kappa}'} \hat{\mathbf{v}} \cdot \hat{\mathbf{n}} f \, d\hat{s} \quad \forall f \in P_{l-1} \quad \text{on each side } \hat{\kappa}' \text{ of } \hat{\kappa};$$

(4.62) (ii) the interior moments of order $l-2$ for $\hat{\mathbf{v}}$:

$$\int_{\hat{\kappa}} \hat{\mathbf{v}} \cdot \mathbf{f} \, d\hat{x} \quad \forall \mathbf{f} \in \begin{cases} P_{l-2}^2 & \text{if } \hat{\kappa} \text{ is a triangle,} \\ Q_{l-2,l-1} \times Q_{l-1,l-2} & \text{if } \hat{\kappa} \text{ is a square.} \end{cases}$$

It is a matter of routine to check that (4.62) defines a unique vector $\hat{\mathbf{v}}$ of $\hat{D}$ and that the restriction of $\hat{\mathbf{v}} \cdot \hat{\mathbf{n}}$ on each side $\hat{\kappa}'$ depends only upon the l values prescribed on this side.

To switch from $\hat{\kappa}$ to an arbitrary element κ, we introduce the following contravariant transformation between the vector function $\mathbf{v} = (v_1, v_2)$ defined on κ and $\hat{\mathbf{v}} = (\hat{v}_1, \hat{v}_2)$ defined on $\hat{\kappa}$:

$$\mathbf{v} = \mathscr{F}_\kappa \hat{\mathbf{v}} \quad \text{where} \quad \mathbf{v} \circ F_\kappa = (1/J_F) DF_\kappa \cdot \hat{\mathbf{v}}. \tag{4.63}$$

The choice of $\mathscr{F}_\kappa$ is justified by the fact that, roughly speaking, it preserves the divergence, **curl** and normal component:

$$(\operatorname{div} \mathbf{v}) \circ F_\kappa = (1/J_F) \operatorname{div}(\mathscr{F}_\kappa^{-1} \mathbf{v}),$$

$$\mathscr{F}_\kappa^{-1}(\mathbf{curl}\, \psi) = \mathbf{curl}(\psi \circ F_\kappa),$$

$$\int_{\partial\kappa} \phi \mathbf{v} \cdot \mathbf{n} \, ds = \int_{\partial\hat{\kappa}} (\phi \circ F_\kappa)(\mathscr{F}_\kappa^{-1} \mathbf{v}) \cdot \hat{\mathbf{n}} \, d\hat{s}.$$

Then we *fix l distinct points* on each segment κ' of Γ_h and we set:

$$D_{0h} = \{\mathbf{v}_h \in L^2(\Omega)^2;\ \mathbf{v}_h|_\kappa = \mathscr{F}_\kappa \hat{\mathbf{v}},\ \hat{\mathbf{v}} \in \hat{D} \ \ \forall \kappa \in \mathscr{T}_h \ \ \mathbf{v}_h \cdot \mathbf{n} \text{ is continuous (resp. 0) at the } l \text{ points of each interior (resp. boundary) segment } \kappa' \text{ of } \Gamma_h\}. \tag{4.64}$$

Once the space D_{0h} is constructed, the choice of the pressure space Q_h is dictated by the above requirements and considerations. Clearly, we must choose

(4.65) $$Q_h = \{q_h \in L_0^2(\Omega);\ q_h|_\kappa \in P_{l-1} \text{ or } Q_{l-1}(\kappa) \text{ according that } \kappa \text{ is a triangle or a quadrilateral}\}.$$

It follows immediately that:

$$D_{0h} \subset H_0(\text{div}; \Omega),$$

$$\{\mathbf{v}_h \in D_{0h} \text{ and } (q_h, \text{div}\,\mathbf{v}_h) = 0 \quad \forall q_h \in Q_h\} \Rightarrow \text{div}\,\mathbf{v}_h = 0$$

$$\Rightarrow \mathbf{v}_h = \mathbf{curl}\,\phi_h \quad \text{for a unique } \phi_h \text{ in } \Phi_h.$$

Remark 4.5. When $l = 1$, the functions of Q_h are constants in each κ while the functions of $\hat{D}$ have the form:

$$\hat{\mathbf{v}} = (c_1 + c_0\hat{x}_1, c_2 + c_0\hat{x}_2) \quad \text{if } \hat{\kappa} \text{ is the unit triangle},$$

$$\hat{\mathbf{v}} = (c_1 + c_3\hat{x}_1, c_2 + c_4\hat{x}_2) \quad \text{if } \hat{\kappa} \text{ is the unit square}.$$

From the degrees of freedom (4.62) we deduce a straightforward restriction operator $\hat{\pi}$ from $H^1(\hat{\kappa})^2$ onto $\hat{D}$:

$\hat{\pi}\hat{\mathbf{v}}$ is the unique polynomial of $\hat{D}$ that has the same degrees of freedom (4.62) on $\hat{\kappa}$ as $\hat{\mathbf{v}}$.

Then the transformation $\mathscr{F}_\kappa$ yields the following restriction operator

$$\pi_h \in \mathscr{L}(H_0(\text{div}; \Omega) \cap H^1(\Omega)^2; D_{0h}):$$

(4.66) $$\pi_h \mathbf{v}|_\kappa = \mathscr{F}_\kappa(\hat{\pi}\hat{\mathbf{v}}) \qquad \forall \kappa \in \mathscr{T}_h.$$

The operator π_h satisfies Lemma II.1.1. More precisely, we have the following crucial result.

Theorem 4.4. *If the triangulation* $\mathscr{T}_h$ *is regular, the operator* π_h *defined by* (4.66) *satisfies for all* $\mathbf{v} \in H^1(\Omega)^2$:

(4.67) $$\left\{\sum_{\kappa \in \mathscr{T}_h} |\pi_h \mathbf{v} - \mathbf{v}|_{1,\kappa}^2\right\}^{1/2} + h^{-2/s}\|\pi_h \mathbf{v} - \mathbf{v}\|_{0,s,\Omega} \leqslant C(s)|\mathbf{v}|_{1,\Omega} \qquad \forall s \geqslant 2,$$

(4.68) $$\|\text{div}(\pi_h \mathbf{v} - \mathbf{v})\|_{0,\Omega} \leqslant C\|\text{div}\,\mathbf{v}\|_{0,\Omega}.$$

In addition, whatever the triangulation, we have:

(4.69) $$(q_h, \text{div}(\pi_h \mathbf{v} - \mathbf{v})) = 0 \qquad \forall q_h \in Q_h.$$

Finally, when $\mathscr{T}_h$ *is made exclusively of triangles, the inequality* (4.68) *holds with* $C = 1$ *and no regularity requirement on* $\mathscr{T}_h$.

Proof. The properties (4.68) and (4.69) are an easy consequence of the definition of $\hat{\pi}$.

Next, let us sketch the proof of (4.67). Consider the case where κ is a quadrilateral. Take $\mathbf{v}$ in $H^1(\kappa)^2$, $\hat{\mathbf{v}} = \mathscr{F}_\kappa^{-1}\mathbf{v}$ and let us split $\hat{\mathbf{v}}$ as follows:

$$\hat{\mathbf{v}} = \mathbf{grad}\, \hat{q} + \mathbf{curl}\, \hat{\phi} \tag{4.70}$$

where $\hat{q}$ is the solution of

$$\Delta \hat{q} = \operatorname{div} \hat{\mathbf{v}} \quad \text{in } \hat{\kappa}, \qquad \hat{q}|_{\partial\hat{\kappa}} = 0.$$

Since $\hat{\kappa}$ is convex, Theorem I.1.8 says that $\hat{q} \in H^2(\hat{\kappa})$ with

$$\|\hat{q}\|_{2,\hat{\kappa}} \leqslant C_1 \|\operatorname{div} \hat{\mathbf{v}}\|_{0,\hat{\kappa}}.$$

As a consequence we can find $\hat{\phi}$ in $H^2(\hat{\kappa})$ that satisfies (4.70) and

$$\|\partial^2\hat{\phi}/\partial\hat{x}_1^2\|_{0,\hat{\kappa}}^2 + \|\partial^2\hat{\phi}/\partial\hat{x}_2^2\|_{0,\hat{\kappa}}^2$$
$$\leqslant C_2\{\|\partial\hat{v}_1/\partial\hat{x}_2\|_{0,\hat{\kappa}}^2 + \|\partial\hat{v}_2/\partial\hat{x}_1\|_{0,\hat{\kappa}}^2 + \|\operatorname{div}\hat{\mathbf{v}}\|_{0,\hat{\kappa}}^2\}.$$

Thus a straightforward application of Theorem A.3 yields:

$$|\hat{\pi}\hat{\mathbf{v}} - \hat{\mathbf{v}}|_{1,\hat{\kappa}} \leqslant C_3\{\|\partial\hat{v}_1/\partial\hat{x}_2\|_{0,\hat{\kappa}}^2 + \|\partial\hat{v}_2/\partial\hat{x}_1\|_{0,\hat{\kappa}}^2 + \|\operatorname{div}\hat{\mathbf{v}}\|_{0,\hat{\kappa}}^2\}^{1/2}, \tag{4.71}$$

and a similar upper bound (with a different constant) for $\|\hat{\pi}\hat{\mathbf{v}} - \hat{\mathbf{v}}\|_{0,s,\hat{\kappa}}$. In view of (4.63) and the regularity of $\mathscr{T}_h$, a simple calculation now leads to (4.67). □

Theorem 4.4 gives us the following inf-sup condition:

for each $q_h \in Q_h$, *there exists* $\mathbf{v}_h \in D_{0h}$ *such that*

$$(q_h, \operatorname{div} \mathbf{v}_h) = \|q_h\|_{0,\Omega}^2,$$

and

$$\left\{\sum_{\kappa\in\mathscr{T}_h} |\mathbf{v}_h|_{1,\kappa}^2\right\}^{1/2} + \|\mathbf{v}_h\|_{0,s,\Omega} + \|\operatorname{div}\mathbf{v}_h\|_{0,\Omega} \leqslant C(s)\|q_h\|_{0,\Omega} \qquad \forall s \geqslant 2.$$

With the spaces D_{0h} and Q_h we propose the following discretization of Problem (4.8):

Find a function p_h *in* Q_h *satisfying*

$$\int_\Omega p_h \operatorname{div}\mathbf{v}_h\,dx = -\int_\Omega \mathbf{f}\cdot\mathbf{v}_h\,dx$$
$$+ \nu \sum_{\kappa\in\mathscr{T}_h}\left\{\int_\kappa (\lambda_h)_{ij}(\partial v_{hi}/\partial x_j)\,dx - \int_{\partial\kappa} M_{nt}(\lambda_h)\mathbf{v}_h\cdot\mathbf{t}\,ds\right\} \tag{4.72}$$
$$\forall \mathbf{v}_h \in D_{0h},$$

where the tensor λ_h *is related by* (4.10) *to the solution* $\sigma_h \in \Sigma_h$ *of* (4.20).

Since Problem (4.20) can be solved independently of Problem (4.72), then owing to the above inf-sup condition Problem (4.72) has a unique solution p_h in Q_h. Moreover, we have the following error estimate:

Theorem 4.5. *Assume that Ω is a bounded polygon and suppose $\mathscr{T}_h$ is a uniformly regular triangulation of $\bar{\Omega}$. If the solution* $(\mathbf{u}, p)$ *of the Stokes Problem* (2.1) *has the regularity:*

$$\mathbf{u} \in H^{k+1}(\Omega)^2, \quad p \in H^k(\Omega) \cap L_0^2(\Omega) \quad \text{for some } k \in [1, l],$$

then the solution p_h of Problem (4.72) *satisfies the error estimate:*

$$\|p - p_h\|_{0,\Omega} \leqslant Ch^k\{|\mathbf{u}|_{k+1,\Omega} + |p|_{k,\Omega}\}. \tag{4.73}$$

Proof. In view of (4.10), for each q_h in Q_h we have:

$$\left|\int_\Omega (q_h - p_h)\operatorname{div}\mathbf{v}_h\,dx\right| \leqslant \|q_h - p\|_{0,\Omega}\|\operatorname{div}\mathbf{v}_h\|_{0,\Omega} + \nu\|\sigma - \sigma_h\|_{0,\Omega}\left(\sum_{K \in \mathscr{T}_h}|\mathbf{v}_h|_{1,K}^2\right)^{1/2} + \nu\|M_n(\sigma - \sigma_h)\|_{0,\Gamma_h}\|S(\mathbf{v}_h \cdot \mathbf{t})\|_{0,\Gamma_h}.$$

Then, according to the inf-sup condition, we can choose $\mathbf{v}_h$ in D_{0h} such that

$$\|q_h - p_h\|_{0,\Omega}^2 \leqslant C_1\{\|q_h - p\|_{0,\Omega} + \nu\|\sigma - \sigma_h\|_{0,\Omega}\}\|q_h - p_h\|_{0,\Omega} + \nu\|M_n(\sigma - \sigma_h)\|_{0,\Gamma_h}\|S(\mathbf{v}_h \cdot \mathbf{t})\|_{0,\Gamma_h}.$$

It remains to estimate $S(\mathbf{v}_h \cdot \mathbf{t})$. To this end, we use the fact that

$$\mathbf{v}_h = \pi_h \mathbf{v}$$

with $\quad \operatorname{div}\mathbf{v} = q_h - p_h, \quad |\mathbf{v}|_{1,\Omega} \leqslant C_2\|q_h - p_h\|_{0,\Omega}, \quad \mathbf{v} \in H_0^1(\Omega)^2.$

Since $S(\mathbf{v} \cdot \mathbf{t}) = 0$ we can write

$$S(\mathbf{v}_h \cdot \mathbf{t}) = S((\pi_h \mathbf{v} - \mathbf{v}) \cdot \mathbf{t}).$$

Hence the argument of Theorem 4.4 gives:

$$\|S(\mathbf{v}_h \cdot \mathbf{t})\|_{0,K'} \leqslant C_3(\rho_K)^{-1/2}\{\|\partial\hat{v}_1/\partial\hat{x}_2\|_{0,\hat{K}}^2 + \|\partial\hat{v}_2/\partial\hat{x}_1\|_{0,\hat{K}}^2 + \|\operatorname{div}\hat{\mathbf{v}}\|_{0,\hat{K}}^2\}^{1/2}.$$

Therefore

$$\|S(\mathbf{v}_h \cdot \mathbf{t})\|_{0,\Gamma_h} \leqslant C_4 h^{1/2}\|q_h - p_h\|_{0,\Omega}$$

and consequently,

$$\|q_h - p_h\|_{0,\Omega} \leqslant C_5\{\|q_h - p\|_{0,\Omega} + \nu\|\sigma - \sigma_h\|_{0,h}\}.$$

Then (4.73) follows from (4.43) and Lemma A.5 or (A.51).

Observe that just (4.43) requires the uniformity of $\mathscr{T}_h$. □

Remark 4.6. It is also possible to associate this discontinuous approximation of the pressure with the "stream function-vorticity" scheme studied in § 2. The discrete version of the first equation (2.13) is:

Find $p_h \in Q_h$, defined by (4.65), *solution of*

$$(p_h, \operatorname{div}\mathbf{v}_h) = \nu(\mathbf{curl}\,\omega_h, \mathbf{v}_h) - (\mathbf{f}, \mathbf{v}_h) \quad \forall \mathbf{v}_h \in D_{0h}. \tag{4.74}$$

This scheme has the same order of convergence as (2.36) but the proof requires more regularity of the solution (cf. Girault & Raviart [33]).

Remark 4.7. Of course, we can use here a continuous approximation of the pressure analogous to (2.36), but the error analysis of the corresponding scheme is more delicate.

§5. A "Vector Potential-Vorticity" Scheme in Three Dimensions

It is not easy to adapt the schemes developed in the previous paragraphs to the three-dimensional Stokes problem. The obvious reason is that the conditions determining the vector potential are more intricate than those defining the two-dimensional stream function. Therefore we shall only attempt to extend to the homogeneous Stokes problem in a very simple region of $\mathbb{R}^3$ the "stream function-vorticity" scheme of §2.

Throughout this paragraph, we shall assume that Ω is a *bounded, simply-connected open subset of* $\mathbb{R}^3$ *with a polyhedral connected bounded* Γ. Leaving the approximation of the pressure to the last section, our first object is to relax the regularity of the function spaces related to the biharmonic problems of Section I.5.3. The reader will discover that it suffices to work with functions in $H(\mathbf{curl}; \Omega)$. This approach will lead to the construction of finite-dimensional subspaces of conforming finite elements in $H(\mathbf{curl}; \Omega)$, which are not subspaces of $H^1(\Omega)^3$.

Finally, since discontinuous elements are used it is reasonable to use a discontinuous approximation of the pressure, very similar to that of Section 4.4.

5.1. A Mixed Formulation of the Three-Dimensional Stokes Problem

Let $\mathbf{f}$ be a given vector of $L^2(\Omega)^3$ and consider the *homogeneous* Stokes Problem:

Find $(\mathbf{u}, p)$ *in* $H^1(\Omega)^3 \times L_0^2(\Omega)$ *satisfying*:

$$\text{(5.1)} \qquad \begin{cases} \left.\begin{aligned} -\nu\Delta\mathbf{u} + \mathbf{grad}\, p &= \mathbf{f}, \\ \operatorname{div}\mathbf{u} &= 0 \end{aligned}\right\} \quad \textit{in } \Omega, \\ \mathbf{u} = \mathbf{0} \quad \textit{on } \Gamma. \end{cases}$$

We have seen in Section I.5.3 that this problem can be interpreted as a biharmonic problem for the vector potential $\boldsymbol{\psi}$ of $\mathbf{u}$ (recall that $\mathbf{u} = \mathbf{curl}\, \boldsymbol{\psi}$) where $\boldsymbol{\psi}$ belongs to the space:

$$\text{(5.2)} \qquad \Psi = \{\boldsymbol{\phi} \in L^2(\Omega)^3;\ \operatorname{div}\boldsymbol{\phi} \in H^1(\Omega),\ \mathbf{curl}\,\boldsymbol{\phi} \in H_0^1(\Omega)^3,\ \boldsymbol{\phi} \times \mathbf{n}|_\Gamma = \mathbf{0}\}.$$

In order to derive a mixed formulation of Problem (5.1), let us multiply both sides of (5.1) with $\mathbf{curl}\,\boldsymbol{\phi}$ and determine exactly what properties we require of $\boldsymbol{\phi}$.

As $\operatorname{div} \mathbf{u} = 0$ we obtain first:

$$\nu \langle \mathbf{curl}\, \mathbf{curl}\, \mathbf{u}, \mathbf{curl}\, \boldsymbol{\phi} \rangle = (\mathbf{f}, \mathbf{curl}\, \boldsymbol{\phi}) \qquad \forall \boldsymbol{\phi} \in H_0(\mathbf{curl}; \Omega),$$

where $\langle .,. \rangle$ denotes the duality between $H_0(\mathbf{curl}; \Omega)$ and its dual space. Now, let us set

$$\boldsymbol{\omega} = \mathbf{curl}\, \mathbf{u} \tag{5.3}$$

and *assume that* $\boldsymbol{\omega} \in H(\mathbf{curl}; \Omega)$; then we have

$$\nu (\mathbf{curl}\, \boldsymbol{\omega}, \mathbf{curl}\, \boldsymbol{\phi}) = (\mathbf{f}, \mathbf{curl}\, \boldsymbol{\phi}) \qquad \forall \boldsymbol{\phi} \in H_0(\mathbf{curl}; \Omega).$$

Finally, since

$$\mathbf{u} = \mathbf{curl}\, \boldsymbol{\psi} \quad \text{with} \quad \operatorname{div} \boldsymbol{\psi} = 0 \quad \text{and} \quad \boldsymbol{\psi} \in \boldsymbol{\Psi},$$

equation (5.3) can be written equivalently as:

$$(\mathbf{curl}\, \mathbf{curl}\, \boldsymbol{\psi}, \boldsymbol{\mu}) = (\boldsymbol{\omega}, \boldsymbol{\mu}) \qquad \forall \boldsymbol{\mu} \in L^2(\Omega)^3.$$

And by restricting $\boldsymbol{\mu}$ to $H(\mathbf{curl}; \Omega)$ this becomes:

$$(\mathbf{curl}\, \boldsymbol{\psi}, \mathbf{curl}\, \boldsymbol{\mu}) = (\boldsymbol{\omega}, \boldsymbol{\mu}) \qquad \forall \boldsymbol{\mu} \in H(\mathbf{curl}; \Omega).$$

Hence, summing up we see that the following *Problem* $(\tilde{\mathrm{Q}})$:

Find a pair $(\boldsymbol{\psi}, \boldsymbol{\omega}) \in H_0(\mathbf{curl}; \Omega) \times H(\mathbf{curl}; \Omega)$ *such that*:

$$\nu (\mathbf{curl}\, \boldsymbol{\omega}, \mathbf{curl}\, \boldsymbol{\phi}) = (\mathbf{f}, \mathbf{curl}\, \boldsymbol{\phi}) \qquad \forall \boldsymbol{\phi} \in H_0(\mathbf{curl}; \Omega), \tag{5.4}$$

$$(\mathbf{curl}\, \boldsymbol{\psi}, \mathbf{curl}\, \boldsymbol{\mu}) = (\boldsymbol{\omega}, \boldsymbol{\mu}) \qquad \forall \boldsymbol{\mu} \in H(\mathbf{curl}; \Omega), \tag{5.5}$$

$$\operatorname{div} \boldsymbol{\psi} = 0, \tag{5.6}$$

has at least one solution $(\boldsymbol{\psi}, \boldsymbol{\omega} = -\Delta \boldsymbol{\psi})$ where $\mathbf{u} = \mathbf{curl}\, \boldsymbol{\psi}$ and $\boldsymbol{\psi} \in \boldsymbol{\Psi}$. Conversely, it is easy to check that this problem has at most one solution. Indeed, if

$$(\mathbf{curl}\, \boldsymbol{\omega}, \mathbf{curl}\, \boldsymbol{\phi}) = 0 \qquad \forall \boldsymbol{\phi} \in H_0(\mathbf{curl}; \Omega),$$

then choosing $\boldsymbol{\phi} = \boldsymbol{\psi}$ and $\boldsymbol{\mu} = \boldsymbol{\omega}$ in (5.5) we obtain successively $\boldsymbol{\omega} = \mathbf{0}$ and $\mathbf{curl}\, \boldsymbol{\psi} = \mathbf{0}$. Since $\operatorname{div} \boldsymbol{\psi} = 0$ and Γ has only one connected component this implies that $\boldsymbol{\psi} = \mathbf{0}$ (cf. Remark I.3.9). Therefore, we have proved the following result:

Theorem 5.1. *Assume that the solution* $\mathbf{u}$ *of Problem* (5.1) *satisfies*

$$\mathbf{curl}\, \mathbf{u} \in H(\mathbf{curl}; \Omega).$$

Then Problem (5.4) (5.5) (5.6) *has the unique solution*:

$$(\boldsymbol{\psi}, -\Delta \boldsymbol{\psi}) \quad \textit{where } \mathbf{curl}\, \boldsymbol{\psi} = \mathbf{u}.$$

Now we want to insert Problem $(\tilde{\mathrm{Q}})$ into the framework of Section 1.1. We set

$$\Phi_0 = \{\boldsymbol{\phi} \in H_0(\mathbf{curl}; \Omega);\ \operatorname{div} \boldsymbol{\phi} = 0\}.$$

Clearly, Problem $(\tilde{\mathrm{Q}})$ is equivalent to:

Find a pair $(\boldsymbol{\psi}, \boldsymbol{\omega})$ *in* $\Phi_0 \times H(\mathbf{curl}; \Omega)$ *satisfying*:

$$(5.7)\qquad \begin{cases} \nu(\mathbf{curl}\,\boldsymbol{\omega}, \mathbf{curl}\,\boldsymbol{\phi}) = (\mathbf{f}, \mathbf{curl}\,\boldsymbol{\phi}) & \forall \boldsymbol{\phi} \in \Phi_0, \\ (\mathbf{curl}\,\boldsymbol{\psi}, \mathbf{curl}\,\boldsymbol{\mu}) = (\boldsymbol{\omega}, \boldsymbol{\mu}) & \forall \boldsymbol{\mu} \in H(\mathbf{curl}; \Omega). \end{cases}$$

Next, we introduce:

$$\tilde{X} = \{\mathbf{curl}\,\boldsymbol{\phi};\, \boldsymbol{\phi} \in \Phi_0\} \times L^2(\Omega)^3, \qquad \tilde{M} = H(\mathbf{curl}; \Omega),$$

$$\tilde{a}(u, v) = \nu(\boldsymbol{\omega}, \boldsymbol{\theta}) \qquad \forall u = (\mathbf{curl}\,\boldsymbol{\psi}, \boldsymbol{\omega}), \qquad v = (\mathbf{curl}\,\boldsymbol{\phi}, \boldsymbol{\theta}) \in \tilde{X},$$

$$\tilde{b}(v, \boldsymbol{\mu}) = (\mathbf{curl}\,\boldsymbol{\phi}, \mathbf{curl}\,\boldsymbol{\mu}) - (\boldsymbol{\theta}, \boldsymbol{\mu}) \qquad \forall v = (\mathbf{curl}\,\boldsymbol{\phi}, \boldsymbol{\theta}) \in \tilde{X}, \qquad \boldsymbol{\mu} \in \tilde{M},$$

$$\langle l, v\rangle = (\mathbf{f}, \mathbf{curl}\,\boldsymbol{\phi}) \qquad \forall v = (\mathbf{curl}\,\boldsymbol{\phi}, \boldsymbol{\theta}) \in \tilde{X}.$$

With this notation, Problem $(\tilde{\mathrm{Q}})$ takes the more familiar form:

Find a pair $(u, \boldsymbol{\lambda})$ *in* $\tilde{X} \times \tilde{M}$ *such that*:

$$(5.8)\qquad \begin{cases} \tilde{a}(u, v) + \tilde{b}(v, \boldsymbol{\lambda}) = \langle l, v\rangle & \forall v \in \tilde{X}, \\ \tilde{b}(u, \boldsymbol{\mu}) = 0 & \forall \boldsymbol{\mu} \in \tilde{M}. \end{cases}$$

As usual, the space $\tilde{V}$ is defined by:

$$\tilde{V} = \{v \in \tilde{X};\, \tilde{b}(v, \boldsymbol{\mu}) = 0 \quad \forall \boldsymbol{\mu} \in \tilde{M}\},$$

i.e.

$$\tilde{V} = \{v = (\mathbf{curl}\,\boldsymbol{\phi}, \boldsymbol{\theta}) \in \tilde{X};\, (\mathbf{curl}\,\boldsymbol{\phi}, \mathbf{curl}\,\boldsymbol{\mu}) = (\boldsymbol{\theta}, \boldsymbol{\mu}) \quad \forall \boldsymbol{\mu} \in H(\mathbf{curl}; \Omega)\}.$$

Since the mapping $\boldsymbol{\phi} \to \|\mathbf{curl}\,\boldsymbol{\phi}\|_{0,\Omega}$ is a norm on Φ_0 equivalent to the norm of $H(\mathbf{curl}; \Omega)$ (cf. Lemma I.3.4):

$$(5.9)\qquad \|\boldsymbol{\phi}\|_{0,\Omega} \leqslant C \|\mathbf{curl}\,\boldsymbol{\phi}\|_{0,\Omega} \qquad \forall \boldsymbol{\phi} \in \Phi_0,$$

it follows that on the one hand we can choose the following norm on $\tilde{X}$:

$$\|v\|_{\tilde{X}} = (\|\mathbf{curl}\,\boldsymbol{\phi}\|_{0,\Omega}^2 + \|\boldsymbol{\theta}\|_{0,\Omega}^2)^{1/2} \qquad \forall v = (\mathbf{curl}\,\boldsymbol{\phi}, \boldsymbol{\theta}) \in \tilde{X};$$

and on the other hand we have:

$$(5.10)\qquad \|\mathbf{curl}\,\boldsymbol{\phi}\|_{0,\Omega} \leqslant C \|\boldsymbol{\theta}\|_{0,\Omega} \qquad \forall v = (\mathbf{curl}\,\boldsymbol{\phi}, \boldsymbol{\theta}) \in \tilde{V}.$$

Hence the *mapping* $v = (\mathbf{curl}\,\boldsymbol{\phi}, \boldsymbol{\theta}) \to \|\boldsymbol{\theta}\|_{0,\Omega}$ *is a norm on* $\tilde{V}$ *equivalent to the norm of* $\tilde{X}$:

$$\|v\|_{\tilde{X}} \leqslant (C^2 + 1)^{1/2} \|\boldsymbol{\theta}\|_{0,\Omega}$$

and we set:

$$|v| = \|\boldsymbol{\theta}\|_{0,\Omega}.$$

As a consequence, the form $\tilde{a}(.,.)$ is $\tilde{V}$-elliptic:

$$\tilde{a}(v, v) = \nu |v|^2 \geqslant \tilde{\alpha} \|v\|_{\tilde{X}}^2 \qquad \forall v \in \tilde{V},$$

with $\tilde{\alpha} = \nu/(C^2 + 1)$.

Likewise, it is easy to check that the form $\tilde{b}(.,.)$ satisfies the weak inf-sup condition:

$$\sup_{v \in \tilde{X}} (\tilde{b}(v, \boldsymbol{\mu})/\|v\|_{\tilde{X}}) \geqslant \|\boldsymbol{\mu}\|_{0,\Omega} \qquad \forall \boldsymbol{\mu} \in \tilde{M}.$$

Finally, we readily derive that here the Lagrange multiplier $\boldsymbol{\lambda}$ satisfies $\boldsymbol{\lambda} = \nu\boldsymbol{\omega}$.

Remark 5.1. Observe the analogy between Problem (5.7) and Problem (2.20) in two dimensions.

Remark 5.2. Note that the first equation of (5.8) holds on a larger space than $\tilde{X}$:

$$\tilde{a}(u, v) + \tilde{b}(v, \boldsymbol{\lambda}) = \langle l, v \rangle \qquad \forall v = (\mathbf{curl}\, \boldsymbol{\phi}, \boldsymbol{\theta})$$
$$\text{with } (\boldsymbol{\phi}, \boldsymbol{\theta}) \in H_0(\mathbf{curl}; \Omega) \times L^2(\Omega)^3.$$

5.2. Mixed Approximation in $H(\mathbf{curl}; \Omega)$

The statement of Problem $(\tilde{\mathrm{Q}})$ induces us to define its approximation in finite-dimensional subspaces of $H(\mathbf{curl}; \Omega)$. Thus, we introduce three finite-dimensional spaces:

$$(5.11) \qquad \Phi_h \subset H_0(\mathbf{curl}; \Omega), \qquad M_h \subset H(\mathbf{curl}; \Omega), \qquad \Theta_h \subset H_0^1(\Omega)$$

and we assume that

$$\Phi_h \subset M_h.$$

Since Φ_h is not necessarily contained in $H(\mathrm{div}; \Omega)$, the divergence-free condition is expressed by:

$$(5.12) \qquad (\boldsymbol{\phi}_h, \mathbf{grad}\, q_h) = 0 \qquad \forall q_h \in \Theta_h.$$

In other words, the space Φ_0 is approximated by:

$$(5.13) \qquad \Phi_{h0} = \{\boldsymbol{\phi}_h \in \Phi_h;\ \boldsymbol{\phi}_h \text{ satisfies (5.12)}\},$$

which, in general, is not contained in Φ_0. Nevertheless, it is reasonable to ask that the functions of Φ_{h0} satisfy the same equivalence of norms as Φ_0, namely:

there exists a positive constant $C^* > 0$ *such that*:

$$(5.14) \qquad \|\boldsymbol{\phi}_h\|_{0,\Omega} \leqslant C^* \|\mathbf{curl}\, \boldsymbol{\phi}_h\|_{0,\Omega} \qquad \forall \boldsymbol{\phi}_h \in \Phi_{h0}.$$

With these spaces, we propose the following approximation of Problem (5.7) called *Problem* (Q_h):

Find a pair $(\boldsymbol{\psi}_h, \boldsymbol{\omega}_h) \in \Phi_{h0} \times M_h$ *such that*:

$$(5.15) \qquad \begin{cases} \nu(\mathbf{curl}\, \boldsymbol{\omega}_h, \mathbf{curl}\, \phi_h) = (\mathbf{f}, \mathbf{curl}\, \boldsymbol{\phi}_h) & \forall \boldsymbol{\phi}_h \in \Phi_{h0}, \\ (\mathbf{curl}\, \boldsymbol{\psi}_h, \mathbf{curl}\, \boldsymbol{\mu}_h) = (\boldsymbol{\omega}_h, \boldsymbol{\mu}_h) & \forall \boldsymbol{\mu}_h \in M_h. \end{cases}$$

From a practical point of view, this problem is of little use because of the constraint (5.12) on the test functions $\boldsymbol{\phi}_h$. Therefore we shall use instead the following approximation of Problem $(\tilde{Q})$ which works with the entire space Φ_h:

Find a pair $(\boldsymbol{\psi}_h, \boldsymbol{\omega}_h) \in \Phi_{h0} \times M_h$ *satisfying*:

$$\text{(5.16)} \qquad \begin{cases} \nu(\mathbf{curl}\,\boldsymbol{\omega}_h, \mathbf{curl}\,\boldsymbol{\phi}_h) = (\mathbf{f}, \mathbf{curl}\,\boldsymbol{\phi}_h) & \forall \boldsymbol{\phi}_h \in \Phi_h, \\ (\mathbf{curl}\,\boldsymbol{\psi}_h, \mathbf{curl}\,\boldsymbol{\mu}_h) = (\boldsymbol{\omega}_h, \boldsymbol{\mu}_h) & \forall \boldsymbol{\mu}_h \in M_h. \end{cases}$$

Obviously, it is desirable that these two problems be equivalent; but this requires an additional hypothesis:

for each function $\boldsymbol{\phi}_h$ *of* Φ_h *there exists a function* $\boldsymbol{\phi}_{h0}$ *of* Φ_{h0} *with*

$$\text{(5.17)} \qquad \mathbf{curl}\,\boldsymbol{\phi}_h = \mathbf{curl}\,\boldsymbol{\phi}_{h0}.$$

Clearly, (5.17) implies the equivalence between Problems (5.15) and (5.16).

As far as the solution of these problems is concerned, uniqueness implies existence, for (5.15) is a square system of linear equations. Like in the continuous case, we readily infer this existence from (5.14) and the inclusion $\Phi_h \subset M_h$. Hence we have the following result.

Lemma 5.1. *Let the spaces* Φ_h, M_h *and* Θ_h *satisfy* (5.11) *with* $\Phi_h \subset M_h$ *and let* Φ_{h0} *be defined by* (5.13). *Under the hypothesis* (5.14), *Problem* (Q_h) *has a unique solution* $(\boldsymbol{\psi}_h, \boldsymbol{\omega}_h)$ *in* $\Phi_{h0} \times M_h$. *If, in addition the hypothesis* (5.17) *holds, then Problem* (5.16) *is equivalent to Problem* (Q_h).

It is possible to derive directly an error bound for the solution of (Q_h), but it is easier and more satisfactory to place this problem into the setting of Section 1.2 and use Theorem 1.2. Here, there is a slight difficulty because the natural discretization of $\tilde{X}$: $(\mathbf{curl}\,\Phi_{h0}) \times M_h$ is not contained in $\tilde{X}$. However we can make use of Remark 5.2 and observe that the crucial spaces of Problem (5.8) are in fact $H_0(\mathbf{curl}; \Omega) \times L^2(\Omega)^3$ and $\tilde{V}$. Thus, we take

$$\text{(5.18)} \qquad X_h = \{\mathbf{curl}\,\boldsymbol{\phi}_h;\ \boldsymbol{\phi}_h \in \Phi_{h0}\} \times M_h \subset \{\mathbf{curl}\,\boldsymbol{\phi};\ \boldsymbol{\phi} \in H_0(\mathbf{curl}; \Omega)\} \times L^2(\Omega)^3,$$

$$\text{(5.19)} \qquad V_h = \{(\mathbf{curl}\,\boldsymbol{\phi}_h, \boldsymbol{\theta}_h) \in X_h;\ (\mathbf{curl}\,\boldsymbol{\phi}_h, \mathbf{curl}\,\boldsymbol{\mu}_h) = (\boldsymbol{\theta}_h, \boldsymbol{\mu}_h) \quad \forall \boldsymbol{\mu}_h \in M_h\}.$$

Of course, V_h is generally not included in $\tilde{V}$ but if $\Phi_h \subset M_h$ and if (5.14) holds then we have the analogue of (5.10):

$$\text{(5.20)} \qquad \|\mathbf{curl}\,\boldsymbol{\phi}_h\|_{0,\Omega} \leqslant C^* \|\boldsymbol{\theta}_h\|_{0,\Omega} \qquad \forall v_h = (\mathbf{curl}\,\boldsymbol{\phi}_h, \boldsymbol{\theta}_h) \in V_h,$$

which means that the mapping $v_h = (\mathbf{curl}\,\boldsymbol{\phi}_h, \boldsymbol{\theta}_h) \to \|\boldsymbol{\theta}_h\|_{0,\Omega} = |v_h|$ is an equivalent norm on V_h.

With this notation, *Problem* (Q_h) becomes:

Find a pair $(u_h, \boldsymbol{\lambda}_h) \in X_h \times M_h$ *such that*

$$\tilde{a}(u_h, v_h) + \tilde{b}(v_h, \boldsymbol{\lambda}_h) = \langle l, v_h \rangle \qquad \forall v_h \in X_h,$$

$$\tilde{b}(u_h, \boldsymbol{\mu}_h) = 0 \qquad \forall \boldsymbol{\mu}_h \in M_h.$$

Then we can easily apply the argument of Theorem 1.2 and derive the next result.

Theorem 5.2. *We retain all the hypotheses of Lemma* 5.1 *except* (5.17). *Then the solution* $u_h = (\mathbf{curl}\,\psi_h, \omega_h)$ *of Problem* (Q_h) *satisfies the error estimate*:

$$(5.21)\qquad \|\omega - \omega_h\|_{0,\Omega} \leqslant 2 \inf_{v_h \in V_h} |u - v_h| + (1 + C^{*2})^{1/2} \inf_{\mu_h \in M_h} \|\omega - \mu_h\|_{H(\mathbf{curl};\,\Omega)}$$

$$(5.22)\qquad \begin{aligned} &\|\mathbf{curl}(\psi - \psi_h)\|_{0,\Omega} \leqslant (1 + C^{*2})^{1/2} \\ &\quad \times \left\{ \inf_{v_h \in V_h} \|u - v_h\|_{\tilde{X}} + C^* \inf_{\mu_h \in M_h} \|\omega - \mu_h\|_{H(\mathbf{curl};\,\Omega)} \right\}, \end{aligned}$$

where C^* *is the constant of* (5.14).

Like in the two-dimensional case, we are now faced with the evaluation of the approximation error of V_h: $\inf_{v_h \in V_h} \|u - v_h\|_{\tilde{X}}$. It is easy to see that the statement of Lemma 3.1 is still valid here.

Lemma 5.2. *With the notations of Lemma* 5.1 *we have the upper bound for all* $v = (\mathbf{curl}\,\phi, \theta) \in \tilde{V}$:

$$\inf_{w_h \in V_h} |v - w_h| \leqslant \inf_{v_h = (\mathbf{curl}\,\phi_h, \theta_h) \in X_h} \left\{ 2|v - v_h| + \sup_{\mu_h \in M_h} \frac{(\mathbf{curl}(\phi - \phi_h), \mathbf{curl}\,\mu_h)}{\|\mu_h\|_{0,\Omega}} \right\},$$
(5.23)

and a similar upper bound for $\|v - w_h\|_{\tilde{X}}$ *with the norm* $|.|$ *replaced by* $\|.\|_{\tilde{X}}$ *in the right-hand side side of* (5.23).

5.3. A Family of Conforming Finite Elements in $H(\mathbf{curl};\Omega)$

In this section, we present a space of finite elements developed by Nédélec [59, 60]. Its construction is by no means straightforward, inasmuch as it requires exactly the continuity of the tangential components at element interfaces. This implies that we must work with incomplete spaces of polynomials of (say) degree l, for some integer $l \geqslant 1$.

Let $\tilde{P}_l$ denote the space of homogeneous polynomials of degree l in $\mathbb{R}^3$ and consider the following subspaces of P_l^3:

$$(5.24)\qquad \begin{cases} S_l = \{\mathbf{p} \in \tilde{P}_l^3;\ \mathbf{p}(x)\cdot\mathbf{x} \equiv 0,\ \mathbf{x} = (x_1, x_2, x_3)\}, \\ R_l = P_{l-1}^3 \oplus S_l. \end{cases}$$

Examples. Let us exhibit S_1 and S_2. Clearly, all homogeneous polynomial vectors of degree one that satisfy $\mathbf{p}(x)\cdot\mathbf{x} = 0$ must necessarily be of the form:

$$\mathbf{p}(x) = \boldsymbol{\alpha} \times \mathbf{x}$$

where $\boldsymbol{\alpha}$ is an arbitrary vector of $\mathbb{R}^3$. Thus, S_1 has the basis:

$$\mathbf{p}_1(x) = (0, -x_3, x_2), \qquad \mathbf{p}_2(x) = (x_3, 0, -x_1),$$
$$\mathbf{p}_3(x) = (-x_2, x_1, 0).$$

Likewise, it is easy to see that all polynomials of S_2 must necessarily be of the form:

$$\mathbf{p}(x) = \sum_{i,j=1}^{3} \alpha_{ij} x_j \mathbf{p}_i(x) \quad \text{with the above } \mathbf{p}_i.$$

But the nine polynomials $x_j \mathbf{p}_i(x)$ are not all linearly independent, for they are linked by one relation:

$$\sum_{i=1}^{3} x_i \mathbf{p}_i(x) = \mathbf{x} \times \mathbf{x} = \mathbf{0}.$$

Thus we can suppress one of these polynomials and it can be readily checked that the remaining eight are linearly independent. For example, we can take for S_2 the following eight basis functions:

$$x_1\mathbf{p}_1, \quad x_2\mathbf{p}_1, \quad x_3\mathbf{p}_1, \quad x_1\mathbf{p}_2, \quad x_2\mathbf{p}_2, \quad x_3\mathbf{p}_2, \quad x_1\mathbf{p}_3, \quad x_2\mathbf{p}_3.$$

The space R_l has the following attractive property.

Lemma 5.3. *If the vector field* $\mathbf{u} \in R_l$ *satisfies* $\mathbf{curl}\,\mathbf{u} = \mathbf{0}$ *then*

$$\mathbf{u} = \mathbf{grap}\, p \quad \textit{with } p \in P_l.$$

Proof. First observe that each $f \in \tilde{P}_l$ satisfies

$$lf = \mathbf{grad}\, f \cdot \mathbf{x}.$$

Now, we know that $\mathbf{u} = \mathbf{grad}\, p$ with $p \in P_{l+1}$. Therefore, the term in $\mathbf{grad}\, p$ that belongs to S_l vanishes according to the definition (5.24). Hence p has no term of degree $l + 1$. □

Remark 5.3. The definition of R_l can obviously be extended to an arbitrary dimension N. Then the statement of Lemma 5.3 is also valid for all dimensions. □

Definition 5.1. Let κ be a tetrahedron in $\mathbb{R}^3$ with edges denoted by e and faces by f and let $\mathbf{u}$ be a function in $W^{1,s}(\kappa)^3$ for some $s > 2$. We define the three sets of moments of $\mathbf{u}$ on κ:

$$M_e(\mathbf{u}) = \left\{ \int_e (\mathbf{u} \cdot \boldsymbol{\tau}) q \, de \quad \forall q \in P_{l-1}(e) \text{ for all edges } e \text{ of } \kappa \right\},$$

where $\boldsymbol{\tau}$ denotes the unit vector of e;

$$M_f(\mathbf{u}) = \left\{ \int_f (\mathbf{u} \times \mathbf{n}) \cdot \mathbf{q} \, ds \quad \forall \mathbf{q} \in P_{l-2}^2(f) \text{ for all faces } f \text{ of } \kappa \right\};$$

$$M_\kappa(\mathbf{u}) = \left\{ \int_\kappa \mathbf{u} \cdot \mathbf{q} \, dx \quad \forall \mathbf{q} \in P_{l-3}^3(\kappa) \right\}.$$

Remark 5.4. In the above definition, it is necessary to suppose that $\mathbf{u}$ has a little more regularity than $H^1(\kappa)^3$ because $M_e(\mathbf{u})$ makes no sense when $\mathbf{u}$ is only in $H^1(\kappa)^3$.

These definitions will enable us to construct conforming finite elements in $H(\mathbf{curl}; \Omega)$ provided that on the one hand, the above set of moments is unisolvent on R_l and on the other hand, the moments M_e and M_f determine entirely the tangential components of polynomials of R_l. This is achieved in the next lemmas.

Lemma 5.4. *The total number of moments in Definition* 5.1 *is equal to* N_l, *the dimension of* R_l:

$$N_l = (1/2)l(l+2)(l+3).$$

Proof. In view of Definition 5.1, we have:

$$\operatorname{card}(M_e(\mathbf{u})) = 6\dim(P_{l-1} \text{ in } \mathbb{R}) = 6l,$$
$$\operatorname{card}(M_f(\mathbf{u})) = 8\dim(P_{l-2} \text{ in } \mathbb{R}^2) = 4l(l-1),$$
$$\operatorname{card}(M_\kappa(\mathbf{u})) = 3\dim(P_{l-3} \text{ in } \mathbb{R}^3) = (1/2)l(l-1)(l-2).$$

On summing these three quantities we obtain $(1/2)l(l+2)(l+3)$ moments.

On the other hand, observe that the product of an arbitrary polynomial of $\tilde{P}_l^3$ by $\mathbf{x}$: $\mathbf{p}(x)\cdot\mathbf{x}$ yields an arbitrary polynomial of $\tilde{P}_{l+1}$. Hence the identity $\mathbf{p}(x)\cdot\mathbf{x} \equiv 0$ amounts to $\dim(\tilde{P}_{l+1})$ independent conditions. Therefore

$$\begin{aligned}\dim(R_l) &= 3\dim(P_l \text{ in } \mathbb{R}^3) - \dim(\tilde{P}_{l+1})\\ &= (1/2)(l+3)(l+2)(l+1) - (1/2)(l+3)(l+2)\\ &= (1/2)(l+3)(l+2)l.\end{aligned}$$

□

The equality in Lemma 5.4 means that the polynomials of R_l are uniquely determined by their three sets of moments if and only if the zero moments define only the zero polynomial. But this unisolvence is not easily established on an arbitrary tetrahedron κ. Therefore, we shall first prove that the zero moments are preserved by an affine transformation and subsequently work on the reference tetrahedron $\hat{\kappa}$ whenever it is convenient.

As usual, we denote by F_κ the affine invertible transformation from $\hat{\kappa}$ onto κ:

$$x = F_\kappa(\hat{x}) = B_\kappa \hat{x} + \mathbf{b}_\kappa.$$

Scalar functions defined on κ are transformed by a composition with F_κ:

$$\hat{\phi} = \phi \circ F_\kappa \quad \forall \phi \text{ defined on } \kappa, \tag{5.25}$$

while vector functions defined on κ are transformed like gradients:

$$\hat{\mathbf{u}} = B_\kappa^T(\mathbf{u} \circ F_\kappa) \quad \forall \mathbf{u} \text{ defined on } \kappa. \tag{5.26}$$

Recall that the unit normal and unit tangent vectors are transformed respectively by

(5.27) $$\mathbf{n} \circ F_\kappa = [(B_\kappa^{-1})^T \cdot \hat{\mathbf{n}}] / \| (B_\kappa^{-1})^T \cdot \hat{\mathbf{n}} \|,$$

(5.28) $$\tau \circ F_\kappa = [B_\kappa \cdot \hat{\tau}] / \| B_\kappa \cdot \hat{\tau} \|.$$

The main reason for adopting the transformation (5.26) is that it preserves the **curl** in a certain sense. Indeed, let us introduce the matrices

(5.29) $$\begin{aligned} C &= (c_{ij})_{i,j} = (\partial u_j / \partial x_i - \partial u_i / \partial x_j)_{i,j}, \\ \hat{C} &= (\hat{c}_{ij})_{i,j} = (\partial \hat{u}_j / \partial \hat{x}_i - \partial \hat{u}_i / \partial \hat{x}_j)_{i,j}. \end{aligned}$$

Then by expanding the formula (5.26) we easily derive that the matrices C and $\hat{C}$ are related by:

(5.30) $$C \circ F_\kappa = (B_\kappa^{-1})^T \hat{C} (B_\kappa^{-1}).$$

As a consequence, **curl u** and **curl û** vanish always simultaneously.

Besides that, the transformation (5.26) preserves the space R_l.

Lemma 5.5. *The space R_l is invariant under the transformation* (5.26).

Proof. Clearly (5.26) preserves the space P_k^3 for arbitrary k; hence we need only consider $\mathbf{u}$ in S_l. Formula (5.26) reads:

$$\begin{aligned} \hat{\mathbf{u}}(\hat{x}) &= B_\kappa^T \mathbf{u}(B_\kappa \hat{x} + \mathbf{b}_\kappa) \\ &= B_\kappa^T \mathbf{u}(B_\kappa \hat{x}) + \mathbf{p}(\hat{x}) \end{aligned}$$

where the degree of $\mathbf{p}$ is strictly less than l and $B_\kappa^T \mathbf{u}(B_\kappa \hat{x}) \in \tilde{P}_l^3$. Now,

$$B_\kappa^T \mathbf{u}(B_\kappa \hat{x}) \cdot \hat{x} = \mathbf{u}(B_\kappa \hat{x}) \cdot (B_\kappa \hat{x}) = 0$$

since $\mathbf{u} \in S_l$. Hence $\hat{\mathbf{u}} \in R_l$ on $\hat{\kappa}$. Conversely, the same argument shows that if $\hat{\mathbf{u}} \in R_l$ on $\hat{\kappa}$ then $\mathbf{u} \in R_l$ on κ. □

Lemma 5.6. *The three sets of moments of a function* **u** *given by Definition* 5.1 *vanish on κ iff the moments of* **û** *vanish on $\hat{\kappa}$.*

Proof. In view of (5.26) we have:

$$\int_\kappa \mathbf{u} \cdot \mathbf{q} \, dx = |\det(B_\kappa)| \int_{\hat{\kappa}} \hat{\mathbf{u}} \cdot (B_\kappa^{-1})(\mathbf{q} \circ F_\kappa) \, d\hat{x}.$$

Hence

$$\int_\kappa \mathbf{u} \cdot \mathbf{q} \, dx = 0 \quad \forall \mathbf{q} \in P_{l-3}^3(\kappa) \Leftrightarrow \int_{\hat{\kappa}} \hat{\mathbf{u}} \cdot \hat{\mathbf{q}} \, d\hat{x} = 0 \quad \forall \hat{\mathbf{q}} \in P_{l-3}^3(\hat{\kappa}).$$

Next, observe that every vector $\mathbf{q}$ of $\mathbb{R}^3$ satisfies

$$\mathbf{u} \times \mathbf{n} \cdot \mathbf{q} = -\mathbf{q} \times \mathbf{n} \cdot \mathbf{u}.$$

Furthermore, all tangent vectors $\mathbf{q}$ to the affine variety f with normal $\mathbf{n}$ (i.e. $\mathbf{q}$ is characterized by $\mathbf{q} \cdot \mathbf{n} = 0$) are of the form $\mathbf{q} = \mathbf{p} \times \mathbf{n}$ for arbitrary $\mathbf{p}$ of $\mathbb{R}^3$. Hence

$$M_f(\mathbf{u}) = \{0\} \Leftrightarrow \int_f \mathbf{u} \cdot \mathbf{q} \, ds = 0 \quad \forall \mathbf{q} \in P_{l-2}^3(\kappa) \quad \text{such that} \quad \mathbf{q} \cdot \mathbf{n} = 0.$$

Therefore, applying (5.26) and (5.27) we have:

$$M_f(\mathbf{u}) = \{0\} \Leftrightarrow \int_{\hat{f}} \hat{\mathbf{u}} \cdot (B_\kappa^{-1})(\mathbf{q} \circ F_\kappa) \, d\hat{s} = 0 \quad \forall \mathbf{q} \in P_{l-2}^3(\kappa)$$

$$\text{such that} \quad (B_\kappa^{-1})(\mathbf{q} \circ F_\kappa) \cdot \hat{\mathbf{n}} = 0$$

$$\Leftrightarrow M_{\hat{f}}(\hat{\mathbf{u}}) = \{0\}.$$

Likewise, owing to (5.28) we readily derive that

$$M_e(\mathbf{u}) = \{0\} \Leftrightarrow M_{\hat{e}}(\hat{\mathbf{u}}) = \{0\}. \qquad \square$$

Now we turn to the unisolvence. Let us start with a boundary result.

Lemma 5.7. *A vector* $\mathbf{u}$ *of* R_l *has all its moments zero on a given face* f *of* κ *iff the tangential components of* $\mathbf{u}$ *vanish on* f.

Proof. As all conditions involved are preserved by an affine transformation, we can assume that the face f lies on the plane $x_3 = 0$. Then the tangential components $\mathbf{u}_T$ of $\mathbf{u}$ on f reduce to its first two components:

$$\mathbf{u}_T(x_1, x_2) = (u_1(x_1, x_2, 0), u_2(x_1, x_2, 0)).$$

Moreover, the conditions $M_f(\mathbf{u}) = \{0\}$ and $M_e(\mathbf{u}) = \{0\}$ are respectively equivalent to:

$$\int_f \mathbf{u}_T \cdot \mathbf{q} \, dx_1 \, dx_2 = 0 \quad \forall \mathbf{q} \in P_{l-2}^2(f), \tag{5.31}$$

$$\int_e \mathbf{u}_T \cdot \boldsymbol{\tau} q \, de = 0 \quad \forall q \in P_{l-1}(e). \tag{5.32}$$

Hence Green's formula (I.2.22) in two dimensions gives:

$$\int_f \operatorname{curl} \mathbf{u}_T q \, dx_1 \, dx_2 = 0 \quad \forall q \in P_{l-1}(f),$$

i.e.

$$\operatorname{curl} \mathbf{u}_T = 0 \quad \text{on } f.$$

Now, it is easy to verify that $\mathbf{u}_T$ belongs to the two-dimensional analogue of R_l. Therefore it follows from Lemma 5.3 and its Remark that

$$\mathbf{u}_T = \mathbf{grad}\, p \quad \text{with } p \in P_l(f).$$

As a consequence, (5.32) implies that p is constant on the boundary ∂f of f; thus we can take

$$p = 0 \quad \text{on } \partial f,$$

i.e.

$$p = \lambda_1 \lambda_2 \lambda_3 r \quad \text{with } r \in P_{l-3}(f),$$

where λ_1, λ_2, λ_3 are the three barycentric coordinates of f. Then (5.31) readily yields that $r = 0$. □

Lemma 5.8. *If the moments of the vector* $\mathbf{u}$ *of* R_l *are all zero on* κ *then* $\mathbf{u}$ *is identically zero.*

Proof. On the one hand, Lemma 5.7 shows that

$$\mathbf{u} \times \mathbf{n} = \mathbf{0} \quad \text{on } \partial\kappa; \tag{5.33}$$

on the other hand, we have

$$\int_\kappa \mathbf{u} \cdot \mathbf{q} \, dx = 0 \qquad \forall \mathbf{q} \in P_{l-3}^3(\kappa). \tag{5.34}$$

Again, since these conditions are preserved by an affine transformation, we can switch to the reference element. Then Green's formula gives:

$$\int_{\hat{\kappa}} \mathbf{curl}\, \hat{\mathbf{u}} \cdot \hat{\mathbf{q}} \, d\hat{x} = 0 \qquad \forall \hat{\mathbf{q}} \in P_{l-2}^3(\hat{\kappa})$$

and it stems from (5.33) that

$$\mathbf{curl}\, \hat{\mathbf{u}} \cdot \hat{\mathbf{n}} = 0 \quad \text{on } \partial\hat{\kappa}.$$

Now, *taking advantage of the geometry of* $\hat{\kappa}$, it is easy to prove that these conditions (together with the fact that $\mathbf{curl}\, \hat{\mathbf{u}} \in P_{l-1}^3(\hat{\kappa})$) imply

$$\mathbf{curl}\, \hat{\mathbf{u}} = \mathbf{0} \quad \text{in } \hat{\kappa}.$$

Hence it follows from (5.30) that

$$\mathbf{curl}\, \mathbf{u} = \mathbf{0} \quad \text{in } \kappa.$$

Therefore, owing to Lemma 5.3,

$$\mathbf{u} = \mathbf{grad}\, p$$

with $p \in P_l$ and $p|_{\partial\kappa} = 0$ because $\mathbf{u} \times \mathbf{n} = \mathbf{0}$ on $\partial\kappa$. As a consequence, $p = \lambda_1\lambda_2\lambda_3\lambda_4 r$ with $r \in P_{l-4}(\kappa)$ and (5.34) implies that $r = 0$. □

Remark 5.5. By applying the arguments of Lemmas 5.7 and 5.8 it can also be proved that every vector $\mathbf{u}$ of P_l^3 with zero moments in κ satisfies

$$\mathbf{curl}\, \mathbf{u} = \mathbf{0}.$$

First, observe that Lemma 5.7 shows that

$$\mathbf{curl}\, \mathbf{u} \cdot \mathbf{n} = 0 \quad \text{on the particular face } x_3 = 0.$$

But since this property is preserved by an affine transformation, it holds on each face of κ. Then the argument of Lemma 5.8 yields

$$\mathbf{curl}\, \mathbf{u} = \mathbf{0} \quad \text{in } \kappa.$$

By combining Lemmas 5.7 and 5.8, we derive the result announced at the beginning of this section.

Theorem 5.3. *A vector field* $\mathbf{u}$ *of* R_l *is entirely determined in a tetrahedron* κ *by its three sets of moments:* $M_e(\mathbf{u})$, $M_f(\mathbf{u})$, $M_\kappa(\mathbf{u})$. *Moreover the tangential components of* $\mathbf{u}$ *on a given face* f *of* κ *depend only upon the moments* $M_f(\mathbf{u})$ *and* $M_e(\mathbf{u})$ *defined on that face.*

This theorem induces a natural interpolation operator in κ.

Definition 5.2. Let κ be an arbitrary tetrahedron of $\mathbb{R}^3$ and $\mathbf{u} \in W^{1,s}(\kappa)^3$ for some $s > 2$. Its interpolant $r_\kappa \mathbf{u}$ is the unique polynomial of R_l that has the same moments as $\mathbf{u}$ on κ.

In other words, $r_\kappa \mathbf{u}$ is determined by:

$$M_\kappa(r_\kappa \mathbf{u} - \mathbf{u}) = \{0\}, \qquad M_f(r_\kappa \mathbf{u} - \mathbf{u}) = \{0\}, \qquad M_e(r_\kappa \mathbf{u} - \mathbf{u}) = \{0\}.$$

Clearly, it follows from the invariance Lemmas 5.5 and 5.6 that

$$\widehat{r_\kappa \mathbf{u}} = r_{\hat{\kappa}} \hat{\mathbf{u}}, \tag{5.35}$$

i.e.

$$B_\kappa^T[(r_\kappa \mathbf{u}) \circ F_\kappa] = r_{\hat{\kappa}}[B_\kappa^T(\mathbf{u} \circ F_\kappa)].$$

Remark 5.6. When $\mathbf{u} \in W^{1,s}(\kappa)^3$ satisfies $\mathbf{curl}\,\mathbf{u} = \mathbf{0}$, the argument of Lemma 5.8 shows that $\mathbf{curl}(r_\kappa \mathbf{u}) = \mathbf{0}$ in κ.

Likewise, when $\mathbf{u} \in P_l^3$, Remark 5.5 establishes that

$$\mathbf{curl}(\mathbf{u} - r_\kappa \mathbf{u}) = \mathbf{0} \quad \text{in } \kappa.$$

Now we are in a position to define the finite element spaces M_h and Φ_h. As a matter of convenience, we assume that Ω is a *bounded polyhedron.* Let $\mathscr{T}_h$ be a triangulation of $\bar{\Omega}$ consisting of polyhedra κ with diameters bounded by h. For each integer $l \geqslant 1$, we set:

$$M_h = \{\boldsymbol{\mu}_h \in H(\mathbf{curl}; \Omega);\ \boldsymbol{\mu}_h|_\kappa \in R_l \quad \forall \kappa \in \mathscr{T}_h\}, \tag{5.36a}$$

$$\Phi_h = M_h \cap H_0(\mathbf{curl}; \Omega) \tag{5.36b}$$

and we define the interpolation operator r_h on M_h by:

$$r_h \mathbf{u}|_\kappa = r_\kappa \mathbf{u} \quad \text{on } \kappa \quad \forall \kappa \in \mathscr{T}_h \tag{5.37}$$

for all $\mathbf{u} \in W^{1,s}(\Omega)^3$ for some $s > 2$. The next lemma asserts that M_h (resp. Φ_h) is a conforming approximation of $H(\mathbf{curl}; \Omega)$ (resp. $H_0(\mathbf{curl}; \Omega)$).

Lemma 5.9. *If* $\mathbf{u} \in W^{1,s}(\Omega)^3$, *then* $r_h \mathbf{u} \in M_h$. *Similarly, when* $\mathbf{u} \in W^{1,s}(\Omega)^3$ *with* $\mathbf{u} \times \mathbf{n}|_\Gamma = \mathbf{0}$ *then* $r_h \mathbf{u} \in \Phi_h$.

We skip the proof as it is a straightforward consequence of Lemma 5.7.

The following theorem establishes the approximation properties of M_h when the triangulation $\mathscr{T}_h$ is regular as h tends to zero (cf. Definition A.2):

$$h_\kappa/\rho_\kappa = \sigma_\kappa \leqslant \sigma \qquad \forall\kappa \in \mathscr{T}_h, \qquad \sigma > 0 \quad \text{independent of } h.$$

Theorem 5.4. *Let $\mathscr{T}_h$ be a regular family of triangulations of $\bar{\Omega}$ and let M_h and r_h be defined by* (5.36) *and* (5.37) *for some integer $l \geqslant 1$. We have the upper bound for all $\mathbf{u} \in H^{l+1}(\Omega)^3$:*

$$\|\mathbf{u} - r_h\mathbf{u}\|_{H(\mathbf{curl};\,\Omega)} \leqslant C_1 h^l\{|\mathbf{u}|_{l,\Omega} + |\mathbf{u}|_{l+1,\Omega}\}. \tag{5.38}$$

Moreover, the operator r_h satisfies the following stability estimate:

$$\|\mathbf{u} - r_h\mathbf{u}\|_{0,\Omega} + h\|\mathbf{curl}(\mathbf{u} - r_h\mathbf{u})\|_{0,\Omega} \leqslant C_2 h|\mathbf{u}|_{1,s,\Omega} \tag{5.39}$$

for all $\mathbf{u} \in W^{1,s}(\Omega)^3$ with $s > 2$, where the positive constants C_1 and C_2 are independent of h and $\mathbf{u}$.

Proof. Let us first prove (5.38). By virtue of (5.26) we have:

$$\|\mathbf{u} - r_h\mathbf{u}\|_{0,\kappa} \leqslant |\det(B_\kappa)|^{1/2}\|B_\kappa^{-1}\|\,\|\hat{\mathbf{u}} - r_{\hat{\kappa}}\hat{\mathbf{u}}\|_{0,\hat{\kappa}}.$$

But since the operator $r_{\hat{\kappa}}$ preserves the polynomials of P_{l-1}^3, Corollary A.1 implies that:

$$\|\hat{\mathbf{u}} - r_{\hat{\kappa}}\hat{\mathbf{u}}\|_{0,\hat{\kappa}} \leqslant C_1 \begin{cases} |\hat{\mathbf{u}}|_{l,\hat{\kappa}} & \text{if } l \geqslant 2, \\ |\hat{\mathbf{u}}|_{1,\hat{\kappa}} + |\hat{\mathbf{u}}|_{2,\hat{\kappa}} & \text{if } l = 1. \end{cases}$$

Next, combining formulas (A.7) and (5.26) we derive:

$$|\hat{\mathbf{u}}|_{k,\hat{\kappa}} \leqslant C_2\|B_\kappa\|^{k+1}|\det(B_\kappa)|^{-1/2}|\mathbf{u}|_{k,\kappa}. \tag{5.40}$$

Therefore, these three inequalities yield:

$$\begin{cases} \|\mathbf{u} - r_h\mathbf{u}\|_{0,\kappa} \leqslant C_3\|B_\kappa^{-1}\|\,\|B_\kappa\|^{l+1}|\mathbf{u}|_{l,\kappa} & \text{when } l \geqslant 2, \\ \|\mathbf{u} - r_h\mathbf{u}\|_{0,\kappa} \leqslant C_3\|B_\kappa^{-1}\|\,\|B_\kappa\|^2(|\mathbf{u}|_{1,\kappa} + \|B_\kappa\|\,|\mathbf{u}|_{2,\kappa}) & \text{when } l = 1. \end{cases} \tag{5.41}$$

Next, let us examine $\mathbf{curl}(\mathbf{u} - r_h\mathbf{u})$. According to (5.30), we have:

$$\|\mathbf{curl}(\mathbf{u} - r_h\mathbf{u})\|_{0,\kappa} \leqslant C_4|\det(B_\kappa)|^{1/2}\|B_\kappa^{-1}\|^2\|\mathbf{curl}(\hat{\mathbf{u}} - r_{\hat{\kappa}}\hat{\mathbf{u}})\|_{0,\hat{\kappa}}.$$

As mentioned in Remark 5.6, the linear mapping $\hat{\mathbf{u}} \to \mathbf{curl}(\hat{\mathbf{u}} - r_{\hat{\kappa}}\hat{\mathbf{u}})$ vanishes on the space P_l^3. Therefore, a simple application of Theorem A.1 yields:

$$\|\mathbf{curl}(\hat{\mathbf{u}} - r_{\hat{\kappa}}\hat{\mathbf{u}})\|_{0,\hat{\kappa}} \leqslant C_5|\hat{\mathbf{u}}|_{l+1,\hat{\kappa}}.$$

Hence

$$\|\mathbf{curl}(\mathbf{u} - r_h\mathbf{u})\|_{0,\kappa} \leqslant C_6\|B_\kappa^{-1}\|^2\|B_\kappa\|^{l+2}|\mathbf{u}|_{l+1,\kappa}. \tag{5.42}$$

Finally (5.38) stems from (5.41) and (5.42) together with (A.2) and the regularity of $\mathscr{T}_h$.

The proof of the stability estimate (5.39) is a trifle more intricate. Taking into account the facts that $r_{\hat{\kappa}}$ preserves the constant polynomials and belongs to

$\mathscr{L}(W^{1,s}(\hat{\kappa})^3; R_l)$, Corollary A.1 yields:

$$\|\hat{\mathbf{u}} - r_{\hat{\kappa}}\hat{\mathbf{u}}\|_{0,\hat{\kappa}} \leqslant C_7 |\hat{\mathbf{u}}|_{1,s,\hat{\kappa}}.$$

Hence

$$\|\mathbf{u} - r_h\mathbf{u}\|_{0,\kappa} \leqslant C_8(\text{meas}(\kappa))^{1/2-1/s} \|B_\kappa^{-1}\| \, \|B_\kappa\|^2 |\mathbf{u}|_{1,s,\kappa}.$$

Then Hölder's inequality and the regularity of $\mathscr{T}_h$ imply that

$$\|\mathbf{u} - r_h\mathbf{u}\|_{0,\Omega} \leqslant C_9 h(\text{meas}(\Omega))^{1/2-1/s} |\mathbf{u}|_{1,s,\Omega}.$$

Likewise, we have

$$\|\mathbf{curl}(\hat{\mathbf{u}} - r_{\hat{\kappa}}\hat{\mathbf{u}})\|_{0,\hat{\kappa}} \leqslant C_{10} |\hat{\mathbf{u}}|_{1,s,\hat{\kappa}}.$$

Therefore, we infer from the above inequalities that:

$$\|\mathbf{curl}(\mathbf{u} - r_h\mathbf{u})\|_{0,\Omega} \leqslant C_{11}(\text{meas}(\Omega))^{1/2-1/s} |\mathbf{u}|_{1,s,\Omega}. \qquad \square$$

5.4. Error Analysis for Finite Elements of Degree l

The spaces M_h and Φ_h have already been defined in (5.36) and it remains to define the space Θ_h. Here, we simply take the standard finite element space:

$$\Theta_h = \{q_h \in \mathscr{C}^0(\bar{\Omega});\ q_h|_\kappa \in P_l \quad \forall \kappa \in \mathscr{T}_h,\ q_h|_\Gamma = 0\}. \tag{5.43}$$

Recall that the functions of Φ_{h0} satisfy

$$(\boldsymbol{\phi}_h, \mathbf{grad}\, q_h) = 0 \qquad \forall q_h \in \Theta_h. \tag{5.12}$$

The next three results check the hypotheses (5.14) and (5.17). They will lead in particular to an interesting decomposition of our discrete finite element spaces. Before proving that the space Φ_{h0} satisfies (5.14), let us show the following preliminary result.

Lemma 5.10. *Let* $\mathbf{u}$ *be a function of the form*:

$$\mathbf{u} = \mathbf{grad}\, p \quad \text{with } p \in H_0^1(\Omega)$$

and assume that $\mathbf{u}$ *is such that* $r_h\mathbf{u}$ *is well defined. Then there exists* p_h *in* Θ_h *such that*

$$r_h\mathbf{u} = \mathbf{grad}\, p_h.$$

Proof. As $\mathbf{curl}\,\mathbf{u} = \mathbf{0}$, Remark 5.6 implies that

$$\mathbf{curl}\, r_h\mathbf{u} = \mathbf{0} \quad \text{in each } \kappa.$$

But since p is constant on Γ, we also have $\mathbf{u} \times \mathbf{n} = \mathbf{0}$ on Γ. Therefore, it follows from Lemma 5.9 that $r_h\mathbf{u} \in H_0(\mathbf{curl}; \Omega)$; this means that

$$\mathbf{curl}\, r_h\mathbf{u} = \mathbf{0} \quad \text{in } \Omega, \qquad r_h\mathbf{u} \times \mathbf{n} = \mathbf{0} \quad \text{on } \Gamma.$$

Hence

$$r_h\mathbf{u} = \mathbf{grad}\, q \quad \text{for some } q \text{ in } H_0^1(\Omega).$$

On the other hand, Lemma 5.3 implies that

$$q|_\kappa \in P_l \quad \text{for each } \kappa.$$

Therefore $q \in \Theta_h$. □

Remark 5.7. Lemma 5.10 shows that for each function $\boldsymbol{\phi}_h$ in Φ_h that satisfies $\mathbf{curl}\,\boldsymbol{\phi}_h = \mathbf{0}$ in Ω there exists a (unique) element p_h of Θ_h such that

$$\boldsymbol{\phi}_h = \mathbf{grad}\, p_h.$$

Thus
$$\{\mathbf{grad}\, p_h;\ p_h \in \Theta_h\} = \{\boldsymbol{\phi}_h \in \Phi_h;\ \mathbf{curl}\,\boldsymbol{\phi}_h = \mathbf{0}\}.$$

It follows from this last remark that Φ_{h0} satisfies (5.14). But if we want to check that (5.14) holds *uniformly*, we shall require below a uniformly regular triangulation, i.e. a regular triangulation $\mathscr{T}_h$ that also satisfies for some $\tau > 0$ independent of h:

$$\tau h \leqslant h_\kappa \leqslant \sigma \rho_\kappa \qquad \forall \kappa \in \mathscr{T}_h.$$

Proposition 5.1. *Let Ω be an open, bounded and convex region of $\mathbb{R}^3$ with a polyhedral boundary Γ. If $\mathscr{T}_h$ is a uniformly regular triangulation of $\bar{\Omega}$, there exists a constant C^*, independent of h, such that:*

$$\|\boldsymbol{\phi}_h\|_{0,\Omega} \leqslant C^* \|\mathbf{curl}\,\boldsymbol{\phi}_h\|_{0,\Omega} \qquad \forall \boldsymbol{\phi}_h \in \Phi_{h0}. \tag{5.14}$$

Proof. The idea is to write $\boldsymbol{\phi}_h$ as the sum of a gradient and a divergence-free function $\mathbf{w}$, smooth enough to satisfy an inequality similar to (5.14). First, let $p \in H_0^1(\Omega)$ be the unique solution of the problem:

$$(\mathbf{grad}\, p, \mathbf{grad}\, q) = (\boldsymbol{\phi}_h, \mathbf{grad}\, q) \qquad \forall q \in H_0^1(\Omega).$$

Clearly the difference

$$\mathbf{w} = \boldsymbol{\phi}_h - \mathbf{grad}\, p$$

satisfies
$$\mathbf{curl}\,\mathbf{w} = \mathbf{curl}\,\boldsymbol{\phi}_h, \qquad \operatorname{div}\mathbf{w} = 0, \qquad \mathbf{w} \times \mathbf{n}|_\Gamma = \mathbf{0}.$$

In addition, $\mathbf{curl}\,\boldsymbol{\phi}_h$ belongs to $L^\gamma(\Omega)^3$ for all γ. Therefore, since Ω is convex, it follows from Remark I.3.14 that there exists a real $s > 2$ such that:

$$\mathbf{w} \in W^{1,s}(\Omega)^3$$

and

$$\|\mathbf{w}\|_{1,\alpha,\Omega} \leqslant C_1(\alpha)\, \|\mathbf{curl}\,\mathbf{w}\|_{0,\alpha,\Omega} \quad \text{for all } \alpha \text{ with } 2 \leqslant \alpha \leqslant s. \tag{5.44}$$

Hence, the interpolate $r_h\mathbf{w}$ is well defined. As $\boldsymbol{\phi}_h$ belongs to Φ_h, this in turn implies that $r_h(\mathbf{grad}\, p)$ is also well defined and owing to Lemma 5.10, there exists p_h in Θ_h such that

$$r_h(\mathbf{grad}\, p) = \mathbf{grad}\, p_h.$$

Therefore, $\boldsymbol{\phi}_h$ has the following decomposition:

$$\boldsymbol{\phi}_h = r_h \mathbf{w} + \mathbf{grad}\, p_h.$$

Then, applying (5.12) with $q_h = p_h$ we easily derive

$$\|\boldsymbol{\phi}_h\|_{0,\Omega} \leqslant \|r_h \mathbf{w}\|_{0,\Omega}.$$

Thus, (5.14) will be established if we show that

$$\|r_h \mathbf{w}\|_{0,\Omega} \leqslant C^* \|\mathbf{curl}\, \boldsymbol{\phi}_h\|_{0,\Omega}. \tag{5.45}$$

Now (5.39) and (5.44) yield:

$$\|\mathbf{w} - r_h \mathbf{w}\|_{0,\Omega} \leqslant C_2 h \|\mathbf{curl}\, \boldsymbol{\phi}_h\|_{0,s,\Omega}.$$

But since $\mathscr{T}_h$ is uniformly regular and $\boldsymbol{\phi}_h$ is a polynomial on each κ, we easily obtain from (5.30) and (A.34):

$$\|\mathbf{curl}\, \boldsymbol{\phi}_h\|_{0,s,\Omega} \leqslant C_3 h^{3(1/s-1/2)} \|\mathbf{curl}\, \boldsymbol{\phi}_h\|_{0,\Omega}.$$

Therefore

$$\|\mathbf{w} - r_h \mathbf{w}\|_{0,\Omega} \leqslant C_4 h^{\alpha} \|\mathbf{curl}\, \boldsymbol{\phi}_h\|_{0,\Omega}$$

with a non negative exponent α as long as $2 < s \leqslant 6$. This proves (5.45). □

Corollary 5.1. *Let Ω be an open, bounded polyhedron of $\mathbb{R}^3$. For each function $\mathbf{w}_h$ of Φ_h there exists a unique function $\mathbf{v}_h$ in Φ_{h0} and p_h in Θ_h such that:*

$$\begin{aligned} \mathbf{w}_h &= \mathbf{v}_h + \mathbf{grad}\, p_h, \\ |p_h|_{1,\Omega} &\leqslant \|\mathbf{w}_h\|_{0,\Omega}. \end{aligned} \tag{5.46}$$

Moreover, under the assumptions of Proposition 5.1, $\mathbf{v}_h$ *is bounded as follows:*

$$\|\mathbf{v}_h\|_{H(\mathbf{curl};\,\Omega)} \leqslant (1 + C^{*2})^{1/2} \|\mathbf{curl}\, \mathbf{w}_h\|_{0,\Omega}. \tag{5.47}$$

Proof. Let us take for p_h the unique solution in Θ_h of

$$(\mathbf{grad}\, p_h, \mathbf{grad}\, q_h) = (\mathbf{w}_h, \mathbf{grad}\, q_h) \qquad \forall q_h \in \Theta_h.$$

Then the difference

$$\mathbf{v}_h = \mathbf{w}_h - \mathbf{grad}\, p_h$$

belongs to Φ_{h0} and (5.47) follows immediately from Proposition 5.1. □

Observe that the first part of this corollary establishes (5.17).

From Lemmas 5.1 and 5.2, Theorems 5.1, 5.2 and 5.4, Proposition 5.1 and Corollary 5.1, we derive the major result of this section.

Theorem 5.5. *Let Ω be a bounded polyhedron in $\mathbb{R}^3$. Then Problems* (5.15) *and* (5.16) *associated with the choice of finite element spaces* (5.36) *and* (5.43) *are equivalent and have a unique solution $u_h = (\mathbf{curl}\, \boldsymbol{\psi}_h, \boldsymbol{\omega}_h)$.*

Assume in addition that Ω is convex and that the solution $\mathbf{u} = \mathbf{curl}\,\boldsymbol{\psi}$ *of the Stokes problem* (5.1) *satisfies*:

$$\boldsymbol{\psi} \in H^{l+1}(\Omega)^3, \qquad \boldsymbol{\omega} = -\Delta\boldsymbol{\psi} \in H^{l+1}(\Omega)^3$$

for some integer $l \geqslant 1$. *Then, if* $\mathscr{T}_h$ *is a uniformly regular family of triangulations of* $\bar{\Omega}$, u_h *satisfies the error estimates*:

$$(5.48)\quad \begin{cases} \|\boldsymbol{\omega} - \boldsymbol{\omega}_h\|_{0,\Omega} \leqslant C_1(h^{l-1}|\boldsymbol{\psi}|_{l+1,\Omega} + h^l\|\boldsymbol{\omega}\|_{l+1,\Omega}), \\ \|\mathbf{curl}(\boldsymbol{\psi} - \boldsymbol{\psi}_h)\|_{0,\Omega} \leqslant C_2\{(h^{l-1} + h^l)|\boldsymbol{\psi}|_{l+1,\Omega} + h^l\|\boldsymbol{\omega}\|_{l+1,\Omega}\}, \end{cases}$$

with positive constants C_1 *and* C_2 *independent of* h, $\boldsymbol{\omega}$ *and* $\boldsymbol{\psi}$.

Remark 5.8. Like in the two-dimensional case, we observe a loss of one power of h arising from the term (cf. Lemma 5.2):

$$(5.49)\qquad \inf_{\boldsymbol{\phi}_h \in \Phi_{h0}} \sup_{\boldsymbol{\mu}_h \in M_h} \frac{|(\mathbf{curl}(\boldsymbol{\psi} - \boldsymbol{\phi}_h), \mathbf{curl}\,\boldsymbol{\mu}_h)|}{\|\boldsymbol{\mu}_h\|_{0,\Omega}}.$$

If it were known that the projection $\mathring{P}_h\boldsymbol{\psi}$ (for $\boldsymbol{\psi}$ in Φ_0):

$$\mathring{P}_h\boldsymbol{\psi} \in \Phi_{h0} \qquad (\mathbf{curl}(\mathring{P}_h\boldsymbol{\psi} - \boldsymbol{\psi}), \mathbf{curl}\,\boldsymbol{\phi}_h) = 0 \qquad \forall \boldsymbol{\phi}_h \in \Phi_{h0}$$

satisfied the L^p-estimate:

$$(5.50)\qquad \|\mathring{P}_h\boldsymbol{\psi} - \boldsymbol{\psi}\|_{0,p,\Omega} + h\|\mathbf{curl}(\mathring{P}_h\boldsymbol{\psi} - \boldsymbol{\psi})\|_{0,p,\Omega} \leqslant Ch^{s+1}\|\boldsymbol{\psi}\|_{s+1,p,\Omega}$$

for all $p \in [2, \infty]$ and $s \in [1, l]$, then the argument of Section 3.1 could be applied to derive a sharper estimate than (5.48) and regain part of the missing power of h. In particular this would enable us to obtain an acceptable rate of convergence when using first degree elements, which Theorem 5.5 fails to show.

Although (5.50) is still a conjecture, it does not sound unreasonable and it is hoped that this problem will be solved in a near future.

5.5. Discontinuous Approximation of the Pressure

This section briefly describes and analyzes a finite element method that solves for the pressure term underlying Problems (5.15) and (5.16). Since the situation is fairly similar to that in Section 4.4 we shall state nearly all results without proof. The reader will easily fill in the blanks.

It is clear that here we must construct subspaces D_h of $H(\mathrm{div};\Omega)$ such that, on the one hand, $\mathbf{curl}\,\boldsymbol{\mu}_h$ belongs to D_h for $\boldsymbol{\mu}_h$ in M_h and on the other hand, $\mathrm{div}\,\mathbf{v}_h$ belongs to the discrete pressure space for $\mathbf{v}_h$ in D_h. The following definition generalizes the polynomial space D defined by (4.61a).

Definition 5.3. 1°) For each integer $l \geqslant 1$, let

$$D_l = P_{l-1}^3 \oplus \{p(x)\mathbf{x};\, p \in \tilde{P}_{l-1}\}.$$

2°) Let κ be a tetrahedron in $\mathbb{R}^3$ with faces denoted by f and let $\mathbf{u} \in H^1(\kappa)^3$. We define the two sets of moments of $\mathbf{u}$ on κ by:

$$N_f(\mathbf{u}) = \int_f \mathbf{u} \cdot \mathbf{n}\, q\, ds \qquad \forall q \in P_{l-1}(f),$$

$$N_\kappa(\mathbf{u}) = \int_\kappa \mathbf{u} \cdot \mathbf{q}\, dx \qquad \forall \mathbf{q} \in P_{l-2}^3(\kappa).$$

We can immediately check that for $\mathbf{u}$ in D_l, $\mathbf{u} \cdot \mathbf{n}$ belongs to P_{l-1} on each face f of κ. In addition, it easy to see that when $\mathbf{u}$ is a *divergence-free* vector field of D_l then $\mathbf{u}$ belongs to P_{l-1}^3.

As usual, let $\hat{\kappa}$ denote the unit reference tetrahedron. Instead of (5.26), let us transform vector functions defined on κ by the contravariant transformation:

(5.51) $$\hat{\mathbf{u}} = B_\kappa^{-1}(\mathbf{u} \circ F_\kappa) \qquad \forall \mathbf{u} \text{ defined on } \kappa.$$

It coincides with the contravariant transformation (4.63), up to the multiplicative factor J_F which is constant here:

$$J_F = |\det(B_\kappa)|.$$

As a consequence (5.51) preserves entirely the divergence:

(5.52) $$(\operatorname{div} \mathbf{u}) \circ F_\kappa = \operatorname{div} \hat{\mathbf{u}}.$$

In addition, we have the analogue of Lemmas 5.5, 5.6 and Theorem 5.3.

Proposition 5.2. 1°) *The space D_l is invariant under the transformation* (5.51) *and the moments of* $\mathbf{u}$ *given by Definition* 5.3 *vanish on κ iff the same moments of* $\hat{\mathbf{u}}$ *vanish on $\hat{\kappa}$.*

2°) *A vector field* $\mathbf{u}$ *of D_l is entirely determined in a tetrahedron κ by its two sets of moments: $N_f(\mathbf{u})$, $N_\kappa(\mathbf{u})$. Moreover, the normal component of* $\mathbf{u}$ *on a given face f of κ depends only upon the moments $N_f(\mathbf{u})$ defined on that face.*

Definition 5.4. Let $\mathbf{u} \in H^1(\kappa)^3$, where κ is an arbitrary tetrahedron. Its interpolant $\omega_\kappa \mathbf{u}$ is the unique polynomial of D_l that has the same moments as $\mathbf{u}$ on κ.

Thus $\omega_\kappa \mathbf{u}$ is determined by the conditions:

$$N_\kappa(\mathbf{u} - \omega_\kappa \mathbf{u}) = \{0\}, \qquad N_f(\mathbf{u} - \omega_\kappa \mathbf{u}) = \{0\}.$$

Again, the invariance in Proposition 5.2 implies that

$$\widehat{\omega_\kappa \mathbf{u}} = \omega_{\hat{\kappa}} \hat{\mathbf{u}}.$$

Moreover, we observe that

$$\operatorname{div}(\mathbf{u} - \omega_\kappa \mathbf{u}) = 0 \quad \text{on } \kappa \qquad \forall \mathbf{u} \in P_l^3$$

and that $\operatorname{div} \mathbf{u} = 0$ on κ implies that $\operatorname{div}(\omega_\kappa \mathbf{u}) = 0$.

Now, we suppose that Ω is a *bounded polyhedron* and we take a family $\mathscr{T}_h$ of triangulations of $\bar{\Omega}$. For each integer $l \geqslant 1$, we define:

$$(5.53)\qquad \begin{cases} D_h = \{\mathbf{u}_h \in H(\operatorname{div}; \Omega);\ \mathbf{u}_h|_\kappa \in D_l \quad \forall \kappa \in \mathscr{T}_h\}, \\ D_{0h} = D_h \cap H_0(\operatorname{div}; \Omega), \\ Q_h = \{p_h \in L_0^2(\Omega);\ p_h|_\kappa \in P_{l-1} \quad \forall \kappa \in \mathscr{T}_h\}, \end{cases}$$

together with the interpolation operator ω_h:

$$\omega_h \mathbf{u}|_\kappa = \omega_k \mathbf{u} \quad \text{on } \kappa \qquad \forall \kappa \in \mathscr{T}_h,$$

and for all $\mathbf{u} \in H^1(\Omega)^3$. Clearly, div $\mathbf{u}_h$ belongs to Q_h for all $\mathbf{u}_h$ in D_{0h} and $\mathbf{curl}\,\boldsymbol{\mu}_h$ belongs to D_{0h} for all $\boldsymbol{\mu}_h$ in Φ_h. Furthermore, we have the analogue of Lemma 5.9:

$$\mathbf{u} \in H^1(\Omega)^3 \quad \text{implies} \quad \omega_h \mathbf{u} \in D_h,$$

$$\mathbf{u} \in H^1(\Omega)^3 \quad \text{with } \mathbf{u}\cdot\mathbf{n} = 0 \quad \text{implies} \quad \omega_h \mathbf{u} \in D_{0h}.$$

The following proposition states the approximation properties of D_h.

Proposition 5.3. *Let $\mathscr{T}_h$ be a regular family of triangulations of $\bar{\Omega}$ and let D_h be defined by* (5.53) *for an integer $l \geqslant 1$. We have the estimates:*

$$(5.54)\qquad \begin{cases} \|\mathbf{u} - \omega_h \mathbf{u}\|_{0,\Omega} \leqslant C_1 h^l |\mathbf{u}|_{l,\Omega} \qquad \forall \mathbf{u} \in H^l(\Omega)^3, \\ \|\operatorname{div}(\mathbf{u} - \omega_h \mathbf{u})\|_{0,\Omega} \leqslant C_2 h^l |\mathbf{u}|_{l+1,\Omega} \qquad \forall \mathbf{u} \in H^{l+1}(\Omega)^3. \end{cases}$$

The next lemma establishes the desired relationship between the spaces D_h and M_h.

Lemma 5.11. *Let Ω be an open, bounded polyhedron in $\mathbb{R}^3$ and let Γ_i, $0 \leqslant i \leqslant p$, denote the connected components of its boundary. A function $\mathbf{u}_h$ of D_h (resp. D_{0h}) satisfies:*

$$\operatorname{div} \mathbf{u}_h = 0 \quad \text{in } \Omega, \qquad \int_{\Gamma_i} \mathbf{u}_h \cdot \mathbf{n}\, ds = 0 \quad \text{for } 0 \leqslant i \leqslant p$$

iff there exists a function $\boldsymbol{\phi}_h$ in M_h (resp. Φ_h) such that:

$$\mathbf{u}_h = \mathbf{curl}\,\boldsymbol{\phi}_h.$$

Proof. We already know that $\mathbf{curl}\,\boldsymbol{\phi}_h$ belongs to D_h (resp. D_{0h}) whenever $\boldsymbol{\phi}_h$ belongs to M_h (resp. Φ_h).

Conversely, Theorem I.3.4 asserts that there exists $\boldsymbol{\phi} \in H^1(\Omega)^3$ such that

$$\mathbf{u}_h = \mathbf{curl}\,\boldsymbol{\phi}.$$

Furthermore, the fact that $\mathbf{u}_h$ belongs to $H^\alpha(\Omega)^3$ for all α with $0 < \alpha < 1/2$ implies that $\boldsymbol{\phi}$ belongs to $H^{1+\alpha}(\Omega)^3$ (cf. Remark I.3.12). Thus, the interpolate of $\boldsymbol{\phi}$, $r_h\boldsymbol{\phi}$, is well defined. Let us prove that:

$$\mathbf{u}_h = \mathbf{curl}\, r_h \boldsymbol{\phi},$$

i.e.

$$\mathbf{curl}(\boldsymbol{\phi} - r_h \boldsymbol{\phi}) = \mathbf{0} \quad \text{in } \Omega.$$

On the one hand, observe that $\mathbf{curl}(\boldsymbol{\phi} - r_h \boldsymbol{\phi})|_\kappa \in P_{l-1}^3(\kappa)$. On the other hand, we have

$$\int_\kappa \mathbf{curl}(\boldsymbol{\phi} - r_\kappa \boldsymbol{\phi}) \cdot \mathbf{q}\, dx = 0 \qquad \forall \mathbf{q} \in P_{l-2}^3(\kappa),$$

$$\mathbf{curl}(\boldsymbol{\phi} - r_\kappa \boldsymbol{\phi}) \cdot \mathbf{n} = 0 \quad \text{on each face } f \text{ of } \kappa.$$

Hence like in Lemma 5.8, we deduce:

$$\mathbf{curl}(\boldsymbol{\phi} - r_\kappa \boldsymbol{\phi}) = \mathbf{0} \quad \text{in each } \kappa \text{ of } \mathscr{T}_h$$

and since $\boldsymbol{\phi} - r_h \boldsymbol{\phi}$ belongs to $H(\mathbf{curl}; \Omega)$, its **curl** vanishes on the whole of Ω.

It remains to establish that when $\mathbf{u}_h \cdot \mathbf{n}$ vanishes on Γ then $\boldsymbol{\phi}_h$ may be chosen such that $\boldsymbol{\phi}_h \times \mathbf{n} = \mathbf{0}$ on Γ. The proof follows the lines of Theorem I.3.6. We take an open ball $\mathcal{O}$ containing $\bar{\Omega}$; then, it is easy to construct a function q in $H^2(\mathcal{O})$ such that

$$\mathbf{grad}\, q \times \mathbf{n} = \boldsymbol{\phi} \times \mathbf{n} \quad \text{on } \Gamma.$$

Note that this requires no regularity on Ω since $\mathbf{grad}\, q$ need not be divergence-free. As q belongs to $H^2(\Omega)$, $r_h\, \mathbf{grad}\, q$ is well defined and therefore $r_h(\boldsymbol{\phi} - \mathbf{grad}\, q)$ is the desired potential vector of $\mathbf{u}_h$ in Φ_h. □

Remark 5.9. According to Corollary 5.1, for each *divergence-free* vector field $\mathbf{u}_h$ in D_{0h} there exists a unique vector potential $\boldsymbol{\phi}_h$ in Φ_{h0} such that:

$$\mathbf{u}_h = \mathbf{curl}\, \boldsymbol{\phi}_h.$$

In addition, under the hypotheses of Proposition 5.1, we have:

$$\|\boldsymbol{\phi}_h\|_{H(\mathbf{curl}; \Omega)} \leqslant C \|\mathbf{u}_h\|_{0,\Omega}.$$

With this lemma and the statement of Problem (5.16), we can formulate the corresponding problem in $\mathbf{u}_h$, $\boldsymbol{\omega}_h$, p_h:

Find a pair $(\mathbf{u}_h, \boldsymbol{\omega}_h)$ *in* $D_{0h} \times M_h$ *and a function* p_h *in* Q_h *such that*:

$$(5.55) \quad \begin{cases} \nu(\mathbf{curl}\, \boldsymbol{\omega}_h, \mathbf{v}_h) - (p_h, \operatorname{div} \mathbf{v}_h) = (\mathbf{f}, \mathbf{v}_h) & \forall \mathbf{v}_h \in D_{0h}, \\ (\mathbf{u}_h, \mathbf{curl}\, \boldsymbol{\mu}_h) = (\boldsymbol{\omega}_h, \boldsymbol{\mu}_h) & \forall \boldsymbol{\mu}_h \in M_h, \\ \operatorname{div} \mathbf{u}_h = 0 \quad \text{in } \Omega. \end{cases}$$

The existence and uniqueness of p_h is a consequence of the following lemma which establishes the inf-sup condition relative to the space $D_{0h} \times M_h$ and Q_h. Then Lemma 5.11 implies the equivalence between this problem and (5.15) or (5.16). Therefore when Ω is a bounded polyhedron, Problem (5.55) has a unique solution $(\mathbf{u}_h, \boldsymbol{\omega}_h, p_h)$.

Lemma 5.12. *Let Ω be a bounded polyhedron in $\mathbb{R}^3$. Then for each p_h in Q_h there exists a function $\mathbf{v}_h$ in D_{0h} and a function $\boldsymbol{\theta}_h$ in M_h such that:*

$$\operatorname{div} \mathbf{v}_h = p_h \quad in\ \Omega,$$

$$(\boldsymbol{\theta}_h, \boldsymbol{\mu}_h) = (\mathbf{v}_h, \mathbf{curl}\, \boldsymbol{\mu}_h) \qquad \forall \boldsymbol{\mu}_h \in M_h.$$

In addition, when $\mathscr{T}_h$ is a regular family of triangulations of $\bar{\Omega}$, we have:

$$\|\boldsymbol{\theta}_h\|_{0,\Omega} + \|\mathbf{v}_h\|_{H(\mathrm{div};\,\Omega)} \leqslant C \|p_h\|_{0,\Omega}, \tag{5.56}$$

where the positive constant C is independent of h and p_h.

The above lemmas lead to the expected estimate for the error $p - p_h$.

Theorem 5.6. *Let Ω be a bounded polyhedron in $\mathbb{R}^3$. Then Problem (5.55) has a unique solution $(\mathbf{u}_h, \boldsymbol{\omega}_h)$ in $D_{0h} \times M_h$ and p_h in Q_h where $(\mathbf{u}_h = \mathbf{curl}\, \boldsymbol{\psi}_h, \boldsymbol{\omega}_h)$ is the solution of Problem (5.15).*

Moreover, under the hypotheses of Theorem 5.5 and if the exact pressure p belongs to $H^l(\Omega)$ for $l \geqslant 1$, the following error estimate holds:

$$\|p - p_h\|_{0,\Omega} \leqslant C(h^l |p|_{l,\Omega} + h^l \|\boldsymbol{\omega}\|_{l+1,\Omega} + h^{l-1} |\boldsymbol{\psi}|_{l+1,\Omega}), \tag{5.57}$$

with a positive constant C independent of h, p, $\boldsymbol{\omega}$ and $\boldsymbol{\psi}$.

Chapter IV. Theory and Approximation of the Navier-Stokes Problem

§ 1. A Class of Nonlinear Problems

In this paragraph, we study a nonlinear generalization of the abstract variational problem analyzed in Paragraph I.4. This family of nonlinear problems contains in particular the Navier-Stokes problem.

We retain the notations of Paragraph I.4. Namely, we consider two (real) Hilbert spaces X and M normed by $\|.\|_X$ and $\|.\|_M$ respectively and a bilinear continuous form

$$b(.,.): (v, \mu) \in X \times M \to b(v, \mu) \in \mathbb{R}.$$

The nonlinearity is introduced by means of a form

$$a(.;.,.): (u, v, w) \in X \times X \times X \to a(w; u, v) \in \mathbb{R}$$

where, for $w \in X$, the mapping $(u, v) \to a(w; u, v)$ is a bilinear continuous form on $X \times X$.

Then, we consider the following problem, also called *Problem* (Q):

Given $l \in X'$, find a pair $(u, \lambda) \in X \times M$ satisfying

$$a(u; u, v) + b(v, \lambda) = \langle l, v \rangle \quad \forall v \in X, \tag{1.1}$$

$$b(u, \mu) = 0 \quad \forall \mu \in M. \tag{1.2}$$

Let us introduce the linear operators $A(w) \in \mathscr{L}(X; X')$ for w in X, and $B \in \mathscr{L}(X; M')$ defined by:

$$\langle A(w)u, v \rangle = a(w; u, v) \quad \forall u, v \in X,$$

$$\langle Bv, \mu \rangle = b(v, \mu) \quad \forall v \in X, \quad \forall \mu \in M.$$

With these notations, Problem (Q) becomes:

Find $(u, \lambda) \in X \times M$ such that

$$A(u)u + B'\lambda = l \quad \textit{in } X', \tag{1.3}$$

$$Bu = 0 \quad \textit{in } M'. \tag{1.4}$$

As in the linear case we set $V = \text{Ker}(B)$ and we associate with Problem (Q) the following problem, called *Problem* (P):

Find $u \in V$ *such that*

$$a(u; u, v) = \langle l, v \rangle \quad \forall v \in V, \tag{1.5}$$

or equivalently such that

$$\pi A(u)u = \pi l \quad \textit{in } V', \tag{1.6}$$

where the linear operator $\pi \in \mathscr{L}(X'; V')$ is again defined by:

$$\langle \pi l, v \rangle = \langle l, v \rangle \quad \forall v \in V.$$

Of course, if (u, λ) is a solution of Problem (Q), then u is a solution of Problem (P). The converse property may be easily established as in the linear case provided the inf-sup condition holds. Therefore, the real difficulty here lies in solving the *nonlinear* Problem (P). To begin with, we need to derive a simple consequence of the following classical fixed-point theorem due to Brouwer, which we state without proof.

Theorem 1.1. *Let C denote a non-void, convex and compact subset of a finite-dimensional space and let F be a continuous mapping from C into C. Then, F has at least one fixed point.*

Corollary 1.1. *Let H be a finite-dimensional Hilbert space whose scalar product is denoted by $(.,.)$ and the corresponding norm by $|.|$. Let Φ be a continuous mapping from H into H with the following property:*

there exists $\mu > 0$ such that

$$(\Phi(f), f) \geqslant 0 \quad \textit{for all} \quad f \in H \quad \textit{with} \quad |f| = \mu. \tag{1.7}$$

Then, there exists an element f in H such that

$$\Phi(f) = 0, \quad |f| \leqslant \mu. \tag{1.8}$$

Proof. The proof proceeds by contradiction. Suppose that $\Phi(f) \neq 0$ in the closed sphere $S = \{f \in H; |f| \leqslant \mu\}$. Then, the mapping

$$f \to -\mu\Phi(f)/|\Phi(f)|$$

is continuous from S into S. As the dimension of H is finite and since the set S is obviously non-void, convex and compact, we may apply Theorem 1.1:

there exists an $f \in S$ such that

$$f = -\mu\Phi(f)/|\Phi(f)|.$$

Thus, we have exhibited an $f \in H$ such that $|f| = \mu$ and

$$(\Phi(f), f) = -\mu|\Phi(f)| < 0.$$

This contradicts (1.7). □

Now, we are in a position to establish the following existence result for Problem (P).

Theorem 1.2. *Assume that the following hypotheses hold:*

(i) *there exists a constant $\alpha > 0$ such that*

$$a(v; v, v) \geqslant \alpha \|v\|_X^2 \quad \forall v \in V; \tag{1.9}$$

(ii) *the space V is separable and, for all $v \in V$, the mapping*

$$u \to a(u; u, v)$$

is sequentially weakly continuous on V, i.e.,

$$\text{weak} \lim_{m\to\infty} u_m = u \quad \text{in } V \quad \text{implies} \quad \lim_{m\to\infty} a(u_m; u_m, v) = a(u; u, v) \qquad \forall v \in V. \tag{1.10}$$

Then, Problem (P) *has at least one solution $u \in V$.*

Proof. 1°) We begin by constructing a sequence of approximate solutions by *Galerkin's method*. Since the space V is separable, there exists a sequence $(w_m)_{m\geqslant 1}$ in V such that:

(i) for all $m \geqslant 1$, the elements $w_1, \ldots, w_m$ are linearly independent;

(ii) the finite linear combinations of the w_i, $\sum_i v_i w_i$, are dense in V.

Such a sequence $(w_m)_{m\geqslant 1}$ is called a "basis" of the separable space V.

We denote by V_m the subspace of V spanned by $w_1, \ldots, w_m$. Then, we approximate Problem (P) by the following problem, called *Problem* (P_m):

Find $u_m \in V_m$ satisfying

$$a(u_m; u_m, v) = \langle l, v\rangle \qquad \forall v \in V_m. \tag{1.11}$$

If we set

$$u_m = \sum_{i=1}^{m} v_i w_i,$$

we find that Problem (P_m) amounts to solve a system of m nonlinear equations in the m unknowns v_i.

Let us now show that, for each m, Problem (P_m) has at least one solution. We introduce the mapping Φ_m: $V_m \to V_m$ defined by

$$(\Phi_m(v), w_i) = a(v; v, w_i) - \langle l, w_i\rangle, \qquad 1 \leqslant i \leqslant m,$$

where $(.,.)$ is the scalar product in X. Hence, $u_m \in V_m$ is a solution of Problem (P_m) if and only if $\Phi_m(u_m) = 0$. Since

$$(\Phi_m(v), v) = a(v; v, v) - \langle l, v\rangle \qquad \forall v \in V_m,$$

it follows from the hypothesis (1.9) that

$$(\Phi_m(v), v) \geqslant (\alpha\|v\|_X - \|\pi l\|_{V'})\|v\|_X.$$

Hence, choosing $\mu = (1/\alpha)\,\|\pi l\|_{V'}$, we get for all $v \in V_m$ with $\|v\|_X = \mu$

$$(\Phi_m(v), v) \geqslant 0.$$

Moreover, Φ_m is continuous in V_m by virtue of the hypothesis (1.10). The space V_m being finite-dimensional, we may apply Corollary 1.1:

there exists at least one solution $u_m \in V_m$ of Problem (P_m). Furthermore, we have for all solution u_m

$$0 = (\Phi_m(u_m), u_m) \geqslant (\alpha\,\|u_m\|_X - \|\pi l\|_{V'})\,\|u_m\|_X$$

so that

$$\|u_m\|_X \leqslant (1/\alpha)\,\|\pi l\|_{V'}. \tag{1.12}$$

2°) Next, we construct a sequence (u_m) in V by taking, for each m, one arbitrary solution of Problem (P_m). We want to establish that we can extract a subsequence which converges towards a solution of Problem (P). It follows from (1.12) that the sequence (u_m) is bounded in V. Therefore, we can extract a subsequence (u_{m_p}) such that

$$u_{m_p} \to u^* \quad \text{weakly in } V \text{ as } p \to +\infty.$$

Then, the hypothesis (1.10) implies that

$$\lim_{p\to\infty} a(u_{m_p}; u_{m_p}, v) = a(u^*; u^*, v) \qquad \forall v \in V.$$

Taking in (1.11) $v = w_i$ and $m = m_p \geqslant i$, this yields:

$$a(u^*; u^*, w_i) = \langle l, w_i \rangle, \qquad i \geqslant 1.$$

Since the finite linear combinations of the w_i are dense in V, we get

$$a(u^*; u^*, v) = \langle l, v \rangle \qquad \forall v \in V,$$

so that u^* is a solution of Problem (P). □

Remark 1.1. Clearly, Theorem 1.2 is valid in a more general context: V is a separable Hilbert space; $(u, v) \to a(u, v)$ is a mapping from $V \times V$ into $\mathbb{R}$ such that

(i) $a(v, v) \geqslant \alpha\,\|v\|_V^2 \quad \forall v \in V, \alpha > 0$;

(ii) for each $u \in V$, $v \to a(u, v)$ is a linear continuous form on V;

(iii) for each $v \in V$, $u \to a(u, v)$ is sequentially weakly continuous on V.

Then, given $l \in V'$, there exists at least one element $u \in V$ such that

$$a(u, v) = \langle l, v \rangle \qquad \forall v \in V.$$

Now, we turn to the uniqueness of the solution of Problem (P). This requires stronger hypotheses than (1.9) and (1.10). Namely, we assume that

(i) the bilinear form $a(w; .,.)$ is uniformly V-elliptic with respect to w, i.e., there exists a constant $\alpha > 0$ such that

$$a(w; v, v) \geqslant \alpha\,\|v\|_X^2 \qquad \forall v, w \in V; \tag{1.13}$$

(ii) the mapping $w \to \pi A(w)$ is locally Lipschitz-continuous in V, i.e., there exists a continuous and monotonically increasing function $L: \mathbb{R}_+ \to \mathbb{R}_+$ such that for all $\mu > 0$

$$(1.14)\quad \begin{aligned} &|a(w_1; u, v) - a(w_2; u, v)| \leqslant L(\mu) \|u\|_X \|v\|_X \|w_1 - w_2\|_X \\ &\forall u, v \in V, \quad \forall w_1, w_2 \in S_\mu = \{w \in V;\ \|w\|_X \leqslant \mu\}. \end{aligned}$$

Theorem 1.3. *Assume that the hypotheses* (1.13) *and* (1.14) *hold. Then, under the condition*

$$(1.15)\qquad [\|\pi l\|_{V'}/(\alpha^2)]L(\|\pi l\|_{V'}/\alpha) < 1,$$

Problem (P) *has a unique solution* $u \in V$.

Proof. According to the hypothesis (1.13) and the Lax § Milgram's Theorem I.1.7, the operator $\pi A(w) \in \mathscr{L}(V; V')$ is invertible for each $w \in V$. Moreover, $T(w) = (\pi A(w))^{-1}$ belongs to $\mathscr{L}(V'; V)$ and satisfies

$$(1.16)\qquad \|T(w)\|_{\mathscr{L}(V';V)} \leqslant 1/\alpha.$$

With these notations, Problem (P) becomes

$$u = T(u)\pi l \quad \text{in } V.$$

Let us show that $v \to T(v)\pi l$ maps V into S_μ and is a strict contraction mapping in S_μ, where $\mu = (1/\alpha)\|\pi l\|_{V'}$. For all $v \in V$, we have

$$\|T(v)\pi l\|_X \leqslant \|T(v)\|_{\mathscr{L}(V';V)} \|\pi l\|_{V'} \leqslant (1/\alpha)\|\pi l\|_{V'} = \mu,$$

so that $T(v)\pi l$ belongs to S_μ. Next, we evaluate $T(u) - T(v)$ for $u, v \in S_\mu$. By virtue of the identity

$$T(u) - T(v) = T(u)[\pi A(v) - \pi A(u)]T(v)$$

and (1.16), we find

$$\|T(u) - T(v)\|_{\mathscr{L}(V';V)} \leqslant (1/\alpha^2)\|\pi A(v) - \pi A(u)\|_{\mathscr{L}(V;V')}.$$

Therefore, (1.14) yields

$$\|(T(u) - T(v))\pi l\|_X \leqslant (1/\alpha^2)\|\pi l\|_{V'} L(\mu)\|u - v\|_X.$$

Hence, by (1.15), the mapping $v \to T(v)\pi l$ is a strict contraction in S_μ and has a unique fixed point $u \in V$ which is the unique solution of Problem (P). □

Remark 1.2. Since $v \to T(v)\pi l$ is a strict contraction mapping in S_μ, its fixed point u can be computed by the method of successive approximations. More precisely, starting from any $u_0 \in S_\mu$, we construct the sequence (u_m) defined by

$$u_{m+1} = T(u_m)\pi l,$$

or equivalently by

$$(1.17)\qquad a(u_m; u_{m+1}, v) = \langle l, v\rangle \quad \forall v \in V.$$

Then

$$\lim_{m\to\infty} \| u_m - u \|_X = 0.$$

Note that u_0 need not be picked in S_μ since, for any $u_0 \in V$, $u_1 = T(u_0)\pi l$ is necessarily in S_μ.

We end this paragraph by solving Problem (Q). As mentioned in the beginning, the analysis parallels that of the linear case.

Theorem 1.4. *Assume that the bilinear form $b(.,.)$ satisfies the inf-sup condition*

$$\inf_{\mu \in M} \sup_{v \in X} \frac{b(v,\mu)}{\| v \|_X \| \mu \|_M} \geqslant \beta > 0. \tag{1.18}$$

Then, for each solution u of Problem (P), *there exists a unique $\lambda \in M$ such that the pair (u, λ) is a solution of Problem* (Q).

Proof. Assume that $u \in V$ is a solution of Problem (P). We must find $\lambda \in M$ which satisfies the equation (1.3). But $l - A(u)u$ belongs to the polar set V^0 of V. Moreover, the inf-sup condition (1.18) and Lemma I.4.1 imply that B' is an isomorphism from M onto V^0. Thus, there exists a unique $\lambda \in M$ such that (u, λ) is a solution of Problem (Q). □

Remark 1.3. We can now extend the iterative scheme of Remark 1.2 in order to solve Problem (Q). Starting from any $u_0 \in V$, we construct the sequence (u_m, λ_m) in $X \times M$ by solving

$$\begin{cases} a(u_m; u_{m+1}, v) + b(v, \lambda_{m+1}) = \langle l, v \rangle & \forall v \in X, \\ b(u_{m+1}, \mu) = 0 & \forall \mu \in M. \end{cases} \tag{1.19}$$

Let us check that, under the assumptions (1.13), (1.14), (1.15) and (1.18), we have for any u_0 in V:

$$\lim_{m\to\infty} \{ \| u_m - u \|_X + \| \lambda_m - \lambda \|_M \} = 0. \tag{1.20}$$

In fact, u_m belongs to the space V and satisfies (1.17). Hence by Remark 1.2, we get

$$\lim_{m\to\infty} \| u_m - u \|_X = 0.$$

Moreover, using (1.1) and (1.19), we obtain

$$b(v, \lambda_m - \lambda) = a(u; u, v) - a(u_{m-1}; u_m, v) \qquad \forall v \in X$$

and by (1.18)

$$\| \lambda_m - \lambda \|_M \leqslant (1/\beta) \sup_{v \in X} \left\{ \frac{a(u; u, v) - a(u_{m-1}; u_m, v)}{\| v \|_X} \right\}.$$

The desired result then follows from the convergence of u_m to u in X and the continuity properties of $a(.;.,.)$.

§ 2. Theory of the Steady-State Navier-Stokes Equations

As we have seen in Paragraph I.5, the stationary Navier-Stokes equations may be written in the form

$$\text{(2.1)} \qquad \left\{ \begin{aligned} -\nu\Delta\mathbf{u} + \sum_{j=1}^{N} u_j \partial\mathbf{u}/\partial x_j + \mathbf{grad}\, p = \mathbf{f}, \\ \operatorname{div}\mathbf{u} = 0, \end{aligned} \right\} \quad \text{in } \Omega$$

$$\mathbf{f} \text{ given in } H^{-1}(\Omega)^N,$$

where again Ω is a bounded domain of $\mathbb{R}^N$ ($N = 2, 3$) with a Lipschitz-continuous boundary Γ. In this paragraph, we want to derive from the abstract material of §1 existence and uniqueness results for various formulations of the Navier-Stokes equations.

2.1. The Dirichlet Problem in the Velocity-Pressure Formulation

We first consider the case of the *homogeneous Dirichlet boundary condition*

$$\text{(2.2)} \qquad \mathbf{u} = \mathbf{0} \quad \text{on } \Gamma.$$

In order to write Problem (2.1), (2.2) in a variational form, we introduce the trilinear form

$$\text{(2.3)} \qquad a_1(\mathbf{w}; \mathbf{u}, \mathbf{v}) = \sum_{i,j=1}^{N} \int_\Omega w_j (\partial u_i/\partial x_j) v_i \, dx.$$

The next two lemmas state useful properties of the trilinear form $a_1(.; ., .)$.

Lemma 2.1. *For $N \leqslant 4$, the trilinear form $a_1(.; ., .)$ is continuous on $(H^1(\Omega)^N)^3$.*

Proof. According to the Sobolev Imbedding Theorem I.1.3, the space $H^1(\Omega)$ is continuously imbedded in $L^4(\Omega)$ for $N \leqslant 4$. Then by Hölder's inequality, we have if $\mathbf{u}, \mathbf{v}, \mathbf{w} \in H^1(\Omega)^N$:

$$w_j(\partial u_i/\partial x_j) v_i \in L^1(\Omega), \qquad 1 \leqslant i, j \leqslant N,$$

with

$$\left| \int_\Omega w_j (\partial u_i/\partial x_j) v_i \, dx \right| \leqslant \| w_j \|_{0,4,\Omega} \| \partial u_i/\partial x_j \|_{0,\Omega} \| v_i \|_{0,4,\Omega}$$

$$\leqslant C_1 \| w_j \|_{1,\Omega} | u_i |_{1,\Omega} \| v_i \|_{1,\Omega}.$$

Thus, the form $a_1(.; ., .)$ is well defined and continuous on $(H^1(\Omega)^N)^3$ and

$$|a_1(\mathbf{w}; \mathbf{u}, \mathbf{v})| \leqslant C_2 |\mathbf{u}|_{1,\Omega} \|\mathbf{v}\|_{1,\Omega} \|\mathbf{w}\|_{1,\Omega}. \qquad \square$$

Lemma 2.2. *Let* $\mathbf{u}, \mathbf{v} \in H^1(\Omega)^N$ *and let* $\mathbf{w} \in H^1(\Omega)^N$ *with* $\operatorname{div} \mathbf{w} = 0$ *and* $\mathbf{w} \cdot \mathbf{n}|_\Gamma = 0$. *Then, we have*:

$$a_1(\mathbf{w}; \mathbf{u}, \mathbf{v}) + a_1(\mathbf{w}; \mathbf{v}, \mathbf{u}) = 0, \tag{2.4}$$

$$a_1(\mathbf{w}; \mathbf{v}, \mathbf{v}) = 0. \tag{2.5}$$

Proof. Clearly, the properties (2.4) and (2.5) are equivalent and it suffices to check (2.5). Let $\mathbf{v} \in \mathscr{D}(\bar{\Omega})^N$ and $\mathbf{w} \in H^1(\Omega)^N$; we may write:

$$a_1(\mathbf{w}; \mathbf{v}, \mathbf{v}) = (1/2) \sum_{i,j=1}^{N} \int_\Omega w_j \partial(v_i^2)/\partial x_j \, dx$$

and by Green's formula (I.2.17):

$$a_1(\mathbf{w}; \mathbf{v}, \mathbf{v}) = -(1/2) \sum_{i,j=1}^{N} \left\{ \int_\Omega \operatorname{div} \mathbf{w} v_i \, dx + \int_\Gamma \mathbf{w} \cdot \mathbf{n} v_i \, ds \right\}.$$

When $\mathbf{w}$ satisfies $\operatorname{div} \mathbf{w} = 0$ and $\mathbf{w} \cdot \mathbf{n}|_\Gamma = 0$, this implies

$$a_1(\mathbf{w}; \mathbf{v}, \mathbf{v}) = 0.$$

Then the lemma follows by using the density of $\mathscr{D}(\bar{\Omega})$ into $H^1(\Omega)$ (cf. Theorem I.1.2). □

Now, recall the following spaces:

$$\mathscr{V} = \{\mathbf{v} \in \mathscr{D}(\Omega)^N; \operatorname{div} \mathbf{v} = 0\},$$
$$V = \{\mathbf{v} \in H_0^1(\Omega)^N; \operatorname{div} \mathbf{v} = 0\}.$$

We set

$$\begin{aligned} a_0(\mathbf{u}, \mathbf{v}) &= \nu(\mathbf{grad}\, \mathbf{u}, \mathbf{grad}\, \mathbf{v}) \\ &= 2\nu \sum_{i,j=1}^{N} (D_{ij}(\mathbf{u}), D_{ij}(\mathbf{v})) \quad \text{if either } \mathbf{u} \text{ or } \mathbf{v} \in \mathbf{V}, \end{aligned} \tag{2.6}$$

and

$$a(\mathbf{w}; \mathbf{u}, \mathbf{v}) = a_0(\mathbf{u}, \mathbf{v}) + a_1(\mathbf{w}; \mathbf{u}, \mathbf{v}). \tag{2.7}$$

Then Problem (2.1) (2.2) has the equivalent form:

Find a pair $(\mathbf{u}, p) \in V \times L_0^2(\Omega)$ *such that*:

$$a(\mathbf{u}; \mathbf{u}, \mathbf{v}) - (p, \operatorname{div} \mathbf{v}) = \langle \mathbf{f}, \mathbf{v} \rangle \quad \forall \mathbf{v} \in H_0^1(\Omega)^N. \tag{2.8}$$

Theorem 2.1. *Let* $N \leqslant 4$ *and let* Ω *be a bounded domain of* $\mathbb{R}^N$ *with a Lipschitz-continuous boundary* Γ. *Given* $\mathbf{f} \in H^{-1}(\Omega)^N$, *there exists at least one pair* $(\mathbf{u}, p) \in V \times L_0^2(\Omega)$ *which satisfies* (2.8) *or equivalently* (2.1) (2.2).

Proof. We apply the material of Paragraph 1 as follows. We set

$$X = H_0^1(\Omega)^N \quad \text{normed by } |.|_{1,\Omega}, \qquad M = L_0^2(\Omega),$$
$$b(\mathbf{v}, q) = -(q, \operatorname{div} \mathbf{v}), \qquad \langle l, \mathbf{v}\rangle = \langle \mathbf{f}, \mathbf{v}\rangle.$$

Hence (2.8) is a particular case of Problem (Q). Thus, it remains to check the hypotheses of Theorems 1.2 and 1.4. First, using (2.5), we get for all $\mathbf{v}$, $\mathbf{w} \in V$

$$a(\mathbf{w}; \mathbf{v}, \mathbf{v}) = a_0(\mathbf{v}, \mathbf{v}) = \nu |\mathbf{v}|_{1,\Omega}^2.$$

Therefore, the form $a(.;.,.)$ satisfies the property (1.13) (and thus (1.9)).

Next, let $\mathbf{u}$ be a function of V and $(\mathbf{u}_m)$ be a sequence in V such that

$$\mathbf{u}_m \to \mathbf{u} \quad \text{weakly in } V \text{ as } m \to \infty.$$

Then, the compactness of the imbedding of $H_0^1(\Omega)$ into $L^2(\Omega)$ (cf. Theorem I.1.3) implies that:

$$\mathbf{u}_m \to \mathbf{u} \quad \text{strongly in } L^2(\Omega)^N \text{ as } m \to \infty.$$

Now, let $\mathbf{v}$ be in $\mathscr{V}$ and let us take the limit of $a(\mathbf{u}_m; \mathbf{u}_m, \mathbf{v})$. According to (2.4), we have

$$\begin{aligned} a_1(\mathbf{u}_m; \mathbf{u}_m, \mathbf{v}) &= -a_1(\mathbf{u}_m; \mathbf{v}, \mathbf{u}_m) \\ &= -\sum_{i,j=1}^{N} \int_\Omega u_{mi} u_{mj} (\partial v_i/\partial x_j)\, dx. \end{aligned}$$

As $\partial v_i/\partial x_j \in L^\infty(\Omega)$ and $\lim_{m\to\infty} u_{mi} u_{mj} = u_i u_j$ in $L^1(\Omega)$, it follows that

$$\begin{aligned} \lim_{m\to\infty} a_1(\mathbf{u}_m; \mathbf{u}_m, \mathbf{v}) &= -\sum_{i,j=1}^{N} \int_\Omega u_i u_j (\partial v_i/\partial x_j)\, dx \\ &= -a_1(\mathbf{u}; \mathbf{v}, \mathbf{u}) = a_1(\mathbf{u}; \mathbf{u}, \mathbf{v}). \end{aligned}$$

Since it is clear that

$$\lim_{m\to\infty} a_0(\mathbf{u}_m, \mathbf{v}) = a_0(\mathbf{u}, \mathbf{v}),$$

we get

$$\lim_{m\to\infty} a(\mathbf{u}_m; \mathbf{u}_m, \mathbf{v}) = a(\mathbf{u}; \mathbf{u}, \mathbf{v})$$

for all $\mathbf{v} \in \mathscr{V}$ and therefore for all $\mathbf{v} \in V$ by virtue of the density of $\mathscr{V}$ in V (cf. Corollary I.2.5) and the continuity of the mapping $\mathbf{v} \to a(\mathbf{u}; \mathbf{u}, \mathbf{v})$.

Finally, we have already seen that the bilinear form $b(.,.)$ satisfies the inf-sup condition of Theorem 1.4 (see inequality (I.5.14)).

Thus, the hypotheses of Theorems 1.2 and 1.4 are fulfilled. Hence, there exists at least one function $\mathbf{u} \in V$ such that

$$a(\mathbf{u}; \mathbf{u}, \mathbf{v}) = \langle \mathbf{f}, \mathbf{v}\rangle \quad \forall \mathbf{v} \in V. \tag{2.9}$$

Moreover, for each solution $\mathbf{u}$ of (2.9), there exists a unique $p \in L_0^2(\Omega)$ such that $(\mathbf{u}, p)$ is a solution of Problem (2.8). □

Now, we turn to the uniqueness of the solution $(\mathbf{u}, p)$ of Problem (2.8). For this, we introduce the norm of the trilinear form $a_1(.;.,.)$ in V^3:

(2.10) $$\mathcal{N} = \sup_{\mathbf{u},\mathbf{v},\mathbf{w} \in V} \frac{a_1(\mathbf{w};\mathbf{u},\mathbf{v})}{|\mathbf{u}|_{1,\Omega}|\mathbf{v}|_{1,\Omega}|\mathbf{w}|_{1,\Omega}}.$$

We also set:

(2.11) $$\|\mathbf{f}\|_{V'} = \sup_{\mathbf{v} \in V} \frac{\langle \mathbf{f}, \mathbf{v} \rangle}{|\mathbf{v}|_{1,\Omega}}.$$

Theorem 2.2. *Under the hypotheses of Theorem* 2.1 *and if in addition*

(2.12) $$(\mathcal{N}/\nu^2)\,\|\mathbf{f}\|_{V'} < 1,$$

then Problem (2.8) *has a unique solution* $(\mathbf{u}, p)$ *in* $V \times L_0^2(\Omega)$.

Proof. Here, we make use of Theorem 1.3. We have already proved the property (1.13) with $\alpha = \nu$ and it suffices to establish (1.14). Let $\mathbf{u}$, $\mathbf{v}$, $\mathbf{w}_1$ and $\mathbf{w}_2$ be in V; we have:

$$\begin{aligned}|a(\mathbf{w}_1;\mathbf{u},\mathbf{v}) - a(\mathbf{w}_2;\mathbf{u},\mathbf{v})| &= |a(\mathbf{w}_1 - \mathbf{w}_2;\mathbf{u},\mathbf{v})| \\ &\leqslant \mathcal{N}|\mathbf{u}|_{1,\Omega}|\mathbf{v}|_{1,\Omega}|\mathbf{w}_1 - \mathbf{w}_2|_{1,\Omega}.\end{aligned}$$

Therefore, the form $a(.;.,.)$ satisfies the hypothesis (1.14) with $L(\mu) = \mathcal{N}$ for all μ. Then, the condition (2.12) coincides precisely with (1.15). Hence, the conclusion of Theorem 1.3 is valid. □

Let us next consider the general case of a *nonhomogeneous Dirichlet boundary condition*

(2.13) $$\mathbf{u} = \mathbf{g} \quad \text{on } \Gamma.$$

Denote again by Γ_i, $0 \leqslant i \leqslant p$, the connected components of the boundary Γ like in Figure 2. We shall assume in all the sequel that

(2.14) $$\int_{\Gamma_i} \mathbf{g}\cdot\mathbf{n}\, ds = 0, \qquad 0 \leqslant i \leqslant p.$$

Now, we need the following important technical result due to Hopf [45].

Lemma 2.3. *Suppose* $N \leqslant 3$ *and* Ω *is like in Theorem* 2.1. *Then, given a function* $\mathbf{g} \in H^{1/2}(\Gamma)^N$ *satisfying the conditions* (2.14), *there exists for any* $\varepsilon > 0$ *a function* $\mathbf{u}_0 = \mathbf{u}_0(\varepsilon) \in H^1(\Omega)^N$ *such that*

(2.15) $$\operatorname{div} \mathbf{u}_0 = 0, \qquad \mathbf{u}_0|_\Gamma = \mathbf{g},$$

(2.16) $$|a_1(\mathbf{v}; \mathbf{u}_0, \mathbf{v})| \leqslant \varepsilon |\mathbf{v}|_{1,\Omega}^2 \qquad \forall \mathbf{v} \in V.$$

Before proving Lemma 2.3, we check two preliminary lemmas. For any point $x \in \Omega$, we denote by $d(x; \Gamma)$ the distance of x to the boundary Γ. Then, we have:

Lemma 2.4. *Let Ω be like in Theorem 2.1. For all $\varepsilon > 0$, there exists a function $\theta_\varepsilon \in \mathscr{C}^2(\bar{\Omega})$ such that*

(2.17) $$\begin{cases} \theta_\varepsilon = 1 & \text{in a neighborhood of } \Gamma, \\ \theta_\varepsilon(x) = 0 & \text{if } d(x;\Gamma) \geqslant 2\delta(\varepsilon), \quad \delta(\varepsilon) = \exp(-1/\varepsilon), \\ |\partial\theta_\varepsilon(x)/\partial x_i| \leqslant \varepsilon/d(x;\Gamma) & \text{if } d(x;\Gamma) \leqslant 2\delta(\varepsilon), \quad 1 \leqslant i \leqslant N. \end{cases}$$

Proof. Let us consider the function $\mu \to \phi_\varepsilon(\mu)$ defined for $\mu \geqslant 0$ by

$$\phi_\varepsilon(\mu) = \begin{cases} 1 & \text{if } 0 \leqslant \mu \leqslant \delta(\varepsilon)^2, \\ \varepsilon \operatorname{Log}(\delta(\varepsilon)/\mu) & \text{if } \delta(\varepsilon)^2 \leqslant \mu \leqslant \delta(\varepsilon), \\ 0 & \text{if } \mu \geqslant \delta(\varepsilon). \end{cases}$$

Clearly, $\phi_\varepsilon \in W^{1,\infty}(\mathbb{R}_+)$. We set:

$$\chi_\varepsilon(x) = \phi_\varepsilon(d(x;\Gamma)), \qquad x \in \Omega.$$

Since Γ is Lipschitz-continuous, the function d belongs to $W^{1,\infty}(\Omega)$ and we have

$$|\partial d(x;\Gamma)/\partial x_i| \leqslant 1.$$

Hence the function χ_ε belongs to $W^{1,\infty}(\Omega)$ and satisfies

$$|\partial \chi_\varepsilon(x)/\partial x_i| \leqslant \varepsilon/d(x;\Gamma) \quad \text{if } d(x;\Gamma) \leqslant \delta(\varepsilon).$$

By regularizing χ_ε, we obtain a function $\theta_\varepsilon \in \mathscr{C}^2(\bar{\Omega})$ with the properties (2.17). □

Lemma 2.5. *Let Ω be like in Theorem 2.1. There exists a constant $C = C(\Omega) > 0$ such that*

(2.18) $$\|\phi/d(.;\Gamma)\|_{0,\Omega} \leqslant C|\phi|_{1,\Omega} \qquad \forall \phi \in H_0^1(\Omega).$$

Proof. By introducing a partition of unity subordinate to a covering of Γ and systems of local coordinates near Γ, we need only to investigate the case where Ω is the half-space

$$\mathbb{R}_+^N = \{x = (x', x_N);\, x_N > 0\}$$

and $d(x;\Gamma) = x_N$. Hence, for proving (2.18), it is sufficient to check that

$$\int_{\mathbb{R}_+^N} [\phi(x)/x_N]^2\, dx \leqslant C_1 \int_{\mathbb{R}_+^N} |\partial\phi(x)/\partial x_N|^2\, dx \qquad \forall \phi \in \mathscr{D}(\mathbb{R}_+^N).$$

This in turn will be a consequence of the following Hardy inequality:

$$\int_0^\infty |\phi(t)/t|^2\,dt \leqslant 4\int_0^\infty |\phi'(t)|^2\,dt \qquad \forall \phi\in\mathscr{D}(0,\infty), \tag{2.19}$$

which can be established as follows. By writing

$$\phi(t)=\int_0^t \phi'(s)\,ds$$

and setting $t=e^\tau$, $s=e^\sigma$, we have

$$\begin{aligned}\int_0^\infty |\phi(t)/t|^2\,dt &= \int_0^\infty (1/t)\left[\int_0^t \phi'(s)\,ds\right]^2 (dt/t)\\ &= \int_{-\infty}^{+\infty} e^{-\tau}\left[\int_{-\infty}^{\tau} \phi'(e^\sigma)e^\sigma\,d\sigma\right]^2 d\tau\\ &= \int_{-\infty}^{+\infty}\left[\int_{-\infty}^{+\infty} H(\tau-\sigma)e^{-(\tau-\sigma)/2}\phi'(e^\sigma)e^{\sigma/2}\,d\sigma\right]^2 d\tau,\end{aligned}$$

where H is the classical Heaviside function. Now, using the standard convolution inequality:

$$\|f*g\|_{L^2}\leqslant \|f\|_{L^1}\|g\|_{L^2},$$

we get

$$\begin{aligned}\int_0^\infty |\phi(t)/t|^2\,dt &\leqslant \left[\int_{-\infty}^{+\infty} H(\tau)e^{-\tau/2}\,d\tau\right]^2\int_{-\infty}^{+\infty}\phi'(e^\sigma)^2e^\sigma\,d\sigma\\ &= 4\int_0^\infty \phi'(s)^2\,ds\end{aligned}$$

which is the desired inequality. □

Proof of Lemma 2.3. Let the function $\mathbf{g}\in H^{1/2}(\Gamma)^N$ satisfy the conditions (2.14). We already know from Lemma I.2.2 that there exists a function $\mathbf{w}_0\in H^1(\Omega)^N$ such that

$$\operatorname{div}\mathbf{w}_0=0,\qquad \mathbf{w}_0|_\Gamma=\mathbf{g}.$$

Moreover, it follows from (2.14), Theorem I.3.1 or I.3.4 and Corollary I.3.3, that we can find a stream function $\psi_0\in H^2(\Omega)$ if the dimension $N=2$ or a vector potential $\boldsymbol{\psi}_0\in H^2(\Omega)^3$ if the dimension $N=3$, such that

$$\mathbf{w}_0=\begin{cases}\mathbf{curl}\,\psi_0, & N=2,\\ \mathbf{curl}\,\boldsymbol{\psi}_0, & N=3.\end{cases}$$

Consider the case $N=3$, the case $N=2$ being exactly similar. For all $\mu>0$, we introduce the function

$$\mathbf{u}_{0\mu} = \mathbf{curl}(\theta_\mu \boldsymbol{\psi}_0)$$

where θ_μ is defined as in Lemma 2.4. Clearly $\mathbf{u}_{0\mu} \in H^1(\Omega)^3$ and

$$\operatorname{div} \mathbf{u}_{0\mu} = 0, \qquad \mathbf{u}_{0\mu}|_\Gamma = \mathbf{g}.$$

Now, using Lemma 2.4, we have if $d(x;\Gamma) \leqslant 2\delta(\mu)$

$$\left|\frac{\partial(\theta_\mu \psi_{0i})}{\partial x_j}(x)\right| \leqslant \frac{\mu}{d(x;\Gamma)}|\psi_{0i}(x)| + \left|\frac{\partial \psi_{0i}}{\partial x_j}(x)\right|$$

so that

$$\|\mathbf{u}_{0\mu}(x)\| \leqslant C_1[(\mu/d(x;\Gamma))\,\|\boldsymbol{\psi}_0(x)\| + \|D\boldsymbol{\psi}_0(x)\|]$$

where $\|\,.\,\|$ denotes as usual the Euclidean norm of $\mathbb{R}^N$ and

$$\|D\boldsymbol{\psi}_0(x)\| = \left(\sum_{i,j=1}^{N} |\partial \psi_{0i}(x)/\partial x_j|^2\right)^{1/2}.$$

Let $\mathbf{v}$ belong to V. Since $H^2(\Omega) \hookrightarrow \mathscr{C}^0(\bar{\Omega})$, we obtain

$$\|v_i u_{0\mu j}\|_{0,\Omega} \leqslant C_2\left\{\mu\|v_i/d(.\,;\Gamma)\|_{0,\Omega} + \left(\int_{d(x;\Gamma)\leqslant 2\delta(\mu)} [v_i\|D\boldsymbol{\psi}_0(x)\|]^2\,dx\right)^{1/2}\right\}.$$

Applying Lemma 2.5 gives

$$\|v_i/d(.\,;\Gamma)\|_{0,\Omega} \leqslant C_3|v_i|_{1,\Omega}.$$

Moreover, since $H^1(\Omega) \hookrightarrow L^6(\Omega)$, we have by Hölder's inequality

$$\left(\int_{d(x;\Gamma)\leqslant 2\delta(\mu)} [v_i\|D\boldsymbol{\psi}_0\|]^2\,dx\right)^{1/2} \leqslant C_4|v_i|_{1,\Omega} \times \left(\int_{d(x;\Gamma)\leqslant 2\delta(\mu)} \|D\boldsymbol{\psi}_0(x)\|^3\,dx\right)^{1/3}.$$

Setting

$$\phi(\mu) = \left(\int_{d(x;\Gamma)\leqslant 2\delta(\mu)} \|D\boldsymbol{\psi}_0(x)\|^3\,dx\right)^{1/3},$$

we get

$$\|v_i u_{0\mu j}\|_{0,\Omega} \leqslant C_5(\mu + \phi(\mu))|v_i|_{1,\Omega}.$$

Therefore, using Lemma 2.2, we have

$$|a_1(\mathbf{v};\mathbf{u}_{0\mu},\mathbf{v})| = |a_1(\mathbf{v};\mathbf{v},\mathbf{u}_{0\mu})| = \left|\sum_{i,j=1}^{N}\int_\Omega v_i u_{0\mu j}\,\partial v_j/\partial x_i\,dx\right|$$
$$\leqslant C_6(\mu + \phi(\mu))|\mathbf{v}|_{1,\Omega}^2.$$

Thus, given $\varepsilon > 0$ and since $\lim_{\mu\to 0}\phi(\mu) = 0$, we may choose $\mu = \mu(\varepsilon)$ small enough so that

$$C_6(\mu + \phi(\mu)) \leqslant \varepsilon.$$

The corresponding function $\mathbf{u}_0(\varepsilon) = \mathbf{u}_{0\mu}$ satisfies the requirements (2.15) and (2.16). □

Now, a variational form of Problem (2.1), (2.13) consists in finding a pair $(\mathbf{u}, p) \in H^1(\Omega)^N \times L_0^2(\Omega)$ solution of the equations

$$\text{(2.20)} \qquad \begin{cases} a(\mathbf{u}; \mathbf{u}, \mathbf{v}) - (p, \operatorname{div} \mathbf{v}) = \langle \mathbf{f}, \mathbf{v} \rangle \quad \forall \mathbf{v} \in H_0^1(\Omega)^N, \\ \operatorname{div} \mathbf{u} = 0 \quad \text{in } \Omega, \\ \mathbf{u} = \mathbf{g} \quad \text{on } \Gamma. \end{cases}$$

Theorem 2.3. *Let $N \leqslant 3$ and let Ω be a bounded domain of $\mathbb{R}^N$ with a Lipschitz-continuous boundary Γ. Given $\mathbf{f} \in H^{-1}(\Omega)^N$ and $\mathbf{g} \in H^{1/2}(\Gamma)^N$ satisfying the conditions* (2.14), *there exists at least one pair $(\mathbf{u}, p) \in H^1(\Omega)^N \times L_0^2(\Omega)$ solution of* (2.20) *or equivalently solution of Problem* (2.1) (2.13).

Proof. Let $\mathbf{u}_0$ be a function of $H^1(\Omega)^N$ such that

$$\operatorname{div} \mathbf{u}_0 = 0, \qquad \mathbf{u}_0|_\Gamma = \mathbf{g}.$$

We set $\mathbf{u} = \mathbf{u}_0 + \mathbf{w}$. Since

$$a(\mathbf{u}_0 + \mathbf{w}; \mathbf{u}_0 + \mathbf{w}, \mathbf{v}) = a(\mathbf{w}; \mathbf{w}, \mathbf{v}) + a_1(\mathbf{u}_0; \mathbf{w}, \mathbf{v}) + a_1(\mathbf{w}; \mathbf{u}_0, \mathbf{v}) + a(\mathbf{u}_0; \mathbf{u}_0, \mathbf{v}),$$

Problem (2.20) may be equivalently stated as follows:

Find a pair $(\mathbf{w}, p) \in V \times L_0^2(\Omega)$ such that

$$\text{(2.21)} \qquad \tilde{a}(\mathbf{w}; \mathbf{w}, \mathbf{v}) - (p, \operatorname{div} \mathbf{v}) = \langle \mathbf{f}, \mathbf{v} \rangle - a(\mathbf{u}_0; \mathbf{u}_0, \mathbf{v}) \qquad \forall \mathbf{v} \in H_0^1(\Omega)^N,$$

where

$$\tilde{a}(\mathbf{w}; \mathbf{u}, \mathbf{v}) = a(\mathbf{w}; \mathbf{u}, \mathbf{v}) + a_1(\mathbf{u}_0; \mathbf{u}, \mathbf{v}) + a_1(\mathbf{u}; \mathbf{u}_0, \mathbf{v}).$$

Observe that Problem (2.21) fits into the framework of Paragraph 1 if we take

$$X = H_0^1(\Omega)^N, \qquad M = L_0^2(\Omega),$$

$$a(.; ., .) \text{ replaced by } \tilde{a}(.; ., .), \quad b(\mathbf{v}, q) = -(q, \operatorname{div} \mathbf{v}),$$

$$\langle l, \mathbf{v} \rangle = \langle \mathbf{f}, \mathbf{v} \rangle - a(\mathbf{u}_0; \mathbf{u}_0, \mathbf{v}).$$

Again, we have to check the hypotheses of Theorems 1.2 and 1.4. First, using (2.5), we have for all $\mathbf{v}, \mathbf{w} \in V$

$$\tilde{a}(\mathbf{w}; \mathbf{v}, \mathbf{v}) = \nu |\mathbf{v}|_{1,\Omega}^2 + a_1(\mathbf{v}; \mathbf{u}_0, \mathbf{v}).$$

It follows from Lemma 2.3 that we may choose the function $\mathbf{u}_0$ so that

$$|a_1(\mathbf{v}; \mathbf{u}_0, \mathbf{v})| \leqslant \varepsilon |\mathbf{v}|_{1,\Omega}^2 \qquad \forall \mathbf{v} \in V, \qquad \varepsilon < \nu$$

and therefore

$$\tilde{a}(\mathbf{w}; \mathbf{v}, \mathbf{v}) \geqslant (\nu - \varepsilon) |\mathbf{v}|_{1,\Omega}^2 \qquad \forall \mathbf{v}, \mathbf{w} \in V.$$

Hence, the form $\tilde{a}(.;.,.)$ satisfies the property (1.13) (and also (1.9)). Besides that, the property (1.10) is established exactly as in the proof of Theorem 2.1. Since the inf-sup condition (1.18) holds, we obtain that Problem (2.21) has at least one solution $(\mathbf{w}, p) \in V \times L_0^2(\Omega)$, which proves the theorem. □

Next, we derive a uniqueness result. For any function $\mathbf{u}_0 \in H^1(\Omega)^N$, we set

$$\rho(\mathbf{u}_0) = \sup_{\mathbf{v} \in V} \frac{a_1(\mathbf{v}; \mathbf{u}_0, \mathbf{v})}{|\mathbf{v}|_{1,\Omega}^2}, \tag{2.22}$$

$$\| l(\mathbf{f}; \mathbf{u}_0) \|_{V'} = \sup_{\mathbf{v} \in V} \frac{\langle l, \mathbf{v} \rangle}{|\mathbf{v}|_{1,\Omega}}, \qquad \langle l, \mathbf{v} \rangle = \langle \mathbf{f}, \mathbf{v} \rangle - a(\mathbf{u}_0; \mathbf{u}_0, \mathbf{v}). \tag{2.23}$$

Then, we define $\nu_0 = \nu_0(\Omega; \mathbf{f}, \mathbf{g})$ by

$$\nu_0 = \inf\{\rho(\mathbf{u}_0) + (\mathcal{N} \| l(\mathbf{f}; \mathbf{u}_0) \|_{V'})^{1/2};\ \mathbf{u}_0 \in H^1(\Omega)^N \text{ satisfies (2.15)}\}. \tag{2.24}$$

Given any number $\nu > 0$, it follows from Lemma 2.3 and the continuity of the mapping $\mathbf{g} \to \mathbf{u}_0$ (cf. (I.2.12)) that we have $\nu_0 < \nu$ for $\|\mathbf{f}\|_{V'}$ and $\|\mathbf{g}\|_{1/2,\Gamma}$ small enough.

Theorem 2.4. *Assume the hypotheses of Theorem* 2.3. *Then, for* $\nu > \nu_0(\Omega; \mathbf{f}, \mathbf{g})$, *Problem* (2.20) *has a unique solution* $(\mathbf{u}, p) \in H^1(\Omega)^N \times L_0^2(\Omega)$.

Proof. Let us choose a function $\mathbf{u}_0 \in H^1(\Omega)^N$ which satisfies (2.15) and $\rho(\mathbf{u}_0) < \nu$. We want to apply Theorem 1.3 to Problem (2.21). We have for all $\mathbf{v}, \mathbf{w} \in V$

$$\tilde{a}(\mathbf{w}; \mathbf{v}, \mathbf{v}) \geqslant (\nu - \rho(\mathbf{u}_0)) |\mathbf{v}|_{1,\Omega}^2,$$

so that the property (1.13) holds with $\alpha = \nu - \rho(\mathbf{u}_0)$. Hence, it remains only to check (1.14). Let $\mathbf{u}$, $\mathbf{v}$, $\mathbf{w}_1$, $\mathbf{w}_2$ be in V; using (2.10), we get

$$\begin{aligned} |\tilde{a}(\mathbf{w}_1; \mathbf{u}, \mathbf{v}) - \tilde{a}(\mathbf{w}_2; \mathbf{u}, \mathbf{v})| &= |a_1(\mathbf{w}_1 - \mathbf{w}_2; \mathbf{u}, \mathbf{v})| \\ &\leqslant \mathcal{N} |\mathbf{u}|_{1,\Omega} |\mathbf{v}|_{1,\Omega} |\mathbf{w}_1 - \mathbf{w}_2|_{1,\Omega}. \end{aligned}$$

Hence, the form $\tilde{a}(.;.,.)$ satisfies the property (1.14) with $L(\mu) = \mathcal{N}$. Now, the condition (1.15) becomes in our case

$$\frac{\mathcal{N} \| l(\mathbf{f}; \mathbf{u}_0) \|_{V'}}{(\nu - \rho(\mathbf{u}_0))^2} < 1$$

or equivalently

$$\nu > \rho(\mathbf{u}_0) + (\mathcal{N} \| l(\mathbf{f}; \mathbf{u}_0) \|_{V'})^{1/2}.$$

By taking the infimum over all the admissible functions $\mathbf{u}_0$, we obtain that, for $\nu > \nu_0$, Problem (2.21) has a unique solution $(\mathbf{w}, p) \in H_0^1(\Omega)^N \times L_0^2(\Omega)$ and therefore Problem (2.20) has a unique solution $(\mathbf{u}_0 + \mathbf{w}, p) \in H^1(\Omega)^N \times L_0^2(\Omega)$. □

Remark 2.1. Let $\mathbf{u}_0$ be a function of $H^1(\Omega)^N$ such that

$$\operatorname{div}\mathbf{u}_0 = 0, \qquad \mathbf{u}_0|_\Gamma = \mathbf{g}, \qquad \nu > \rho(\mathbf{u}_0).$$

Then, we have for all solution $(\mathbf{u} = \mathbf{u}_0 + \mathbf{w}, \mathbf{p})$ of Problem (2.20)

$$|\mathbf{w}|_{1,\Omega} \leqslant \frac{\|l(\mathbf{f};\mathbf{u}_0)\|_{V'}}{\nu - \rho(\mathbf{u}_0)}.$$

Remark 2.2. Assuming the hypotheses of Theorem 2.4, we may apply the results of Remark 1.3 to Problem (2.21). Starting from an arbitrary $\mathbf{w}^0 \in V$, the iterative scheme

$$\tilde{a}(\mathbf{w}^m;\mathbf{w}^{m+1},\mathbf{v}) - (p^{m+1},\operatorname{div}\mathbf{v}) = \langle\mathbf{f},\mathbf{v}\rangle - a(\mathbf{u}_0;\mathbf{u}_0,\mathbf{v}) \qquad \forall\mathbf{v}\in H_0^1(\Omega)^N$$

uniquely defines a sequence $(\mathbf{w}^m, p^m)$ in $V \times L_0^2(\Omega)$ which converges towards the solution $(\mathbf{w}, p)$ of Problem (2.21).

Equivalently, starting from an arbitrary function $\mathbf{u}^0 \in H^1(\Omega)^N$ such that $\operatorname{div}\mathbf{u}^0 = 0$ and $\mathbf{u}^0|_\Gamma = \mathbf{g}$, the iterative scheme

$$(2.25)\qquad \begin{cases} -\nu\Delta\mathbf{u}^{m+1} + \sum_{j=1}^{N} u_j^m(\partial\mathbf{u}^{m+1}/\partial x_j) + \mathbf{grad}\, p^{m+1} = \mathbf{f} & \text{in } \Omega,\\ \operatorname{div}\mathbf{u}^{m+1} = 0 & \text{in } \Omega,\\ \mathbf{u}^{m+1} = \mathbf{g} & \text{on } \Gamma \end{cases}$$

uniquely defines a sequence $(\mathbf{u}^m, p^m)$ in $H^1(\Omega)^N \times L_0^2(\Omega)$ such that

$$\lim_{m\to\infty} \{\|\mathbf{u}^m - \mathbf{u}\|_{1,\Omega} + \|p^m - p\|_{0,\Omega}\} = 0.$$

2.2. The Stream Function Formulation of the Homogeneous Problem

For the sake of simplicity, we restrict the discussion exclusively to the homogeneous boundary condition:

$$(2.2)\qquad \mathbf{u} = \mathbf{0} \quad \text{on } \Gamma.$$

Let us first consider the *two-dimensional case*: $N = 2$. Since $\mathbf{u} \in H_0^1(\Omega)^2$ is divergence-free, we know from Section I.3.1 that

$$\mathbf{u} = \mathbf{curl}\,\psi$$

for a unique stream function ψ in the space

$$\Psi = \{\chi \in H^2(\Omega);\, \chi|_{\Gamma_0} = 0,\, \chi|_{\Gamma_i} \text{ is constant for } 1 \leqslant i \leqslant p,\, \partial\chi/\partial n|_\Gamma = 0\}.$$

To express the nonlinear term $a_1(\mathbf{u};\mathbf{u},\mathbf{v})$ in terms of stream functions, observe that

$$a_1(\mathbf{u};\mathbf{u},\mathbf{v}) = \int_\Omega \operatorname{curl}\mathbf{u}(u_1 v_2 - u_2 v_1)\,dx \tag{2.26}$$

$$\forall \mathbf{u} = (u_1, u_2) \in H_0^1(\Omega)^2, \qquad \mathbf{v} = (v_1, v_2) \in V.$$

This result stems from the identities:

$$\begin{cases} u_1 \dfrac{\partial u_1}{\partial x_1} + u_2 \dfrac{\partial u_1}{\partial x_2} = (1/2)\dfrac{\partial}{\partial x_1}(u_1^2 + u_2^2) - u_2 \operatorname{curl}\mathbf{u}, \\[2ex] u_1 \dfrac{\partial u_2}{\partial x_1} + u_2 \dfrac{\partial u_2}{\partial x_2} = (1/2)\dfrac{\partial}{\partial x_2}(u_1^2 + u_2^2) + u_1 \operatorname{curl}\mathbf{u}, \end{cases} \tag{2.27}$$

followed by an integration by parts to eliminate $\mathbf{grad}(\|\mathbf{u}\|^2)$. Therefore the nonlinear term reads:

$$a_1(\mathbf{u};\mathbf{u},\mathbf{v}) = \int_\Omega \Delta\psi \left(\frac{\partial \psi}{\partial x_2}\frac{\partial \phi}{\partial x_1} - \frac{\partial \psi}{\partial x_1}\frac{\partial \phi}{\partial x_2}\right) dx \tag{2.28}$$

$$\forall \mathbf{u} = \mathbf{curl}\,\psi, \qquad \mathbf{v} = \mathbf{curl}\,\phi \quad \text{with } \psi, \phi \quad \text{in } \Psi.$$

In addition, we have proved in Theorem I.5.5 that

$$(\mathbf{grad}\,\mathbf{u}, \mathbf{grad}\,\mathbf{v}) = (\Delta\psi, \Delta\phi) \quad \text{with the notations of (2.28).}$$

Thus, the Navier-Stokes Problem (2.1) (2.2) has the *equivalent formulation*:

Find a function ψ in Ψ such that

$$\nu(\Delta\psi, \Delta\phi) + \int_\Omega \Delta\psi \left(\frac{\partial \psi}{\partial x_2}\frac{\partial \phi}{\partial x_1} - \frac{\partial \psi}{\partial x_1}\frac{\partial \phi}{\partial x_2}\right) dx = \langle \mathbf{f}, \mathbf{curl}\,\phi\rangle \quad \forall \phi \in \Psi. \tag{2.29}$$

Owing to this equivalence, all the existence and uniqueness results of the preceding section carry over to Problem (2.29).

It remains to interpret Problem (2.29). We easily derive that ψ satisfies the following nonlinear biharmonic equations:

$$\nu\Delta^2\psi - \frac{\partial}{\partial x_1}\left(\Delta\psi \frac{\partial \psi}{\partial x_2}\right) + \frac{\partial}{\partial x_2}\left(\Delta\psi \frac{\partial \psi}{\partial x_1}\right) = \operatorname{curl}\mathbf{f} \quad \text{in } \Omega,$$

$$\psi|_{\Gamma_0} = 0, \qquad \psi|_{\Gamma_i} = \text{a constant } c_i, \qquad 1 \leqslant i \leqslant p,$$

$$\partial\psi/\partial n|_\Gamma = 0,$$

$$\int_{\Gamma_i} (\nu\partial(\Delta\psi)/\partial n - \mathbf{f}\cdot\boldsymbol{\tau})\,ds = 0, \qquad 1 \leqslant i \leqslant p.$$

Now, we turn to the *three-dimensional case*: $N = 3$. Here, we consider only *simply-connected domains* Ω. In order to express Problem (2.1) (2.2) in terms of vector potentials, we use extensively the material of Section I.5.3. Let us take the space of vector potentials:

$$\Psi = \left\{ \boldsymbol{\phi} \in L^2(\Omega)^3;\ \operatorname{div} \boldsymbol{\phi} \in H^1(\Omega),\ \mathbf{curl}\, \boldsymbol{\phi} \in H_0^1(\Omega)^3,\ \boldsymbol{\phi} \times \mathbf{n}|_\Gamma = \mathbf{0}, \right.$$

$$\left. \int_{\Gamma_i} \boldsymbol{\phi} \cdot \mathbf{n}\, ds = 0,\ 0 \leqslant i \leqslant p \right\}$$

with the norm

$$\|\boldsymbol{\phi}\| = \{ \|\boldsymbol{\phi}\|_{0,\Omega}^2 + \|\operatorname{div} \boldsymbol{\phi}\|_{1,\Omega}^2 + \|\mathbf{curl}\, \boldsymbol{\phi}\|_{1,\Omega}^2 \}^{1/2}.$$

Recall that this norm is equivalent to $\|\Delta \boldsymbol{\phi}\|_{0,\Omega}$ (cf. Lemma I.5.2) and for the sake of brevity we denote: $|\boldsymbol{\phi}| = \|\Delta \boldsymbol{\phi}\|_{0,\Omega}$.

Since $\operatorname{div} \mathbf{u} = 0$ and Ω is simply-connected, $\mathbf{u}$ has a unique vector potential $\boldsymbol{\psi}$ in Ψ that satisfies

$$\mathbf{u} = \mathbf{curl}\, \boldsymbol{\psi}, \qquad \operatorname{div} \boldsymbol{\psi} = 0.$$

Then, taking into account the identity:

$$\text{(2.30)} \qquad \sum_{i,j=1}^{3} u_j (\partial u_i / \partial x_j) v_i = (\mathbf{curl}\, \mathbf{u} \times \mathbf{u}) \cdot \mathbf{v} + (1/2)\, \mathbf{grad}(\|\mathbf{u}\|^2) \cdot \mathbf{v},$$

we derive exactly like in Section I.5.3 that this vector $\boldsymbol{\psi}$ is a solution of the nonlinear biharmonic problem:

Find $\boldsymbol{\psi}$ in Ψ such that

$$\text{(2.31)} \quad \nu(\Delta \boldsymbol{\psi}, \Delta \boldsymbol{\phi}) - \int_\Omega (\Delta \boldsymbol{\psi} \times \mathbf{curl}\, \boldsymbol{\psi}) \cdot \mathbf{curl}\, \boldsymbol{\phi}\, dx = \langle \mathbf{f}, \mathbf{curl}\, \boldsymbol{\phi} \rangle \qquad \forall \boldsymbol{\phi} \in \Psi.$$

Therefore, it follows from Theorem 2.1 that, for each $\mathbf{f}$ in $H^{-1}(\Omega)^3$, Problem (2.31) has at least one divergence-free solution $\boldsymbol{\psi}$ in Ψ.

But conversely, we cannot establish in general that every solution $\boldsymbol{\psi}$ of (2.31) is a vector potential of a solution $\mathbf{u}$ of the Navier-Stokes Problem (2.1) (2.3). Indeed, setting

$$\mathbf{w} = \mathbf{curl}\, \boldsymbol{\psi},$$

we infer from (2.31) that $\mathbf{w}$ satisfies: $\mathbf{w} \in V$,

$$\nu(\mathbf{grad}\, \mathbf{w}, \mathbf{grad}\, \mathbf{v}) + \int_\Omega \mathbf{curl}\, \mathbf{w} \times \mathbf{w} \cdot \mathbf{v}\, dx = \langle \mathbf{f}, \mathbf{v} \rangle + \int_\Omega \mathbf{grad}(\operatorname{div} \boldsymbol{\psi}) \times \mathbf{w} \cdot \mathbf{v}\, dx \qquad \forall \mathbf{v} \in V.$$

Unless $\int_\Omega \mathbf{grad}(\operatorname{div} \boldsymbol{\psi}) \times \mathbf{w} \cdot \mathbf{v}\, dx = 0$ (and this property does not stem from (2.31)), we see that $\mathbf{w}$ does not satisfy the Navier-Stokes equations.

The trilinear form naturally attached to Problem (2.31) is

$$\text{(2.32)} \qquad a_1(\boldsymbol{\phi}; \boldsymbol{\psi}, \boldsymbol{\chi}) = - \int_\Omega (\Delta \boldsymbol{\phi} \times \mathbf{curl}\, \boldsymbol{\psi}) \cdot \mathbf{curl}\, \boldsymbol{\chi}\, dx.$$

Clearly, this form satisfies (2.4) and (2.5). In addition, like in Lemma 2.1, $a_1(.;.,.)$

is continuous on Ψ^3 by virtue of the continuity of the imbedding of $H^1(\Omega)$ into $L^4(\Omega)$ (cf. Theorem I.1.3). More precisely, we have:

$$|a_1(\boldsymbol{\phi}; \boldsymbol{\psi}, \boldsymbol{\chi})| \leqslant \|\Delta \boldsymbol{\phi}\|_{0,\Omega} \|\mathbf{curl}\, \boldsymbol{\psi}\|_{0,4,\Omega} \|\mathbf{curl}\, \boldsymbol{\chi}\|_{0,4,\Omega}.$$
$$\leqslant C_1 \|\Delta \boldsymbol{\phi}\|_{0,\Omega} \|\mathbf{curl}\, \boldsymbol{\psi}\|_{1,\Omega} \|\mathbf{curl}\, \boldsymbol{\chi}\|_{1,\Omega}.$$

Hence

$$|a_1(\boldsymbol{\phi}; \boldsymbol{\psi}, \boldsymbol{\chi})| \leqslant C_1 \|\boldsymbol{\phi}\| \, \|\boldsymbol{\psi}\| \, \|\boldsymbol{\chi}\| \qquad \forall \boldsymbol{\phi}, \boldsymbol{\psi}, \boldsymbol{\chi} \in \Psi. \tag{2.33}$$

As far as the uniqueness of the solution is concerned, the equivalence of norms:

$$\|\boldsymbol{\phi}\| \cong \|\Delta \boldsymbol{\phi}\|_{0,\Omega} = |\boldsymbol{\phi}| \qquad \forall \boldsymbol{\phi} \in \Psi$$

guarantees the ellipticity of the form

$$a(\boldsymbol{\phi}; \boldsymbol{\psi}, \boldsymbol{\chi}) = \nu(\Delta \boldsymbol{\psi}, \Delta \boldsymbol{\chi}) + a_1(\boldsymbol{\phi}; \boldsymbol{\psi}, \boldsymbol{\chi}).$$

On the other hand, setting

$$\tilde{\mathcal{N}} = \sup_{\boldsymbol{\phi}, \boldsymbol{\psi}, \boldsymbol{\chi} \in \Psi} \frac{a_1(\boldsymbol{\phi}; \boldsymbol{\psi}, \boldsymbol{\chi})}{|\boldsymbol{\phi}| |\boldsymbol{\psi}| |\boldsymbol{\chi}|}, \tag{2.34}$$

we get

$$|a_1(\boldsymbol{\phi}_1; \boldsymbol{\psi}, \boldsymbol{\chi}) - a_1(\boldsymbol{\phi}_2; \boldsymbol{\psi}, \boldsymbol{\chi})| = |a_1(\boldsymbol{\phi}_1 - \boldsymbol{\phi}_2; \boldsymbol{\psi}, \boldsymbol{\chi})|$$
$$\leqslant \tilde{\mathcal{N}} |\boldsymbol{\phi}_1 - \boldsymbol{\phi}_2| |\boldsymbol{\psi}| |\boldsymbol{\chi}|.$$

Therefore the trilinear form $a_1(.;.,.)$ satisfies (1.14) with $L(\mu) = \tilde{\mathcal{N}}$, independent of μ (the space Ψ being equipped with the norm $|.|$). Thus when ν is large enough or $\|\mathbf{f}\|_{V'}$ is sufficiently small the solution $\boldsymbol{\psi}$ of Problem (2.31) is unique. As the vector potential of every solution $\mathbf{u}$ of Problem (2.1) (2.2) is a solution of (2.31), this means that Problem (2.1) (2.2) has also a unique solution $\mathbf{u}$ and

$$\mathbf{u} = \mathbf{curl}\, \boldsymbol{\psi}$$

with $\boldsymbol{\psi}$ solution of (2.31). As a consequence $\boldsymbol{\psi}$ satisfies necessarily

$$\operatorname{div} \boldsymbol{\psi} = 0.$$

These results are summed up in the following theorem.

Theorem 2.5. *Let Ω be a bounded, simply-connected domain of $\mathbb{R}^3$ with a Lipschitz-continuous boundary Γ. Then Problem* (2.31) *has at least one solution $\boldsymbol{\psi}$ in Ψ. In addition, if*

$$\tilde{\mathcal{N}}(1/\nu^2) \sup_{\boldsymbol{\phi} \in \Psi} \left\{ \frac{\langle \mathbf{f}, \mathbf{curl}\, \boldsymbol{\phi} \rangle}{|\boldsymbol{\phi}|} \right\} < 1 \tag{2.35}$$

then the solution $\boldsymbol{\psi}$ is unique, there exists p in $L_0^2(\Omega)$ such that $(\mathbf{u} = \mathbf{curl}\, \boldsymbol{\psi}, p)$ *is the unique solution of Problem* (2.1) (2.2) *and*

$$\operatorname{div} \boldsymbol{\psi} = 0 \quad \textit{in } \Omega.$$

Remark 2.3. The same analysis with analogous conclusions can be carried out with vector potentials in the space:

$$\Psi_1 = \{\boldsymbol{\phi} \in L^2(\Omega)^3;\ \operatorname{div} \boldsymbol{\phi} \in H^1(\Omega),\ \mathbf{curl}\, \boldsymbol{\phi} \in H_0^1(\Omega)^3,\ \boldsymbol{\phi} \cdot \mathbf{n}|_\Gamma = \mathbf{0}\}.$$

§ 3. Approximation of Branches of Nonsingular Solutions

As proved in the preceding paragraph, the Navier-Stokes equations have in general more than one solution, unless the data (namely, the viscosity and external forces) satisfy very stringent requirements. However, it can also be shown that in many practical examples these solutions are mostly isolated, i.e. there exists a neighborhood in which each solution is unique. Furthermore, it can be established that the solutions depend continuously on the viscosity. Thus, as the viscosity varies along an interval, each solution of the Navier-Stokes equations describes an isolated branch. In particular, this means that the bifurcation phenomenon is rare. This situation, very frequently encountered in practice, is expressed mathematically by the notion of branches of nonsingular solutions.

This paragraph proposes and analyzes several approximations of branches of nonsingular solutions pertaining to a wide class of nonlinear problems, including the Navier-Stokes problem. The analysis, based on a general form of the implicit function theorem, is a variant of a broader theory developed by Brezzi, Rappaz & Raviart [14]. The version that we present here is due to Crouzeix [22].

3.1. An Abstract Framework

Let X and $\mathscr{X}$ be two Banach spaces and Λ a compact interval of the real line $\mathbb{R}$. We are given a $\mathscr{C}^p$-mapping $(p \geq 1)$

$$F: (\lambda, u) \in \Lambda \times X \to F(\lambda, u) \in \mathscr{X}$$

and we want to solve the equation

$$F(\lambda, u) = 0, \tag{3.1}$$

i.e. we want to find pairs $(\lambda, u) \in \Lambda \times X$ solutions of (3.1).

Let $\{(\lambda, u(\lambda)); \lambda \in \Lambda\}$ be a branch of solutions of equation (3.1). This means that

$$\lambda \to u(\lambda) \text{ is a continuous function from } \Lambda \text{ into } X; \tag{3.2}$$

$$F(\lambda, u(\lambda)) = 0. \tag{3.3}$$

Moreover, we suppose that these solutions are *nonsingular* in the sense that:

$$D_u F(\lambda, u(\lambda)) \text{ is an isomorphism from } X \text{ onto } \mathscr{X} \text{ for all } \lambda \in \Lambda. \tag{3.4}$$

As an immediate consequence of (3.4), it follows from the implicit function theorem that $\lambda \to u(\lambda)$ is a $\mathscr{C}^p$-function from Λ into X.

As our fundamental example, let us show that the Dirichlet problem for the Navier-Stokes equations in the velocity-pressure formulation (2.20) fits into the above framework. We first set:

$$X = \mathscr{X} = H^1(\Omega)^N \times L_0^2(\Omega), \tag{3.5}$$

and we introduce the intermediate space

$$Y = H^{-1}(\Omega)^N \times \left\{\mathbf{g} \in H^{1/2}(\Gamma)^N;\ \int_{\Gamma_i} \mathbf{g}\cdot\mathbf{n}\,ds = 0,\ 0 \leqslant i \leqslant p\right\}. \tag{3.6}$$

Next we define a linear operator T as follows: given $(\mathbf{f}_*, \mathbf{g}_*) \in Y$, we denote by $(\mathbf{u}_*, p_*) = T(\mathbf{f}_*, \mathbf{g}_*) \in X$ the solution of the Dirichlet problem for the Stokes equations:

$$\begin{cases} \left.\begin{aligned} -\Delta \mathbf{u}_* + \mathbf{grad}\, p_* &= \mathbf{f}_* \\ \operatorname{div} \mathbf{u}_* &= 0 \end{aligned}\right\} \text{ in } \Omega, \\ \mathbf{u}_*|_\Gamma = \mathbf{g}_*. \end{cases} \tag{3.7}$$

Finally, with the data $(\mathbf{f}, \mathbf{g}) \in Y$ we associate a $\mathscr{C}^\infty$-mapping G from $\mathbb{R}_+ \times X$ into Y defined by

$$G\colon (\lambda, v = (\mathbf{v}, q)) \to G(\lambda, v) = \left(\lambda\left(\sum_{j=1}^N v_j \delta\mathbf{v}/\partial x_j - \mathbf{f}\right), -\mathbf{g}\right) \tag{3.8}$$

and we set

$$F(\lambda, v) = v + TG(\lambda, v). \tag{3.9}$$

Now we may state:

Lemma 3.1. *The pair* $(\mathbf{u}, p) \in H^1(\Omega)^N \times L_0^2(\Omega)$ *is a solution of Problem* (2.1) (2.13) *if and only if* $u = (\mathbf{u}, p/v)$ *is a solution of* (3.1) *where* $\lambda = 1/v$, *the spaces* X *and* $\mathscr{X}$ *are defined by* (3.5) *and the compound mapping* F *is defined by* (3.9).

Proof. We observe that the equations (2.1) may be equivalently written in the form

$$-\Delta \mathbf{u} + \mathbf{grad}(p/v) = (1/v)\left(\mathbf{f} - \sum_{j=1}^N u_j \partial\mathbf{u}/\partial x_j\right) \quad \text{in } \Omega,$$

$$\operatorname{div} \mathbf{u} = 0 \quad \text{in } \Omega,$$

$$\mathbf{u} = \mathbf{g} \quad \text{on } \Gamma;$$

or by (3.7)

$$(\mathbf{u}, p/v) = T\left[(1/v)\left(\mathbf{f} - \sum_{j=1}^N u_j \partial\mathbf{u}/\partial x_j\right), \mathbf{g}\right].$$

Hence the lemma is established using (3.8) and (3.9). □

Next, we note that for all $v = (\mathbf{v}, q) \in X$

$$D_u G(\lambda, u) \cdot v = \left(\lambda \sum_{j=1}^{N} (u_j \partial \mathbf{v}/\partial x_j + v_j \partial \mathbf{u}/\partial x_j), \mathbf{0} \right).$$

Hence in our example, $u = (\mathbf{u}, p/v)$ is a nonsingular solution of (3.1), or equivalently $D_u F(1/v, u)$ is an isomorphism of X if and only if, for each $w = (\mathbf{w}, \sigma) \in X$ there exists a unique $v = (\mathbf{v}, q)$ in X such that

$$-\Delta \mathbf{v} + \mathbf{grad}\, q + (1/v) \sum_{j=1}^{N} (u_j \partial \mathbf{v}/\partial x_j + v_j \partial \mathbf{u}/\partial x_j) = -\Delta \mathbf{w} + \mathbf{grad}\, \sigma \quad \text{in } \Omega,$$

$$\operatorname{div} \mathbf{v} = \operatorname{div} \mathbf{w}$$

$$\mathbf{v} = \mathbf{w} \quad \text{on } \Gamma.$$

Now, this problem can be simplified because on the one hand, an obvious extension of Theorem I.5.4, proved in Cattabriga [18], guarantees that for every $\mathbf{f} \in H^{-1}(\Omega)^N$, $\mu \in L^2(\Omega)$ and $\mathbf{g} \in H^{1/2}(\Gamma)^N$ fulfilling the compatibility condition

$$\int_\Omega \mu \, dx = \int_\Gamma \mathbf{g} \cdot \mathbf{n} \, ds$$

there exists a unique $w = (\mathbf{w}, \sigma)$ in X solution of

$$\left. \begin{aligned} -\Delta \mathbf{w} + \mathbf{grad}\, \sigma &= \mathbf{f} \\ \operatorname{div} \mathbf{w} &= \mu \end{aligned} \right\} \quad \text{in } \Omega,$$

$$\mathbf{w} = \mathbf{g} \quad \text{on } \Gamma.$$

On the other hand, when μ and $\mathbf{g}$ satisfy the above conditions an extension of Lemma I.2.2 shows that there exists $\mathbf{v}_0$ in $H^1(\Omega)^N$ such that

$$\operatorname{div} \mathbf{v}_0 = \mu, \qquad \mathbf{v}_0|_\Gamma = \mathbf{g}.$$

Therefore, applying these two results we can prove that $u = (\mathbf{u}, p/v)$ is a nonsingular solution of (3.1) if and only if the homogeneous linearized problem

$$(3.10) \qquad \left. \begin{aligned} -v\Delta \mathbf{v} + \sum_{j=1}^{N} (u_j \partial \mathbf{v}/\partial x_j + v_j \partial \mathbf{u}/\partial x_j) + \mathbf{grad}\, q &= \mathbf{f}_* \\ \operatorname{div} \mathbf{v} &= 0 \end{aligned} \right\} \quad \text{in } \Omega,$$

$$\mathbf{v} = \mathbf{0} \quad \text{on } \Gamma$$

has a unique solution $v = (\mathbf{v}, q) \in X$ for each $\mathbf{f}_* \in H^{-1}(\Omega)^N$, the mapping $\mathbf{f}_* \to v$ being continuous from $H^{-1}(\Omega)^N$ into X.

Although we do not want to give here a precise statement, it can be proved that the solutions of the Dirichlet problem for the Navier-Stokes equations are "in general" nonsingular. We shall only derive the following simple result.

Lemma 3.2. *Assume the hypotheses of Theorem* 2.4. *Then, if* $(\mathbf{u}, p)$ *is the unique solution of Problem* (2.20), $u = (\mathbf{u}, p/\nu)$ *is a nonsingular solution of* (3.1).

Proof. First, using the notations of Paragraph 2.1, we note that a variational form of Problem (3.10) consists in finding a pair $(\mathbf{v}, q) \in H^1(\Omega)^N \times L_0^2(\Omega)$ solution of

$$(3.11) \qquad \begin{cases} a(\mathbf{u}; \mathbf{v}, \mathbf{z}) + a_1(\mathbf{v}; \mathbf{u}, \mathbf{z}) - (q, \operatorname{div} \mathbf{z}) = \langle \mathbf{f}_*, \mathbf{z} \rangle & \forall \mathbf{z} \in H_0^1(\Omega)^N, \\ \operatorname{div} \mathbf{v} = 0 \quad \text{in } \Omega, \\ \mathbf{v} = \mathbf{g}_* \quad \text{on } \Gamma. \end{cases}$$

In order to check that the equations (3.11) have a unique solution, it is sufficient to prove that the bilinear form

$$(\mathbf{v}, \mathbf{z}) \to c(\mathbf{v}, \mathbf{z}) = a(\mathbf{u}; \mathbf{v}, \mathbf{z}) + a_1(\mathbf{v}; \mathbf{u}, \mathbf{z})$$

is V-elliptic. But it follows from (2.5) that

$$c(\mathbf{v}, \mathbf{v}) = \nu |\mathbf{v}|_{1,\Omega}^2 + a_1(\mathbf{v}; \mathbf{u}, \mathbf{v}) \qquad \forall \mathbf{v} \in V.$$

Now, assume that $\nu > \nu_0$ where ν_0 is defined by (2.24). Then, there exists a function $\mathbf{u}_0 \in H^1(\Omega)^N$ such that

$$\operatorname{div} \mathbf{u}_0 = 0, \qquad \mathbf{u}_0|_\Gamma = \mathbf{g},$$

$$\nu > \rho(\mathbf{u}_0) + (\mathcal{N} \| l(\mathbf{f}; \mathbf{u}_0) \|_{V'})^{1/2}.$$

Next, setting $\mathbf{u} = \mathbf{u}_0 + \mathbf{w}$, we have by (2.10) and (2.22)

$$|a_1(\mathbf{v}; \mathbf{u}, \mathbf{v})| \leqslant |a_1(\mathbf{v}; \mathbf{u}_0, \mathbf{v})| + |a_1(\mathbf{v}; \mathbf{w}, \mathbf{v})|$$
$$\leqslant (\rho(\mathbf{u}_0) + \mathcal{N} |\mathbf{w}|_{1,\Omega}) |\mathbf{v}|_{1,\Omega}^2.$$

Since (cf. Remark 2.1)

$$|\mathbf{w}|_{1,\Omega} \leqslant \frac{\| l(\mathbf{f}; \mathbf{u}_0) \|_{V'}}{\nu - \rho(\mathbf{u}_0)},$$

we obtain

$$c(\mathbf{v}, \mathbf{v}) \geqslant \left(\nu - \rho(\mathbf{u}_0) - \frac{\mathcal{N} \| l(\mathbf{f}; \mathbf{u}_0) \|_{V'}}{\nu - \rho(\mathbf{u}_0)} \right) |\mathbf{v}|_{1,\Omega}^2$$
$$= (\nu - \rho(\mathbf{u}_0)) \left(1 - \frac{\mathcal{N} \| l(\mathbf{f}; \mathbf{u}_0) \|_{V'}}{(\nu - \rho(\mathbf{u}_0))^2} \right) |\mathbf{v}|_{1,\Omega}^2,$$

so that the ellipticity property holds. □

In the subsequent paragraphs, we shall be essentially concerned with the approximation of branches of nonsingular solutions $\{(\lambda, u(\lambda) = (\mathbf{u}(\lambda), \lambda p(\lambda))); \lambda \in \Lambda\}$ of the Dirichlet problem for the Navier-Stokes equations where the parameter $\lambda = 1/\nu$ plays the role of the Reynolds number.

3.2. Approximation of Branches of Nonsingular Solutions

Let us go back to the general abstract problem (3.1) in order to introduce the method of approximation. For each value of a real parameter $h > 0$ which will tend to zero, we are given a $\mathscr{C}^p$-mapping F_h, presumably an approximation of F, defined on $\Lambda \times X$ with values in $\mathscr{X}$. The problem now is to find pairs $(\lambda, u_h) \in \Lambda \times X$, solutions of

$$F_h(\lambda, u_h) = 0. \tag{3.12}$$

Let us assume that $\{(\lambda, u(\lambda)); \lambda \in \Lambda\}$ is a branch of *nonsingular solutions of* (3.1). We want to find sufficient conditions ensuring the existence and uniqueness of a branch $\{(\lambda, u_h(\lambda)); \lambda \in \Lambda\}$ of solutions of (3.12) in a suitable neighborhood of the branch of solutions of (3.1).

In a first stage, we fix λ in Λ and we propose to approximate the solution of (3.1) by the solution $u_h(\lambda)$ of (3.12). Let $\tilde{u}_h$ $(= \tilde{u}_h(\lambda))$ be an arbitrary element of X and let us investigate under what conditions the mapping $D_u F_h(\lambda, \tilde{u}_h)$ is invertible. To this end, we introduce the two quantities:

$$\gamma(\lambda) = \| \{D_u F(\lambda, u(\lambda))\}^{-1} \|_{\mathscr{L}(\mathscr{X}; X)}, \tag{3.13}$$

$$\mu_h(\lambda) = \| D_u F(\lambda, u(\lambda)) - D_u F_h(\lambda, \tilde{u}_h) \|_{\mathscr{L}(X; \mathscr{X})}. \tag{3.14}$$

Lemma 3.3. *Under the condition*

$$\gamma(\lambda)\mu_h(\lambda) < 1, \tag{3.15}$$

the mapping $D_u F_h(\lambda, \tilde{u}_h)$ is an isomorphism from X onto $\mathscr{X}$.

Proof. We set

$$B = \{D_u F(\lambda, u(\lambda))\}^{-1} \{D_u F(\lambda, u(\lambda)) - D_u F_h(\lambda, \tilde{u}_h)\}.$$

Then

$$D_u F_h(\lambda, \tilde{u}_h) = D_u F(\lambda, u(\lambda)) \{I - B\}.$$

But in view of (3.15), we have

$$\| B \|_{\mathscr{L}(X; X)} \leqslant \gamma(\lambda)\mu_h(\lambda) < 1;$$

therefore $I - B$ is an isomorphism of X and

$$\| (I - B)^{-1} \|_{\mathscr{L}(X; X)} \leqslant 1/(1 - \gamma(\lambda)\mu_h(\lambda)).$$

As a consequence $D_u F_h(\lambda, \tilde{u}_h)$ is also an isomorphism from X onto $\mathscr{X}$ and

$$\| \{D_u F_h(\lambda, \tilde{u}_h)\}^{-1} \|_{\mathscr{L}(\mathscr{X}; X)} \leqslant \gamma(\lambda)/(1 - \gamma(\lambda)\mu_h(\lambda)). \tag{3.16}$$ □

Remark 3.1. The above lemma relies on the fact that both F and F_h map $\Lambda \times X$ into $\mathscr{X}$. This hypothesis makes the forthcoming theory simpler than when F_h and

F are not defined on the same space, but it is not a fundamental tool in the fixed-point argument used below. The reader will find in Section 3.4 an example of a situation where F and F_h are not defined in the same space.

Now, we assume that $D_u F_h(\lambda, \tilde{u}_h)$ is an isomorphism from X onto $\mathscr{X}$ and we introduce the following notations:

$$(3.17)\quad \begin{cases} \varepsilon_h(\lambda) = \|F_h(\lambda, \tilde{u}_h)\|_{\mathscr{X}}, \\ \gamma_h(\lambda) = \|\{D_u F_h(\lambda, \tilde{u}_h)\}^{-1}\|_{\mathscr{L}(\mathscr{X};X)}, \\ S(u;\alpha) = \{v \in X;\ \|v - u\|_X \leqslant \alpha\}, \\ L_h(\lambda;\alpha) = \sup\limits_{v \in S(\tilde{u}_h;\alpha)} \|D_u F_h(\lambda, \tilde{u}_h) - D_u F_h(\lambda, v)\|_{\mathscr{L}(X;\mathscr{X})}. \end{cases}$$

Clearly, the function $\alpha \to L_h(\lambda;\alpha)$: $\mathbb{R}_+ \to \mathbb{R}_+$ is monotonically increasing.

The next theorem solves Problem (3.12) and gives a fundamental error estimate. In addition, it shows that Problem (3.12) has no other solution in a suitable neighborhood of $\tilde{u}_h$.

Theorem 3.1. *Under the following assumptions*:

$$(3.18)\qquad D_u F_h(\lambda, \tilde{u}_h) \text{ is an isomorphism of } X \text{ onto } \mathscr{X},$$

$$(3.19)\qquad 2\gamma_h(\lambda) L_h(\lambda; 2\gamma_h(\lambda)\varepsilon_h(\lambda)) < 1,$$

then Problem (3.12) *has a solution* (λ, u_h) *with*

$$(3.20)\qquad u_h \in S(\tilde{u}_h; 2\gamma_h(\lambda)\varepsilon_h(\lambda)),$$

$D_u F_h(\lambda, u_h)$ *is an isomorphism of* X *onto* $\mathscr{X}$ *and*

$$(3.21)\qquad \|\{D_u F_h(\lambda, u_h)\}^{-1}\|_{\mathscr{L}(\mathscr{X};X)} \leqslant 2\gamma_h(\lambda).$$

Furthermore, we have the following uniqueness result: u_h *is the only solution in every ball* $S(\tilde{u}_h;\alpha)$ *whose radius* α *satisfies*:

$$(3.22)\qquad \gamma_h(\lambda) L_h(\lambda;\alpha) < 1$$

and we have the estimate:

$$(3.23)\quad \|u_h - v\|_X \leqslant [\gamma_h(\lambda)/(1 - \gamma_h(\lambda)L_h(\lambda;\alpha))]\,\|F_h(\lambda, v)\|_{\mathscr{X}} \qquad \forall v \in S_h(u_h;\alpha).$$

Proof. Let us define the function

$$\Phi_h(v) = v - [D_u F_h(\lambda, \tilde{u}_h)]^{-1} F_h(\lambda, v) \qquad \forall v \in X.$$

Clearly, the pair (λ, u_h) is a solution of (3.12) if and only if u_h is a fixed point of Φ_h. First, we are going to establish that Φ_h is a strict contraction of $S = S(\tilde{u}_h; 2\gamma_h(\lambda)\varepsilon_h(\lambda))$ into itself.

1°) Let $v \in S$; then

$$\Phi_h(v) - \tilde{u}_h = [D_u F_h(\lambda, \tilde{u}_h)]^{-1}[D_u F_h(\lambda, \tilde{u}_h)\cdot(v - \tilde{u}_h) - (F_h(\lambda, v) - F_h(\lambda, \tilde{u}_h)) - F_h(\lambda, \tilde{u}_h)].$$

But

$$F_h(\lambda, v) - F_h(\lambda, \tilde{u}_h) = \int_0^1 D_u F_h(\lambda, \tilde{u}_h + \theta(v - \tilde{u}_h))\cdot(v - \tilde{u}_h)\,d\theta.$$

Thus

$$\begin{aligned}&\| D_u F_h(\lambda, \tilde{u}_h)\cdot(v - \tilde{u}_h) - (F_h(\lambda, v) - F_h(\lambda, \tilde{u}_h))\| \\ &= \left\| \int_0^1 \{D_u F_h(\lambda, \tilde{u}_h) - D_u F_h(\lambda, \tilde{u}_h + \theta(v - \tilde{u}_h))\}\cdot(v - \tilde{u}_h)\,d\theta \right\| \\ &\leqslant L_h(\lambda; 2\gamma_h(\lambda)\varepsilon_h(\lambda))2\gamma_h(\lambda)\varepsilon_h(\lambda)\end{aligned}$$

because $v \in S$ (cf. (3.17)). Hence:

$$\| \Phi_h(v) - \tilde{u}_h \|_X \leqslant \gamma_h(\lambda)\varepsilon_h(\lambda)\{L_h(\lambda; 2\gamma_h(\lambda)\varepsilon_h(\lambda))2\gamma_h(\lambda) + 1\} < 2\gamma_h(\lambda)\varepsilon_h(\lambda),$$

by virtue of (3.19). Therefore $\Phi_h(v)$ belongs to S.

2°) Let v and w belong to S. Like above, we can write:

$$\begin{aligned}\Phi_h(v) - \Phi_h(w) &= [D_u F_h(\lambda, \tilde{u}_h)]^{-1}[D_u F_h(\lambda, \tilde{u}_h)\cdot(v - w) - (F_h(\lambda, v) - F_h(\lambda, w))] \\ &= [D_u F_h(\lambda, \tilde{u}_h)]^{-1} \int_0^1 \{D_u F_h(\lambda, \tilde{u}_h) \\ &\qquad - D_u F_h(\lambda, w + \theta(v - w))\}\cdot(v - w)\,d\theta.\end{aligned}$$

Again (3.17) and (3.19) yield:

$$\| \Phi_h(v) - \Phi_h(w) \|_X \leqslant \gamma_h(\lambda) L_h(\lambda; 2\gamma_h(\lambda)\varepsilon_h(\lambda)) \| v - w \|_X < (1/2) \| v - w \|_X.$$

Hence the mapping Φ_h is a strict contraction of S into itself. As a consequence, Φ_h has a unique fixed point in S.

Let $u_h = u_h(\lambda)$ denote the fixed point of Φ_h in S. It follows from the above considerations that the pair (λ, u_h) is the (unique) solution of (3.12) in S. In addition, observe that

$$\begin{aligned}\| D_u F_h(\lambda, u_h) - D_u F_h(\lambda, \tilde{u}_h) \|_{\mathscr{L}(X; \mathscr{X})} &\leqslant L_h(\lambda; 2\gamma_h(\lambda)\varepsilon_h(\lambda)) \\ &< 1/(2\gamma_h(\lambda)) \quad \text{owing to (3.19).}\end{aligned}$$

Thus the assumption (3.18) and Lemma 3.3 imply that $D_u F_h(\lambda, u_h)$ is also an isomorphism of X onto $\mathscr{X}$ and (3.21) follows from (3.16).

Next, let us prove that Φ_h has no other fixed point than u_h in the larger ball $S(\tilde{u}_h; \alpha)$ with α prescribed by (3.22). Indeed, let v be another fixed point of Φ_h in $S(\tilde{u}_h; \alpha)$. The argument of part 2°) gives:

$$\| v - u_h \|_X = \| \Phi_h(v) - \Phi_h(u_h) \|_X \leqslant \gamma_h(\lambda) L_h(\lambda; \alpha) \| v - u_h \|_X.$$

Thus $v = u_h$ when α satisfies (3.22).

Finally, if v is an arbitrary element of $S(\tilde{u}_h; \alpha)$, we can write

$$
\begin{aligned}
v - u_h &= [D_u F_h(\lambda, \tilde{u}_h)]^{-1}[D_u F_h(\lambda, \tilde{u}_h)\cdot(v - u_h) \\
&\quad - (F_h(\lambda, v) - F_h(\lambda, u_h)) + F_h(\lambda, v)] \\
&= [D_u F_h(\lambda, \tilde{u}_h)]^{-1}\left[\int_0^1 \{D_u F_h(\lambda, \tilde{u}_h)\right. \\
&\quad \left. - D_u F_h(\lambda, u_h + \theta(v - u_h))\}\cdot(v - u_h)\,d\theta + F_h(\lambda, v)\right]. \qquad (3.24)
\end{aligned}
$$

Hence

$$\|v - u_h\|_X \leqslant \gamma_h(\lambda)\{L_h(\lambda; \alpha)\,\|v - u_h\|_X + \|F_h(\lambda, v)\|_{\mathscr{X}}\},$$

thus proving (3.23). □

Now we allow λ to vary in the compact interval Λ. We replace the fixed element $\tilde{u}_h$ by a $\mathscr{C}^0$-mapping $\lambda \to \tilde{u}_h(\lambda)$: $\Lambda \to X$. The following lemma generalizes Lemma 3.3.

Lemma 3.4. *Under the condition*:

$$\lim_{h\to 0}\sup_{\lambda\in\Lambda} \mu_h(\lambda) = 0, \qquad (3.25)$$

there exists a real $h_0 > 0$ such that for all $\lambda \in \Lambda$ and all $h \leqslant h_0$, $D_u F_h(\lambda, \tilde{u}_h(\lambda))$ is an isomorphism of X onto $\mathscr{X}$ and we have the bound:

$$\gamma_h(\lambda) = \|[D_u F_h(\lambda, \tilde{u}_h(\lambda))]^{-1}\|_{\mathscr{L}(\mathscr{X}; X)} \leqslant 2\gamma(\lambda). \qquad (3.26)$$

Proof. Since Λ is compact, the function $\gamma(\lambda)$ is bounded above on Λ. Therefore (3.25) implies that for some $h_0 > 0$, we have

$$\sup_{\lambda\in\Lambda}\{\gamma(\lambda)\mu_h(\lambda)\} \leqslant 1/2 \qquad \forall h \leqslant h_0.$$

Hence Lemma 3.3 asserts that $D_u F_h(\lambda, \tilde{u}_h(\lambda))$ is an isomorphism of X onto $\mathscr{X}$ for all $\lambda \in \Lambda$ and all $h \leqslant h_0$. Moreover the bound (3.16) yields directly (3.26). □

Combined with Theorem 3.1, this lemma shows that Problem (3.12) has a unique branch of nonsingular solutions in a neighborhood of $\tilde{u}_h$.

Theorem 3.2. *Let $\{(\lambda, u(\lambda)); \lambda \in \Lambda\}$ be a branch of nonsingular solutions of* (3.1) *and let $\lambda \to \tilde{u}_h(\lambda)$ be a given function in $\mathscr{C}^0(\Lambda; X)$ that satisfies*

$$\lim_{h\to 0}\sup_{\lambda\in\Lambda} \mu_h(\lambda) = 0.$$

If, besides that,

$$\lim_{h \to 0} \sup_{\lambda \in \Lambda} \varepsilon_h(\lambda) = 0 \tag{3.27}$$

and

$$\lim_{\alpha \to 0} \left\{ \sup_{\lambda \in \Lambda} L_h(\lambda; \alpha) \right\} = 0 \quad \textit{uniformly for all } h \leqslant h_0, \tag{3.28}$$

where h_0 is the parameter of Lemma 3.4, *then there exist two real constants $\alpha > 0$ and $h_1 > 0$ and a function $u_h \in \mathscr{C}^0(\Lambda; X)$ such that for all $h \leqslant h_1$:*

$$\{(\lambda, u_h(\lambda)); \lambda \in \Lambda\} \quad \textit{is a branch of nonsingular solutions of } (3.12), \tag{3.29}$$

for each $\lambda \in \Lambda$, $u_h(\lambda)$ is the only solution of (3.12) *in the ball $S(\tilde{u}_h(\lambda); \alpha)$ and we have the bound*:

$$\| u_h(\lambda) - v \|_X \leqslant 4\gamma(\lambda) \| F_h(\lambda, v) \|_{\mathscr{X}} \qquad \forall v \in S(\tilde{u}_h(\lambda); \alpha). \tag{3.30}$$

Proof. According to Lemma 3.4, we know that $\gamma_h(\lambda)$ is bounded above for all λ in Λ and for all $h \leqslant h_0$. Therefore (3.28) implies that we can find an $\alpha > 0$ such that

$$\gamma_h(\lambda) L_h(\lambda; \alpha) < 1/2 \qquad \forall \lambda \in \Lambda, \qquad \forall h \leqslant h_0. \tag{3.31}$$

Likewise, (3.27) implies that there exists an $h_0' > 0$ such that

$$2\gamma_h(\lambda) L_h(\lambda; 2\gamma_h(\lambda)\varepsilon_h(\lambda)) < 1 \qquad \forall \lambda \in \Lambda, \qquad \forall h \leqslant h_0'.$$

Take $h_1 = \min(h_0, h_0')$. Then it follows from Theorem 3.1 that for all $\lambda \in \Lambda$ and all $h \leqslant h_1$, there exists a $u_h(\lambda)$ in X, unique in $S(\tilde{u}_h(\lambda); \alpha)$, such that:

$$F_h(\lambda, u_h(\lambda)) = 0$$

and $D_u F_h(\lambda, u_h(\lambda))$ is an isomorphism of X onto $\mathscr{X}$. In addition Lemma 3.4 and (3.21) yield:

$$\| [D_u F_h(\lambda, u_h(\lambda))]^{-1} \|_{\mathscr{L}(\mathscr{X}; X)} \leqslant 4\gamma(\lambda). \tag{3.32}$$

Similarly, (3.23), (3.26) and (3.31) give (3.30).

Finally, the continuity of u_h with respect to λ is an easy consequence of (3.30), the continuity of $\tilde{u}_h$ and the continuity of F_h. □

Remark 3.2. An interesting feature of this proof is that the function $\tilde{u}_h$ is arbitrary. The most obvious choice of function in the present case where F and F_h are defined on the same space X is of course $\tilde{u}_h(\lambda) = u(\lambda)$, and this will be used below. When F_h is defined on a space X_h different from X, like in Section 3.4, we shall take for $\tilde{u}_h$ an adequate approximation of u.

Remark 3.3. Theorem 3.2 does not use fully the regularity of F and F_h with respect to λ. When $\lambda \to \tilde{u}_h(\lambda)$ is a $\mathscr{C}^p$-mapping from Λ into X, it is established in Crouzeix [22] that u_h is also in $\mathscr{C}^p(\Lambda; X)$.

3.3. Application to a Class of Nonlinear Problems

Let us apply the preceding theoretical approach to solve the following class of problems:

$$F(\lambda, u) \equiv u + TG(\lambda, u) = 0, \tag{3.33}$$

where $T \in \mathscr{L}(Y; X)$, G is a $\mathscr{C}^2$-mapping from $\Lambda \times X$ into Y, X and Y are two Banach spaces and Λ is a compact interval of $\mathbb{R}$. As a consequence, here we have $\mathscr{X} = X$. We have seen in Section 3.1 that the Dirichlet problem for the Navier-Stokes equations can be put into the form (3.33).

To approximate Problem (3.33) we introduce an operator $T_h \in \mathscr{L}(Y; X)$ intended to approximate T and we set:

$$F_h(\lambda, u) = u + T_h G(\lambda, u). \tag{3.34}$$

The approximate problem reads:

Find $u_h \in X$ *such that*

$$F_h(\lambda, u_h) = 0. \tag{3.35}$$

Now, suppose that Problem (3.33) has a branch of nonsingular solutions $\{(\lambda, u(\lambda)); \lambda \in \Lambda\}$. In order to apply Theorem 3.2 we need to make additional assumptions. First, we suppose there exists another Banach space Z contained in Y, with continuous imbedding, such that

$$D_u G(\lambda, u) \in \mathscr{L}(X; Z) \quad \forall \lambda \in \Lambda, \quad \forall u \in X. \tag{3.36}$$

Next, concerning the approximation properties of the operator T_h, we assume that:

$$\lim_{h \to 0} \|(T_h - T)g\|_X = 0 \quad \forall g \in Y \tag{3.37}$$

and

$$\lim_{h \to 0} \|T_h - T\|_{\mathscr{L}(Z; X)} = 0. \tag{3.38}$$

Remark 3.4. When the range of the operator T_h is a *finite-dimensional* subspace X_h of X (which is the case in finite element approximation), the assumption (3.38) implies that the operator T is compact from Z into X as the uniform limit of the sequence of compact operators T_h. Hence, in that case, the operator $TD_u G(\lambda, u) \in \mathscr{L}(X; X)$ is *compact* and

$$D_u F(\lambda, u) = I + TD_u G(\lambda, u)$$

is a *compact perturbation of the identity.*

Note also that (3.38) is a consequence of (3.37) when the imbedding of Z into Y is compact.

Theorem 3.3. *Assume that G is a $\mathscr{C}^2$-mapping from $\Lambda \times X$ into Y and the mapping $D^2 G$ is bounded on all bounded subsets of $\Lambda \times X$. Assume in addition that the conditions* (3.36), (3.37) *and* (3.38) *hold and that $\{(\lambda, u(\lambda)); \lambda \in \Lambda\}$ is a branch of nonsingular solutions of* (3.33). *Then there exists a neighborhood $\mathcal{O}$ of the origin in X and for $h \leqslant h_0$ small enough a unique $\mathscr{C}^2$-function $\lambda \in \Lambda \to u_h(\lambda) \in X$ such that:*

(3.39) $\{(\lambda, u_h(\lambda)); \lambda \in \Lambda\}$ *is a branch of nonsingular solutions of* (3.35),

(3.40) $$u_h(\lambda) - u(\lambda) \in \mathcal{O} \quad \text{for all } \lambda \in \Lambda.$$

Furthermore, there exists a constant $K > 0$ independent of h and λ with:

(3.41) $$\| u_h(\lambda) - u(\lambda) \|_X \leqslant K \| (T_h - T) G(\lambda, u(\lambda)) \|_X \qquad \forall \lambda \in \Lambda.$$

Proof. We are going to apply Theorem 3.2 with $\tilde{u}_h(\lambda) = u(\lambda)$. Since by assumption the mapping $(\lambda, u) \to G(\lambda, u)$ is $\mathscr{C}^2$, then $(\lambda, u) \to F(\lambda, u)$ is also $\mathscr{C}^2$ and so is $\lambda \to u(\lambda)$.

Let us check the condition (3.25). We have

$$\mu_h(\lambda) = \| (T - T_h) D_u G(\lambda, u(\lambda)) \|_{\mathscr{L}(X;X)}.$$

Then (3.36), (3.38) and the continuity of the mapping $\lambda \to D_u G(\lambda, u(\lambda))$ yield (3.25). Next, we turn to (3.27). Since

$$F(\lambda, u(\lambda)) = 0$$

we can write:

$$\begin{aligned} \varepsilon_h(\lambda) &= \| F_h(\lambda, u(\lambda)) - F(\lambda, u(\lambda)) \|_X \\ &= \| (T_h - T) G(\lambda, u(\lambda)) \|_X. \end{aligned}$$

Hence the continuity of the mapping $\lambda \to G(\lambda, u(\lambda))$ and (3.37) imply (3.27). Finally, consider (3.28). Using (3.27) and the uniform-boundedness theorem we obtain

$$\| T_h \|_{\mathscr{L}(Y;X)} \leqslant C.$$

Therefore,

$$L_h(\lambda; \alpha) \leqslant C \sup_{v \in S(u(\lambda); \alpha)} \| D_u G(\lambda, u(\lambda)) - D_u G(\lambda, v) \|_{\mathscr{L}(X;Y)}.$$

Thus, by the mean-value theorem:

$$L_h(\lambda; \alpha) \leqslant \alpha C L(\alpha)$$

where
$$L(\alpha) = \sup_{\lambda \in \Lambda,\, v \in S(u(\lambda); \alpha)} \| D^2_{uu} G(\lambda, v) \|_{\mathscr{L}_2(X;Y)}.$$

As $D^2 G$ is bounded on all bounded subsets of $\Lambda \times X$, we immediately derive (3.28).

Therefore, the conclusion of Theorem 3.2 holds, (3.41) follows readily from

(3.30), and the $\mathscr{C}^2$-regularity of u_h is a consequence of the regularity of G (cf. Remark 3.3). □

Remark 3.5. The conclusion of Theorem 3.3 (apart from the $\mathscr{C}^2$-regularity of u_h) can be obtained by replacing the $\mathscr{C}^2$-regularity of G by the Lipschitz-continuity of $D_u G$:

there exists a function $\mu \to L(\mu)$: $\mathbb{R}_+ \to \mathbb{R}_+$, locally bounded, such that for all v in $S(u(\lambda); \mu)$ and all λ in Λ:

$$\| D_u G(\lambda, u(\lambda)) - D_u G(\lambda, v) \|_{\mathscr{L}(X;Y)} \leqslant L(\mu) \| u(\lambda) - v \|_X. \tag{3.42}$$

Remark 3.6. On the other hand, when G is a $\mathscr{C}^p$-mapping (with $p \geqslant 2$) and $D^p G$ is bounded on all bounded subsets of $\Lambda \times X$, then the argument of Brezzi, Rappaz & Raviart [14] shows that $u_h(\lambda)$ is a $\mathscr{C}^p$-mapping from Λ into X and gives the following bound for each m with $0 \leqslant m \leqslant p - 1$:

$$\| d^m(u_h(\lambda) - u(\lambda))/d\lambda^m \|_X \leqslant C_m \sum_{l=0}^{m} \| (T_h - T) d^l G(\lambda, u(\lambda))/d\lambda^l \|_X. \tag{3.43}$$

As a first application of Theorem 3.3, we generalize to the Navier-Stokes equations (2.1) (2.13) the regularization method or penalty method introduced in Section I.5.1 for the Stokes equations. We consider the following problem:

given $\varepsilon > 0$ *find* $(\mathbf{u}^\varepsilon, p^\varepsilon) \in H^1(\Omega)^N \times L_0^2(\Omega)$ *solution of*

$$\begin{cases} \left.\begin{aligned} -\nu \Delta \mathbf{u}^\varepsilon + \sum_{j=1}^{N} u_j^\varepsilon \partial \mathbf{u}^\varepsilon/\partial x_j + \mathbf{grad}\, p^\varepsilon = \mathbf{f} \\ p^\varepsilon = -(1/\varepsilon) \operatorname{div} \mathbf{u}^\varepsilon \end{aligned}\right\} \text{ in } \Omega, \\ \mathbf{u}^\varepsilon = \mathbf{g} \quad \text{on } \Gamma, \end{cases} \tag{3.44}$$

or equivalently

find $\mathbf{u}^\varepsilon \in H^1(\Omega)^N$ *such that*

$$\begin{cases} -\nu \Delta \mathbf{u}^\varepsilon - (1/\varepsilon) \mathbf{grad}(\operatorname{div} \mathbf{u}^\varepsilon) + \sum_{j=1}^{N} u_j^\varepsilon \partial \mathbf{u}^\varepsilon/\partial x_j = \mathbf{f} \quad \text{in } \Omega, \\ \mathbf{u}^\varepsilon = \mathbf{g} \quad \text{on } \Gamma. \end{cases} \tag{3.45}$$

In order to study the convergence of this regularization method, we consider a branch of nonsingular solutions $\{(\lambda, u(\lambda) = (\mathbf{u}(\lambda), \lambda p(\lambda))); \lambda = 1/\nu \in \Lambda\}$ of the equations (2.1) (2.13) in a compact interval Λ of $\mathbb{R}_+ - \{0\}$. This means that for all $(\mathbf{u}, \lambda p) = (\mathbf{u}(\lambda), \lambda p(\lambda))$, $\lambda = 1/\nu \in \Lambda$, the linearized problem (3.10) is well posed.

Theorem 3.4. *Let* $N \leqslant 4$ *and let* $\{(\lambda, u(\lambda) = (\mathbf{u}(\lambda), \lambda p(\lambda))); \lambda = 1/\nu \in \Lambda\}$ *be a branch of nonsingular solutions of* (2.1) (2.13). *Then there exists a neighborhood* $\mathcal{O}$ *of the origin in* $H^1(\Omega)^N \times L_0^2(\Omega)$ *and for* $\varepsilon \leqslant \varepsilon_0$ *small enough a unique* $\mathscr{C}^\infty$ *branch*

$\{(\lambda, u^\varepsilon(\lambda) = (\mathbf{u}^\varepsilon(\lambda), \lambda p^\varepsilon(\lambda))); \lambda \in \Lambda\}$ *of solutions of* (3.44) *such that*

$$u^\varepsilon(\lambda) \in u(\lambda) + \mathcal{O} \qquad \forall \lambda \in \Lambda.$$

Moreover, we get the estimate

$$\sup_{\lambda \in \Lambda} (\|\mathbf{u}^\varepsilon(\lambda) - \mathbf{u}(\lambda)\|_{1,\Omega} + \|p^\varepsilon(\lambda) - p(\lambda)\|_{0,\Omega}) \leqslant C\varepsilon, \tag{3.46}$$

where the constant C is independent of ε.

Proof. If we define the spaces X and Y by (3.5) and (3.6) and the mappings T and G by (3.7) and (3.8) respectively, we have already seen that Problem (2.1) (2.13) fits into the framework of Section 3.1. Then, in order to apply Theorem 3.3 we take $Z = Y$ and we define the operator T^ε of $\mathscr{L}(Y; X)$ as follows:

given $(\mathbf{f}_*, \mathbf{g}_*) \in Y$ *let* $(\mathbf{u}_*^\varepsilon, p_*^\varepsilon) = T^\varepsilon(\mathbf{f}_*, \mathbf{g}_*)$ *denote the solution of the regularized Stokes problem*

$$\begin{cases} \left.\begin{aligned} -\Delta \mathbf{u}_*^\varepsilon + \mathbf{grad}\, p_*^\varepsilon &= \mathbf{f}_* \\ p_*^\varepsilon &= -(1/\varepsilon) \operatorname{div} \mathbf{u}_*^\varepsilon \end{aligned}\right\} \quad \text{in } \Omega, \\ \mathbf{u}_*^\varepsilon = \mathbf{g}_* \quad \text{on } \Gamma. \end{cases}$$

Clearly, $(\mathbf{u}^\varepsilon, p^\varepsilon)$ is a solution of (3.44) if and only if

$$u^\varepsilon + T^{\varepsilon\nu} G(\lambda, u^\varepsilon) = 0,$$

where $u^\varepsilon = (\mathbf{u}^\varepsilon, p^\varepsilon/\nu)$. Moreover, Theorem I.5.3 gives for all $\varepsilon \leqslant \varepsilon_0$ sufficiently small:

$$\|T^{\varepsilon\nu} - T\|_{\mathscr{L}(Y;X)} \leqslant C_1 \varepsilon\nu \leqslant C_2 \varepsilon,$$

with a constant C_2 independent of λ owing to the compactness of Λ.

Now since

$$G(\lambda, v) = \left(\lambda\left(\sum_{j=1}^N v_j \partial \mathbf{v}/\partial x_j - \mathbf{f}\right), -\mathbf{g}\right),$$

it follows that $D^2 G$ is independent of v:

$$D^2 G(\lambda, v) \cdot (u, w) = \lambda \sum_{j=1}^N (w_j \partial \mathbf{u}/\partial x_j + u_j \partial \mathbf{w}/\partial x_j).$$

Thus the mapping $D^2 G$ is bounded on all bounded subsets of $\Lambda \times X$ for $N \leqslant 4$, by virtue of the Sobolev's Imbedding Theorem I.1.3. (And more generally, G is $\mathscr{C}^\infty$ and $D^p G$ is zero for all $p \geqslant 2$). Therefore the fact that $\{(\lambda, u(\lambda)); \lambda \in \Lambda\}$ is a branch of nonsingular solutions of

$$F(\lambda, u) \equiv u + TG(\lambda, u) = 0$$

permits to apply Theorem 3.3. In other words, if ε_0 is small enough there exist a real $a > 0$ and a unique branch $\{(\lambda, u^\varepsilon(\lambda) = (\mathbf{u}^\varepsilon(\lambda), \lambda p^\varepsilon(\lambda))); \lambda \in \Lambda\}$ of nonsingular solutions of the problem

$$F^\varepsilon(\lambda, u^\varepsilon) \equiv u^\varepsilon + T^{\varepsilon\nu} G(\lambda, u^\varepsilon) = 0 \qquad \nu = 1/\lambda$$

such that

$$\| u^\varepsilon(\lambda) - u(\lambda) \|_X \leqslant a.$$

Moreover we have

$$\begin{aligned} \| u^\varepsilon(\lambda) - u(\lambda) \|_X &= \| \mathbf{u}^\varepsilon(\lambda) - \mathbf{u}(\lambda) \|_{1,\Omega} + \lambda \| p^\varepsilon(\lambda) - p(\lambda) \|_{0,\Omega} \\ &\leqslant C_3 \| (T^{\varepsilon\nu} - T) G(\lambda, u(\lambda)) \|_X \\ &\leqslant C_2 C_3 \varepsilon \| G(\lambda, u(\lambda)) \|_X \leqslant C_4 \varepsilon. \end{aligned}$$

In addition Remark 3.6 implies that the mapping $\lambda \to u^\varepsilon(\lambda)$ is $\mathscr{C}^\infty$ from Λ into X. □

Let H be a Banach space such that

$$X \hookrightarrow H$$

where as usual the sign $\hookrightarrow$ means that the imbedding is continuous. Now, we want to derive a sharper estimate for $\| u_h(\lambda) - u(\lambda) \|_H$. To this end, we assume that there exists another Banach space W with

$$W \hookrightarrow X$$

such that the following property holds:

(3.47) $\begin{cases} \text{for all } v \in W, \text{ the operator } D_u G(\lambda, v) \text{ may be extended as a linear operator of } \mathscr{L}(H; Y), \\ \text{the mapping } v \to D_u G(\lambda, v) \text{ being continuous from } W \text{ into } \mathscr{L}(H; Y). \end{cases}$

Hence, for $v \in W$, both $D_u F(\lambda, v)$ and $D_u F_h(\lambda, v)$ may be extended as operators of $\mathscr{L}(H; H)$. Next, we suppose that:

(3.48) $$\lim_{h \to 0} \| T_h - T \|_{\mathscr{L}(Y; H)} = 0.$$

Remark 3.7. Note that (3.48) is again a consequence of (3.37) when the imbedding of X into H is compact.

Then we can prove the following result.

Theorem 3.5. *We retain the hypotheses of Theorem* 3.3 *together with* (3.47) *and* (3.48). *Assume in addition that*:

(3.49) *for each* $\lambda \in \Lambda$, $u(\lambda) \in W$ *and the function* $\lambda \to u(\lambda) \in \mathscr{C}^0(\Lambda; W)$;

(3.50) *for each* $\lambda \in \Lambda$, $D_u F(\lambda, u(\lambda))$ *is an isomorphism of* H.

Then, for $h \leqslant h_1$ small enough, there exists a constant $K' > 0$, independent of h and λ, such that:

$$\|u_h(\lambda) - u(\lambda)\|_H \leqslant K'\{\|(T - T_h)G(\lambda, u(\lambda))\|_H + \|u_h(\lambda) - u(\lambda)\|_X^2\}. \tag{3.51}$$

Proof. First, in view of (3.47) and (3.49), observe that:

$$\begin{aligned}\|D_u F(\lambda, u(\lambda)) - D_u F_h(\lambda, u(\lambda))\|_{\mathscr{L}(H;H)} &= \|(T - T_h)D_u G(\lambda, u(\lambda))\|_{\mathscr{L}(H;H)} \\ &\leqslant \|T - T_h\|_{\mathscr{L}(Y;H)}\|D_u G(\lambda, u(\lambda))\|_{\mathscr{L}(H;Y)}.\end{aligned}$$

Therefore, (3.48) implies that

$$\limsup_{h \to 0} \sup_{\lambda \in \Lambda} \|D_u F(\lambda, u(\lambda)) - D_u F_h(\lambda, u(\lambda))\|_{\mathscr{L}(H;H)} = 0.$$

Hence it stems from (3.50) and Lemma 3.3 that for all sufficiently small h and all λ in Λ, $D_u F_h(\lambda, u(\lambda))$ is an isomorphism of H with:

$$\|[D_u F_h(\lambda, u(\lambda))]^{-1}\|_{\mathscr{L}(H;H)} \leqslant C_1.$$

Then, like in Theorem 3.1, we can write:

$$\begin{aligned}u(\lambda) - u_h(\lambda) = [D_u F_h(\lambda, u(\lambda))]^{-1}[T_h\{D_u G(\lambda, u(\lambda))\cdot(u(\lambda) - u_h(\lambda)) - (G(\lambda, u(\lambda)) \\ - G(\lambda, u_h(\lambda)))\} + F_h(\lambda, u(\lambda))].\end{aligned}$$

But

$$\begin{aligned}&G(\lambda, u(\lambda)) - G(\lambda, u_h(\lambda)) - D_u G(\lambda, u(\lambda))\cdot(u(\lambda) - u_h(\lambda)) \\ &= -\int_0^1 (1 - t)D_{uu}^2 G(\lambda, (1 - t)u(\lambda) + tu_h(\lambda))\, dt\cdot(u(\lambda) - u_h(\lambda))^2.\end{aligned} \tag{3.52}$$

Therefore the boundedness assumption on $D^2 G$ together with (3.48) yield:

$$\|u(\lambda) - u_h(\lambda)\|_H \leqslant C_1[C_2\|u(\lambda) - u_h(\lambda)\|_X^2 + \|F_h(\lambda, u(\lambda)) - F(\lambda, u(\lambda))\|_H],$$

which proves (3.51). □

3.4. Non-Differentiable Approximation of Branches of Nonsingular Solutions

So far, we have assumed that the approximate mapping F_h retained the smoothness properties of the mapping F, because the approximation was performed on the linear operator T alone and not on G. But it is sometimes necessary to approximate G by a mapping G_h which is no longer differentiable. This occurs, for example, when an upwind discretization of the convective terms is used in the Navier-Stokes equations. This situation is analyzed by the following theory, which is an easy variant of the one elaborated in Sections 3.2 and 3.3. The reader can refer to Rappaz [67] for a more general approach that encompasses all the material of this paragraph.

The situation is somewhat more general than that of Section 3.2. We introduce a closed subspace X_h of X and a closed subspace $\mathscr{X}_h$ of $\mathscr{X}$ and we assume that the mapping F_h is defined and continuous on $\Lambda \times X_h$ with values in $\mathscr{X}_h$. Then, for a given element $\tilde{u}_h$ of X_h and a given $\lambda \in \Lambda$, we replace the notion of differentiability by the following assumptions:

(3.53) $\begin{cases} \text{there exists an operator } \nabla_u F_h(\lambda, \tilde{u}_h) \in \mathscr{L}(X_h; \mathscr{X}_h) \text{ which is an isomorphism from } X_h \text{ onto } \mathscr{X}_h; \end{cases}$

there exists a continuous, monotonically increasing function

(3.54) $\begin{cases} \mu \to L_h(\lambda; \mu): \mathbb{R}_+ \to \mathbb{R}_+ \text{ such that:} \\ \| F_h(\lambda, v) - F_h(\lambda, w) - \nabla_u F_h(\lambda, \tilde{u}_h)\cdot(v - w)\|_{\mathscr{X}} \leqslant L_h(\lambda; \alpha)\, \| v - w\|_X \\ \qquad \text{for all } v \text{ and } w \in S(\tilde{u}_h; \alpha) \cap X_h. \end{cases}$

We retain the following notations:

$$\begin{cases} \varepsilon_h(\lambda) = \| F_h(\lambda, \tilde{u}_h)\|_{\mathscr{X}}, \\ \gamma_h(\lambda) = \| [\nabla_u F_h(\lambda, \tilde{u}_h)]^{-1} \|_{\mathscr{L}(\mathscr{X}_h; X_h)}, \end{cases} \tag{3.55}$$

where the norm $\| B \|_{\mathscr{L}(\mathscr{X}_h; X_h)}$ stands for $\sup_{v \in \mathscr{X}_h} (\| Bv \|_X / \| v \|_{\mathscr{X}})$. Then we have the analogue of Theorem 3.1.

Theorem 3.6. *If* (3.53) *and* (3.54) *hold and if, in addition*:

$$2\gamma_h(\lambda) L_h(\lambda; 2\gamma_h(\lambda)\varepsilon_h(\lambda)) < 1 \tag{3.19'}$$

then Problem (3.12) *has a unique solution* $(\lambda, u_h(\lambda))$ *such that*:

$$u_h(\lambda) \in S(\tilde{u}_h; 2\gamma_h(\lambda)\varepsilon_h(\lambda)) \cap X_h. \tag{3.20'}$$

In addition, $u_h(\lambda)$ *is the only solution of* (3.12) *in the larger ball* $S(\tilde{u}_h; \alpha) \cap X_h$ *for all* $\alpha \geqslant 2\gamma_h(\lambda)\varepsilon_h(\lambda)$ *that satisfy*

$$\gamma_h(\lambda) L_h(\lambda; \alpha) < 1 \tag{3.22'}$$

and we have the estimate:

$$\| u_h(\lambda) - v_h \|_X \leqslant [\gamma_h(\lambda)/(1 - \gamma_h(\lambda) L_h(\lambda; \alpha))] \, \| F_h(\lambda, v_h) \|_{\mathscr{X}} \qquad \forall v_h \in S(\tilde{u}_h; \alpha) \cap X_h. \tag{3.23'}$$

We skip the proof since it is very similar to that of Theorem 3.1. Likewise, we can easily prove the following counterpart of Theorem 3.2.

Theorem 3.7. *Let* $\tilde{u}_h$ *be a given function in* $\mathscr{C}^0(\Lambda; X_h)$ *that satisfies*:

$$\sup_{h \leqslant h_0} \left(\sup_{\lambda \in \Lambda} \gamma_h(\lambda) \right) = \gamma \quad \textit{for some } h_0 > 0, \tag{3.26'}$$

(3.27′) $$\lim_{h\to 0}\left(\sup_{\lambda\in\Lambda}\varepsilon_h(\lambda)\right)=0.$$

If, in addition,

(3.56) $$\sup_{\lambda\in\Lambda} L_h(\lambda;\mu)=L_h(\mu)\qquad \forall h\leqslant h_0,$$

where the function $L_h(\mu)$ is monotonically increasing with respect to μ and h, continuous at $\mu=0$, and such that:

(3.57) $$\lim_{h\to 0} L_h(0)=0$$

then there exists two real constants $\alpha>0$ and $h_1>0$ and a function $u_h\in\mathscr{C}^0(\Lambda;X_h)$ such that for all $h\leqslant h_1$:

$$\{(\lambda,u_h(\lambda));\ \lambda\in\Lambda\}\quad \textit{is a branch of solutions of } (3.12),$$

for each $\lambda\in\Lambda$, $u_h(\lambda)$ is the only solution of (3.12) *in $S(\tilde{u}_h(\lambda);\alpha)\cap X_h$ and we have the estimate:*

(3.30′) $$\|u_h(\lambda)-v_h\|_X\leqslant 2\gamma\|F_h(\lambda,v_h)\|_{\mathscr{X}}\qquad \forall v_h\in S(\tilde{u}_h(\lambda);\alpha)\cap X_h.$$

Remark 3.8. The assumption that $L_h(\mu)$ be monotonically increasing with respect to h is not necessary. It can be replaced by the condition that $L_h(\mu)$ be uniformly continuous with respect to h at $\mu=0$.

For the sake of simplicity, we are going to apply Theorem 3.7 to solve a narrower range of problems than (3.33), but the reader will easily extend the forthcoming analysis to the general case. More precisely, with the notations of Section 3.3 we propose to solve the problem:

(3.33′) $$F(\lambda,u)\equiv u+\lambda TG(u)=0$$

where G is a $\mathscr{C}^1$-mapping from X into Y, $\lambda\in\Lambda$, a compact interval of $\mathbb{R}$, and T is unchanged but we assume that $T\in\mathscr{L}(Y;V)$ where V is a closed subspace of X. This may amount to a regularity assumption on T. Furthermore, we suppose that the problem

$$F(\lambda,u(\lambda))=0$$

has a branch of nonsingular solutions $\lambda\to u(\lambda)$ from Λ into X with $u\in\mathscr{C}^0(\Lambda;V)$.

In view of the approximation, we introduce a closed subspace V_h of X, equipped with the norm of X and a space Y_h that *contains* Y with continuous imbedding. To avoid confusion, we denote the norm of Y_h by $\|\,.\,\|_h$. The mapping G is approximated by a $\mathscr{C}^0$-mapping G_h from V_h into Y_h and T is approximated by an operator $T_h\in\mathscr{L}(Y_h;V_h)$. Then we set

(3.58) $$F_h(\lambda,u_h)=u_h+\lambda T_hG_h(u_h)$$

which is a $\mathscr{C}^0$-mapping from $\mathbb{R} \times V_h$ into V_h. The following theorem solves Problem (3.33′) without involving the differentiability of G_h.

Theorem 3.8. *Under the following hypotheses:*

(i)

$$\| T_h \|_{\mathscr{L}(Y_h; V_h)} \leqslant C, \tag{3.59}$$

$$\lim_{h \to 0} \| T - T_h \|_{\mathscr{L}(Y; X)} = 0. \tag{3.60}$$

(ii) *There exists an operator* $\pi_h \in \mathscr{L}(V; V_h)$ *such that:*

$$\lim_{h \to 0} \| v - \pi_h v \|_X = 0 \qquad \forall v \in V \tag{3.61}$$

and

$$\limsup_{h \to 0} \sup_{\lambda \in \Lambda} \| G_h(\pi_h u(\lambda)) - G(u(\lambda)) \|_h = 0. \tag{3.62}$$

(iii) *For all* $u_h \in V_h$, *there exists an operator* $\nabla G_h(u_h) \in \mathscr{L}(V_h; Y_h)$ *such that*

$$\limsup_{h \to 0} \sup_{\lambda \in \Lambda} \| \nabla G_h(\pi_h u(\lambda)) - DG(u(\lambda)) \|_{\mathscr{L}(V_h; Y_h)} = 0 \tag{3.63}$$

and

$$\| G_h(u_h) - G_h(u_h^*) - \nabla G_h(u_h^0) \cdot (u_h - u_h^*) \|_h \leqslant L_h(\mu; \| u_h^0 \|_X) \| u_h - u_h^* \|_X \tag{3.64}$$

for all $u_h^0 \in V_h$, u_h *and* $u_h^* \in S(u_h^0; \mu) \cap V_h$, *where* $L_h\colon \mathbb{R}_+ \times \mathbb{R}_+ \to \mathbb{R}_+$ *is a continuous, monotonically increasing function with respect to each variable and with respect to h, which satisfies:*

$$\lim_{h \to 0} L_h(0; \mu) = 0 \qquad \forall \mu \in \mathbb{R}_+. \tag{3.65}$$

Then there exists a neighborhood $\mathcal{O}$ *of the origin in* X *and for* $h \leqslant h_0$ *small enough, a unique* $\mathscr{C}^0$*-function* $u_h\colon \Lambda \to V_h$ *such that*

$$F_h(\lambda, u_h(\lambda)) = 0, \qquad u_h(\lambda) \in u(\lambda) + \mathcal{O} \qquad \forall \lambda \in \Lambda. \tag{3.66}$$

In addition, the following error estimate holds:

$$\begin{aligned} \| u_h(\lambda) - u(\lambda) \|_X \leqslant C\{ & \| (I - \pi_h) u(\lambda) \|_X + \| \lambda (T - T_h) G(u(\lambda)) \|_X \\ & + \| G_h(\pi_h u(\lambda)) - G(u(\lambda)) \|_h \}. \end{aligned} \tag{3.67}$$

Proof. Let us apply Theorem 3.7 with $\tilde{u}_h = \tilde{u}_h(\lambda) = \pi_h u(\lambda)$, V_h playing the role of X_h. Since by assumption, u belongs to $\mathscr{C}^0(\Lambda; V)$ then $\tilde{u}_h$ belongs to $\mathscr{C}^0(\Lambda; V_h)$. Next, let us prove (3.27′). We have:

$$\varepsilon_h(\lambda) = \| F_h(\lambda, \tilde{u}_h(\lambda)) - F(\lambda, u(\lambda)) \|_X,$$

i.e.

$$\varepsilon_h(\lambda) \leqslant \| \tilde{u}_h(\lambda) - u(\lambda) \|_X + |\lambda| \, \| T_h\{G_h(\tilde{u}_h(\lambda)) - G(u(\lambda))\} \|_X \\ + |\lambda| \, \| (T_h - T) G(u(\lambda)) \|_X. \tag{3.68}$$

But on the one hand, (3.61) and Ascoli's Lemma imply that

$$\lim_{h \to 0} \sup_{\lambda \in \Lambda} \| \tilde{u}_h(\lambda) - u(\lambda) \|_X = 0.$$

On the other hand, (3.59) and (3.62) yield:

$$\lim_{h \to 0} \sup_{\lambda \in \Lambda} \| T_h\{G_h(\tilde{u}_h(\lambda)) - G(u(\lambda))\} \|_X = 0.$$

In addition, it stems from the continuity of the mapping G and (3.60) that

$$\lim_{h \to 0} \sup_{\lambda \in \Lambda} \| (T_h - T) G(u(\lambda)) \|_X = 0.$$

These three limits give immediately (3.27').

Now, we turn to (3.26'). For each $u_h \in V_h$, it is natural to define $\nabla_u F_h(\lambda, u_h) \in \mathscr{L}(V_h; V_h)$ by:

$$\nabla_u F_h(\lambda, u_h) = I + \lambda T_h \nabla G_h(u_h). \tag{3.69}$$

Then we can write:

$$\nabla_u F_h(\lambda, \tilde{u}_h(\lambda)) = \lambda T_h\{\nabla G_h(\tilde{u}_h(\lambda)) - DG(u(\lambda))\} + I + \lambda T_h DG(u(\lambda)). \tag{3.70}$$

But observe that the operator $I + \lambda T_h DG(u(\lambda))$ belongs to $\mathscr{L}(X; X) \cap \mathscr{L}(V_h; V_h)$. Furthermore,

$$I + \lambda T_h DG(u(\lambda)) - D_u F(\lambda, u(\lambda)) = \lambda (T_h - T) DG(u(\lambda))$$

and owing to (3.60) and the differentiability of G, we have

$$\lim_{h \to 0} \sup_{\lambda \in \Lambda} \| (T_h - T) DG(u(\lambda)) \|_{\mathscr{L}(X; X)} = 0.$$

Hence, applying Lemma 3.3 we find that there exists a real $h_0 > 0$ such that for all $h \leqslant h_0$ and all $\lambda \in \Lambda$, $I + \lambda T_h DG(u(\lambda))$ is an isomorphism of X as well as an isomorphism of V_h. Besides that, we have the bound:

$$\| (I + \lambda T_h DG(u(\lambda)))^{-1} \|_{\mathscr{L}(V_h; V_h)} \leqslant \| (I + \lambda T_h DG(u(\lambda)))^{-1} \|_{\mathscr{L}(X; X)} \\ \leqslant 2\gamma(\lambda) \quad \forall h \leqslant h_0 \quad \text{and} \quad \forall \lambda \in \Lambda, \tag{3.71}$$

where $\gamma(\lambda)$ is defined by (3.13). Likewise, it follows from (3.59) and (3.63) that

$$\lim_{h \to 0} \sup_{\lambda \in \Lambda} \| T_h\{\nabla_h G_h(\tilde{u}_h(\lambda)) - DG(u(\lambda))\} \|_{\mathscr{L}(V_h; V_h)} = 0.$$

Therefore, applying again Lemma 3.3 and using (3.70) and (3.71), we see that there exists another h_1, with $0 < h_1 \leqslant h_0$, such that for all $h \leqslant h_1$ and all $\lambda \in \Lambda$, $\nabla_u F_h(\lambda, \tilde{u}_h(\lambda))$ is an isomorphism of V_h and

$$\| [\nabla_u F_h(\lambda, \tilde{u}_h(\lambda))]^{-1} \|_{\mathscr{L}(V_h; V_h)} \leqslant 4\gamma(\lambda) \qquad \forall h \leqslant h_1, \qquad \forall \lambda \in \Lambda.$$

As $\gamma(\lambda)$ is bounded above, this establishes (3.26').

It remains to fulfill the conditions on the function L_h defined by (3.54). In view of (3.59) and (3.64), we have the upper bound:

$$\| \nabla_u F_h(\lambda, \tilde{u}_h(\lambda)) \cdot (v_h - w_h) - F_h(\lambda, v_h) + F_h(\lambda, w_h) \|_X$$
$$\leqslant C |\lambda| L_h(\alpha; \| \tilde{u}_h(\lambda) \|_X) \| v_h - w_h \|_X$$
$$\forall \lambda \in \Lambda, \qquad \forall v_h, w_h \in S(\tilde{u}_h(\lambda); \alpha) \cap V_h.$$

Let

$$K = \sup_{h \leqslant h_1} \sup_{\lambda \in \Lambda} \| \tilde{u}_h(\lambda) \|_X.$$

The monotonicity of L_h implies that

$$L_h(\alpha; \| \tilde{u}_h(\lambda) \|_X) \leqslant L_h(\alpha; K).$$

By assumption, the mapping $\mu \to L_h(\mu; K)$ is monotonically increasing with respect to h and μ and continuous at $\mu = 0$. In addition, by virtue of (3.65), it satisfies the condition (3.57).

As a consequence, we can apply the conclusion of Theorem 3.7. It yields (3.66); then (3.67) stems from (3.30′) with $v_h = \tilde{u}_h(\lambda)$, (3.68) and (3.59). □

Remark 3.9. The space V, which enters only in the assumption (3.61), plays a minor part in this proof. It is useful when the restriction operator π_h requires more regularity than the space X can provide.

§ 4. Numerical Analysis of Centered Finite Element Schemes

In this paragraph, we propose to apply to the homogeneous Navier-Stokes equations most of the finite element methods developed in Chapters II and III to solve the Stokes problem. The reader will find that in nearly every case, the error analysis carries over successfully to Navier-Stokes equations. One of the few exceptions is the three-dimensional mixed method elaborated in Paragraph III.5. This method can be tried to solve the Navier-Stokes system but at the present stage, its numerical analysis is still an open problem.

4.1. Formulation in Primitive Variables: Methods Using Discontinuous Pressures

The situation is that of Section II.1.3. For each value of the parameter $h > 0$, we are given two finite-dimensional spaces:

$$W_h \subset H^1(\Omega)^N, \qquad Q_h \subset L^2(\Omega)$$

and we assume that Q_h contains the constant functions. Then we define:

$$W_{0h} = W_h \cap H_0^1(\Omega)^N, \qquad M_h = Q_h \cap L_0^2(\Omega).$$

As usual, Ω is a bounded, connected, open subset of $\mathbb{R}^N$ with a Lipschitz-continuous boundary Γ.

With the above spaces, the homogeneous Navier-Stokes Problem:

Given $\mathbf{f}$ *in* $H^{-1}(\Omega)^N$, *find* $(\mathbf{u}, p)$ *in* $H_0^1(\Omega)^N \times L_0^2(\Omega)$ *satisfying:*

$$\text{(4.1)} \qquad \begin{cases} a(\mathbf{u}; \mathbf{u}, \mathbf{v}) - (p, \operatorname{div} \mathbf{v}) = \langle \mathbf{f}, \mathbf{v} \rangle & \forall \mathbf{v} \in H_0^1(\Omega)^N, \\ \operatorname{div} \mathbf{u} = 0 \quad in\ \Omega \end{cases}$$

is discretized by the following *Problem* (Q_h)

Find $(\mathbf{u}_h, p_h) \in W_{0h} \times M_h$ *solution of*

$$\text{(4.2)} \qquad \begin{cases} a(\mathbf{u}_h; \mathbf{u}_h, \mathbf{v}_h) - (p_h, \operatorname{div} \mathbf{v}_h) = \langle \mathbf{f}, \mathbf{v}_h \rangle & \forall \mathbf{v}_h \in W_{0h}, \\ (q_h, \operatorname{div} \mathbf{u}_h) = 0 \quad \forall q_h \in M_h. \end{cases}$$

Recall the three hypotheses introduced in Section II.1.3 and related to the approximation of the Stokes system:

Hypothesis H1 (Approximation property of W_{0h}). *There exist an operator* $r_h \in \mathscr{L}([H^2(\Omega) \cap H_0^1(\Omega)]^N; W_{0h})$ *and an integer* l *such that:*

$$\text{(4.3)} \qquad \|\mathbf{v} - r_h \mathbf{v}\|_{1,\Omega} \leqslant C h^m \|\mathbf{v}\|_{m+1,\Omega} \quad \forall \mathbf{v} \in H^{m+1}(\Omega)^N, \quad 1 \leqslant m \leqslant l.$$

Hypothesis H2 (Approximation property of Q_h). *There exists an operator* $s_h \in \mathscr{L}(L^2(\Omega); Q_h)$ *such that:*

$$\text{(4.4)} \qquad \|q - s_h q\|_{0,\Omega} \leqslant C h^m \|q\|_{m,\Omega} \quad \forall q \in H^m(\Omega), \quad 0 \leqslant m \leqslant l.$$

Hypothesis H3 (Uniform inf-sup condition). *For each* $q_h \in M_h$ *there exists a* $\mathbf{v}_h \in W_{0h}$ *such that:*

$$\text{(4.5)} \qquad (q_h, \operatorname{div} \mathbf{v}_h) = \|q_h\|_{0,\Omega}^2, \quad |\mathbf{v}_h|_{1,\Omega} \leqslant C \|q_h\|_{0,\Omega},$$

with a constant $C > 0$ *independent of* h, q_h *and* $\mathbf{v}_h$.

In view of these assumptions, let us apply the material of Section 3.3.

Theorem 4.1. *Suppose* $N \leqslant 3$ *and assume that the hypotheses* H1, H2 *and* H3 *hold. Let* $\{(\lambda, (\mathbf{u}(\lambda), \lambda p(\lambda))); \lambda = 1/\nu \in \Lambda\}$ *be a branch of nonsingular solutions of the Navier-Stokes Problem* (4.1). *Then there exists a neighborhood* $\mathcal{O}$ *of the origin in* $H_0^1(\Omega)^N \times L_0^2(\Omega)$ *and for* $h \leqslant h_0$ *sufficiently small a unique* $\mathscr{C}^\infty$ *branch* $\{(\lambda, (\mathbf{u}_h(\lambda), \lambda p_h(\lambda))); \lambda \in \Lambda\}$ *of nonsingular solutions of Problem* (4.2) *such that:*

$$(\mathbf{u}_h(\lambda), \lambda p_h(\lambda)) \in (\mathbf{u}(\lambda), \lambda p(\lambda)) + \mathcal{O} \qquad \forall \lambda \in \Lambda.$$

Moreover, we have the convergence property:

$$\text{(4.6)} \qquad \lim_{h \to 0} \sup_{\lambda \in \Lambda} \{|\mathbf{u}_h(\lambda) - \mathbf{u}(\lambda)|_{1,\Omega} + \|p_h(\lambda) - p(\lambda)\|_{0,\Omega}\} = 0.$$

In addition, if the mapping $\lambda \to (\mathbf{u}(\lambda), p(\lambda))$ is continuous from Λ into $H^{m+1}(\Omega)^N \times H^m(\Omega)$ for some integer m with $1 \leqslant m \leqslant l$, we have for all $\lambda \in \Lambda$:

$$|\mathbf{u}_h(\lambda) - \mathbf{u}(\lambda)|_{1,\Omega} + \|p_h(\lambda) - p(\lambda)\|_{0,\Omega} \leqslant Kh^m. \tag{4.7}$$

Proof. The idea is to apply Theorem 3.3 with the following choice:

$$X = H_0^1(\Omega)^N \times L_0^2(\Omega), \qquad Y = H^{-1}(\Omega)^N.$$

The operator $T \in \mathscr{L}(Y; X)$ is the Stokes operator:

for $\mathbf{f}$ given in Y, $T\mathbf{f} = (\mathbf{v}, q) \in X$ is the solution of the Stokes problem:

$$\begin{cases} -\Delta \mathbf{v} + \mathbf{grad}\, q = \mathbf{f} & \\ \qquad \operatorname{div} \mathbf{v} = 0 & \text{in } \Omega, \\ \qquad \mathbf{v} = \mathbf{0} & \text{on } \Gamma. \end{cases} \tag{4.8}$$

For a fixed $\mathbf{f}$, the $\mathscr{C}^\infty$-mapping G: $\mathbb{R}_+ \times X \to Y$ is defined like in (3.8):

$$G(\lambda, v) = \lambda\left(\sum_{j=1}^N v_j \partial\mathbf{v}/\partial x_j - \mathbf{f}\right), \qquad v = (\mathbf{v}, q) \in X,$$

and

$$D_u G(\lambda, v) \cdot w = \lambda \sum_{j=1}^N (v_j \partial\mathbf{w}/\partial x_j + w_j \partial\mathbf{v}/\partial x_j), \qquad w = (\mathbf{w}, r) \in X.$$

Let us determine the space Z. In view of Theorem I.1.3, we know that

$$H_0^1(\Omega) \hookrightarrow L^6(\Omega) \quad \text{for } N \leqslant 3$$

and the imbedding of $H_0^1(\Omega)$ into $L^p(\Omega)$ is compact for $p < 6$. Next, applying Corollary I.1.1 observe that for $\mathbf{v}$ and $\mathbf{w}$ in $H_0^1(\Omega)^N$, we have

$$\sum_{j=1}^N (v_j \partial\mathbf{w}/\partial x_j + w_j \partial\mathbf{v}/\partial x_j) \in L^{3/2}(\Omega)^N.$$

Finally, applying again Theorem I.1.3 we find that $L^q(\Omega)$ is compactly imbedded in $H^{-1}(\Omega)$ whenever $q > 6/5$. It stems from these remarks that we can choose

$$Z = L^{3/2}(\Omega)^N \hookrightarrow Y \quad \text{with a compact imbedding}$$

and (3.36) holds.

Now, let $X_h = W_{0h} \times M_h$ and let $T_h \in \mathscr{L}(Y; X_h)$ be the approximate Stokes operator defined by:

$$T_h\mathbf{f} = (\mathbf{v}_h, q_h) \in X_h \quad \text{is the solution of}$$

$$\begin{cases} (\mathbf{grad}\, \mathbf{v}_h, \mathbf{grad}\, \mathbf{w}_h) - (q_h, \operatorname{div} \mathbf{w}_h) = \langle \mathbf{f}, \mathbf{w}_h \rangle & \forall \mathbf{w}_h \in W_{0h}, \\ \qquad (r_h, \operatorname{div} \mathbf{v}_h) = 0 & \forall r_h \in M_h. \end{cases}$$

It follows from Theorem II.1.8 that, owing to the hypotheses H1, H2 and H3 each $\mathbf{f}$ in Y determines a unique $T_h\mathbf{f}$ in X_h and

$$\lim_{h\to 0}\{|\mathbf{v}_h - \mathbf{v}|_{1,\Omega} + \|q_h - q\|_{0,\Omega}\} = 0, \tag{4.9}$$

i.e.

$$\lim_{h\to 0}\|(T_h - T)\mathbf{f}\|_X = 0 \qquad \forall \mathbf{f}\in Y.$$

Furthermore, when $(\mathbf{v}, q)$ belongs to $H^{m+1}(\Omega)^N \times H^m(\Omega)$ for some integer m in $[1, l]$, this same theorem asserts that:

$$|\mathbf{v}_h - \mathbf{v}|_{1,\Omega} + \|q_h - q\|_{0,\Omega} \leqslant Ch^m(\|\mathbf{v}\|_{m+1,\Omega} + \|q\|_{m,\Omega}), \tag{4.10}$$

i.e.
$$\|(T_h - T)\mathbf{f}\|_X \leqslant Ch^m\|T\mathbf{f}\|_{H^{m+1}(\Omega)^N\times H^m(\Omega)}.$$

As mentioned at the end of Remark 3.4, the compactness of the imbedding of Z into Y together with (4.9) imply that

$$\lim_{h\to 0}\|T_h - T\|_{\mathscr{L}(Z;X)} = 0.$$

Thus (3.37) and (3.38) hold.

Finally, since

$$a(\mathbf{u}_h;\mathbf{u}_h,\mathbf{v}_h) = \nu(\mathbf{grad}\,\mathbf{u}_h, \mathbf{grad}\;\mathbf{v}_h) + \left(\sum_{j=1}^N u_{hj}\partial\mathbf{u}_h/\partial x_j, \mathbf{v}_h\right),$$

we can express Problem (4.2) as follows:

$$(\mathbf{grad}\,\mathbf{u}_h, \mathbf{grad}\,\mathbf{v}_h) - (1/\nu)(p_h, \operatorname{div}\mathbf{v}_h) = (1/\nu)\left\langle \mathbf{f} - \sum_{j=1}^N u_{hj}\partial\mathbf{u}_h/\partial x_j, \mathbf{v}_h\right\rangle \qquad \forall \mathbf{v}_h\in W_{oh},$$

$$(q_h, \operatorname{div}\mathbf{u}_h) = 0 \qquad \forall q_h\in M_h.$$

In other words, $u_h = (\mathbf{u}_h, (1/\nu)p_h)$ satisfies:

$$u_h = -T_hG(1/\nu, u_h)$$

i.e.
$$F_h(\lambda, u_h) \equiv u_h + T_hG(\lambda, u_h) = 0 \quad \text{with } \lambda = 1/\nu.$$

Consequently, we can apply the conclusion of Theorem 3.3: for $h \leqslant h_0$ sufficiently small there exists *a unique branch of nonsingular solutions of* (4.2): $\{(\lambda, u_h(\lambda) = (\mathbf{u}_h(\lambda), \lambda p_h(\lambda))); \lambda\in\Lambda\}$,

i.e.
$$u_h(\lambda) + T_hG(\lambda, u_h(\lambda)) = 0 \qquad \forall\lambda\in\Lambda,$$

and a real number $a > 0$, independent of h, such that:

$$\|u_h(\lambda) - u(\lambda)\|_X \leqslant a \qquad \forall\lambda\in\Lambda.$$

In addition, according to Remark 3.6, the mapping $\lambda\to u_h(\lambda)$ is $\mathscr{C}^\infty$ because the mapping G is also $\mathscr{C}^\infty$ with bounded derivatives of all order on every bounded subsets of $\Lambda\times X$.

As far as the convergence and error bounds are concerned, (3.41) implies that

$$|\mathbf{u}_h(\lambda) - \mathbf{u}(\lambda)|_{1,\Omega} + |\lambda| \, \|p_h(\lambda) - p(\lambda)\|_{0,\Omega} \leqslant K \|(T_h - T)G(\lambda, u(\lambda))\|_X.$$

Hence (4.6) follows from (4.9). Furthermore, since

$$u(\lambda) = (\mathbf{u}(\lambda), \lambda p(\lambda)) \in H^{m+1}(\Omega)^N \times H^m(\Omega)$$

is the solution of the Stokes system:

$$u(\lambda) = -TG(\lambda, u(\lambda)),$$

the error estimate (II.1.46) gives:

$$\|(T_h - T)G(\lambda, u(\lambda))\|_X \leqslant Ch^m\{\|\mathbf{u}(\lambda)\|_{m+1,\Omega} + \|p(\lambda)\|_{m,\Omega}\}.$$

Thus (4.7) stems from the continuity of the mapping $\lambda \to u(\lambda)$ from Λ into $H^{m+1}(\Omega)^N \times H^m(\Omega)$. □

It is also possible to derive an L^2-estimate for the velocity. Like in the linear case we must assume that the associated Stokes problem is *regular* (cf. Definition II.1.1):

(4.11) $\left\{ \begin{array}{l} \text{the mapping } (\boldsymbol{\phi}, \mu) \to -\nu\Delta\boldsymbol{\phi} + \mathbf{grad}\,\mu \text{ is an isomorphism from } [H^2(\Omega)^N \cap \\ V] \times [H^1(\Omega) \cap L_0^2(\Omega)] \text{ onto } L^2(\Omega)^N. \end{array} \right.$

Theorem 4.2. *We retain the hypotheses of Theorem* 4.1 *and we assume that* (4.11) *holds. If the mapping* $\lambda \to (\mathbf{u}(\lambda), p(\lambda))$ *is continuous from* Λ *into* $H^{m+1}(\Omega)^N \times H^m(\Omega)$ *for some integer* m *in* $[1, l]$, *then we have the following* L^2*-estimate for all* λ *in* Λ:

(4.12) $$\|\mathbf{u}_h(\lambda) - \mathbf{u}(\lambda)\|_{0,\Omega} \leqslant Kh^{m+1}.$$

Proof. Let us apply Theorem 3.5. Since we are only interested in the velocity $\mathbf{u}$, we are going to drop entirely the pressure p. Thus we take

$$X = H_0^1(\Omega)^N, \qquad Y = H^{-1}(\Omega)^N$$

and for $\mathbf{f}$ in Y we define $\mathbf{v} = T\mathbf{f}$ by:

$$\mathbf{v} \in V, \qquad (\mathbf{grad}\,\mathbf{v}, \mathbf{grad}\,\mathbf{w}) = \langle \mathbf{f}, \mathbf{w} \rangle \qquad \forall \mathbf{w} \in V.$$

As the mapping $G(\lambda, v)$ depends only on the velocity $\mathbf{v}$ (and not on q) we can leave it as such. Thus, $\mathbf{u}$ is a solution of the Navier-Stokes system (4.1) if and only if:

$$F(\lambda, \mathbf{u}) \equiv \mathbf{u} + TG(\lambda, \mathbf{u}) = \mathbf{0}.$$

Next, we choose

$$H = L^2(\Omega)^N, \qquad W = [H_0^1(\Omega) \cap H^2(\Omega)]^N.$$

From Theorem I.1.3, we know that X (resp. W) is compactly imbedded into H (resp. X). Now, take $\mathbf{u}$ in W and let us check that $D_u G(\lambda, \mathbf{u}) \in \mathscr{L}(H; Y) = \mathscr{L}(L^2(\Omega)^N; H^{-1}(\Omega)^N)$. Indeed, if $\mathbf{v} \in L^2(\Omega)^N$ and $\boldsymbol{\phi} \in \mathscr{D}(\Omega)^N$, we can write:

$$\langle D_u G(\lambda, \mathbf{u})\cdot \mathbf{v}, \boldsymbol{\phi}\rangle = \lambda \sum_{j=1}^{N} \{\langle u_j \partial \mathbf{v}/\partial x_j, \boldsymbol{\phi}\rangle + \langle v_j \partial \mathbf{u}/\partial x_j, \boldsymbol{\phi}\rangle\}.$$

But
$$\langle u_j \partial \mathbf{v}/\partial x_j, \boldsymbol{\phi}\rangle = -\langle \mathbf{v}, \partial(u_j \boldsymbol{\phi})/\partial x_j\rangle;$$
hence
$$|\langle u_j \partial \mathbf{v}/\partial x_j, \boldsymbol{\phi}\rangle| \leqslant C_1 \|\mathbf{v}\|_{0,\Omega} \|u_j\|_{2,\Omega} \|\boldsymbol{\phi}\|_{1,\Omega},$$
with a similar bound for $\langle v_j \partial \mathbf{u}/\partial x_j, \boldsymbol{\phi}\rangle$. As a consequence, $D_u G(\lambda, \mathbf{u})$ can be extended to a continuous linear operator from $L^2(\Omega)^N$ into $H^{-1}(\Omega)^N$, provided $\mathbf{u}$ belongs to $\mathrm{H}^2(\Omega)^N$. This settles (3.47).

As far as (3.48) is concerned, we make use of Remark 3.7 and the compactness of X into H.

It remains to verify (3.50): $D_u F(\lambda, \mathbf{u}(\lambda))$ is an isomorphism of H. By assumption, we already know that:

$$D_u F(\lambda, \mathbf{u}(\lambda)) = I + T D_u G(\lambda, \mathbf{u}(\lambda))$$

is an isomorphism of X. In addition, we have just seen that $T D_u G(\lambda, \mathbf{u}(\lambda)) \in \mathscr{L}(L^2(\Omega)^N; V)$ whenever $\mathbf{u}(\lambda) \in H^2(\Omega)^N$. Therefore, the compactness of the imbedding of $H^1(\Omega)$ into $L^2(\Omega)$ implies that $T D_u G(\lambda, \mathbf{u}(\lambda))$ is a compact operator from $L^2(\Omega)^N$ into itself. Hence we can apply to $D_u F(\lambda, \mathbf{u}(\lambda))$ Fredholm's alternative, namely $D_u F(\lambda, \mathbf{u}(\lambda))$ is an isomorphism of H iff the equation

(4.13) $$D_u F(\lambda, \mathbf{u}(\lambda))\cdot \mathbf{v} = \mathbf{0} \quad \text{with } \mathbf{v} \quad \text{in } H$$

has only the zero solution. But if $\mathbf{v}$ in H satisfies (4.13), then $\mathbf{v}$ belongs to V since $\mathbf{v} = -T D_u G(\lambda, \mathbf{u}(\lambda))\cdot \mathbf{v}$. Therefore $\mathbf{v} = \mathbf{0}$ because $D_u F(\lambda, \mathbf{u}(\lambda))$ is an isomorphism of X. This proves (3.50).

Consequently, we can apply the conclusion of Theorem 3.5: if the mapping $\lambda \to \mathbf{u}(\lambda)$ is continuous from Λ into $[H_0^1(\Omega) \cap H^2(\Omega)]^N$, then for all h sufficiently small, we have:

$$\|\mathbf{u}_h(\lambda) - \mathbf{u}(\lambda)\|_{0,\Omega} \leqslant C_2 \{\|(T - T_h) G(\lambda, \mathbf{u}(\lambda))\|_{0,\Omega} + |\mathbf{u}_h(\lambda) - \mathbf{u}(\lambda)|_{1,\Omega}^2\}.$$

Thus, when $(\mathbf{u}(\lambda), p(\lambda))$ belongs to $H^{m+1}(\Omega)^N \times H^m(\Omega)$, the bound (II.1.50) gives:

$$\|(T - T_h) G(\lambda, \mathbf{u}(\lambda))\|_{0,\Omega} \leqslant C_3 h^{m+1} \{\|\mathbf{u}(\lambda)\|_{m+1,\Omega} + \|p(\lambda)\|_{m,\Omega}\}$$

and (4.12) stems from this estimate and (4.7). □

Roughly speaking, Theorems 4.1 and 4.2 can be summarized by saying that all the function spaces introduced in Paragraphs II.2 and II.3 to solve the Stokes equations can be also applied to approximate branches of nonsingular solutions of the Navier-Stokes problem with a *similar accuracy*. For instance, assume that the two-dimensional Navier-Stokes equations (4.1) have a branch of nonsingular solutions with the regularity:

(4.14) $\lambda \to (\mathbf{u}(\lambda), p(\lambda))$ is continuous from Λ into $H^2(\Omega)^2 \times H^1(\Omega)$.

Assume that Ω is a bounded polygon triangulated by $\mathscr{T}_h$ and for each $\kappa \in \mathscr{T}_h$ take the polynomial space defined by (II.2.2):

$$\mathscr{P}_1(\kappa) = P_1^2 \oplus \text{span}\{\mathbf{n}_i\lambda_j\lambda_k;\ 1 \leqslant i, j, k \leqslant 3, i \neq j \neq k\}.$$

Then, if the triangulation $\mathscr{T}_h$ is regular the finite element scheme (4.2) with the spaces:

$$W_h = \{\mathbf{w} \in \mathscr{C}^0(\bar{\Omega})^2;\ \mathbf{w}|_\kappa \in \mathscr{P}_1(\kappa) \quad \forall \kappa \in \mathscr{T}_h\},$$

$$Q_h = \{q \in L^2(\Omega);\ q|_\kappa \in P_0 \quad \forall \kappa \in \mathscr{T}_h\}$$

has a unique branch of solutions $\{(\lambda, (\mathbf{u}_h(\lambda), \lambda p_h(\lambda)));\ \lambda \in \Lambda\}$ that satisfies the error bound:

$$|\mathbf{u}_h(\lambda) - \mathbf{u}(\lambda)|_{1,\Omega} + \|p_h(\lambda) - p(\lambda)\|_{0,\Omega} \leqslant Ch(|\mathbf{u}(\lambda)|_{2,\Omega} + |p(\lambda)|_{1,\Omega}) \qquad \forall \lambda \in \Lambda.$$

In addition, when Ω is convex we have the L^2-estimate:

$$\|\mathbf{u}_h(\lambda) - \mathbf{u}(\lambda)\|_{0,\Omega} \leqslant Ch^2(|\mathbf{u}(\lambda)|_{2,\Omega} + |p(\lambda)|_{1,\Omega}) \qquad \forall \lambda \in \Lambda.$$

Another interesting example is the scheme derived from the controversial $Q_1 - P_0$ element discussed in Sections II.3.3 and II.3.4 in the case of the square $(-1, 1) \times (-1, 1)$. Assume that $\mathscr{T}_h$ is a square grid with mesh-size $h = 1/(4n)$ so that the conclusion of Theorem II.3.4 be valid. Then take

$$W_{0h} = \tilde{V}_h \quad \text{and} \quad M_h = \tilde{M}_h$$

with $\tilde{V}_h$ and $\tilde{M}_h$ defined respectively by (II.3.33) and (II.3.32). It is established in Sections II.3.3 and II.3.4 that this choice of spaces satisfies the three hypotheses H1, H2 and H3 (cf. (II.3.39), (II.3.40) and Theorem II.3.4). Therefore Theorems 4.1 and II.3.5 imply that, under the condition (4.14), the finite element scheme (4.2) has a unique branch of solutions

$$\{(\lambda, (\mathbf{u}_h(\lambda), \lambda \tilde{p}_h(\lambda)));\ \lambda \in \Lambda\}$$

and

$$|\mathbf{u}_h(\lambda) - \mathbf{u}(\lambda)|_{1,\Omega} + \|\tilde{p}_h(\lambda) - p(\lambda)\|_{0,\Omega} \leqslant Ch\{|\mathbf{u}(\lambda)|_{2,\Omega} + |p(\lambda)|_{1,\Omega}\}.$$

Likewise, we derive immediately from Theorem 4.2 and Corollary II.3.2 that:

$$\|\mathbf{u}_h(\lambda) - \mathbf{u}(\lambda)\|_{0,\Omega} \leqslant Ch^2\{|\mathbf{u}(\lambda)|_{2,\Omega} + |p(\lambda)|_{1,\Omega}\}.$$

Remark 4.1. To solve the scheme (4.2) with the $Q_1 - P_0$ element, it is convenient (like in the linear case) to use the basis of V_h described at the end of Section II.3.4.

4.2. Formulation in Primitive Variables: the Case of Continuous Pressures

Going back to Chapter II, we can plainly see that the previous analysis applies readily to the "mini" element discussed in Section II.4.1 as well as the "Hood-Taylor" elements of Section II.4.2. Indeed, all these elements satisfy the hypotheses

H1, H2 and H3 and besides that, a continuous approximation of the pressure has no direct influence on the nonlinear term of the Navier-Stokes equations. Hence in the neighborhood of a branch of nonsingular solutions with the regularity (4.14) and if the triangulation $\mathcal{T}_h$ is regular, the finite element scheme (4.2) with the spaces

$$W_h = \{\mathbf{w} \in \mathscr{C}^0(\bar{\Omega})^2;\ \mathbf{w}|_\kappa \in \mathscr{P}_1(\kappa) \quad \forall \kappa \in \mathcal{T}_h\},$$
$$\mathscr{P}_1(\kappa) = [P_1 \oplus \mathrm{span}(\lambda_1 \lambda_2 \lambda_3)]^2,$$
$$Q_h = \{q \in \mathscr{C}^0(\bar{\Omega});\ q|_\kappa \in P_1 \quad \forall \kappa \in \mathcal{T}_h\}$$

has a unique branch of solutions: $\{(\lambda, (\mathbf{u}_h(\lambda), \lambda p_h(\lambda)));\ \lambda \in \Lambda\}$ and there exists a constant C independent of λ such that:

$$|\mathbf{u}(\lambda) - \mathbf{u}_h(\lambda)|_{1,\Omega} + \|p(\lambda) - p_h(\lambda)\|_{0,\Omega} \leqslant Ch\{|\mathbf{u}(\lambda)|_{2,\Omega} + |p(\lambda)|_{1,\Omega}\}.$$

When Ω is convex we have the L^2-estimate:

$$\|\mathbf{u}(\lambda) - \mathbf{u}_h(\lambda)\|_{0,\Omega} \leqslant Ch^2\{|\mathbf{u}(\lambda)|_{2,\Omega} + |p(\lambda)|_{1,\Omega}\}.$$

Likewise the error estimates of Theorems II.4.3 and II.4.4 for the two "Hood-Taylor" elements carry over to the scheme (4.2) when $\mathcal{T}_h$ and the branch of nonsingular solutions of the Navier-Stokes equations have the adequate regularity.

Unfortunately, the "Glowinski-Pironneau" scheme does not fit so neatly into the preceding framework and has to be analyzed separately. Recall that the velocity and pressure spaces are those of the "Hood-Taylor" scheme:

$$X_h = \{\mathbf{v} \in \mathscr{C}^0(\bar{\Omega})^2;\ \mathbf{v}|_\kappa \in P_2^2 \quad \forall \kappa \in \mathcal{T}_h,\ \mathbf{v}|_\Gamma = 0\},$$
$$Q_h = \{q \in \mathscr{C}^0(\bar{\Omega});\ q|_\kappa \in P_1 \quad \forall \kappa \in \mathcal{T}_h\},$$

with the additional space

$$\Phi_h = Q_h \cap H_0^1(\Omega)$$

for the auxiliary potential. When adapted to the Navier-Stokes problem, the "Glowinski-Pironneau" method, in the version given by Remark II.4.5, reads:

Find a triple $(\mathbf{u}_h, p_h, \mu_h) \in X_h \times Q_h \times \Phi_h$ *satisfying*:

$$(4.15)\quad \begin{cases} \nu(\mathbf{grad}\,\mathbf{u}_h, \mathbf{grad}\,\mathbf{v}_h) + \left(\sum_{j=1}^{2} (u_h)_j \partial \mathbf{u}_h/\partial x_j, \mathbf{v}_h - \mathbf{grad}\, q_h\right) \\ \quad + (\mathbf{grad}\, p_h, \mathbf{v}_h - \mathbf{grad}\, q_h) = (\mathbf{f}, \mathbf{v}_h - \mathbf{grad}\, q_h) \quad \forall (\mathbf{v}_h, q_h) \in X_h \times \Phi_h \\ (\mathbf{u}_h - \mathbf{grad}\, \mu_h, \mathbf{grad}\, q_h) = 0 \quad \forall q_h \in Q_h. \end{cases}$$

Let us see under what conditions the exact Navier-Stokes problem admits a formulation similar to (4.15). To solve the Stokes problem in Section II.4.3 we took the right-hand side $\mathbf{f}$ in $L^2(\Omega)^2$ and chose $H_0^1(\Omega)$ for space of potentials. Here this is not realistic because the nonlinear term

$$\sum_{j=1}^{2} u_j \partial \mathbf{u}/\partial x_j,$$

which is considered part of the right-hand side, belongs only to $L^{2-\varepsilon}(\Omega)^2$ for any $\varepsilon > 0$ whenever $\mathbf{u}$ belongs to $H^1(\Omega)^2$ (cf. Corollary I.1.1 with $N = 2$). This remark suggests to *fix* a real r in the interval $(1, 2)$ and take $W_0^{1,s}(\Omega)$ for space of potentials with $1/s + 1/r = 1$. Then consider the problem:

For $\mathbf{f} \in L^r(\Omega)^2$ *find a triple* $(\mathbf{u}, p, \mu) \in H_0^1(\Omega)^2 \times W^{1,r}(\Omega) \times W_0^{1,s}(\Omega)$ *such that*

$$(4.16) \quad \begin{cases} \nu(\mathbf{grad}\,\mathbf{u}, \mathbf{grad}\,\mathbf{v}) + \left(\sum_{j=1}^{2} u_j \partial \mathbf{u}/\partial x_j, \mathbf{v} - \mathbf{grad}\, q\right) + (\mathbf{grad}\, p, \mathbf{v} - \mathbf{grad}\, q) \\ \quad = (\mathbf{f}, \mathbf{v} - \mathbf{grad}\, q) \quad \forall (\mathbf{v}, q) \in H_0^1(\Omega)^2 \times W_0^{1,s}(\Omega) \\ \qquad (\mathbf{u} - \mathbf{grad}\, \mu, \mathbf{grad}\, q) = 0 \quad \forall q \in W^{1,r}(\Omega). \end{cases}$$

It is a matter of routine to check that this problem is equivalent to the Navier-Stokes problem (4.1) in the following sense:

if $(\mathbf{u}, p)$ is a solution of (4.1) *with* p *in* $W^{1,r}(\Omega)$ then the triple $(\mathbf{u}, p, \mu = 0)$ is a solution of (4.16). Conversely, each solution of (4.16) is of the form $(\mathbf{u}, p, \mu = 0)$ where the pair $(\mathbf{u}, p)$ satisfies (4.1).

Obviously, the same conclusion applies to the Stokes problem with right-hand side $\mathbf{f}$ in $L^r(\Omega)^2$:

if the pressure solution p of the Stokes problem belongs to $W^{1,r}(\Omega)$ then this problem is equivalent to:

Find $(\mathbf{u}, p, \mu) \in H_0^1(\Omega)^2 \times W^{1,r}(\Omega) \times W_0^{1,s}(\Omega)$ *such that*

$$(4.17) \quad \begin{cases} (\mathbf{grad}\,\mathbf{u}, \mathbf{grad}\,\mathbf{v}) + (\mathbf{grad}\, p, \mathbf{v} - \mathbf{grad}\, q) = (\mathbf{f}, \mathbf{v} - \mathbf{grad}\, q) \\ \qquad\qquad \forall (\mathbf{v}, q) \in H_0^1(\Omega)^2 \times W_0^{1,s}(\Omega), \\ \qquad (\mathbf{u} - \mathbf{grad}\, q) = 0 \quad \forall q \in W^{1,r}(\Omega). \end{cases}$$

Therefore, in order to express the Stokes operator by (4.17) for every right-hand side in $L^r(\Omega)^2$ we must assume that the Stokes problem is regular in a more general sense than (4.11):

$$(4.18) \quad \begin{cases} \text{the mapping } (\boldsymbol{\phi}, \mu) \to -\Delta \boldsymbol{\phi} + \mathbf{grad}\, \mu \text{ is an isomorphism from } [W^{2,r}(\Omega) \cap \\ V]^2 \times [W^{1,r}(\Omega)/\mathbb{R}] \text{ onto } L^r(\Omega)^2 \text{ for all } r \in (1, 2]. \end{cases}$$

Remark 4.2. Here the pressure is taken in the quotient space $W^{1,r}(\Omega)/\mathbb{R}$ instead of $W^{1,r}(\Omega) \cap L_0^2(\Omega)$ because in the practical computation of (4.15), p_h is fixed by the condition $\int_\Gamma p_h \, ds = 0$ (cf. Lemma II.4.3).

Now, we can put Problem (4.16) into the setting of Section 3.3. We take:

$$Y = L^r(\Omega)^2, \qquad X = H_0^1(\Omega)^2 \times [L^2(\Omega)/\mathbb{R}] \times H_0^1(\Omega).$$

Owing to the regularity assumption (4.18), the Stokes operator T can be written in the form (4.17):

$$T \in \mathscr{L}(Y; X), \qquad T\mathbf{f} = (\mathbf{u}, p, \mu = 0) \quad \text{solution of (4.17).}$$

The nonlinearity is embodied by the usual mapping G:

$$G(\lambda, v) = \lambda\left(\sum_{j=1}^{2} v_j \partial \mathbf{v}/\partial x_j - \mathbf{f}\right), \qquad \lambda \in \mathbb{R}_+, \qquad v = (\mathbf{v}, q, \phi) \in X$$

which maps $\mathbb{R}_+ \times X$ into $L^r(\Omega)^2$. With these notations, Problem (4.16) takes the standard form:

$$F(\lambda, u) = 0$$

with $\lambda = 1/\nu$, $u = (\mathbf{u}, \lambda p, 0)$ and $F(\lambda, u) \equiv u + TG(\lambda, u)$.

Remark 4.3. When the Stokes operator has the regularity (4.18), each solution u of Problem (4.16) with right-hand side $\mathbf{f}$ in $L^r(\Omega)^2$ has the regularity $\mathbf{u} \in W^{2,r}(\Omega)^2$, $p \in W^{1,r}(\Omega)$. This is valid for all real $r \in (1, 2]$.

Remark 4.4. It is important to note that if the Stokes operator has the regularity (4.18) then every branch of nonsingular solutions of Problem (4.1) with right-hand side $\mathbf{f}$ in $L^r(\Omega)^2$ is also a branch of nonsingular solutions of Problem (4.16) and conversely.

Next, we set

$$W_h = X_h \times [Q_h/\mathbb{R}] \times \Phi_h \subset X$$

and let $T_h \in \mathscr{L}(Y; W_h)$ be the discrete Stokes operator corresponding to (4.17):

$$T_h\mathbf{f} = (\mathbf{u}_h, p_h, \mu_h) \quad \text{solution of:}$$

$$\text{(4.19)} \quad \begin{cases} (\mathbf{grad}\,\mathbf{u}_h, \mathbf{grad}\,\mathbf{v}_h) + (\mathbf{grad}\,p_h, \mathbf{v}_h - \mathbf{grad}\,q_h) = (\mathbf{f}, \mathbf{v}_h - \mathbf{grad}\,q_h) \\ \qquad\qquad\qquad\qquad\qquad\qquad\qquad\qquad \forall(\mathbf{v}_h, q_h) \in X_h \times \Phi_h, \\ (\mathbf{u}_h - \mathbf{grad}\,\mu_h, \mathbf{grad}\,q_h) = 0 \quad \forall q_h \in Q_h. \end{cases}$$

Therefore, Problem (4.15) has the equivalent formulation:

$$F_h(\lambda, u_h(\lambda)) = 0$$

with $\lambda = 1/\nu$, $u_h(\lambda) = (\mathbf{u}_h(\lambda), \lambda p_h(\lambda), \mu_h(\lambda))$ and $F_h(\lambda, u_h) \equiv u_h + T_h G(\lambda, u_h)$.

We are now in a position to apply Theorem 3.3. We take

$$Z = Y$$

and (3.36) holds automatically. As far as the approximation properties of the operator T_h are concerned, we can apply the material of Section II.4.3. In particular, it is established in Theorem II.4.6 that

$$(4.20)\quad \begin{cases} |\mathbf{u}-\mathbf{u}_h|_{1,\Omega} \leqslant 2(1+\sqrt{2}/\beta^*) \inf\limits_{\mathbf{v}_h\in X_h} |\mathbf{u}-\mathbf{v}_h|_{1,\Omega} + \sqrt{2}\|\bar{p}-P_h\bar{p}\|_{0,\Omega}, \\ |\mu_h|_{1,\Omega} \leqslant \|\mathbf{u}-\mathbf{u}_h\|_{0,\Omega}, \\ \|\bar{p}-\bar{p}_h\|_{0,\Omega} \leqslant (1+\sqrt{2}/\beta^*) \inf\limits_{q_h\in Q_h} \|\bar{p}-q_h\|_{0,\Omega} + (1/\beta^*)|\mathbf{u}-\mathbf{u}_h|_{1,\Omega}, \end{cases}$$

where P_h denotes the H^1-projection on $Q_h \cap L_0^2(\Omega)$ defined by (A.25), $\bar{p}$ (resp. $\bar{p}_h$) denotes the representative of p (resp. p_h) in $L_0^2(\Omega)$ and β^* is the constant of the inf-sup condition. Note that (4.20) requires only the mild assumption (II.4.17) and the regularity of the triangulation $\mathscr{T}_h$. Thus (3.37) follows from (4.20) and a standard density argument.

Finally, (3.38) is a consequence of (4.20) and the regularity assumption (4.18) (which is valid when Ω is convex). More precisely, if (4.18) holds then $\mathbf{u}$ and p have the extra regularity:

$$\mathbf{u}\in \mathbf{w}^{2,r}(\Omega)^2, \qquad p\in w^{1,r}(\Omega).$$

Then just like in Section III.3.1 we derive:

$$(4.21)\qquad \inf_{\mathbf{v}_h\in X_h} |\mathbf{u}-\mathbf{v}_h|_{1,\Omega} \leqslant C_1 h^{2/s}|\mathbf{u}|_{2,r,\Omega}$$

and if in addition Ω is convex and $\mathscr{T}_h$ is uniformly regular, Lemma III.3.4 establishes that

$$(4.22)\qquad \|p-P_h p\|_{0,\Omega} \leqslant C_2 h^{2/s}|\ln h|^{1-2/s}|p|_{1,r',\Omega}.$$

Collecting these inequalities we obtain:

$$(4.23)\qquad |\mathbf{u}-\mathbf{u}_h|_{1,\Omega} + |\mu_h|_{1,\Omega} + \|\bar{p}-\bar{p}_h\|_{0,\Omega} \leqslant C_3 h^{2/s}|\ln h|^{1-2/s}\|\mathbf{f}\|_{0,r,\Omega}$$

which proves (3.38).

Observe also that when Ω is convex and $\mathscr{T}_h$ is uniformly regular, (4.19) and Theorem A.2 imply that:

$$(4.24)\qquad |\mu_h|_{1,\alpha,\Omega} \leqslant C_4(\alpha)\|\operatorname{div}(\mathbf{u}-\mathbf{u}_h)\|_{0,\Omega} \qquad \forall \text{real } \alpha \geqslant 2.$$

Indeed, we can consider that μ_h is the H_0^1-projection, $\mathring{P}_h\mu$, of the solution μ of the Dirichlet problem:

$$(\mathbf{grad}\,\mu, \mathbf{grad}\,q) = -(\operatorname{div}(\mathbf{u}-\mathbf{u}_h), q) \qquad \forall q\in H_0^1(\Omega).$$

Consequently, the conclusion of Theorem 3.3 is valid; it is summarized in the following theorem.

Theorem 4.3. *Let Ω be a bounded, convex polygon and assume that $\mathscr{T}_h$ is a uniformly regular triangulation of $\bar{\Omega}$ that satisfies* (II.4.17). *For $\mathbf{f}\in L^r(\Omega)^2$, $r\in(1,2]$, let $\{(\lambda, (\mathbf{u}(\lambda), \lambda p(\lambda), 0)); \lambda = 1/\nu\in\Lambda\}$, with $p(\lambda)$ chosen in $L_0^2(\Omega)$, be a branch of nonsingular solutions of the Navier-Stokes Problem* (4.16). *Then for $h \leqslant h_0$ sufficiently small, there exists a unique $\mathscr{C}^\infty$ branch $\{(\lambda, (\mathbf{u}_h(\lambda), \lambda p_h(\lambda), \mu_h(\lambda))); \lambda\in\Lambda\}$, with $p_h(\lambda)$ chosen in $L_0^2(\Omega)$, of solutions of Problem* (4.15) *such that:*

$$(4.25)\quad \begin{cases} \sup_{\lambda\in\Lambda}\{|\mathbf{u}_h(\lambda)-\mathbf{u}(\lambda)|_{1,\Omega}+\|p_h(\lambda)-p(\lambda)\|_{0,\Omega}\}\leqslant Ch^{2/s}|\ln h|^{1-2/s}\\ \sup_{\lambda\in\Lambda}|\mu_h(\lambda)|_{1,\alpha,\Omega}\leqslant C(\alpha)h^{2/s}|\ln h|^{1-2/s}\qquad \forall\alpha\geqslant 2,\end{cases}$$

$1/r + 1/s = 1$, *with constants independent of* h *and* λ.

Besides that, if the mapping $\lambda \to (\mathbf{u}(\lambda), p(\lambda))$ *is continuous from* Λ *into* $H^{m+1}(\Omega)^2 \times H^m(\Omega)$ *for* $m = 1$ *or* 2, *we have the estimate*:

$$(4.26)\qquad |\mathbf{u}_h(\lambda)-\mathbf{u}(\lambda)|_{1,\Omega}+\|p_h(\lambda)-p(\lambda)\|_{0,\Omega}+|\mu_h(\lambda)|_{1,\alpha,\Omega}\leqslant Kh^m$$

for all $\lambda\in\Lambda$.

Remark 4.5. It is not yet known whether or not a more accurate L^2-estimate can be obtained for $\mathbf{u} - \mathbf{u}_h + \mathbf{grad}\,\mu_h$, as it is done in Theorem II.4.6 for the Stokes problem. The delicate point is that the proof of Theorem II.4.6 uses explicitly the equality:

$$\Delta p = \operatorname{div}\mathbf{f}$$

which is valid for every Stokes system but is obviously not true for Navier-Stokes equations.

4.3. Mixed Incompressible Methods: the "Stream Function-Vorticity" Formulation

In this section, we investigate exclusively the two-dimensional case. (As mentioned at the beginning of this paragraph, the corresponding analysis of mixed incompressible schemes in three dimensions is still an open problem). We propose to extend to the Navier-Stokes equations the mixed formulation introduced in Section III.2.1. To begin with, recall the stream function-vorticity formulation of the Stokes operator.

Let us *fix* a real $s \geqslant 4$ and let r be its dual exponent:

$$1/r + 1/s = 1.$$

Define the space of stream functions:

$$\Phi = \{\chi\in H^1(\Omega);\ \chi|_{\Gamma_0}=0,\ \chi|_{\Gamma_i}=\text{a constant } c_i,\ 1\leqslant i\leqslant p\}$$

where as usual $\Gamma_0, \ldots, \Gamma_p$ denote the connected components of the boundary Γ with exterior component Γ_0 (cf. Figure 2). We know that the operator **curl** is an isomorphism from

$$(4.27)\qquad \Phi_s = \Phi\cap W^{1,s}(\Omega)$$

onto

$$H\cap L^s(\Omega)^2 = \{\mathbf{v}\in L^s(\Omega)^2;\ \operatorname{div}\mathbf{v}=0,\ \mathbf{v}\cdot\mathbf{n}|_\Gamma=\mathbf{0}\}.$$

It is established in Theorem III.2.4 that when $\mathbf{f} \in L^r(\Omega)^2$, $\operatorname{curl} \mathbf{u} \in W^{1,r}(\Omega)$ and $p \in W^{1,r}(\Omega)$ then the Stokes Problem (4.8) is equivalent to:

Find $\psi \in \Phi_s$ *and* $\omega \in W^{1,r}(\Omega)$ *such that*:

$$\text{(4.28a)} \qquad \begin{cases} (\mathbf{curl}\,\omega, \mathbf{curl}\,\phi) = (\mathbf{f}, \mathbf{curl}\,\phi) & \forall \phi \in \Phi_s, \\ (\mathbf{curl}\,\psi, \mathbf{curl}\,\mu) = (\omega, \mu) & \forall \mu \in W^{1,r}(\Omega), \end{cases}$$

Find $p \in W^{1,r}(\Omega) \cap L_0^2(\Omega)$ *such that*:

$$\text{(4.28b)} \qquad (\mathbf{grad}\, p, \mathbf{grad}\, q) = (\mathbf{f} - \mathbf{curl}\,\omega, \mathbf{grad}\, q) \quad \forall q \in W^{1,s}(\Omega),$$

with $\mathbf{u} = \mathbf{curl}\,\psi$ *and* $\omega = \operatorname{curl} \mathbf{u}$.

Here again, it is necessary to write the Stokes problem in the form (4.28) for every right-hand side $\mathbf{f} \in L^r(\Omega)^2$. Therefore we assume that (4.18) holds. Then setting

$$X = \{\mathbf{curl}\,\phi;\ \phi \in \Phi_s\} \times L^2(\Omega) \times L_0^2(\Omega), \qquad Y = L^r(\Omega)^2,$$

the Stokes operator T is defined by:

$$T \in \mathscr{L}(Y; X), \qquad T\mathbf{f} = (\mathbf{curl}\,\psi, \omega, p) \quad \text{solution of (4.28).}$$

Next, in view of (2.27) the convection term satisfies the identities:

$$\text{(4.29)} \qquad \begin{aligned} \int_\Omega \left(\sum_{j=1}^2 u_j \partial \mathbf{u}/\partial x_j \right) \mathbf{curl}\,\phi\, dx &= \int_\Omega \omega\, \mathbf{grad}\,\psi \cdot \mathbf{curl}\,\phi\, dx \\ &= \int_\Omega \omega \left(\frac{\partial \psi}{\partial x_1} \frac{\partial \phi}{\partial x_2} - \frac{\partial \psi}{\partial x_2} \frac{\partial \phi}{\partial x_1} \right) dx, \end{aligned}$$

$$\text{(4.30)} \qquad \begin{aligned} \int_\Omega \left(\sum_{j=1}^2 u_j \partial \mathbf{u}/\partial x_j \right) \mathbf{grad}\, q\, dx &= \int_\Omega \omega\, \mathbf{grad}\,\psi \cdot \mathbf{grad}\, q\, dx \\ &\quad + (1/2) \int_\Omega \mathbf{grad}(\|\mathbf{u}\|^2) \cdot \mathbf{grad}\, q\, dx, \end{aligned}$$

where $\mathbf{curl}\,\psi = \mathbf{u}$, $\omega = \operatorname{curl} \mathbf{u}$ and $\|.\|$ denotes the Euclidean norm. Hence we introduce the nonlinearity by the mapping

$$\text{(4.31)} \qquad G(\lambda, u) = \lambda(\omega\, \mathbf{grad}\,\psi - \mathbf{f}) \quad \lambda \in \mathbb{R}_+, \quad u = (\mathbf{curl}\,\psi, \omega, p) \in X,$$

and we agree to include the term $(1/2)\|\mathbf{u}\|^2$ in the pressure:

i.e. we work instead with the kinematic pressure $p^* = p + (1/2)\|\mathbf{u}\|^2$. As $s \geqslant 4$, the terms $\omega\, \mathbf{grad}\,\psi$ and $\|\mathbf{u}\|^2$ belong respectively to $L^{4/3}(\Omega)^2$ and $L^2(\Omega)$.

Now consider the following problem for $\mathbf{f}$ in $L^r(\Omega)^2$:

Find $\psi \in \Phi_s$ *and* $\omega \in W^{1,r}(\Omega)$ *such that*

$$\text{(4.32a)} \qquad \begin{cases} \nu(\mathbf{curl}\,\omega, \mathbf{curl}\,\phi) + (\omega\, \mathbf{grad}\,\psi, \mathbf{curl}\,\phi) = (\mathbf{f}, \mathbf{curl}\,\phi) & \forall \phi \in \Phi_s, \\ (\mathbf{curl}\,\psi, \mathbf{curl}\,\mu) = (\omega, \mu) & \forall \mu \in W^{1,r}(\Omega); \end{cases}$$

Find $p^* \in W^{1,r}(\Omega) \cap L_0^2(\Omega)$ such that

(4.32b) $$(\mathbf{grad}\, p^*, \mathbf{grad}\, q) = (\mathbf{f} - \omega\, \mathbf{grad}\, \psi - \nu\, \mathbf{curl}\, \omega, \mathbf{grad}\, q) \quad \forall q \in W^{1,s}(\Omega).$$

With the above notations this problem reads:

$$F(\lambda, u) \equiv u + TG(\lambda, u) = 0$$

with $\lambda = 1/\nu$, $u = (\mathbf{curl}\, \psi, \omega, \lambda p^*) \in X$, G defined by (4.31) and T defined by (4.28).

Again, a routine calculation shows that if $(\mathbf{u}, p)$ is a solution of Problem (4.1) with curl $\mathbf{u}$ and p in $W^{1,r}(\Omega)$, $\mathbf{f}$ in $L^r(\Omega)^2$ then the triple $(\mathbf{curl}\, \psi, \omega, p^*)$ with

(4.33) $$\mathbf{curl}\, \psi = \mathbf{u}, \qquad \mathrm{curl}\, \mathbf{u} = \omega, \qquad p^* = p + (1/2)\, \|\mathbf{u}\|^2$$

is a solution of Problem (4.32). Conversely, each solution $(\mathbf{curl}\, \psi, \omega, p^*)$ of (4.32) is such that $\omega = -\Delta\psi$ and the pair $(\mathbf{u}, p)$ defined by (4.33) satisfies (4.1). In addition, when the Stokes operator has the regularity (4.18), each solution $u = (\mathbf{curl}\, \psi, \omega, p^*)$ of Problem (4.32) with $\mathbf{f}$ in $L^\gamma(\Omega)^2$ for some real $\gamma \in [r, 2]$ has the regularity $\psi \in W^{3,\gamma}(\Omega)$, $\omega \in W^{1,\gamma}(\Omega)$, $p^* \in W^{1,\gamma}(\Omega)$. Furthermore, just like in Remark 4.4, every branch of nonsingular solutions of Problem (4.1) with right-hand side $\mathbf{f}$ in $L^r(\Omega)^2$ is also a branch of nonsingular solutions of Problem (4.32) and conversely.

As far as the approximation is concerned, we assume that Ω is a polygonal domain of $\mathbb{R}^2$ in order to triangulate it entirely. Then let $\mathscr{T}_h$ be a family of triangulations of $\bar{\Omega}$ and $l \geqslant 1$ a fixed integer. We take:

(4.34) $$\begin{cases} \Theta_h = \{\theta_h \in \mathscr{C}^0(\bar{\Omega});\ \theta_h|_\kappa \in P_l \quad \forall \kappa \in \mathscr{T}_h\} \subset W^{1,\infty}(\Omega), \\ \Phi_h = \Theta_h \cap \Phi = \{\phi_h \in \Theta_h;\ \phi_h|_{\Gamma_0} = 0, \\ \qquad \phi_h|_{\Gamma_i} = \text{an arbitrary constant} c_i,\ 1 \leqslant i \leqslant p\}, \\ Q_h = \{q_h \in \mathscr{C}^0(\bar{\Omega}) \cap L_0^2(\Omega);\ q_h|_\kappa \in P_k \quad \forall \kappa \in \mathscr{T}_h\}, \qquad k = \min(1, l-1), \\ X_h = \{\mathbf{curl}\, \phi_h;\ \phi_h \in \Phi_h\} \times \Theta_h \times Q_h \subset X. \end{cases}$$

With these spaces, the Stokes problem is approximated by:

Find ψ_h in Φ_h and ω_h in Θ_h solution of

(4.35a) $$\begin{cases} (\mathbf{curl}\, \omega_h, \mathbf{curl}\, \phi_h) = (\mathbf{f}, \mathbf{curl}\, \phi_h) \quad \forall \phi_h \in \Phi_h, \\ (\mathbf{curl}\, \psi_h, \mathbf{curl}\, \mu_h) = (\omega_h, \mu_h) \quad \forall \mu_h \in \Theta_h; \end{cases}$$

Find p_h in Q_h such that

(4.35b) $$(\mathbf{grad}\, p_h, \mathbf{grad}\, q_h) = (\mathbf{f} - \mathbf{curl}\, \omega_h, \mathbf{grad}\, q_h) \quad \forall q_h \in Q_h.$$

The corresponding operator $T_h \in \mathscr{L}(Y; X_h)$ is defined by

$$T_h\mathbf{f} = (\mathbf{curl}\, \psi_h, \omega_h, p_h) \quad \text{solution of (4.35).}$$

Likewise, the Navier-Stokes problem (4.32) is discretized by:

Find $\psi_h \in \Phi_h$ *and* $\omega_h \in \Theta_h$ *satisfying:*

$$\text{(4.36a)} \quad \begin{cases} \nu(\mathbf{curl}\,\omega_h, \mathbf{curl}\,\phi_h) + (\omega_h\,\mathbf{grad}\,\psi_h, \mathbf{curl}\,\phi_h) = (\mathbf{f}, \mathbf{curl}\,\phi_h) & \forall \phi_h \in \Phi_h, \\ (\mathbf{curl}\,\psi_h, \mathbf{curl}\,\mu_h) = (\omega_h, \mu_h) \quad \forall \mu_h \in \Theta_h; \end{cases}$$

Find $p_h^* \in Q_h$ *such that:*

$$\text{(4.36b)} \qquad (\mathbf{grad}\,p_h^*, \mathbf{grad}\,q_h) = (\mathbf{f} - \omega_h\,\mathbf{grad}\,\psi_h - \nu\,\mathbf{curl}\,\omega_h, \mathbf{grad}\,q_h) \qquad \forall q_h \in Q_h.$$

In other words, this problem can also be written as:

$$F_h(\lambda, u_h) \equiv u_h + T_h G(\lambda, u_h) = 0,$$

with $\lambda = 1/\nu$, $u_h = (\mathbf{curl}\,\psi_h, \omega_h, \lambda p_h^*) \in X_h$, G defined by (4.31) and T_h by (4.35).

Now, let us apply Theorem 3.3 with $Z = Y$. Recall the approximation properties of the operator T_h derived in Section III.3.1 (cf. Theorem III.3.2). When Ω is a convex polygon, the Stokes problem has the regularity (4.18); in other words, if $\mathbf{f}$ belongs to $L^t(\Omega)^2$ with $r \leqslant t \leqslant 2$ the solution of the Stokes problem $(\mathbf{curl}\,\psi, \omega, p)$ belongs to $W^{2,t}(\Omega) \times W^{1,t}(\Omega) \times W^{1,t}(\Omega)$ and

$$\|\psi\|_{3,t,\Omega} + \|\omega\|_{1,t,\Omega} + |p|_{1,t,\Omega} \leqslant C_1 \|\mathbf{f}\|_{0,t,\Omega}.$$

Hence, if $\mathscr{T}_h$ is a uniformly regular family of triangulations of $\bar{\Omega}$, we have the following estimates for the solution $(\mathbf{curl}\,\psi_h, \omega_h, p_h)$ of Problem (4.35):

$$|\psi - \psi_h|_{1,s,\Omega} + \|\omega - \omega_h\|_{0,\Omega} + \|p - p_h\|_{0,\Omega} \leqslant C_2 h^\alpha \|\mathbf{f}\|_{0,t,\Omega},$$

with $r \leqslant t < 2$, $1/\gamma + 1/t = 1$, $\alpha = 1/\gamma$ when $l = 1$ and $\alpha = 2/\gamma$ when $l \geqslant 2$. This settles (3.37) and (3.38). Therefore the conclusion of Theorem 3.3 holds and combined with Theorem III.3.1, it gives the next result.

Theorem 4.4. 1°) *Let* Ω *be a bounded, convex polygon and let* $\mathscr{T}_h$ *be a uniformly regular family of triangulations of* $\bar{\Omega}$. *For* $\mathbf{f} \in L^t(\Omega)^2$, $t \in [r, 2)$, *let* $\{(\lambda, (\mathbf{curl}\,\psi(\lambda), \omega(\lambda), \lambda p^*(\lambda))); \lambda = 1/\nu \in \Lambda\}$ *be a branch of nonsingular solutions of the Navier-Stokes Problem* (4.32). *Then for* $h \leqslant h_0$ *small enough there exists a unique* $\mathscr{C}^\infty$ *branch* $\{(\lambda, (\mathbf{curl}\,\psi_h(\lambda), \omega_h(\lambda), \lambda p_h^*(\lambda))); \lambda = 1/\nu \in \Lambda\}$ *of solutions of Problem* (4.36) *that satisfies:*

$$\sup_{\lambda \in \Lambda} \{|\psi_h(\lambda) - \psi(\lambda)|_{1,s,\Omega} + \|\omega_h(\lambda) - \omega(\lambda)\|_{0,\Omega} + \|p_h^*(\lambda) - p^*(\lambda)\|_{0,\Omega}\}$$

$$\text{(4.37)} \qquad \leqslant C h^\alpha \quad \text{with } \alpha = \begin{cases} 1/\gamma & \text{if } l = 1 \\ 2/\gamma & \text{if } l \geqslant 2, \end{cases}$$

$1/t + 1/\gamma = 1$ *and the constant* C *is independent of* h *or* λ. *This bound is still valid when* $t = 2$ *and either* $l \geqslant 2$ *or* ψ *belongs also to* $W^{2,\infty}(\Omega)$. *When* $l = 1$ *and* $t = 2$, *the left-hand side of* (4.37) *is bounded by*

$$C(\varepsilon) h^{1/2 - \varepsilon} \quad \text{for all } \varepsilon > 0.$$

2°) *Moreover, when the mapping* $\lambda \to (\psi(\lambda), p^*(\lambda))$ *is continuous from* Λ *into* $[H^{m+2}(\Omega) \cap W^{m+3/2,\infty}(\Omega)] \times H^m(\Omega)$ *for some real* $m \in [1, l - 1/2]$ *we have the error estimate for all* λ *in* Λ:

$$(4.38)\quad |\psi_h(\lambda) - \psi(\lambda)|_{1,s,\Omega} + \|\omega_h(\lambda) - \omega(\lambda)\|_{0,\Omega} + \|p_h^*(\lambda) - p^*(\lambda)\|_{0,\Omega} \leqslant Kh^m.$$

Like in the linear case, it is possible to sharpen the above estimate for $\psi_h - \psi$ in the H_0^1 norm. However, the argument of Theorem 3.5 does not seem to apply here because it would have to bear on both ψ and ω whereas the estimate on ω is unlikely to be improved. Let us introduce instead a more direct duality argument. Since we are not concerned by the pressure, we take

$$X = \{\mathbf{curl}\, \phi;\ \phi \in \Phi_s\} \times L^2(\Omega), \qquad H = \{\mathbf{curl}\, \phi;\ \phi \in \Phi\} \times L^2(\Omega)$$

and we suppose that

$$\lambda \to u(\lambda) = (\mathbf{curl}\, \psi(\lambda), \omega(\lambda))$$

is a branch of nonsingular solutions of Problem (4.32a) with $\mathbf{f} \in L^2(\Omega)^2$. Again we assume that (4.18) holds so that $u(\lambda)$ belongs to $H^2(\Omega) \times H^1(\Omega)$. To simplify we denote

$$D = D_u G(\lambda, u(\lambda)).$$

Recall that

$$D \cdot v = \lambda(\omega(\lambda)\, \mathbf{grad}\, \phi + \theta\, \mathbf{grad}\, \psi(\lambda)) \qquad \forall v = (\mathbf{curl}\, \phi, \theta) \in X.$$

Then we introduce the operator $D^* \in \mathscr{L}(H; X')$ defined by:

$$(4.39)\qquad \langle D^* z, v\rangle = (\mathbf{curl}\, \chi, D \cdot v) \qquad \forall z = (\mathbf{curl}\, \chi, \mu) \in H.$$

To relate D and D^* recall that the space V associated with X is

$$V = \{v = (\mathbf{curl}\, \phi, \theta) \in X;\ \tilde{b}(v, \mu) = 0 \quad \forall \mu \in W^{1,r}(\Omega)\},$$

where
$$\tilde{b}(v, \mu) = (\mathbf{curl}\, \phi, \mathbf{curl}\, \mu) - (\theta, \mu).$$

Recall also that V is a Hilbert space for the scalar product

$$\tilde{a}(u, v) = (\omega, \theta) \qquad \forall u = (\mathbf{curl}\, \psi, \omega), \qquad v = (\mathbf{curl}\, \phi, \theta) \in V$$

and that the definition of the Stokes operator can be extended to

$$T \in \mathscr{L}(X'; V), \qquad \tilde{a}(Tl, v) = \langle l, v\rangle \qquad \forall v \in V.$$

Thus it follows readily from (4.39) that

$$\tilde{a}(TD^* z, v) = \tilde{a}(TDv, z) \qquad \forall z, v \in V.$$

In other words, TD^* *is the adjoint of* TD *in* V for the scalar product $\tilde{a}(.,.)$. As a consequence, since by assumption $I + TD$ is an isomorphism of V, then $I + TD^*$ *is also an isomorphism of* V.

Now we are in a position to define the dual linearized Navier-Stokes system:

For $g = \mathbf{g} \in L^2(\Omega)^2$ *find* $z = (\mathbf{curl}\,\chi, \mu) \in V$ *such that*:

(4.40) $$\tilde{a}((I + TD^*)z, v) = (\mathbf{g}, \mathbf{curl}\,\phi) \qquad \forall v = (\mathbf{curl}\,\phi, \theta) \in V;$$

i.e. $$z = (I + TD^*)^{-1}Tg.$$

Lemma 4.1. *Let* Ω *be a bounded, Lipschitz-continuous domain of* $\mathbb{R}^2$ *and suppose the Stokes problem has the regularity* (4.18). *Then the solution* $z = (\mathbf{curl}\,\chi, \mu)$ *of Problem* (4.40) *belongs to* $H^2(\Omega) \times H^1(\Omega)$ *with*

(4.41) $$\|\chi\|_{3,\Omega} + \|\mu\|_{1,\Omega} \leqslant C\{1 + \lambda(\|\psi(\lambda)\|_{3,\Omega} + \|\omega(\lambda)\|_{1,\Omega})\}\,\|\mathbf{g}\|_{0,\Omega}.$$

Proof. First, it stems from (4.40) that

(4.42) $$\|z\|_X \leqslant C_1 \|\mathbf{g}\|_{0,\Omega}.$$

Next we know from Lemma III.2.1 that all functions $v = (\mathbf{curl}\,\phi, \theta) \in V$ satisfy:

$$\mathbf{curl}\,\phi \in H_0^1(\Omega)^2, \qquad \theta = -\Delta\phi, \qquad \|\theta\|_{0,\Omega} = |v| \cong \|v\|_X.$$

Therefore

$$\begin{aligned}\langle D^*z, v\rangle &= \lambda(\mathbf{curl}\,\chi, \omega(\lambda)\,\mathbf{grad}\,\phi - \Delta\phi\,\mathbf{grad}\,\psi(\lambda)) \qquad \forall v \in V\\ &= \lambda(\mathbf{curl}(\mathbf{curl}\,\chi \cdot \mathbf{grad}\,\psi(\lambda)) - \omega(\lambda)\,\mathbf{grad}\,\chi, \mathbf{curl}\,\phi).\end{aligned}$$

Thus D^*z can be written in the form

$$\langle D^*z, v\rangle = (\mathbf{l}, \mathbf{curl}\,\phi) \qquad \forall v = (\mathbf{curl}\,\phi, \theta) \in V$$

where, in view of the regularity of $u(\lambda)$ and Sobolev's Imbedding Theorem I.1.3, $\mathbf{l} \in L^2(\Omega)^2$ with

$$\begin{aligned}\|\mathbf{l}\|_{0,\Omega} &\leqslant C_2\lambda(\|\omega(\lambda)\|_{1,\Omega} + \|\psi(\lambda)\|_{3,\Omega})\,\|\chi\|_{2,\Omega},\\ &\leqslant C_3\lambda(\|\omega(\lambda)\|_{1,\Omega} + \|\psi(\lambda)\|_{3,\Omega})\,\|\mathbf{g}\|_{0,\Omega},\end{aligned}$$

since $z \in V$ and satisfies (4.42). Then (4.41) follows from this last inequality, the regularity assumption (4.18) and the fact that z can also be expressed as $z = T(g - D^*z)$. □

Theorem 4.5. *Let* Ω *and* $\mathcal{T}_h$ *be like in Theorem* 4.4 *and suppose that the Navier-Stokes Problem* (4.42) *has a branch of nonsingular solutions such that the mapping* $\lambda \to \psi(\lambda)$ *is continuous from* Λ *into* $H^{l+1}(\Omega)$ *when* $l \geqslant 2$ *or* $H^3(\Omega)$ *when* $l = 1$. *Then we have the following estimate for all* λ *in* Λ:

(4.43) $$|\psi(\lambda) - \psi_h(\lambda)|_{1,\Omega} \leqslant \begin{cases} C_1 h^l & \text{if } l \geqslant 2,\\ C_2(\varepsilon)h^{1-\varepsilon} & \text{if } l = 1 \quad \forall \varepsilon > 0,\end{cases}$$

with constants independent of h *and* λ.

Proof. Let $\mathbf{g}$ be an arbitrary function of $L^2(\Omega)^2$ and $z = (\mathbf{curl}\,\chi, v) \in V$ the corresponding solution of the linearized problem (4.40):

$$(\mathbf{g}, \mathbf{curl}\,\phi) = \tilde{a}(z, v) + (\mathbf{curl}\,\chi, D\cdot v) + \tilde{b}(\mathbf{v}, v) \qquad \forall v = (\mathbf{curl}\,\phi, \theta) \in X.$$

In particular (dropping for the moment the parameter λ):

$$(4.44)\quad (\mathbf{g}, \mathbf{curl}(\psi - \psi_h)) = \tilde{a}(z, u - u_h) + (\mathbf{curl}\,\chi, D\cdot(u - u_h)) + \tilde{b}(u - u_h, v).$$

But we infer from (4.32a) and (4.36a) that $u - u_h$ satisfies:

$$(4.45)\quad \tilde{a}(u - u_h, z_h) + \tilde{b}(z_h, \omega) = -(G(\lambda, u) - G(\lambda, u_h), \mathbf{curl}\,\chi_h)$$

for every $z_h = (\mathbf{curl}\,\chi_h, v_h) \in V_h$ with

$$V_h = \{v_h = (\mathbf{curl}\,\phi_h, \theta_h);\ \phi_h \in \Phi_h,\ \theta_h \in \Theta_h,\ \tilde{b}(v_h, \mu_h) = 0 \quad \forall \mu_h \in \Theta_h\}.$$

Therefore, combining (4.44) and (4.45) and using the fact that u and z belong to V and u_h and z_h belong to V_h we obtain:

$$\begin{aligned}(\mathbf{g}, \mathbf{curl}(\psi - \psi_h)) &= \tilde{a}(z - z_h, u - u_h) + (\mathbf{curl}(\chi - \chi_h), D\cdot(u - u_h)) \\ &\quad + \tilde{b}(u - u_h, v - \theta_h) + \tilde{b}(z - z_h, \omega - \mu_h) \\ &\quad + (\mathbf{curl}\,\chi_h, D\cdot(u - u_h) - (G(\lambda, u) - G(\lambda, u_h))) \\ &\qquad \forall z_h \in V_h, \quad \forall \mu_h, \quad \theta_h \in \Theta_h.\end{aligned}$$

Let us choose $\theta_h = P_h v$ and $\mu_h = P_h\omega$; formula (A.25) gives for all $\phi_h, \delta_h \in \Phi_h$:

$$\tilde{b}(u - u_h, v - \theta_h) = (\mathbf{curl}(\psi - \phi_h), \mathbf{curl}(v - P_h v)) - (\omega - \omega_h, v - P_h v),$$

$$\tilde{b}(z - z_h, \omega - \mu_h) = (\mathbf{curl}(\chi - \delta_h), \mathbf{curl}(\omega - P_h\omega)) - (v - v_h, \omega - P_h\omega).$$

On the other hand, Taylor's formula (3.52) yields here:

$$\begin{aligned}G(\lambda, u) - G(\lambda, u_h) - D_u G(\lambda, u)\cdot(u - u_h) &= -(1/2) D_{uu}^2 G\cdot(u - u_h)^2 \\ &= -\lambda(\omega - \omega_h)\,\mathbf{grad}(\psi - \psi_h).\end{aligned}$$

Hence for all $z_h \in V_h$, ϕ_h and $\delta_h \in \Phi_h$, we have:

$$(4.46)\quad \begin{aligned}(\mathbf{g}, \mathbf{curl}(\psi - \psi_h)) &= \tilde{a}(z - z_h, u - u_h) + \lambda(\mathbf{curl}(\chi - \chi_h), \omega\,\mathbf{grad}(\psi - \psi_h)) \\ &\quad + \lambda(\mathbf{curl}(\chi - \chi_h), (\omega - \omega_h)\,\mathbf{grad}\,\psi) \\ &\quad + \lambda(\mathbf{curl}\,\chi_h, (\omega - \omega_h)\,\mathbf{grad}(\psi - \psi_h)) \\ &\quad + (\mathbf{curl}(\psi - \phi_h), \mathbf{curl}(v - P_h v)) - (\omega - \omega_h, v - P_h v) \\ &\quad + (\mathbf{curl}(\chi - \delta_h), \mathbf{curl}(\omega - P_h\omega)) - (v - v_h, \omega - P_h\omega),\end{aligned}$$

from which we readily infer that

$$
\begin{aligned}
|(\mathbf{g}, \mathbf{curl}(\psi - \psi_h))| \leqslant {} & |u - u_h|(|z - z_h| + \lambda|\psi|_{1,4,\Omega}|\chi - \chi_h|_{1,4,\Omega} \\
& + \lambda|\chi_h|_{1,4,\Omega}|\psi - \psi_h|_{1,4,\Omega} + \|v - P_h v\|_{0,\Omega}) \\
& + \lambda|u||\psi - \psi_h|_{1,4,\Omega}|\chi - \chi_h|_{1,4,\Omega} \\
& + |v - P_h v|_{1,\Omega}|\psi - \phi_h|_{1,\Omega} \\
& + |\chi - \delta_h|_{1,\Omega}|\omega - P_h\omega|_{1,\Omega} \\
& + |z - z_h|\,\|\omega - P_h\omega\|_{0,\Omega} \\
& \qquad \forall z_h \in V_h \quad \forall \phi_h, \delta_h \in \Phi_h.
\end{aligned} \tag{4.47}
$$

Finally, recall that (cf. Lemma III.3.3 and Remark III.3.2):

$$\inf_{z_h \in V_h} \|z - z_h\|_X \leqslant \begin{cases} C_1 h \|\chi\|_{3,\Omega} & \text{if } l \geqslant 2 \\ C_2(\varepsilon) h^{1/2-\varepsilon} \|\chi\|_{3,\Omega} & \text{if } l = 1 \quad \forall \varepsilon > 0; \end{cases}$$

on the other hand,

$$\inf_{\delta_h \in \Phi_h} |\chi - \delta_h|_{1,\Omega} \leqslant C_3 h^\alpha |\psi|_{\alpha+1,\Omega} \quad \text{with } \alpha = \min(l, 2);$$

and likewise:

$$\inf_{\phi_h \in \Phi_h} |\psi - \phi_h|_{1,\Omega} \leqslant C_4 h^l |\psi|_{l+1,\Omega},$$

$$\|\omega - P_h\omega\|_{0,\Omega} + h|\omega - P_h\omega|_{1,\Omega} \leqslant C_5 h^\beta |\omega|_{\beta,\Omega}$$

with $\beta = 1$ when $l = 1$ and $\beta = l - 1$ when $l \geqslant 2$. By substituting these bounds into (4.47) and applying Theorem 4.4 and Lemma 4.1 we easily derive (4.43). □

4.4. Remarks on the "Stream Function-Gradient of Velocity Tensor" Scheme

With minor modifications, the approach of Section 4.3 can be applied to the "stream function-gradient of velocity tensor" method for the Navier-Stokes equations, at least when Ω is a *plane, convex polygon.* Going back to Paragraph 4, Chapter III, recall the bilinear forms:

$$a_h(\sigma, \tau) = \int_\Omega \sigma_{ij}\tau_{ij}\, dx,$$

$$b_h(\tau, \phi) = -\sum_{\kappa \in \mathscr{T}_h} \int_\kappa \tau_{ij}(\partial^2\phi/\partial x_i \partial x_j)\, dx + \int_{\Gamma_h} M_n(\tau) S(\partial\phi/\partial n)\, ds,$$

and the spaces

$$
\begin{aligned}
\tilde{\Sigma} = \{ & \tau = (\tau_{ij}) \in L^2(\Omega)^4;\ \tau_{12} = \tau_{21},\ \tau|_\kappa \in W^{1,r}(\kappa)^4 \quad \forall \kappa \in \mathscr{T}_h, \\
& M_n(\tau) \text{ is continuous on each segment of } \Gamma_h\},
\end{aligned}
$$

$$\tilde{\Psi} = \{\phi \in \Phi_s;\ \phi|_\kappa \in H^2(\kappa) \quad \forall \kappa \in \mathscr{T}_h\},$$

for a *fixed* $s \geqslant 4$ and $1/r + 1/s = 1$. For $\mathbf{f} \in L^r(\Omega)^2$, we know that the Stokes operator can be expressed by:

$$\begin{cases} T\mathbf{f} = u = (\sigma, \psi) \in \tilde{\Sigma} \times \tilde{\Psi}, \\ b_h(\sigma, \phi) = -(\mathbf{f}, \mathbf{curl}\, \phi) \quad \forall \phi \in \tilde{\Psi}, \\ a_h(\sigma, \tau) + b_h(\tau, \psi) = 0 \quad \forall \tau \in \tilde{\Sigma}. \end{cases}$$

Setting

$$Y = L^r(\Omega)^2,$$

$$X = \{\sigma = (\sigma_{ij}) \in L^2(\Omega)^4;\ \sigma_{12} = \sigma_{21}\} \times \Phi_s,$$

we have

$$T \in \mathscr{L}(Y; X)$$

and since the Stokes problem is regular we also have

$$T \in \mathscr{L}(Y; W^{1,r}(\Omega)^4 \times W^{3,r}(\Omega)).$$

In view of (4.31) we introduce the nonlinear convection term by:

$$G(\lambda, u) = -\lambda(\operatorname{tr}(\sigma)\, \mathbf{grad}\, \psi + \mathbf{f})$$

which is a $\mathscr{C}^\infty$-mapping from $\Lambda \times X$ into $L^r(\Omega)^2$. With these notations, the Navier-Stokes equations take the standard form:

$$\text{(4.48)} \qquad u(\lambda) \in X, \quad F(\lambda, u(\lambda)) \equiv u(\lambda) + TG(\lambda, u(\lambda)) = 0 \quad \forall \lambda \in \Lambda.$$

As far as the approximation is concerned, we take

$$X_h = \Sigma_h \times \Phi_h \subset X,$$

where

$$\Sigma_h = \{\tau \in \tilde{\Sigma};\ \tau|_\kappa \in P_{l-1}^4 \quad \forall \kappa \in \mathscr{T}_h\},$$

$$\Phi_h = \{\phi \in \Phi;\ \phi|_\kappa \in P_l \quad \forall \kappa \in \mathscr{T}_h\}$$

for some integer $l \geqslant 1$. The Stokes operator is discretized by

$$T_h\mathbf{f} = u_h = (\sigma_h, \psi_h) \in X_h,$$

$$\left.\begin{aligned} b_h(\sigma_h, \phi_h) &= -(\mathbf{f}, \mathbf{curl}\, \phi_h) \\ a_h(\sigma_h, \tau_h) + b_h(\tau_h, \psi_h) &= 0 \end{aligned}\right\} \quad \forall v_h = (\tau_h, \phi_h) \in X_h;$$

and the Navier-Stokes equations are approximated by:

$$\text{(4.49)} \qquad u_h(\lambda) \in X_h, \quad F_h(\lambda, u_h(\lambda)) \equiv u_h(\lambda) + T_hG(\lambda, u_h(\lambda)) = 0 \quad \forall \lambda \in \Lambda.$$

The results of Section III.4.3 give the following estimate for $T - T_h$:

$$\begin{aligned} \|(T - T_h)\mathbf{f}\|_X &\equiv \|\sigma - \sigma_h\|_{0,\Omega} + |\psi - \psi_h|_{1,s,\Omega} \\ &\leqslant Ch^{2(1-1/r)}|\psi|_{3,r,\Omega}. \end{aligned}$$

Hence applying Theorems 3.3 and III.4.2 and Corollary III.4.2 we obtain the following error bound for $u - u_h$ with u (resp. u_h) satisfying (4.48) (resp. (4.49)):

$$\| u(\lambda) - u_h(\lambda) \|_X \leqslant C h^k$$

provided $\psi(\lambda) \in H^{k+2}(\Omega)$ for some $k \in [1, l]$.

Finally if $\psi(\lambda) \in H^{l+1}(\Omega)$ when $l \geqslant 2$ or $\psi(\lambda) \in H^3(\Omega)$ when $l = 1$, a duality argument similar to that of Theorem 4.5 yields the optimal estimate:

$$|\psi(\lambda) - \psi_h(\lambda)|_{1,\Omega} \leqslant C h^l.$$

By applying the material of Section III.4.4, we can derive error bounds of the same order for the pressure.

The proofs are left as exercises.

§ 5. Numerical Analysis of Upwind Schemes

When the viscosity ν is small, or equivalently when the Reynolds number Re is large compared to the other parameters of the fluid, there arises a boundary layer in the neighborhood of Γ where the viscosity predominates while it is negligible in the interior of Ω. At the same time, the flow becomes turbulent. Thus the solutions of the Navier-Stokes equations are seriously discontinuous at high Reynolds number.

It is not our purpose here to modelize turbulent flows, but all the same, it is worthwhile to examine one possible discretization of discontinuous solutions of the Navier-Stokes equations. Instead of the centered schemes studied so far, we propose upwind schemes which are better adapted to describe discontinuous flows. The forthcoming analysis shows that these schemes are nearly optimal.

The reader can also refer to Johnson & Saranen [47] for an alternate method, the streamline diffusion method, that applies to Euler and Navier-Stokes equations.

5.1. Upwinding in the Stream Function-Vorticity Scheme

The upwind scheme discussed in this section was developed by Fortin [31], inspired by a numerical method advocated in Lesaint & Raviart [51] to solve a neutron transport equation. It stems from the following heuristic remarks. When a smooth vector field $\mathbf{u}$ is divergence-free, the convection term satisfies (with the usual summation convention that a repeated index represents a sum):

$$u_j(\partial u_i / \partial x_j) = \partial(u_j u_i)/\partial x_j.$$

Now, suppose Ω is the union of two bounded regions Ω_1 and Ω_2 separated by an interface S like in Figure 21. Denote by Γ the boundary of Ω and by $\mathbf{n}_i$ the

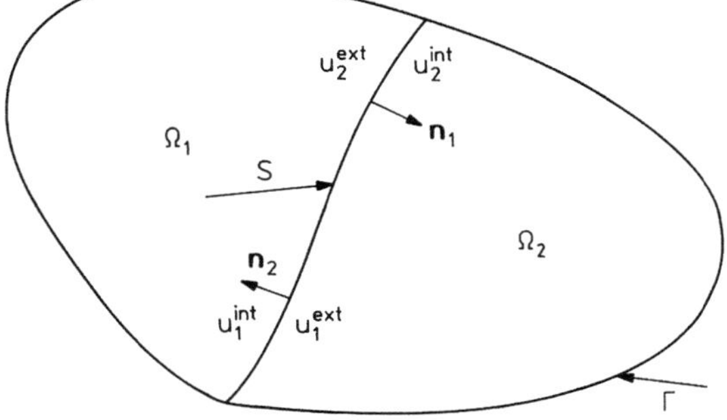

Figure 21

unit exterior normal to Ω_i on S. Assume that the vector field $\mathbf{u}$ is no longer globally smooth in Ω but that instead,

$$\mathbf{u}|_{\Omega_k} \in H^1(\Omega_k)^N, \quad \mathbf{u}|_{\Gamma} = \mathbf{0}, \quad \mathbf{u} \in H(\mathrm{div}; \Omega) \quad \text{and} \quad \mathrm{div}\, \mathbf{u} = 0.$$

Then the product $u_j u_i$ is no longer differentiable in Ω but its distributional derivative has the form:

$$[\partial(u_j u_i)/\partial x_j] = \partial(u_j u_i)/\partial x_j|_{\Omega_1 \cup \Omega_2}$$
+ a surface distribution corresponding to the jump of the field $\mathbf{u}$ across the interface S, namely:
$$\mathbf{u} \cdot \mathbf{n}(u_i^{ext} - u_i^{int})\, ds$$

where $\mathbf{n}$ and the notions of exterior and interior refer to the same region. The upwinding relies on the following principle:

the flow in Ω_k depends exclusively upon the flux entering through S.

By virtue of this principle, we introduce the notation:

$$\partial\_\Omega_k = \{x \in S;\, \mathbf{u} \cdot \mathbf{n}_k(x) < 0\} \tag{5.1}$$

for the portion of the interface S where the fluid enters Ω_k and we take as a *definition* the following expression for the convection term:

$$\int_\Omega u_j(\partial u_i/\partial x_j) v_i\, dx \overset{\text{def}}{=} \tag{5.2}$$
$$\sum_{k=1}^{2} \int_{\Omega_k} u_j(\partial u_i/\partial x_j) v_i\, dx + \sum_{k=1}^{2} \int_{\partial\_\Omega_k} \mathbf{u} \cdot \mathbf{n}_k (u_i^{ext} - u_i^{int}) v_i^{int}\, ds$$

for all $\mathbf{u}$ in $H(\mathrm{div}; \Omega)$ with $\mathrm{div}\, \mathbf{u} = 0$, $\mathbf{u}|_{\Omega_k} \in H^1(\Omega_k)^N$ and $\mathbf{v}|_{\Omega_k} \in H^1(\Omega_k)^N$. Clearly, when $\mathbf{u}$ and $\mathbf{v}$ belong to $H^1(\Omega)^N$, the surface term vanishes and we recover the familiar expression for the convection. But when $\mathbf{u}$ and $\mathbf{v}$ are not globally in H^1 we have to interpret (5.2) as a definition.

We propose to extend the above considerations to the case where Ω is triangulated, each triangle being considered as a subregion of Ω. But, beforehand, let us recall the setting of the stream function-vorticity formulation and approximation of the Navier-Stokes system developed in Section 4.3.

Let Ω be a bounded domain in $\mathbb{R}^2$ with a polygonal boundary Γ, so that it can be entirely triangulated and suppose Ω is *convex* so the regularity condition (4.18) holds. To simplify the discussion we agree to drop the pressure for the moment and consider only the stream function and vorticity. Then we set

$$Y = L^r(\Omega)^2, \qquad X = \{\mathbf{curl}\,\phi;\ \phi \in \Phi_s\} \times L^2(\Omega)$$

with $s \geqslant 4$ and $1/r + 1/s = 1$. The Stokes operator $T \in \mathscr{L}(Y; X)$ is defined by

$$T\mathbf{f} = (\mathbf{curl}\,\psi, \omega) \quad \text{where}$$

$$\text{(5.3)} \qquad \begin{cases} (\mathbf{curl}\,\omega, \mathbf{curl}\,\phi) = (\mathbf{f}, \mathbf{curl}\,\phi) & \forall \phi \in \Phi_s, \\ (\mathbf{curl}\,\psi, \mathbf{curl}\,\mu) = (\omega, \mu) & \forall \mu \in W^{1,r}(\Omega), \end{cases}$$

where ψ and ω are related to the velocity $\mathbf{u}$ by

$$\text{(5.4)} \qquad \mathbf{u} = \mathbf{curl}\,\psi \quad \text{and} \quad \omega = \operatorname{curl} \mathbf{u}.$$

As usual, we introduce the subspace of X

$$\text{(5.5)} \qquad V = \{(\mathbf{curl}\,\phi, \theta) \in X;\ \phi \in H_0^2(\Omega),\ \theta = -\Delta\phi\},$$

(Γ has only one connected component since Ω is convex) and we know from Lemma III.2.1 that $T \in \mathscr{L}(Y; V)$.

The nonlinearity is expressed much like in (4.31):

$$\text{(5.6)} \qquad G(u) = \omega\,\mathbf{grad}\,\psi - \mathbf{f} \qquad u = (\mathbf{curl}\,\psi, \omega) \in X.$$

Then the Navier-Stokes problem is stated like in (4.32):

For $\mathbf{f}$ *in* $L^r(\Omega)^2$ *find* $u = (\mathbf{curl}\,\psi, \omega) \in X$ *such that*:

$$\text{(5.7)} \qquad \begin{cases} \nu(\mathbf{curl}\,\omega, \mathbf{curl}\,\phi) + (\omega\,\mathbf{grad}\,\psi, \mathbf{curl}\,\phi) = (\mathbf{f}, \mathbf{curl}\,\phi) & \forall \phi \in \Phi_s, \\ (\mathbf{curl}\,\psi, \mathbf{curl}\,\mu) = (\omega, \mu) & \forall \mu \in W^{1,r}(\Omega), \end{cases}$$

where ψ and ω are related to $\mathbf{u}$ by (5.4). With T and G defined above and $\lambda = 1/\nu$, this has the compact expression:

$$\text{(5.8)} \qquad F(\lambda, u) \equiv u + \lambda T G(u) = 0.$$

For the purpose of the approximation, we introduce the first two finite-dimensional spaces Θ_h and Φ_h defined by (4.34):

$$\Theta_h = \{\theta_h \in \mathscr{C}^0(\bar{\Omega});\ \theta_h|_\kappa \in P_l \quad \forall \kappa \in \mathscr{T}_h\}, \qquad \Phi_h = \Theta_h \cap H_0^1(\Omega),$$

where $\mathscr{T}_h$ is a triangulation of $\bar{\Omega}$ and the integer $l \geqslant 1$. Naturally we take

$$X_h = \{\mathbf{curl}\,\phi_h;\ \phi_h \in \Phi_h\} \times \Theta_h \subset X$$

and we define the corresponding subspace

(5.9) $$V_h = \{(\mathbf{curl}\,\phi_h, \theta_h) \in X_h;\ (\mathbf{curl}\,\phi_h, \mathbf{curl}\,\mu_h) = (\theta_h, \mu_h) \quad \forall \mu_h \in \Theta_h\}.$$

The discrete Stokes operator can be defined on a wider space than Y, namely:

$$Y_h = \text{the dual space of } \{\mathbf{curl}\,\phi_h;\ \phi_h \in \Phi_h\}$$

equipped with the norm

$$\|l\|_* = \sup_{\phi_h \in \Phi_h} \frac{\langle l, \mathbf{curl}\,\phi_h \rangle}{|\phi_h|_{1,s,\Omega}}.$$

By setting

$$\langle l, v_h \rangle = \langle l, \mathbf{curl}\,\phi_h \rangle \qquad \forall v_h = (\mathbf{curl}\,\phi_h, \theta_h) \in V_h,$$

the space Y_h can also be identified with a subspace of V_h', the dual space of V_h; thus we have the following continuous imbeddings:

$$Y \hookrightarrow Y_h \hookrightarrow V_h'.$$

It will be useful further on to provide Y_h with the norm of V_h', i.e. we put

$$\|l\|_h = \sup_{v_h \in V_h} \frac{\langle l, v_h \rangle}{\|v_h\|_X} \qquad \forall l \in Y_h;$$

clearly we have:

$$\|l\|_h \leqslant \|l\|_* \qquad \forall l \in Y_h.$$

Then we define the approximate Stokes operator $T_h \in \mathscr{L}(Y_h; X_h)$ by:

For $l \in Y_h$, find $T_h l = (\mathbf{curl}\,\psi_h, \omega_h) \in X_h$ solution of

(5.10) $$\begin{cases} (\mathbf{curl}\,\omega_h, \mathbf{curl}\,\phi_h) = \langle l, \mathbf{curl}\,\phi_h \rangle & \forall \phi_h \in \Phi_h, \\ (\mathbf{curl}\,\psi_h, \mathbf{curl}\,\mu_h) = (\omega_h, \mu_h) & \forall \mu_h \in \Theta_h. \end{cases}$$

Obviously the range of the operator T_h is the space V_h.

Now we turn to the upwind discretization of the convection term. Considering that Ω is the union of all the triangles κ of $\mathscr{T}_h$, formula (5.2) can be generalized to yield the following definition:

(5.11) $$\int_\Omega u_j(\partial u_i/\partial x_j) v_i\, dx \stackrel{\text{def}}{=} \sum_{\kappa \in \mathscr{T}_h} \left\{ \int_\kappa u_j(\partial u_i/\partial x_j) v_i\, dx + \int_{\partial_- \kappa} \mathbf{u} \cdot \mathbf{n}(u_i^{ext} - u_i^{int}) v_i^{int}\, ds \right\}.$$

This induces us to split the convection term and replace the single trilinear form by two forms: one for the volume integrals and one for the line integrals. Thus for all $u = (\mathbf{curl}\,\psi, \omega)$ and $v = (\mathbf{curl}\,\phi, \theta)$ in X_h and all $\mathbf{w}$ of the form $w_i = \partial\chi/\partial x_j$ with χ in Φ_h we set:

$$a_1(u; v, \mathbf{w}) = \sum_{\kappa \in \mathcal{T}_h} \int_{\kappa} u_j(\partial v_i/\partial x_j) w_i \, dx \tag{5.12}$$

where $\mathbf{u} = \mathbf{curl}\,\psi$ and $\mathbf{v} = \mathbf{curl}\,\phi$; likewise for $z = (\mathbf{curl}\,v, \mu)$ in X_h we take:

$$a_2^z(u; v, \mathbf{w}) = \sum_{\kappa \in \mathcal{T}_h} \int_{\partial_-\kappa(z)} \mathbf{u}\cdot\mathbf{n}(v_i^{\text{ext}} - v_i^{\text{int}}) w_i^{\text{int}} \, ds, \tag{5.13}$$

where $\mathbf{n}$ denotes always the unit exterior normal to κ, the superscript *ext* (resp. *int*) denotes the external (resp. internal) trace of the function on the boundary of κ and

$$\partial_-\kappa(z) = \{x \in \partial\kappa;\ (\mathbf{z}\cdot\mathbf{n})(x) < 0\}, \qquad \mathbf{z} = \mathbf{curl}\,v, \tag{5.14}$$

i.e. $\partial_-\kappa(z)$ is the portion of the boundary of κ where the fluid with velocity $\mathbf{z}$ enters κ. Finally, we introduce the form

$$\tilde{a}^z(u; v, \mathbf{w}) = a_1(u; v, \mathbf{w}) + a_2^z(u; v, \mathbf{w}) \tag{5.15}$$

and we define the mapping G_h: $v \in X_h \to G_h(v) \in Y_h$ by

$$\langle G_h(v), \mathbf{curl}\,\chi\rangle = \tilde{a}^v(v; v, \mathbf{curl}\,\chi) - (\mathbf{f}, \mathbf{curl}\,\chi) \qquad \forall \chi \in \Phi_h. \tag{5.16}$$

The continuity of G_h will be established subsequently, but right away we observe that owing to the dependence of $a_2^v(v; v, \mathbf{w})$ on $\partial_-\kappa(v)$ the mapping G_h is not differentiable. Note that when $\mathbf{w} \in H^1(\Omega)^2$ the form a satisfies the crucial property justifying definition (5.11):

$$\tilde{a}^z(u; v, \mathbf{w}) = -\int_{\Omega} u_j(\partial w_i/\partial x_j) v_i \, dx, \tag{5.17}$$

whatever the function z, provided of course $\operatorname{div}\mathbf{u} = 0$ in Ω and $\mathbf{u}\cdot\mathbf{n} = 0$ on Γ.

Then the Navier-Stokes Problem (5.7) has the following upwind discretization:

Find $\psi_h \in \Phi_h$ and $\omega_h \in \Theta_h$ solution of

$$\begin{cases} \nu(\mathbf{curl}\,\omega_h, \mathbf{curl}\,\phi_h) + \tilde{a}^{u_h}(u_h; u_h, \mathbf{curl}\,\phi_h) = (\mathbf{f}, \mathbf{curl}\,\phi_h) & \forall \phi_h \in \Phi_h, \\ (\mathbf{curl}\,\psi_h, \mathbf{curl}\,\mu_h) = (\omega_h, \mu_h) \quad \forall \mu_h \in \Theta_h, \end{cases} \tag{5.18}$$

with $\tilde{a}$ defined by (5.12)–(5.15). Keeping in mind the above notations, this problem takes the form of (3.58):

$$F_h(\lambda, u_h) \equiv u_h + \lambda T_h G_h(u_h) = 0, \tag{5.19}$$

where $u_h = (\mathbf{curl}\,\psi_h, \omega_h)$ and $\lambda = 1/\nu$. Note that F_h maps $\Lambda \times X_h$ into V_h.

5.2. Error Analysis of the Upwind Scheme

Before applying the material of Section 3.4 to establish the convergence of (5.18), let us prove some important properties of the form $\tilde{a}(.;.,.)$. The reader will find that this form is not as hard to manipulate as it would seem at first sight. First, it stands out clearly from the expression of $\tilde{a}(.;.,.)$ that it requires a finer norm than the $W^{1,s}$-norm of the space Φ. To be specific, it is desirable to work with the seminorm

$$\left(\sum_{\kappa\in\mathscr{T}_h}|\phi|_{2,\kappa}^2\right)^{1/2}.$$

This is precisely the purpose of the space V_h: it was shown in Chapter III that, under adequate conditions, the functions of V_h satisfy the uniform stability properties:

(5.20) $\quad |\phi_h|_{1,t,\Omega}\leqslant C_t\|\theta_h\|_{0,\Omega}\quad$ (cf. (III.2.41)) $\quad\forall v_h=(\mathbf{curl}\,\phi_h,\theta_h)\in V_h,$

(5.21) $$\left(\sum_{\kappa\in\mathscr{T}_h}|\phi_h|_{2,\kappa}^2\right)^{1/2}\leqslant C\|\theta_h\|_{0,\Omega}\quad\text{(cf. (III.2.65))}$$

with constants independent of h. In other words:

the seminorm of X: $|v_h|\equiv\|\theta_h\|_{0,\Omega}$ is a norm on V_h uniformly equivalent to both

$$|\phi_h|_{1,t,\Omega}+\|\theta_h\|_{0,\Omega}\quad\textit{and}\quad\left(\sum_{\kappa\in\mathscr{T}_h}|\phi_h|_{2,\kappa}^2\right)^{1/2}+\|\theta_h\|_{0,\Omega}\quad\textit{for each } t\geqslant 2.$$

This suggests to adopt the following mesh-dependent norms:

(5.22) $$[\phi]_t=\left(\sum_{\kappa\in\mathscr{T}_h}|\phi|_{2,\kappa}^2\right)^{1/2}+h^{-2/t}|\phi|_{1,t,\Omega}\qquad\forall t\geqslant 2.$$

Next, since the functions of X have little regularity, it will sometimes be useful to work with the following smoothing operator:

$$P\in\mathscr{L}(X;\{\mathbf{curl}\,\phi;\ \phi\in H^2(\Omega)\cap H_0^1(\Omega)\}\times L^2(\Omega))$$

defined by

for $u=(\mathbf{curl}\,\psi,\omega)\in X$, $Pu=(\mathbf{curl}\,\phi,\omega)$ is the solution of the Dirichlet problem

(5.23) $$-\Delta\phi=\omega\quad\text{in }\Omega,\qquad\phi=0\quad\text{on }\Gamma.$$

When Ω is *convex*, this function ϕ belongs indeed to $H^2(\Omega)$ and

(5.24) $$\|\phi\|_{2,\Omega}\leqslant C\|\omega\|_{0,\Omega}.$$

Clearly, the operator P reduces to the identity mapping on V. Moreover, when $u_h=(\mathbf{curl}\,\psi_h,\omega_h)$ belongs to V_h the corresponding function ϕ coincides with the function $\phi(h)$ introduced in the proof of Lemma III.2.5. In other words, $u_h=(\mathbf{curl}\,\psi_h,\omega_h)\in V_h$ and $Pu_h=(\mathbf{curl}\,\phi,\omega_h)$ are related by

$$\psi_h=\mathring{P}_h\phi,$$

the H^1-projection on Φ_h defined by (A.24):

(5.25) $$(\mathbf{curl}(\mathring{P}_h\phi-\phi),\mathbf{curl}\,\mu_h)=0\qquad\forall\mu_h\in\Phi_h.$$

The next lemma gives a useful estimate for $[\phi-\mathring{P}_h\phi]_t$.

Lemma 5.1. *Assume that Ω is a bounded, convex polygon and $\mathcal{T}_h$ a uniformly regular family of triangulations of $\bar{\Omega}$. Let $\phi \in H^{k+1}(\Omega) \cap H_0^1(\Omega)$ for some integer k with $1 \leqslant k \leqslant l$. Then we have the estimate:*

$$(5.26) \qquad [\phi - \mathring{P}_h\phi]_t \leqslant C_t h^{k-1} \|\phi\|_{k+1,\Omega} \qquad \forall t \geqslant 2,$$

with constants C_t independent of h.

This lemma is an easy consequence of Theorem A.2 and

$$(5.27) \qquad [\phi_h]_t \leqslant C(t) h^{-1} |\phi_h|_{1,\Omega} \qquad \forall \phi_h \in \Phi_h.$$

These preliminaries permit to derive some fundamental properties of $\tilde{a}(.;.,.)$. First $\tilde{a}(.;.,.)$ is bounded.

Lemma 5.2. *Let us retain the assumptions of Lemma 5.1. The form $\tilde{a}(.;.,.)$ is bounded as follows,* for all $v_h \in V_h$ *and all* z_h, $u_h = (\mathbf{u}_h = \mathbf{curl}\,\psi_h, \omega_h)$, $w_h = (\mathbf{w}_h = \mathbf{curl}\,\chi_h, \varepsilon_h) \in X_h$:

$$(5.28) \qquad |a_1(u_h; v_h, \mathbf{w}_h)| + |a_2^{z_h}(u_h; v_h, \mathbf{w}_h)| \leqslant C \|\mathbf{u}_h\|_{0,4,\Omega} |v_h| \, \|\mathbf{w}_h\|_{0,4,\Omega}.$$

Proof. Let $v_h = (\mathbf{curl}\,\phi_h, \theta_h)$ belong to V_h. By Hölder's inequality, we have:

$$|a_1(u_h; v_h, \mathbf{w}_h)| \leqslant C_1 |\psi_h|_{1,4,\Omega} |\chi_h|_{1,4,\Omega} \left(\sum_{\kappa \in \mathcal{T}_h} |\phi_h|_{2,\kappa}^2 \right)^{1/2}$$

Thus (5.21) gives:

$$|a_1(u_h; v_h, \mathbf{w}_h)| \leqslant C_2 \|\mathbf{u}_h\|_{0,4,\Omega} |v_h| \, \|\mathbf{w}_h\|_{0,4,\Omega}.$$

Next, consider $a_2^{z_h}(u_h; Pv_h, \mathbf{w}_h)$ with $Pv_h = (\mathbf{curl}\,\phi, \theta_h)$. As ϕ belongs to $H^2(\Omega)$, the surface integrals in $a_2(.;.,.)$ vanish over all interior segments κ' of $\mathcal{T}_h$ and since u_h belongs to X_h the factor $\mathbf{curl}\,\psi_h \cdot \mathbf{n}$ vanishes on all boundary segments of $\mathcal{T}_h$. Hence

$$(5.29) \qquad a_2^{z_h}(u_h; Pv_h, \mathbf{w}_h) = 0,$$

so that

$$|a_2^{z_h}(u_h; v_h, \mathbf{w}_h)| = |a_2^{z_h}(u_h; v_h - Pv_h, \mathbf{w}_h)|.$$

Therefore

$$|a_2^{z_h}(u_h; v_h, \mathbf{w}_h)| \leqslant C_3 \left(\sum_{\kappa \in \mathcal{T}_h} \|\mathbf{curl}\,\psi_h\|_{0,4,\partial\kappa}^4 \right)^{1/4}$$
$$\times \left(\sum_{\kappa \in \mathcal{T}_h} \|\mathbf{curl}\,\chi_h\|_{0,4,\partial\kappa}^4 \right)^{1/4} \left(\sum_{\kappa \in \mathcal{T}_h} \|\mathbf{curl}(\phi_h - \phi)\|_{0,\partial\kappa}^2 \right)^{1/2}$$

On the one hand a routine application of

$$\|\theta\|_{0,t,\partial\kappa} \leqslant C_4 (h_\kappa^{1/t}/\rho_\kappa)(\|\theta\|_{0,\kappa}^2 + h_\kappa^2 |\theta|_{1,\kappa}^2)^{1/2} \qquad \forall \theta \in H^1(\kappa) \qquad \forall t \geqslant 1,$$

yields:

$$\Big(\sum_{\kappa\in\mathscr{T}_h} \|\mathbf{curl}(\phi_h-\phi)\|_{0,\partial\kappa}^2\Big)^{1/2} \leqslant C_5 h^{-1/2}\Big(|\phi_h-\phi|_{1,\Omega} + h\Big(\sum_{\kappa\in\mathscr{T}_h}|\phi_h-\phi|_{2,\kappa}^2\Big)^{1/2}\Big).$$

On the other hand

$$\|\mathbf{curl}\,\theta_h\|_{0,4,\partial\kappa} \leqslant C_6 h^{-1/4}|\theta_h|_{1,4,\kappa} \qquad \forall\kappa\in\mathscr{T}_h \qquad \forall\theta_h\in\Theta_h.$$

Combining these inequalities with (5.22) for $t=2$ we obtain:

$$(5.30)\qquad |a_2^{z_h}(u_h; v_h, \mathbf{w}_h)| \leqslant C_7\|\mathbf{u}_h\|_{0,4,\Omega}\|\mathbf{w}_h\|_{0,4,\Omega}[\phi_h-\phi]_2. \qquad \square$$

The next lemma shows that $a_2^{z_h}(.;.,.)$ is "almost Lipschitz-continuous" with respect to z_h.

Lemma 5.3. *We keep the notations and assumptions of Lemma 5.2 but we suppose that both v_h and w_h belong to V_h. For all pairs $z_h=(\mathbf{z}_h,\zeta_h)$ and $z_h^*=(\mathbf{z}_h^*,\zeta_h^*)$ in X_h, the difference $a_2^{z_h}-a_2^{z_h^*}$ satisfies:*

$$|a_2^{z_h}(u_h; v_h, \mathbf{w}_h) - a_2^{z_h^*}(u_h; v_h, \mathbf{w}_h)|$$

$$(5.31)\qquad \leqslant Ch^{1/2}|v_h||w_h|\begin{cases}\|\mathbf{u}_h\|_{0,4,\Omega} & \text{if } z_h^*\neq u_h\\ \|\mathbf{z}_h-\mathbf{z}_h^*\|_{0,4,\Omega} & \text{if } z_h^*=u_h,\end{cases}$$

$$\forall u_h, z_h, z_h^*\in X_h, \qquad \forall v_h, w_h\in V_h.$$

Proof. This proof is based on the identity:

$$a_2^{z_h}(u_h; v_h, \mathbf{w}_h) - a_2^{z_h^*}(u_h; v_h, \mathbf{w}_h)$$

$$(5.32)\qquad = \sum_{\kappa\in\mathscr{T}_h}\int_{\partial_-\kappa(z_h,-z_h^*)} \mathbf{u}_h\cdot\mathbf{n}(v_{hi}^{\text{ext}}-v_{hi}^{\text{int}})(w_{hi}^{\text{int}}-w_{hi}^{\text{ext}})\,ds$$

where

$$\partial_-\kappa(z,-z^*) = \{x\in\partial\kappa;\ \mathbf{z}\cdot\mathbf{n}(x)<0 \quad\text{and}\quad \mathbf{z}^*\cdot\mathbf{n}(x)>0\}.$$

Thus, in view of (5.29) we can replace not only v_h by $v_h - Pv_h$ but also w_h by $w_h - Pw_h$. This gives:

$$|a_2^{z_h}(u_h; v_h, \mathbf{w}_h) - a_2^{z_h^*}(u_h; v_h, \mathbf{w}_h)| \leqslant C_1\Big(\sum_{\kappa\in\mathscr{T}_h}\|\mathbf{curl}\,\psi_h\|_{0,4,\partial\kappa}^4\Big)^{1/4}$$

$$\times\Big(\sum_{\kappa\in\mathscr{T}_h}\|\mathbf{curl}(\phi_h-\phi)\|_{0,4,\partial\kappa}^4\Big)^{1/4}$$

$$\times\Big(\sum_{\kappa\in\mathscr{T}_h}\|\mathbf{curl}(\chi_h-\chi)\|_{0,\partial\kappa}^2\Big)^{1/2}.$$

Then the techniques of the preceding lemma easily yield:

(5.33) $|a_2^{z_h}(u_h; v_h, \mathbf{w}_h) - a_2^{z_h^*}(u_h; v_h, \mathbf{w}_h)| \leqslant C_2 h^{1/2} \|\mathbf{u}_h\|_{0,4,\Omega}[\phi_h - \phi]_2[\chi_h - \chi]_2.$

Hence (5.31) follows from Lemma 5.1 and (5.24) in the general case.

To handle the particular case where $z_h^* = u_h$, we observe that:

$$|\mathbf{u}\cdot\mathbf{n}(x)| \leqslant |\mathbf{u}\cdot\mathbf{n}(x) - \mathbf{z}\cdot\mathbf{n}(x)| \qquad \forall x \in \partial_-\kappa(z, -u).$$

Therefore the factor $\mathbf{u}_h\cdot\mathbf{n}$ in the right-hand side of (5.32) can be bounded by the difference $|(\mathbf{u}_h - \mathbf{z}_h)\cdot\mathbf{n}| = |(\mathbf{z}_h^* - \mathbf{z}_h)\cdot\mathbf{n}|$. It can be readily checked that the remainder of the above argument is still valid with this modification. □

Finally, we can prove another interesting bound for $\tilde{a}(.;.,.)$.

Lemma 5.4. *We retain the assumptions and notations of Lemma 5.3 except that we take v_h in X_h. Then we have the following estimate for all $z = (\mathbf{curl}\,\alpha, \beta)$ with $\alpha \in H^2(\Omega) \cap H_0^1(\Omega)$ and $\beta \in L^2(\Omega)$:*

$$|\tilde{a}(u_h; v_h - z, \mathbf{w}_h)| \leqslant C h^{1/2} \|\mathbf{u}_h\|_{0,4,\Omega}[\phi_h - \alpha]_4 |w_h|, \tag{5.34}$$
$$\forall u_h, v_h \in X_h, \qquad \forall w_h \in V_h,$$

where the line integrals in $a_2(.;.,.)$ are taken over any portion of $\partial\kappa$.

Proof. Take $Pw_h = (\mathbf{w} = \mathbf{curl}\,\chi, \varepsilon_h)$ and set $\mathbf{z} = \mathbf{curl}\,\alpha$. The linearity of $\tilde{a}(.;.,.)$ with respect to its last argument permits to write:

$$\tilde{a}(u_h; v_h - z, \mathbf{w}_h) = \tilde{a}(u_h; v_h - z, \mathbf{w}_h - \mathbf{w}) + \tilde{a}(u_h; v_h - z, \mathbf{w}).$$

Since $\mathbf{w}$ belongs to $H^1(\Omega)$ we can apply the identity (5.17):

$$\tilde{a}(u_h; v_h - z, \mathbf{w}) = -a_1(u_h; Pw_h, \mathbf{v}_h - \mathbf{z}),$$

so that Lemma 5.2 and (5.24) give:

$$|\tilde{a}(u_h; v_h - z, \mathbf{w})| \leqslant C_1 \|\mathbf{u}_h\|_{0,4,\Omega} |Pw_h|\, \|\mathbf{v}_h - \mathbf{z}\|_{0,4,\Omega}.$$
$$\leqslant C_1 \|\mathbf{u}_h\|_{0,4,\Omega} |w_h|\, \|\mathbf{v}_h - \mathbf{z}\|_{0,4,\Omega}.$$

Likewise, we have

$$|a_1(u_h; v_h - z, \mathbf{w}_h - \mathbf{w})| \leqslant C_2 \|\mathbf{u}_h\|_{0,4,\Omega} \left(\sum_{\kappa \in \mathcal{T}_h} |\phi_h - \alpha|_{2,\kappa}^2\right)^{1/2} \|\mathbf{w}_h - \mathbf{w}\|_{0,4,\Omega}$$
$$\leqslant C_3 h^{1/2} \|\mathbf{u}_h\|_{0,4,\Omega} |w_h| \left(\sum_{\kappa \in \mathcal{T}_h} |\phi_h - \alpha|_{2,\kappa}^2\right)^{1/2}$$

in view of (5.22) and Lemma 5.1. Therefore

$$|a_1(u_h; v_h - z, \mathbf{w}_h - \mathbf{w})| + |\tilde{a}(u_h; v_h - z, \mathbf{w})| \leqslant C_4 h^{1/2} \|\mathbf{u}_h\|_{0,4,\Omega} |w_h|\, [\phi_h - \alpha]_4.$$

Finally, like in Lemma 5.2, we derive

$$|a_2(u_h; v_h - z, \mathbf{w}_h - \mathbf{w})| \leqslant C_5 h^{1/2} \|\mathbf{u}_h\|_{0,4,\Omega} [\phi_h - \alpha]_4 [\chi_h - \chi]_2$$
$$\leqslant C_6 h^{1/2} \|\mathbf{u}_h\|_{0,4,\Omega} |w_h| [\phi_h - \alpha]_4.$$

By collecting these inequalities, we obtain (5.34). □

Remark 5.1. It follows from Lemma 5.2 and the argument of Lemma 5.3 that the mapping G_h is Lipschitz-continuous on V_h:

(5.35) $$\|G_h(u_h) - G_h(u_h^*)\|_* \leqslant C(|u_h| + |u_h^*|)|u_h - u_h^*| \qquad \forall u_h, u_h^* \in V_h,$$

with a constant C independent of h.

Now we are in a position to define the operator $\nabla G_h(u_h)$, i.e. to approach as best as we can the "derivative" of the form $\tilde{a}(.;.,.)$. Clearly, the simplest guess is to linearize the dependence of $\tilde{a}^{z_h}(.;.,.)$ with respect to z_h. Thus we set:

Definition 5.1. For all $u_h \in V_h$, the operator $\nabla G_h(u_h) \in \mathscr{L}(V_h; Y_h)$ is defined by

$$\langle \nabla G_h(u_h) \cdot v_h, \mathbf{w}_h \rangle = \tilde{a}^{u_h}(u_h; v_h, \mathbf{w}_h) + \tilde{a}^{u_h}(v_h; u_h, \mathbf{w}_h) \qquad \forall \mathbf{w}_h = \mathbf{curl}\, \chi_h, \chi_h \in \Phi_h.$$

Obviously $\nabla G_h(u_h)$ is a linear operator from V_h into Y_h and Lemma 5.2 implies that

$$\|\nabla G_h(u_h) \cdot v_h\|_* \leqslant C|u_h||v_h| \qquad \forall u_h, v_h \in V_h,$$

with a constant C independent of h.

Remark 5.2. Note that $\nabla G_h(u_h)$ is "nearly Lipschitz-continuous" with respect to u_h. Indeed,

$$|\langle (\nabla G_h(u_h) - \nabla G_h(u_h^*)) \cdot v_h, \mathbf{w}_h \rangle| = |\tilde{a}^{u_h}(u_h; v_h, \mathbf{w}_h) - \tilde{a}^{u_h^*}(u_h^*; v_h, \mathbf{w}_h)$$
$$+ \tilde{a}^{u_h}(v_h; u_h, \mathbf{w}_h) - \tilde{a}^{u_h^*}(v_h; u_h^*, \mathbf{w}_h)|$$
$$\leqslant |\tilde{a}^{u_h}(u_h - u_h^*; v_h, \mathbf{w}_h)| + |\tilde{a}^{u_h}(u_h^*; v_h, \mathbf{w}_h)$$
$$- \tilde{a}^{u_h^*}(u_h^*; v_h, \mathbf{w}_h)| + |\tilde{a}^{u_h}(v_h; u_h - u_h^*, \mathbf{w}_h)|$$
$$+ |\tilde{a}^{u_h}(v_h; u_h^*, \mathbf{w}_h) - \tilde{a}^{u_h^*}(v_h; u_h^*, \mathbf{w}_h)|$$
$$\leqslant C_1 |v_h||w_h| [|u_h - u_h^*|$$
$$+ h^{1/2}(\|\mathbf{u}_h - \mathbf{u}_h^*\|_{0,4,\Omega} + |u_h^*|)],$$

by applying Lemmas 5.3 and 5.4. Hence

$$\|\nabla G_h(u_h) - \nabla G_h(u_h^*)\|_{\mathscr{L}(V_h; V_h')} \leqslant C_2(|u_h - u_h^*| + h^{1/2}|u_h^*|).$$

Of course, we can derive a similar upper bound with the term $h^{1/2}|u_h|$ in the right-hand side.

Finally in order to apply Theorem 3.8, it is necessary to relate the operator G defined by (5.6) with the operator that corresponds to the trilinear form

$a_1(.;.,.)$. When $\mathbf{u} = \mathbf{curl}\,\psi$ with ψ in $\Phi_s \cap H^2(\Omega)$, formula (2.26) yields:

$$(u_j \partial u_i/\partial x_j, v_i) = (\mathrm{curl}\,\mathbf{u}\,\mathbf{grad}\,\psi, \mathbf{v}) \quad \forall \mathbf{v} = \mathbf{curl}\,\phi, \quad \phi \in \Phi_s.$$

Hence

$$a_1(u; u, \mathbf{v}) = (\omega\,\mathbf{grad}\,\psi, \mathbf{v})$$

provided ψ has the regularity H^2. Of course, when u belongs only to X this equality does not hold but we can extend G to X by regularizing the function u. Thus we define for all u in X:

$$\langle G(u), \mathbf{v}\rangle = a_1(u; Pu, \mathbf{v}) - \langle \mathbf{f}, \mathbf{v}\rangle \quad \forall \mathbf{v} \in L^s(\Omega)^2. \tag{5.36}$$

Since $Pu = u$ for all u in V, this definition coincides with (5.6) when $u \in V$ and $\mathbf{v} = \mathbf{curl}\,\phi$, $\phi \in \Phi_s$. Furthermore, it can be readily checked that (5.36) defines a $\mathscr{C}^\infty$-mapping $G: X \to Y$ whose derivative $DG(u) \in \mathscr{L}(X; Y)$ is given by the expression:

$$\langle DG(u) \cdot v, \mathbf{w}\rangle = a_1(u; Pv, \mathbf{w}) + a_1(v; Pu, \mathbf{w}) \quad \forall \mathbf{w} \in L^s(\Omega)^2. \tag{5.37}$$

Finally, it can be proved that u is a nonsingular solution of (5.7) whenever u is also a nonsingular solution of the Navier-Stokes problem with G defined by (5.36), and conversely.

Let us check the assumptions of Theorem 3.8. It follows readily from (5.10) that

$$\| T_h l \|_X \leqslant (1 + C_s)^2 \| l \|_h \quad \forall l \in Y_h, \tag{5.38}$$

where C_s denotes the constant of (5.20). Next, (3.60) has already been checked in Section 4.3: owing to the regularity condition (4.18) $T_h \mathbf{f} = (\mathbf{curl}\,\psi_h, \omega_h)$ satisfies the error estimate

$$|\psi - \psi_h|_{1,s,\Omega} + \|\omega - \omega_h\|_{0,\Omega} \leqslant C h^\alpha \|\mathbf{f}\|_{0,t,\Omega}, \tag{5.39}$$

where $r \leqslant t < 2$, $1/\gamma + 1/t = 1$, $\alpha = 1/\gamma$ when $l = 1$ and $\alpha = 2/\gamma$ when $l \geqslant 2$. Likewise, the approximation properties of the operator π_h have been derived in Section III.3.1. Indeed, for $v = (\mathbf{curl}\,\phi, \theta) \in V$ we take

$$\pi_h v = (\mathbf{curl}(\mathring{P}_h \phi), \theta_h) \in V_h, \tag{5.40}$$

i.e. θ_h is determined by

$$(\theta_h, \mu_h) = (\mathbf{curl}(\mathring{P}_h \phi), \mathbf{curl}\,\mu_h) \quad \forall \mu_h \in \Theta_h.$$

Then

$$\|\pi_h v - v\|_X \leqslant |\mathring{P}_h \phi - \phi|_{1,s,\Omega} + 2 \inf_{\mu_h \in \Theta_h} \|\mu_h - \theta\|_{0,\Omega} + \sup_{\mu_h \in \Theta_h} \frac{(\mathbf{curl}(\mathring{P}_h \phi - \phi), \mathbf{curl}\,\mu_h)}{\|\mu_h\|_{0,\Omega}}.$$

Therefore Lemma III.3.2, Theorem A.2 and a standard density argument yield

$$\lim_{h \to 0} \|\pi_h v - v\|_X = 0 \quad \forall v \in V.$$

Furthermore, Lemma III.3.3 gives the best estimate when ϕ belongs to $H^{l+3/2}(\Omega)\cap W^{l+1,\infty}(\Omega)$

$$\|\pi_h v - v\|_X \leqslant Ch^{l-1/2}(\|\phi\|_{l+3/2,\Omega} + \|\phi\|_{l+1,\infty,\Omega}).$$

Now we turn to (3.62); we have

$$\langle G_h(\pi_h u) - G(u), \mathbf{v}_h\rangle = \tilde{a}^{\pi_h u}(\pi_h u; \pi_h u, \mathbf{v}_h) - a_1(u; u, \mathbf{v}_h) \qquad \forall \mathbf{v}_h = \mathbf{curl}\,\phi_h,\ \phi_h \in \Phi_h.$$

Using a simple rearrangement of terms we obtain:

$$\langle G_h(\pi_h u) - G(u), \mathbf{v}_h\rangle = \tilde{a}^{\pi_h u}(\pi_h u; \pi_h u - u, \mathbf{v}_h) + a_1(\pi_h u - u; u, \mathbf{v}_h).$$

Therefore Lemma 5.4 and Lemma 5.2 together with (5.22) yield:

$$(5.41) \qquad \|G_h(\pi_h u) - G(u)\|_h \leqslant Ch^{1/2}(|\psi|_{2,\Omega} + |\mathring{P}_h\psi|_{1,4,\Omega})[\mathring{P}_h\psi - \psi]_4.$$

Hence (3.62) follows from Lemma 5.1. Furthermore we infer from this lemma and (5.41) that

$$(5.42) \qquad \begin{aligned} &\|G_h(\pi_h u) - G(u)\|_h \leqslant Ch^{k-1/2}|\psi|_{2,\Omega}\|\psi\|_{k+1,\Omega} \\ &\forall u = (\mathbf{curl}\,\psi, \omega) \in V \quad \text{with } \psi \in H^{k+1}(\Omega), \qquad 1 \leqslant k \leqslant l. \end{aligned}$$

As far as (3.63) is concerned, we apply Definition 5.1 and (5.37):

$$\begin{aligned} \langle (\nabla G_h(\pi_h u) - DG(u))\cdot v_h, \mathbf{w}_h\rangle &= \tilde{a}^{\pi_h u}(\pi_h u; v_h, \mathbf{w}_h) + \tilde{a}^{\pi_h u}(v_h; \pi_h u, \mathbf{w}_h) \\ &\quad - a_1(u; Pv_h, \mathbf{w}_h) - a_1(v_h; u, \mathbf{w}_h) \qquad \forall u \in V \\ &= \tilde{a}^{\pi_h u}(\pi_h u; v_h - Pv_h, \mathbf{w}_h) + \tilde{a}^{\pi_h u}(v_h; \pi_h u - u, \mathbf{w}_h) \\ &\quad + a_1(\pi_h u - u; Pv_h, \mathbf{w}_h) \end{aligned}$$

taking into account the regularity of u and Pv_h. Consequently, it stems from Lemma 5.3 and 5.4 that:

$$\begin{aligned} \|(\nabla G_h(\pi_h u) - DG(u))\cdot v_h\|_h \leqslant C\{&h^{1/2}(|\mathring{P}_h\psi|_{1,4,\Omega}[\phi_h - \phi]_4 + |v_h|[\mathring{P}_h\psi - \psi]_4) \\ &+ |v_h||\mathring{P}_h\psi - \psi|_{1,4,\Omega}\}, \end{aligned}$$

with $Pv_h = (\mathbf{curl}\,\phi, \theta_h)$, $\forall u = (\mathbf{curl}\,\psi, \omega) \in V$, $\forall v_h = (\mathbf{curl}\,\phi_h, \theta_h) \in V_h$.
Thus Lemma 5.1 gives:

$$\|(\nabla G_h(\pi_h u) - DG(u))\cdot v_h\|_h \leqslant Ch^{1/2}(|\psi|_{1,4,\Omega} + \|\psi\|_{2,\Omega})|v_h|,$$

which implies (3.63).

Finally, it remains to verify (3.64). Take u_h, u_h^* and u_h^0 in V_h; by definition we have for all $\mathbf{w}_h = \mathbf{curl}\,\chi_h$, $\chi_h \in \Phi_h$:

$$\begin{aligned} &\langle G_h(u_h) - G_h(u_h^*) - \nabla G_h(u_h^0)\cdot(u_h - u_h^*), \mathbf{w}_h\rangle \\ &\quad = \tilde{a}^{u_h}(u_h; u_h, \mathbf{w}_h) - \tilde{a}^{u_h^*}(u_h^*; u_h^*, \mathbf{w}_h) - \tilde{a}^{u_h^0}(u_h^0; u_h - u_h^*, \mathbf{w}_h) - \tilde{a}^{u_h^0}(u_h - u_h^*; u_h^0, \mathbf{w}_h) \\ &\quad = \tilde{a}^{u_h^0}(u_h - u_h^0; u_h - u_h^*, \mathbf{w}_h) + \tilde{a}^{u_h^0}(u_h - u_h^*; u_h^* - u_h^0, \mathbf{w}_h) + \{a_2^{u_h}(u_h; u_h, \mathbf{w}_h) \\ &\qquad - a_2^{u_h^0}(u_h; u_h, \mathbf{w}_h) - a_2^{u_h^*}(u_h^*; u_h^*, \mathbf{w}_h) + a_2^{u_h^0}(u_h^*; u_h^*, \mathbf{w}_h)\}. \end{aligned}$$

The expression in brackets can be rearranged as follows:

$$\{a_2^{u_h}(u_h - u_h^*; u_h, \mathbf{w}_h) - a_2^{u_h^0}(u_h - u_h^*; u_h, \mathbf{w}_h)\} + \{a_2^{u_h^*}(u_h^*; u_h - u_h^*, \mathbf{w}_h) - a_2^{u_h^0}(u_h^*; u_h - u_h^*, \mathbf{w}_h)\} + \{a_2^{u_h}(u_h^*; u_h, \mathbf{w}_h) - a_2^{u_h^*}(u_h^*; u_h, \mathbf{w}_h)\}.$$

Then using repeatedly Lemma 5.3 this expression can be bounded by:

$$C_1 h^{1/2}\{|u_h||\psi_h - \psi_h^*|_{1,4,\Omega} + |\psi_h^* - \psi_h^0|_{1,4,\Omega}|u_h - u_h^*|\}|w_h|$$
$$\leqslant C_2 h^{1/2}|u_h - u_h^*|\{|u_h - u_h^0| + |u_h^* - u_h^0| + |u_h^0|\}|w_h| \qquad \forall w_h \in V_h.$$

Hence Lemma 5.2 gives the bound

$$\|G_h(u_h) - G_h(u_h^*) - \nabla G_h(u_h^0)\cdot(u_h - u_h^*)\|_h$$
$$\leqslant C_3|u_h - u_h^*|\{|u_h - u_h^0| + |u_h^* - u_h^0| + h^{1/2}|u_h^0|\}.$$

This implies (3.64) with the function

$$L_h(\mu; v) = C_4(\mu + h^{1/2}v).$$

Clearly L_h is continuous, monotonically increasing with respect to each variable and

$$\lim_{h\to 0} L_h(0; v) = \lim_{h\to 0}(h^{1/2}v) = 0 \qquad \forall v \in \mathbb{R}_+.$$

Since all the assumptions of Theorem 3.8 are satisfied, we can apply its conclusion to the upwind scheme (5.18), thus deriving the existence, uniqueness and convergence of its branch of solutions. Furthermore, by comparing the two error estimates (3.67) and (3.41) and taking into account (5.42) we readily deduce that the error of the upwind scheme is bounded exactly like the error of the centered scheme. In other words, in this case *the upwinding does not alter the scheme's order*. These results are summed up in the following theorem.

Theorem 5.1. *Let Ω be a bounded, convex polygon and $\mathcal{T}_h$ a uniformly regular family of triangulations of $\bar{\Omega}$. For $\mathbf{f} \in L^t(\Omega)^2$, $t \in [r, 2)$ let*

$$\{(\lambda, (\mathbf{curl}\,\psi(\lambda), \omega(\lambda))); \lambda = 1/v \in \Lambda\}$$

be a branch of nonsingular solutions of the Navier-Stokes Problem (5.7). *Then for $h \leqslant h_0$ small enough there exists a unique branch*

$$\{(\lambda, (\mathbf{curl}\,\psi_h(\lambda), \omega_h(\lambda))); \lambda = 1/v \in \Lambda\}$$

of $\mathscr{C}^0$-solutions of the upwind scheme (5.18) *satisfying the error estimate:*

$$\sup_{\lambda\in\Lambda}\{|\psi_h(\lambda) - \psi(\lambda)|_{1,s,\Omega} + \|\omega_h(\lambda) - \omega(\lambda)\|_{0,\Omega}\} \leqslant C_1 h^\alpha,$$

(5.43)
$$\alpha = \begin{cases} 1/\gamma & \text{if } l = 1, \\ 2/\gamma & \text{if } l \geqslant 2, \end{cases} \qquad 1/t + 1/\gamma = 1,$$

with a constant C independent of h or λ.

This bound is also valid when $t = 2$ *and either* $l \geqslant 2$ *or* ψ *belongs to* $W^{2,\infty}(\Omega)$. *If* $l = 1$ *and* $t = 2$, *the left-hand side of* (5.43) *is bounded by*

$$C_2(\varepsilon)h^{1/2-\varepsilon} \quad \text{for all } \varepsilon > 0.$$

In addition, when the mapping $\lambda \to \psi(\lambda)$ *is continuous from* Λ *into* $H^{m+2}(\Omega) \cap W^{m+3/2,\infty}(\Omega)$ *for some real* $m \in [1, l - 1/2]$ *we have:*

$$(5.44) \qquad \sup_{\lambda \in \Lambda} \{ |\psi_h(\lambda) - \psi(\lambda)|_{1,s,\Omega} + \|\omega_h(\lambda) - \omega(\lambda)\|_{0,\Omega} \} \leqslant C_3 h^m.$$

Finally, an argument closely resembling that of Section 4.3 permits to refine the error estimate for $|\psi - \psi_h|_{1,\Omega}$ and obtain the same order of convergence as Theorem 4.5.

Theorem 5.2. *Let* Ω *and* $\mathcal{T}_h$ *be like in Theorem* 5.1 *and assume that the branch of nonsingular solutions of the Navier-Stokes Problem* (5.7) *has the regularity*:

$$\lambda \to \psi(\lambda) \in \mathscr{C}^0(\Lambda; H^{l+1}(\Omega)) \quad \textit{for } l \geqslant 2 \quad \text{or} \quad H^3(\Omega) \quad \textit{for } l = 1.$$

Then the approximate solution ψ_h *of the upwind scheme* (5.18) *satisfies the error bound*:

$$(5.45) \qquad |\psi(\lambda) - \psi_h(\lambda)| \leqslant \begin{cases} C_1 h^l & \textit{when } l \geqslant 2, \\ C_2(\varepsilon) h^{1-\varepsilon} & \textit{when } l = 1 \quad \forall \varepsilon > 0, \end{cases}$$

with constants independent of h *and* λ.

Proof. Let $\mathbf{g}$ be any function of $L^2(\Omega)^2$ and let us introduce the following dual linearized Navier-Stokes problem, analogous to (4.40):

$$z = (\mathbf{z} = \mathbf{curl}\,\chi, \nu) \in V \quad \text{such that:}$$

$$(5.46) \qquad (\nu, \theta) + \lambda a_1(v; u(\lambda), \mathbf{z}) - \lambda a_1(u(\lambda); z, \mathbf{v}) = (\mathbf{g}, \mathbf{curl}\,\phi)$$

$$\forall v = (\mathbf{v} = \mathbf{curl}\,\phi, \theta) \in V.$$

We know that on the one hand:

$$a_1(u; z, \mathbf{v}) = -a_1(u; v, \mathbf{z})$$

and on the other hand:

$$a_1(u; v, \mathbf{z}) + a_1(v; u, \mathbf{z}) = (DG(u)\cdot v, \mathbf{z})$$

for all u, v and z in V. Therefore Problem (5.46) is the same as Problem (4.40) and hence its solution enjoys the regularity properties stated in Lemma 4.1.

Now, by reasoning like in Theorem 4.5 we readily obtain:

$$\begin{aligned}(\mathbf{g}, \mathbf{curl}(\psi - \psi_h)) = {} & (\nu - \nu_h, \omega - \omega_h) + \tilde{b}(u - u_h, \nu - \theta_h) + \tilde{b}(z - z_h, \omega - \mu_h) \\ & + \lambda a_1(u - u_h; u, \mathbf{z}) - \lambda a_1(u; z, \mathbf{u} - \mathbf{u}_h) - \lambda a_1(u; u, \mathbf{z}_h) \\ & + \lambda a_1(u_h; u_h, \mathbf{z}_h) + \lambda a_2^{\mathbf{u}_h}(u_h; u_h, \mathbf{z}_h) \\ & \forall z_h = (\mathbf{z}_h = \mathbf{curl}\,\chi_h, \nu_h) \in V_h, \quad \forall \mu_h, \theta_h \in \Theta_h.\end{aligned}$$

Therefore an easy manipulation of terms and (5.17) yield:

$$(5.47)\quad \begin{aligned}(\mathbf{g}, \mathbf{curl}(\psi - \psi_h)) = (v - v_h, \omega - \omega_h) &+ \tilde{b}(u - u_h, v - \theta_h) + \tilde{b}(z - z_h, \omega - \mu_h)\\ &+ \lambda\{a_1(u - u_h; u, \mathbf{z} - \mathbf{z}_h) + \tilde{a}^{u_h}(u_h; u - u_h, \mathbf{z} - \mathbf{z}_h)\\ &- a_1(u - u_h; z, \mathbf{u} - \mathbf{u}_h)\}.\end{aligned}$$

But, the delicate step in this proof is an adequate estimate of the $\tilde{a}(.;.,.)$ term in the factor multiplying λ. Indeed, if we apply Lemma 5.2, we get:

$$|\tilde{a}^{u_h}(u_h; u - u_h, \mathbf{z} - \mathbf{z}_h)| \leqslant Ch^{1/2}|\psi_h|_{1,4,\Omega}[\psi - \psi_h]_2[\chi - \chi_h]_2.$$

And when $l = 1$, this upper bound is useless because $\inf_{\chi_h}[\chi - \chi_h]_2 = O(1)$. Instead, it is better to take advantage of the fact that $\chi \in H^3(\Omega)$ and replace the previous bound by:

$$|\tilde{a}^{u_h}(u_h; u - u_h, \mathbf{z} - \mathbf{z}_h)| \leqslant C_1|\psi_h|_{1,4,\Omega}[\psi - \psi_h]_2 \times \left\{|\chi - \chi_h|_{1,4,\Omega} + h\left(\sum_{\kappa\in\mathscr{T}_h}|\chi - \chi_h|^4_{2,4,\kappa}\right)^{1/4}\right\}.$$

Hence, if we choose $z_h = \pi_h z \in V_h$ defined by (5.40), the fact that $\chi_h = \mathring{P}_h\chi$ implies that

$$|\tilde{a}^{u_h}(u_h; u - u_h, \mathbf{z} - \mathbf{z}_h)| \leqslant C_2 h|\psi_h|_{1,4,\Omega}[\psi - \psi_h]_2\|\chi\|_{2,4,\Omega}.$$

Finally, observe that

$$(5.48)\quad \begin{aligned}[\psi - \psi_h]_2 &\leqslant [\psi - \mathring{P}_h\psi]_2 + [\mathring{P}_h\psi - \psi_h]_2\\ &\leqslant C_3 h^{l-1}\|\psi\|_{l+1,\Omega} + C_4|\pi_h u - u_h|\end{aligned}$$

by virtue of (5.26), (5.21) and (5.40). This gives an upper bound of the form:

$$|\tilde{a}^{u_h}(u_h; u - u_h, \mathbf{z} - \mathbf{z}_h)| \leqslant C_5(\|\psi\|_{2,\Omega})\|\chi\|_{2,4,\Omega}(h^l\|\psi\|_{l+1,\Omega} + h|\pi_h u - u_h|).$$

The other terms in (5.47) are easily estimated and the proof ends exactly like that of Theorem 4.5. □

5.3. Approximating the Pressure with the Upwind Scheme

We have seen in Section 4.3 that the pressure term p underlying the Navier-Stokes system (5.7) is the solution of the problem:

Find $p \in W^{1,r}(\Omega) \cap L^2_0(\Omega)$ such that:

$$(5.49)\quad (\mathbf{grad}\, p, \mathbf{grad}\, q) = \left(\mathbf{f} - v\,\mathbf{curl}\,\omega - \sum_{j=1}^{2} u_j \partial\mathbf{u}/\partial x_j, \mathbf{grad}\, q\right) \quad \forall q \in W^{1,s}(\Omega).$$

Likewise, to recover the pressure p_h associated with the upwind scheme (5.18), we introduce the space Q_h defined by (4.34):

(5.50) $Q_h = \{q_h \in \mathscr{C}^0(\bar{\Omega}) \cap L_0^2(\Omega);\ q_h|_\kappa \in P_k \quad \forall \kappa \in \mathscr{T}_h\} \qquad k = \min(1, l-1)$

and we discretize (5.49) by:

Find $p_h \in Q_h$ satisfying:

$$(\mathbf{grad}\, p_h, \mathbf{grad}\, q_h) = (\mathbf{f} - \nu\, \mathbf{curl}\, \omega_h, \mathbf{grad}\, q_h) - \tilde{a}^{u_h}(u_h; u_h, \mathbf{grad}\, q_h) \qquad \forall q_h \in Q_h. \tag{5.51}$$

Obviously, this problem has a unique solution.

To estimate the error $p - p_h$ we use the same duality argument as in Theorem III.2.7. We introduce the function $\mathbf{v} \in H^1(\Omega)^2$ defined by

$$\operatorname{div} \mathbf{v} = p - p_h, \qquad |\mathbf{v}|_{1,\Omega} \leqslant C_1 \|p - p_h\|_{0,\Omega}$$

which we can split into

$$\mathbf{v} = \mathbf{grad}\, q + \mathbf{curl}\, \phi.$$

Since Ω is assumed to be a convex polygon, both q and ϕ belong to $H^2(\Omega)$ with

$$\|q\|_{2,\Omega} + \|\phi\|_{2,\Omega} \leqslant C_2 |\mathbf{v}|_{1,\Omega} \quad \text{(cf. Lemma III.2.6)}.$$

In addition, if the triangulation $\mathscr{T}_h$ is uniformly regular Lemma III.2.6 shows that

$$|q - P_h q|_{1,s,\Omega} + |\phi - I_h \phi|_{1,s,\Omega} \leqslant C_3 h^{2/s} |\mathbf{v}|_{1,\Omega}. \tag{5.52}$$

Now, like in Theorem III.2.7 we can write:

$$\|p - p_h\|_{0,\Omega}^2 = (\mathbf{grad}(q_h - p), \mathbf{grad}(q - P_h q)) + (\mathbf{grad}(p_h - p), \mathbf{grad}(P_h q)) \qquad \forall q_h \in Q_h.$$

Thus, to estimate $p - p_h$, we must derive a sharp bound for the second term. By subtracting (5.49) from (5.51) we obtain:

$$\begin{aligned}(\mathbf{grad}(p_h - p), \mathbf{grad}(P_h q)) &= -\nu(\mathbf{curl}(\omega_h - \omega), \mathbf{grad}(P_h q)) + a_1(u; u, \mathbf{grad}(P_h q)) \\ &\quad - \tilde{a}^{u_h}(u_h; u_h, \mathbf{grad}(P_h q)) \\ &= \nu(\mathbf{curl}(\omega - \omega_h), \mathbf{v}_h - \mathbf{v}) + \nu(\omega - \omega_h, \operatorname{curl} \mathbf{v}) \\ &\quad + \tilde{a}^{u_h}(u_h; u - u_h, \mathbf{v}_h - \mathbf{v}) + a_1(u - u_h; u, \mathbf{v}_h - \mathbf{v}) \\ &\quad + a_1(u_h - u; v, \mathbf{u}_h) + a_1(u; v, \mathbf{u}_h - \mathbf{u})\end{aligned}$$

where
$$\mathbf{v}_h = \mathbf{grad}(P_h q) + \mathbf{curl}(I_h \phi).$$

Therefore (5.52) together with familiar estimates for the forms $\tilde{a}^{u_h}(.;.,.)$ and $a_1(.;.,.)$ give a result analogous to (III.2.38).

Lemma 5.5. *Let Ω be a bounded, convex polygon and $\mathscr{T}_h$ a uniformly regular triangulation of $\bar{\Omega}$. If p and ω belong to $W^{1,t}(\Omega)$ for some real $t \in [r, 2]$ then the error on p is:*

$$\|p - p_h\|_{0,\Omega} \leqslant C \Big\{ h^{2/\gamma} \inf_{q_h \in Q_h} |p - q_h|_{1,t,\Omega} + \nu \|\omega - \omega_h\|_{0,\Omega}$$

$$(5.53) \qquad + \nu \inf_{\theta_h \in \Theta_h} (\|\omega - \theta_h\|_{0,\Omega} + h^{2/\gamma} |\omega - \theta_h|_{1,t,\Omega})$$

$$+ h^{1/2} [\psi - \psi_h]_4 (|\psi_h|_{1,4,\Omega} + |\psi|_{2,\Omega}) \Big\}$$

where $1/\gamma + 1/t = 1$, $u_h = (\mathbf{u}_h = \mathbf{curl}\, \psi_h, \omega_h)$ *and* $u = (\mathbf{u} = \mathbf{curl}\, \psi, \omega)$ *denote the solutions of* (5.18) *and* (5.7) *respectively.*

It follows readily from (5.53) and (5.48) that the pressure p_h has the same order of convergence as the velocity:

Theorem 5.3. *Under the assumptions of Theorem* 5.1 *the pressure* p_h *defined by Problem* (5.51) *converges to* p *with the same order of accuracy as the velocity. Namely, if* $\mathbf{f} \in L^t(\Omega)^2$ *for* $t \in [r, 2)$ *we have:*

$$\sup_{\lambda \in \Lambda} \|p(\lambda) - p_h(\lambda)\|_{0,\Omega} \leqslant C_1 h^\alpha \qquad \alpha = \begin{cases} 1/\gamma & \text{if } l = 1, \\ 2/\gamma & \text{if } l = 2, \end{cases}$$

where $1/\gamma + 1/t = 1$. *When* ψ *belongs to* $W^{2,\infty}(\Omega)$ *or when* $l \geqslant 2$, *this bound can be extended to* $t = 2$. *When* $l = 1$ *and* $t = 2$ *we have:*

$$\sup_{\lambda \in \Lambda} \|p(\lambda) - p_h(\lambda)\|_{0,\Omega} \leqslant C_2(\varepsilon) h^{1-\varepsilon}.$$

Finally when the mapping $\lambda \to (\psi(\lambda), p(\lambda))$ *is continuous from* Λ *into* $[H^{m+2}(\Omega) \cap W^{m+3/2,\infty}(\Omega)] \times H^m(\Omega)$ *for some real* $m \in [1, l - 1/2]$ *we have*

$$\sup_{\lambda \in \Lambda} \|p(\lambda) - p_h(\lambda)\|_{0,\Omega} \leqslant C_3 h^m.$$

§ 6. Numerical Algorithms

Navier-Stokes equations are difficult to solve in practice because they are nonlinear. We present here a few simple converging algorithms that permit to handle the nonlinearity. Although they are intended to solve the discrete systems of (nonlinear) equations, it is simpler to introduce these algorithms in connection with the continuous problem. The reader will verify easily that the convergence theorems below are also valid for the approximate problems.

6.1. General Methods of Descent and Application to Gradient Methods

Like every nonlinear problem, the Navier-Stokes equations can be put into the framework of an optimization problem. Therefore we shall consider first the minimization of a locally convex functional on an abstract (real) Hilbert space X. Let $\|.\|_X$ and $(.,.)_X$ denote respectively the norm and associated scalar product of X and let J be a $\mathscr{C}^2$-mapping from X into $\mathbb{R}$. We propose to minimize J over an adequate subset D of X, chosen as follows. First, since we are only interested in a minimum of J we restrict the discussion to the set

$$\{v \in X;\ J(v) \leqslant C_0\}$$

for some constant C_0; of course, we suppose that this set is not empty and we take for D one of its connected components. In other words, D is a non empty connected component of:

$$\{v \in X;\ J(v) \leqslant C_0\}. \tag{6.1}$$

Now, we assume that the functional J is *strictly convex in D*, namely:

there exist two constants $\alpha > 0$ and $M > 0$ such that all v in D satisfy

$$\alpha \|w\|_X^2 \leqslant D^2 J(v) \cdot (w, w) \leqslant M \|w\|_X^2 \quad \forall w \in X. \tag{6.2}$$

Then it is well known that the problem

$$\inf_{v \in D} J(v) \tag{6.3}$$

has a unique solution u in D characterized by

$$DJ(u) = 0. \tag{6.4}$$

As usual, we associate with DJ the gradient g defined by:

$$(g(v), w)_X = \langle DJ(v), w \rangle \quad \forall w \in X,$$

where $\langle .,. \rangle$ denotes the duality pairing X and X'.

To solve Problem (6.3) we introduce the following general method of descent *starting from $u^0 \in D$*:

1°) for all $m \geqslant 0$, choose a direction of descent $w^m \in X$ and define $\rho^m \in \mathbb{R}_+$ by

$$J(u^m - \rho^m w^m) = \inf_{\substack{\rho \geqslant 0 \\ u^m - \rho w^m \in D}} J(u^m - \rho w^m), \tag{6.5}$$

2°) set

$$u^{m+1} = u^m - \rho^m w^m \in D. \tag{6.6}$$

As shown below, this algorithm converges under reasonable assumptions.

Theorem 6.1. *Let D be a non empty, connected subset of X satisfying* (6.1) *and let J satisfy* (6.2) *in D. If, for each m, the direction of descent $w^m \in X$ is such that*

$$(g(u^m), w^m)_X \geqslant \beta \|g(u^m)\|_X \|w^m\|_X \tag{6.7}$$

for some constant $\beta > 0$ independent of m, then the algorithm of descent (6.5) (6.6) *defines a sequence $(u^m) \in D$ that converges to the unique solution u of* (6.3).

Proof. First observe that D is a convex subset of X. Indeed, as J is convex in D, we have for all v_1 and v_2 in D:

$$J(\theta v_1 + (1-\theta)v_2) \leqslant \theta J(v_1) + (1-\theta)J(v_2) \qquad \forall\theta\in[0,1]$$
$$\leqslant C_0 \quad \text{by virtue of (6.1).}$$

Since D is connected, this means that $\theta v_1 + (1-\theta)v_2 \in D$.

Next, we set

$$D' = \{v\in D;\ J(v) \leqslant J(u^0)\}$$

and we assume that the starting value u^0 is not the solution of (6.3) (otherwise the algorithm of descent yields the constant value $u^m = u^0$). Again, D' is convex and because J is continuous, D' is obviously a closed subset of X. Let us establish that D' is bounded. Taylor's formula gives:

$$J(v) = J(u^0) + \langle DJ(u^0), v-u^0\rangle + (1/2)D^2J(u^0 + t(v-u^0))\cdot(v-u^0, v-u^0)$$

for some $t\in(0,1)$. To simplify, we denote

$$g^m = g(u^m).$$

Then, for all v in D', the convexity of D' and (6.2) imply

$$J(v) \geqslant J(u^0) - \|g^0\|_X\|v-u^0\|_X + (\alpha/2)\|v-u^0\|_X^2. \tag{6.8}$$

But since $J(v) \leqslant J(u^0)$ this yields:

$$J(u^0) \geqslant J(u^0) - \|g^0\|_X\|v-u^0\|_X + (\alpha/2)\|v-u^0\|_X^2;$$

in other words:

$$\|v-u^0\|_X \leqslant (2/\alpha)\|g^0\|_X. \tag{6.9}$$

This proves the boundedness of D'.

Observe also that J is bounded below in D' because (6.8) and (6.9) imply

$$J(v) \geqslant J(u^0) - (2/\alpha)\|g^0\|_X^2. \tag{6.10}$$

Now we remark that the equation (6.5) defines a unique $\rho^m \geqslant 0$ for each pair u^m and w^m satisfying (6.7). Indeed, since by construction u^m belongs to the interior of D, the mapping $\rho \to J(u^m - \rho w^m)$ is strictly convex for all ρ such that $u^m - \rho w^m \in D$. Hence $J(u^m - \rho w^m)$ has a unique minimum in D and this minimum is realized by a unique interior element of D, $u^m - \rho^m w^m$, with $\rho^m \geqslant 0$. Therefore $u^m - \rho^m w^m$ is characterized by:

$$[dJ(u^m - \rho w^m)/d\rho]\big|_{\rho=\rho^m} = -\langle DJ(u^m - \rho^m w^m), w^m\rangle = 0.$$

With (6.6) this can also be written:

$$\langle DJ(u^{m+1}), w^m\rangle = (g^{m+1}, w^m)_X = 0. \tag{6.11}$$

Note also that (6.11) gives:

$$0 = \langle DJ(u^m - \rho^m w^m), w^m \rangle = \langle DJ(u^m), w^m \rangle - \rho^m D^2 J(u^m - t\rho^m w^m)\cdot(w^m, w^m)$$

for some $t \in (0, 1)$, i.e.

$$(g^m, w^m)_X = \rho^m D^2 J(u^m - t\rho^m w^m)\cdot(w^m, w^m). \tag{6.12}$$

Thus it follows from the convexity of D, (6.2) and (6.7) that

$$\beta \| g^m \|_X \| w^m \|_X \leqslant \rho^m M \| w^m \|_X^2$$

whence we derive the following lower bound for ρ^m:

$$\rho^m \geqslant (\beta/M)(\| g^m \|_X / \| w^m \|_X). \tag{6.13}$$

Next, the sequence $J(u^m)$ is by construction monotonically decreasing and so the sequence (u^m) is contained in D'. We have in particular:

$$J(u^{m+1}) \leqslant J(u^m - \rho w^m) \quad \text{for all } \rho \text{ with } 0 \leqslant \rho \leqslant \rho^m.$$

Therefore

$$J(u^{m+1}) \leqslant J(u^m) - \rho(g^m, w^m)_X + (\rho^2/2) D^2 J(u^m - t\rho w^m)\cdot(w^m, w^m)$$

with $t \in (0, 1)$. Using again (6.7) and (6.2) we obtain:

$$J(u^{m+1}) \leqslant J(u^m) - \rho\beta \| g^m \|_X \| w^m \|_X + M(\rho^2/2) \| w^m \|_X^2.$$

In view of (6.13) we can choose $\rho = (\beta/M)(\| g^m \|_X / \| w^m \|_X)$ thus getting

$$J(u^{m+1}) - J(u^m) \leqslant -(1/2)(\beta^2/M) \| g^m \|_X^2. \tag{6.14}$$

As the sequence $J(u^m)$ is monotonically decreasing and bounded below (cf. (6.10)), it converges. In particular, (6.14) implies that

$$\lim_{m\to\infty} \| g^m \|_X^2 \leqslant 2(M/\beta^2) \lim_{m\to\infty} [J(u^m) - J(u^{m+1})] = 0. \tag{6.15}$$

Besides that we have

$$DJ(u^{m+p}) - DJ(u^m) = D^2 J(u^m + t u^{m+p})\cdot(u^{m+p} - u^m)$$

with $t \in (0, 1)$. Thus (6.2) gives

$$\begin{aligned} \alpha \| u^{m+p} - u^m \|_X^2 &\leqslant \langle DJ(u^{m+p}) - DJ(u^m), u^{m+p} - u^m \rangle \\ &\leqslant (g^{m+p} - g^m, u^{m+p} - u^m)_X, \end{aligned}$$

i.e.

$$\| u^{m+p} - u^m \|_X \leqslant (1/\alpha) \| g^{m+p} - g^m \|_X.$$

Since (g^m) converges in X, this means that (u^m) is a Cauchy sequence in X, and therefore a converging sequence in X. Thus there exists u in D' with

$$\lim_{m\to\infty} u^m = u$$

and furthermore (6.15) and the continuity of DJ imply that

$$g(u) = DJ(u) = 0.$$

Hence

$$J(u) = \underset{v \in D}{\operatorname{Min}} J(v). \qquad \square$$

The simple gradient and conjugate-gradient algorithms are among the most popular applications of the method of descent. The gradient algorithm is obtained by taking $w^m = g^m$ as direction of descent. Obviously this choice satisfies (6.7) and therefore Theorem 6.1 guarantees the local convergence of the simple gradient algorithm (6.5) (6.6) with $w^m = g^m$.

The Polack-Ribiere variant of the conjugate-gradient algorithm is defined by choosing

$$\text{(6.16)} \qquad \begin{cases} w^0 = g^0, \\ w^m = g^m + \sigma^m w^{m-1} \end{cases}$$

with

$$\text{(6.17)} \qquad \sigma^m = \frac{(g^m - g^{m-1}, g^m)_X}{(g^{m-1}, g^{m-1})_X} \qquad \Bigg\} \quad m \geqslant 1.$$

(compare with formula (I.4.68)). Again, Theorem 6.1 implies that this scheme is convergent.

Theorem 6.2. *Let D be defined by* (6.1) *and* (6.2) *and let* u^0 *belong to D. Then the conjugate-gradient algorithm* (6.5) (6.6) (6.16) *and* (6.17) *converges in D.*

Proof. Let us show that w^m defined by (6.16) and (6.17) satisfies (6.7). First observe that the property (6.11) and its consequence (6.12) established in Theorem 6.1 do not require (6.7) but only that $(g^m, w^m)_X$ be positive. Now, this positivity is easily proved by induction as it obviously holds for $m = 0$ and if it is true for $m - 1$ then (6.11) and (6.16) yield:

$$(g^m, w^{m-1})_X = 0, \qquad (g^m, w^m)_X = \|g^m\|_X^2.$$

Hence

$$(g^m, w^m)_X \geqslant 0 \qquad \forall m.$$

Furthermore, (6.12) implies

$$\text{(6.18)} \qquad \rho^m = [\|g^m\|_X^2]/[D^2 J^m \cdot (w^m, w^m)]$$

with

$$D^2 J^m = D^2 J(u^m - t\rho^m w^m), \quad t \text{ defined by (6.12).}$$

Likewise, with the same notation we have

$$\sigma^m = \frac{(g^m - g^{m-1}, g^m)_X}{\|g^{m-1}\|_X^2} = \frac{\langle DJ(u^m) - DJ(u^{m-1}), g^m\rangle}{\|g^{m-1}\|_X^2}$$
$$= -\rho^{m-1}\frac{D^2 J^{m-1}\cdot(w^{m-1}, g^m)}{\|g^{m-1}\|_X^2}$$
$$= -\frac{D^2 J^{m-1}\cdot(w^{m-1}, g^m)}{D^2 J^{m-1}\cdot(w^{m-1}, w^{m-1})}$$

in view of (6.18). Therefore (6.2) yields:

$$|\sigma^m| \leqslant (M/\alpha)(\|g^m\|_X/\|w^{m-1}\|_X),$$

whence

$$\|w^m\|_X = \|g^m + \sigma^m w^{m-1}\|_X \leqslant (1 + M/\alpha)\|g^m\|_X.$$

Thus

$$(g^m, w^m)_X = \|g^m\|_X^2 \geqslant [\|g^m\|_X\|w^m\|_X]/(1 + M/\alpha),$$

proving (6.7) with $\beta = 1/(1 + M/\alpha)$. □

6.2. Least-Squares and Gradient Methods to Solve the Navier-Stokes Equations

We propose first to decouple the divergence-free constraint and the nonlinearity in the Navier-Stokes equations by means of a heuristic alternating direction method introduced by Glowinski [36]. Then we shall solve the resulting nonlinear equations with the gradient methods of the preceding section.

Following the Peaceman-Rachford *alternating directions algorithm* we construct a sequence $(\mathbf{u}^m, p^m)$ starting from an initial pair $(\mathbf{u}^0, p^0)$ by:

$$(6.19)\qquad \begin{cases} -\nu\Delta\mathbf{u}^{m+1/2} + r^m\mathbf{u}^{m+1/2} - \mathbf{grad}\, p^{m+1/2} \\ \quad = \mathbf{f} - \sum_{j=1}^{N} u_j^m \partial\mathbf{u}^m/\partial x_j + r^m\mathbf{u}^m \quad \text{in } \Omega, \\ \operatorname{div}\mathbf{u}^{m+1/2} = 0 \quad \text{in } \Omega, \\ \mathbf{u}^{m+1/2} = \mathbf{0} \quad \text{on } \Gamma; \end{cases}$$

$$(6.20)\qquad \begin{cases} -\nu\Delta\mathbf{u}^{m+1} + \sum_{j=1}^{N} u_j^{m+1}\partial\mathbf{u}^{m+1}/\partial x_j + r^m\mathbf{u}^{m+1} \\ \quad = \mathbf{f} - \mathbf{grad}\, p^{m+1/2} + r^m\mathbf{u}^{m+1/2}, \\ \mathbf{u}^{m+1} = \mathbf{0} \quad \text{on } \Gamma, \end{cases}$$

where the parameters r^m are to be chosen as best as possible. Clearly, Problem (6.19) is like the Stokes problem which has been thoroughly studied in the previous chapters. Therefore we shall concentrate on Problem (6.20) which has no incompressibility constraint and is purely nonlinear.

Problem (6.20) is of the form:

$$
(6.21) \qquad \begin{cases} -\nu\Delta \mathbf{u} + \sum_{j=1}^{N} u_j \partial \mathbf{u}/\partial x_j + c\mathbf{u} = \mathbf{f} & \text{in } \Omega, \\ \mathbf{u} = \mathbf{0} \quad \text{on } \Gamma \end{cases}
$$

with both a linear elliptic and a nonlinear term in the left-hand side. This problem can be easily and conveniently generalized to fit into the abstract setting of Section 6.1.

Let $\|.\|_*$ denote the familiar dual norm of X' and let $A \in \mathscr{L}(X; X')$ be a symmetric and X-elliptic operator in X, i.e.

$$
(6.22) \qquad \langle Av, v\rangle \geqslant \gamma \|v\|_X^2 \quad \forall v \in X, \gamma > 0.
$$

Let G be a $\mathscr{C}^p$-mapping from X into X' (with $p \geqslant 2$) and set

$$
(6.23) \qquad F(v) = Av + G(v)
$$

which is clearly a $\mathscr{C}^p$-mapping from X into X'. Our problem is:

$$
(6.24) \qquad \textit{Find } u \textit{ in } X \textit{ such that } \quad F(u) = 0.
$$

Obviously, Problem (6.21) is a particular case of (6.24) with

$$
X = H_0^1(\Omega)^N, \qquad A = -\nu\Delta + c, \qquad G(\mathbf{v}) = \sum_{j=1}^{N} v_j \partial \mathbf{v}/\partial x_j - \mathbf{f}.
$$

We intend to solve Problem (6.24) by replacing it with an equivalent least-squares problem. To this end, consider the functional:

$$
J(v) = (1/2)\|F(v)\|_{X'}^2
$$

where $\|.\|_{X'}$ is defined by

$$
(6.25) \qquad \|f\|_{X'} = \langle A^{-1}f, f\rangle^{1/2}.
$$

We are going to see below that, because A is symmetric and elliptic, $\|.\|_{X'}$ is a norm on X' equivalent to the dual norm and the corresponding functional J is strictly convex and has a unique minimum.

Lemma 6.1. *The mapping $f \to \langle A^{-1}f, f\rangle^{1/2}$ is a norm on X' equivalent to the dual norm.*

Proof. From (6.22) we infer that

$$
\langle A^{-1}f, f\rangle \geqslant \gamma \|A^{-1}f\|_X^2,
$$

i.e.

$$
\|A^{-1}f\|_X \leqslant (1/\gamma)\|f\|_*.
$$

Hence we have on the one hand

$$\langle A^{-1}f, f\rangle \leqslant (1/\gamma)\,\|f\|_*^2.$$

On the other hand, we set

$$K = \|A\|_{\mathscr{L}(X;X')}.$$

Therefore

$$\|f\|_* \leqslant K\,\|A^{-1}f\|_X$$

and thus

$$\langle A^{-1}f, f\rangle \geqslant (\gamma/K^2)\,\|f\|_*^2.$$

Summing up, we get

$$(6.26)\qquad (\gamma/K^2)\,\|f\|_*^2 \leqslant \langle A^{-1}f, f\rangle \leqslant (1/\gamma)\,\|f\|_*^2. \qquad \square$$

Theorem 6.3. *Let u be a nonsingular solution of Problem* (6.24). *Then the functional J defined by*

$$(6.27)\qquad J(v) = (1/2)\langle A^{-1}(F(v)), F(v)\rangle$$

is strictly convex in a neighborhood of u.

Proof. Taking into account the symmetry of A^{-1}, the first two derivatives of J have the expression:

$$(6.28)\qquad \begin{aligned} DJ(v)\cdot w &= \langle A^{-1}(DF(v)\cdot w), F(v)\rangle \\ &= \langle A^{-1}(F(v)), DF(v)\cdot w\rangle, \end{aligned}$$

$$(6.29)\quad D^2J(v)\cdot(w,z) = \langle A^{-1}(DF(v)\cdot z), DF(v)\cdot w\rangle + \langle A^{-1}(F(v)), D^2F(v)\cdot(w,z)\rangle.$$

Now, recall that u is a nonsingular solution of Problem (6.24) if

$$F(u) = 0 \quad \text{and} \quad DF(u) \quad \text{is an } \textit{isomorphism} \text{ from } X \text{ onto } X'.$$

Hence (6.29) yields:

$$(6.30)\qquad D^2J(u)\cdot(w,w) = \langle A^{-1}(DF(u)\cdot w), DF(u)\cdot w\rangle.$$

Thus (6.26) implies that

$$\begin{aligned} D^2J(u)\cdot(w,w) &\geqslant (\gamma/K^2)\,\|DF(u)\cdot w\|_*^2 \\ &\geqslant \delta\,\|w\|_X^2 \qquad \delta > 0, \end{aligned}$$

since $DF(u)$ is an isomorphism of X onto X'. But the mapping D^2J is continuous in X (because F is a $\mathscr{C}^p$-mapping); therefore there exists $\rho > 0$ such that

$$\|D^2J(u) - D^2J(v)\| \leqslant \delta/2$$

for all $v \in S(u;\rho) = \{v \in X;\ \|v - u\|_X \leqslant \rho\}$. Hence

(6.31) $$D^2 J(v)\cdot(w,w) \geqslant (\delta/2)\,\|w\|_X^2 \qquad \forall v\in S(u;\rho), \qquad \forall w\in X,$$

i.e. J is strictly convex in $S(u;\rho)$. □

As a consequence, Problem (6.24) is equivalent to solve:

(6.32) $$\inf_{v\in S(u;\rho)} (1/2)\langle A^{-1}F(v), F(v)\rangle$$

and this solution can be achieved by the gradient and conjugate-gradient methods of Section 6.1. Indeed, assume that D^2G is bounded on all bounded subsets of X so that F, DF and D^2F are also bounded there, and assume that u is a *nonsingular solution* of (6.24). Then we already know that the first part of (6.2) holds on the ball $S(u;\rho)$, while the second part stems from (6.30), (6.26), the isomorphism property of $DF(u)$ and the continuity of D^2J. In addition, the boundedness of F implies that J is bounded in $S(u;\rho)$. Thus, by choosing a starting value u^0 in $S(u;\rho)$ and setting $J^0 = J(u^0)$ we can take

$$D = \{v\in X;\ J(v) \leqslant J(u^0)\} \cap S(u;\rho)$$

and Theorems 6.1 and 6.2 guarantee the convergence of the gradient and conjugate-gradient algorithms.

Let us examine the practical implementation of the *simple gradient method.* The symmetry and ellipticity of A induce us to equip X with the scalar product:

$$(u,v)_X = \langle Au, v\rangle$$

and associated norm $\|u\|_X = \langle Au,u\rangle^{1/2}$. Hence, in view of (6.28), the gradient $g(v)$ is defined by

$$\begin{aligned}\langle Ag(v), w\rangle &= \langle DJ(v), w\rangle\\ &= \langle A^{-1}F(v), DF(v)\cdot w\rangle\\ &= \langle (DF(v))' A^{-1}F(v), w\rangle,\end{aligned}$$

i.e.

(6.33) $$g(v) = A^{-1}(DF(v))' A^{-1}F(v).$$

Thus one step of the *simple gradient algorithm* can be decomposed into the following operations:

$$\begin{cases} 1^\circ)\ \text{compute} \qquad z^m = A^{-1}F(u^m),\\ \qquad\qquad\qquad\quad g^m = A^{-1}(DF(u^m))' z^m;\\ 2^\circ)\ \text{then minimize } J(u^m-\rho g^m) \text{ with respect to } \rho, \text{ where}\\ \qquad J(u^m-\rho g^m) = (1/2)\langle A^{-1}F(u^m-\rho g^m), F(u^m-\rho g^m)\rangle.\end{cases}$$

Each iteration requires the resolution of two linear problems relative to the operator A plus the determination of ρ^m. As an example, let us explicit the computation of ρ^m for Problem (6.21). First, observe that the mapping $\rho \to J(\mathbf{v} - \rho\mathbf{w})$ is a fourth-degree polynomial because F is a polynomial of degree two. Therefore Taylor's expansion of $J(\mathbf{u}^m - \rho\mathbf{g}^m)$ reduces to:

$$\begin{aligned} J(\mathbf{u}^m - \rho\mathbf{g}^m) = J(\mathbf{u}^m) &- \rho DJ(\mathbf{u}^m)\cdot\mathbf{g}^m + (\rho^2/2)D^2J(\mathbf{u}^m)\cdot(\mathbf{g}^m,\mathbf{g}^m) \\ &-(\rho^3/6)D^3J(\mathbf{u}^m)\cdot(\mathbf{g}^m,\mathbf{g}^m,\mathbf{g}^m) \\ &+(\rho^4/24)D^4J(\mathbf{u}^m)\cdot(\mathbf{g}^m,\mathbf{g}^m,\mathbf{g}^m,\mathbf{g}^m), \end{aligned} \tag{6.34}$$

where the third and fourth derivatives of J have the simple expression:

$$D^3J(\mathbf{u}^m)\cdot(\mathbf{g}^m,\mathbf{g}^m,\mathbf{g}^m) = 3\langle A^{-1}DF(\mathbf{u}^m)\cdot\mathbf{g}^m, D^2F(\mathbf{u}^m)\cdot(\mathbf{g}^m,\mathbf{g}^m)\rangle,$$

$$D^4J(\mathbf{u}^m)\cdot(\mathbf{g}^m,\mathbf{g}^m,\mathbf{g}^m,\mathbf{g}^m) = 3\langle A^{-1}D^2F(\mathbf{u}^m)\cdot(\mathbf{g}^m,\mathbf{g}^m), D^2F(\mathbf{u}^m)\cdot(\mathbf{g}^m,\mathbf{g}^m)\rangle.$$

Summing up, to solve Problem (6.21), each iteration of the *simple gradient algorithm* runs as follows:

1°) given $\mathbf{u}^m \in H_0^1(\Omega)^N$, compute the solution $\mathbf{z}^m \in H^1(\Omega)^N$ of

$$-\nu\Delta\mathbf{z}^m + c\mathbf{z}^m = -\nu\Delta\mathbf{u}^m + c\mathbf{u}^m + \sum_{j=1}^N u_j^m(\partial\mathbf{u}^m/\partial x_j) - \mathbf{f} \quad \text{in } \Omega,$$

$$\mathbf{z}^m = \mathbf{0} \quad \text{on } \Gamma;$$

2°) find the solution $\mathbf{g}^m \in H^1(\Omega)^N$ of

$$\begin{aligned} \nu(\mathbf{grad}\,\mathbf{g}^m, \mathbf{grad}\,\mathbf{v}) + c(\mathbf{g}^m,\mathbf{v}) &= \nu(\mathbf{grad}\,\mathbf{z}^m, \mathbf{grad}\,\mathbf{v}) \\ &+ c(\mathbf{z}^m,\mathbf{v}) + \sum_{j=1}^N (u_j^m\partial\mathbf{v}/\partial x_j + v_j\partial\mathbf{u}^m/\partial x_j, \mathbf{z}^m) \\ &\forall\mathbf{v}\in H_0^1(\Omega)^N, \qquad \mathbf{g}^m = \mathbf{0} \quad \text{on } \Gamma; \end{aligned}$$

3°) compute

$$J(\mathbf{u}^m) = (1/2)\{\nu|\mathbf{z}^m|_{1,\Omega}^2 + c\|\mathbf{z}^m\|_{0,\Omega}^2\}, \tag{6.35}$$

$$DJ(\mathbf{u}^m)\cdot\mathbf{g}^m = (1/2)\{\nu|\mathbf{g}^m|_{1,\Omega}^2 + c\|\mathbf{g}^m\|_{0,\Omega}^2\}; \tag{6.36}$$

4°) find the solution $\mathbf{v}^m \in H^1(\Omega)^N$ of

$$-\nu\Delta\mathbf{v}^m + c\mathbf{v}^m = -\nu\Delta\mathbf{g}^m + c\mathbf{g}^m + \sum_{j=1}^N (u_j^m\partial\mathbf{g}^m/\partial x_j + g_j^m\partial\mathbf{u}^m/\partial x_j) \quad \text{in } \Omega,$$

$$\mathbf{v}^m = \mathbf{0} \quad \text{on } \Gamma;$$

5°) compute

$$\mathbf{t}^m = D^2F(\mathbf{u}^m)\cdot(\mathbf{g}^m,\mathbf{g}^m) = 2\sum_{j=1}^N g_j^m\partial\mathbf{g}^m/\partial x_j,$$

$$D^2J(\mathbf{u}^m)\cdot(\mathbf{g}^m,\mathbf{g}^m) = \nu|\mathbf{v}^m|_{1,\Omega}^2 + c\|\mathbf{v}^m\|_{0,\Omega}^2 + (\mathbf{z}^m,\mathbf{t}^m). \tag{6.37}$$

$$D^3J(\mathbf{u}^m)\cdot(\mathbf{g}^m,\mathbf{g}^m,\mathbf{g}^m) = 3(\mathbf{t}^m,\mathbf{v}^m); \tag{6.38}$$

6°) find the solution $\mathbf{w}^m \in H^1(\Omega)^N$ of

$$-\nu\Delta \mathbf{w}^m + c\mathbf{w}^m = \mathbf{t}^m \quad \text{in } \Omega,$$
$$\mathbf{w}^m = \mathbf{0} \quad \text{on } \Gamma;$$

7°) compute

$$D^4 J(\mathbf{u}^m)\cdot(\mathbf{g}^m, \mathbf{g}^m, \mathbf{g}^m, \mathbf{g}^m) = \nu|\mathbf{w}^m|_{1,\Omega}^2 + c\|\mathbf{w}^m\|_{0,\Omega}^2; \tag{6.39}$$

8°) find the positive root ρ^m of

$$dJ(\mathbf{u}^m - \rho^m \mathbf{g}^m)/d\rho = 0$$

and update $\mathbf{u}^m$ by:

$$\mathbf{u}^{m+1} = \mathbf{u}^m - \rho^m \mathbf{g}^m.$$

Each iteration requires the solution of four Dirichlet problems.

The implementation of the conjugate-gradient algorithm is much like above and is left as an exercise.

6.3. Newton's Method and the Continuation Method

The methods discussed here are intended to solve the complete Navier-Stokes equations: incompressible and nonlinear. More generally, we want to solve equations of the type introduced in Section 3.1, namely:

$$F(\lambda, u) = 0, \tag{6.40}$$

where F is a $\mathscr{C}^p$-mapping ($p \geqslant 1$) defined on $\Lambda \times X$ with values in $\mathscr{X}$, X and $\mathscr{X}$ being two Banach spaces and Λ an interval of $\mathbb{R}$. Let us fix λ for the moment and assume that $u = u(\lambda) \in X$ is a nonsingular solution of (6.40), i.e.

$$F(\lambda, u) = 0, \qquad D_u F(\lambda, u) \text{ is an isomorphism from } X \text{ onto } \mathscr{X}.$$

Then we know from the inverse function theorem (and also from the material of Section 3.2) that there exists a closed ball $S(u;\alpha)$ where the equation (6.40) has no other solution than u.

Since u is an isolated solution of (6.40) and since F is at least differentiable, an efficient way to approximate u is the *Newton's algorithm*:

starting from an initial guess u^0, construct the sequence (u^n) in X by:

$$u^{n+1} = u^n - [D_u F(\lambda, u^n)]^{-1} \cdot F(\lambda, u^n) \qquad n \geqslant 0, \tag{6.41}$$

or equivalently

$$D_u F(\lambda, u^n)\cdot(u^{n+1} - u^n) = -F(\lambda, u^n).$$

As $D_u F(\lambda, u)$ is a linear operator, each step of Newton's method requires the solution of a different linear problem relative to $D_u F(\lambda, u^n)$. If this is too costly, the simplest alternative is to replace (6.41) by:

$$u^{n+1} = u^n - [D_u F(\lambda, u^0)]^{-1} \cdot F(\lambda, u^n) \qquad n \geqslant 0, \tag{6.42}$$

or equivalently

$$D_u F(\lambda, u^0) \cdot (u^{n+1} - u^n) = -F(\lambda, u^n).$$

We are going to prove that both schemes are convergent.

Theorem 6.3. *Assume that $D_u F(\lambda, v)$ is Lipschitz-continuous with respect to v in the ball $S(u; \alpha)$, i.e. there exists a constant $K > 0$ such that*

$$\| D_u F(\lambda, v) - D_u F(\lambda, v^*) \|_{\mathscr{L}(X; \mathscr{X})} \leqslant K \| v - v^* \|_X \qquad \forall v, v^* \in S(u; \alpha). \tag{6.43}$$

Then there exists an α' with $0 < \alpha' \leqslant \alpha$ such that for each initial guess u^0 in $S(u; \alpha')$ the Newton's algorithm (6.41) *determines a unique sequence $(u^n) \subset S(u; \alpha')$ that converges to the solution u of* (6.40). *Furthermore the convergence is quadratic*:

$$\| u^{n+1} - u \|_X \leqslant C \| u^n - u \|_X^2, \qquad C > 0. \tag{6.44}$$

Likewise, there exists an α'' with $0 < \alpha'' \leqslant \alpha$ such that for each initial value u^0 in $S(u; \alpha'')$ the scheme (6.42) *determines a unique sequence $(u^n) \subset S(u; \alpha'')$ that converges to u. But the convergence is only linear*:

$$\| u^{n+1} - u \|_X \leqslant C \| u^n - u \|_X, \qquad C < 1. \tag{6.45}$$

Proof. To begin with, it follows from (6.43) and Lemma 3.3 that there exists an α' with $0 < \alpha' \leqslant \alpha$ such that $D_u F(\lambda, v)$ is an isomorphism of X onto $\mathscr{X}$ for all v in $S(u; \alpha')$. Indeed, if we take $F_h = F$, $\tilde{u}_h = v$, $\gamma = \| [D_u F(\lambda, u)]^{-1} \|_{\mathscr{L}(\mathscr{X}; X)}$, $\mu = \| D_u F(\lambda, u) - D_u F(\lambda, v) \|_{\mathscr{L}(X; \mathscr{X})}$ then Lemma 3.3 says that $D_u F(\lambda, v)$ is an isomorphism of X onto $\mathscr{X}$ provided that $\gamma\mu < 1$. In view of (6.43), this inequality holds if we choose $\alpha' < \operatorname{Min}(\alpha, 1/(\gamma K))$. In particular, we can take

$$\alpha' \leqslant 1/(2\gamma K) \tag{6.46}$$

and formula (3.16) gives the bound

$$\| [D_u F(\lambda, v)]^{-1} \|_{\mathscr{L}(\mathscr{X}; X)} \leqslant 2\gamma. \tag{6.47}$$

Now let us prove that when u^0 belongs to $S(u; \alpha')$ with α' satisfying (6.46) then the scheme (6.41) defines a sequence (u^n) in $S(u; \alpha')$ that converges to u. We proceed by induction: suppose that u^n belongs to $S(u; \alpha')$; then $[D_u F(\lambda, u^n)]^{-1}$ exists and

$$u^{n+1} - u = u^n - u + [D_u F(\lambda, u^n)]^{-1} \cdot (F(\lambda, u) - F(\lambda, u^n)).$$

In other words

$$\begin{aligned} u^{n+1} - u &= [D_u F(\lambda, u^n)]^{-1} [F(\lambda, u) - F(\lambda, u^n) - D_u F(\lambda, u^n) \cdot (u - u^n)] \\ &= [D_u F(\lambda, u^n)]^{-1} \int_0^1 [D_u F(\lambda, u^n + t(u - u^n)) - D_u F(\lambda, u^n)] \cdot (u - u^n) \, dt. \end{aligned}$$

Thus (6.43) and (6.47) imply:

$$\|u^{n+1} - u\|_X \leqslant \gamma K \|u - u^n\|_X^2$$

and since $\alpha'\gamma K \leqslant 1/2$ this yields

$$\|u^{n+1} - u\|_X \leqslant (1/2)\|u^n - u\|_X.$$

Hence u^{n+1} belongs to $S(u;\alpha')$ and these two inequalities show that the sequence (u^n) converges quadratically to u.

Next, consider the scheme (6.42). Like above, we start with u^0 in $S(u;\alpha'')$ for some $\alpha'' \leqslant \alpha'$ that we shall specify subsequently. Then (6.42) determines a unique sequence (u^n) and similarly, we have:

$$u^{n+1} - u = [D_u F(\lambda, u^0)]^{-1} \int_0^1 [D_u F(\lambda, u^n + t(u - u^n)) - D_u F(\lambda, u^0)] \cdot (u - u^n)\, dt.$$

Hence assuming that u^n belongs also to $S(u;\alpha'')$ we derive:

$$\begin{aligned}\|u^{n+1} - u\|_X &\leqslant \gamma K(\|u^n - u^0\|_X + \|u - u^0\|_X)\|u^n - u\|_X\\ &\leqslant 3\alpha''\gamma K \|u^n - u\|_X.\end{aligned}$$

Therefore, by choosing

$$\alpha'' < 1/(3\gamma K)$$

we find that u^{n+1} belongs to $S(u;\alpha'')$ and that the sequence (u^n) converges linearly to u. □

Let us apply Theorem 6.3 to solve the familiar class of problems

$$F(\lambda, u) \equiv u + TG(\lambda, u) = 0 \quad \forall \lambda \in \Lambda, \tag{6.48}$$

where X and Y are two Banach spaces, Λ is a compact interval of $\mathbb{R}$, $T \in \mathscr{L}(Y;X)$ and G is a $\mathscr{C}^2$-mapping from $\Lambda \times X$ into Y with D^2G bounded on all bounded subsets of $\Lambda \times X$. This last property implies the Lipschitz condition (6.43). Therefore, if $\lambda \to u(\lambda)$ is a branch of *nonsingular solutions* of (6.48), Newton's method defines a locally (and quadratically) convergent algorithm:

$$(I + TD_u G(\lambda, u^n)) \cdot u^{n+1} = T[D_u G(\lambda, u^n) \cdot u^n - G(\lambda, u^n)]. \tag{6.49}$$

Likewise, the variant (6.42) is also locally (but linearly) convergent:

$$(I + TD_u G(\lambda, u^0)) \cdot u^{n+1} = T[D_u G(\lambda, u^0) \cdot u^n - G(\lambda, u^n)]. \tag{6.50}$$

As an example, consider the Navier-Stokes equations:

$$\begin{cases} \left.\begin{aligned} -\nu\Delta\mathbf{u} + \sum_{j=1}^{N} u_j(\partial\mathbf{u}/\partial x_j) + \mathbf{grad}\, p &= \mathbf{f},\\ \operatorname{div}\mathbf{u} &= 0, \end{aligned}\right\} \text{ in } \Omega \\ \mathbf{u} = \mathbf{0} \quad \text{on } \Gamma. \end{cases} \tag{6.51}$$

It was shown in Lemma 3.1 that (6.51) enters into that class of problems with the following correspondence:

$$X = H_0^1(\Omega)^N \times L_0^2(\Omega), \qquad Y = H^{-1}(\Omega)^N, \qquad T = \text{the Stokes operator},$$

$$\lambda = 1/\nu, \qquad G(\lambda, u) = \lambda\left(\sum_{j=1}^{N} u_j(\partial \mathbf{u}/\partial x_j) - \mathbf{f}\right).$$

Moreover, $(\mathbf{u}, p)$ is a solution of (6.51) iff $u = (\mathbf{u}, p/\nu)$ is a solution of (6.48). Then Newton's algorithm (6.49) reads:

Find $(\mathbf{u}^{n+1}, p^{n+1}) \in H_0^1(\Omega)^N \times L_0^2(\Omega)$ *such that*

$$(6.52)\quad \begin{cases} -\Delta \mathbf{u}^{n+1} + (1/\nu)\sum_{j=1}^{N} [u_j^n(\partial \mathbf{u}^{n+1}/\partial x_j) + u_j^{n+1}(\partial \mathbf{u}^n/\partial x_j)] + \mathbf{grad}\, p^{n+1} \\ \quad = (1/\nu)\left(\sum_{j=1}^{N} u_j^n(\partial \mathbf{u}^n/\partial x_j) + \mathbf{f}\right), \quad \text{in } \Omega, \\ \operatorname{div} \mathbf{u}^{n+1} = 0 \quad \text{in } \Omega, \\ \mathbf{u}^{n+1} = \mathbf{0} \quad \text{on } \Gamma. \end{cases}$$

Similarly, the simpler variant (6.50) reads:

Find $(\mathbf{u}^{n+1}, p^{n+1}) \in H_0^1(\Omega)^N \times L_0^2(\Omega)$ *such that*

$$(6.53)\quad \begin{cases} -\Delta \mathbf{u}^{n+1} + (1/\nu)\sum_{j=1}^{N} [u_j^0(\partial \mathbf{u}^{n+1}/\partial x_j) + u_j^{n+1}(\partial \mathbf{u}^0/\partial x_j)] + \mathbf{grad}\, p^{n+1} \\ \quad = (1/\nu)\left\{\sum_{j=1}^{N} [(\partial \mathbf{u}^n/\partial x_j)(u_j^0 - u_j^n) + u_j^n(\partial \mathbf{u}^0/\partial x_j)] + \mathbf{f}\right\} \\ \operatorname{div} \mathbf{u}^{n+1} = 0 \quad \text{in } \Omega, \\ \mathbf{u}^{n+1} = \mathbf{0} \quad \text{on } \Gamma. \end{cases}$$

Note that in either case, the next iterate $(\mathbf{u}^{n+1}, p^{n+1})$ is independent of p^n. Also, $D_u F(\lambda, u)$ is obviously Lipschitz-continuous since $D^2 G(\lambda, u)$ is constant. Hence, if $(\mathbf{u}, p)$ is a nonsingular solution of (6.51), starting from an initial guess $\mathbf{u}^0$ sufficiently near $\mathbf{u}$ and an arbitrary p^0, the scheme (6.52) (resp. (6.53)) determines a unique sequence $(\mathbf{u}^n, p^n)$ that converges quadratically (resp. linearly) to $u = (\mathbf{u}, p/\nu)$. Of course, if $u = u(\lambda)$ belongs to a branch of nonsingular solutions on a compact interval Λ, this result stays valid for all λ in Λ with constants independent of λ.

The drawback of Newton's method is that its convergence can only be insured when the first guess u^0 is sufficiently near the solution u. If this solution is part of a branch of nonsingular solutions and if we know the solution at a neighboring point, say $u(\lambda - \Delta\lambda)$ for an adequate increment $\Delta\lambda$, then we can derive from this value the first guess to start Newton's algorithm. This is the *method of continuation*; let us describe it more precisely. Assume that $\lambda \rightarrow u(\lambda)$ is a branch of nonsingular solutions of (6.40). As F is a $\mathscr{C}^p$-mapping ($p \geqslant 2$), so is the mapping $u(\lambda)$ and we can differentiate both sides of (6.40):

$$D_u F(\lambda, u(\lambda)) \cdot (du(\lambda)/d\lambda) + D_\lambda F(\lambda, u(\lambda)) = 0 \quad \forall \lambda \in \Lambda, \tag{6.54}$$

i.e. we find a first order differential equation of the form

$$du(\lambda)/d\lambda = -\phi(\lambda) \tag{6.55}$$

where

$$\phi(\lambda) = [D_u F(\lambda, u(\lambda))]^{-1} D_\lambda F(\lambda, u(\lambda)).$$

The simplest way to solve (6.55) is to use the one-step, explicit, Euler's method; this induces us to choose

$$u^0(\lambda) = u(\lambda - \Delta\lambda) - \phi(\lambda - \Delta\lambda) \cdot \Delta\lambda. \tag{6.56}$$

In other words $u^0(\lambda)$ is defined by

$$\begin{aligned} &D_u F(\lambda - \Delta\lambda, u(\lambda - \Delta\lambda)) \cdot (u^0(\lambda) - u(\lambda - \Delta\lambda)) \\ &\quad = -D_\lambda F(\lambda - \Delta\lambda, u(\lambda - \Delta\lambda)) \cdot \Delta\lambda. \end{aligned} \tag{6.57}$$

Let us estimate the error $u(\lambda) - u^0(\lambda)$. From (6.55) we infer that:

$$u(\lambda) = u(\lambda - \Delta\lambda) - \int_{\lambda - \Delta\lambda}^{\lambda} \phi(\mu)\, d\mu;$$

subtracting (6.56) we obtain

$$\begin{aligned} u(\lambda) - u^0(\lambda) &= -\left[\int_{\lambda - \Delta\lambda}^{\lambda} \phi(\mu)\, d\mu - \phi(\lambda - \Delta\lambda) \cdot \Delta\lambda\right] \\ &= -\int_{\lambda - \Delta\lambda}^{\lambda} \phi'(\theta_\mu) \cdot (\mu - \lambda + \Delta\lambda)\, d\mu. \end{aligned}$$

Hence

$$\|u(\lambda) - u^0(\lambda)\|_X \leqslant [(\Delta\lambda)^2/2] \operatorname*{Max}_{\theta \in (\lambda - \Delta\lambda, \lambda)} \|\phi'(\theta)\|_X.$$

Thus $\|u(\lambda) - u^0(\lambda)\|_X$ is $O((\Delta\lambda)^2)$ and if $\Delta\lambda$ is small enough, $u^0(\lambda)$ defined by (6.56) is an adequate starting value for Newton's algorithm.

As an example, let us explicit formula (6.57) for the Navier-Stokes equation (6.51). To simplify, we set

$$\delta u(\lambda) = u^0(\lambda) - u(\lambda - \Delta\lambda).$$

Then (6.57) amounts to

$$\delta u(\lambda) = -T[D_u G(\lambda - \Delta\lambda, u(\lambda - \Delta\lambda)) \cdot \delta u(\lambda) + D_\lambda G(\lambda - \Delta\lambda, u(\lambda - \Delta\lambda)) \cdot \Delta\lambda],$$

or equivalently

$$\delta u(\lambda) = -T\left\{(\lambda - \Delta\lambda)\sum_{j=1}^{N}[u_j(\lambda - \Delta\lambda)(\partial\delta\mathbf{u}(\lambda)/\partial x_j)\right.$$
$$+ \delta u_j(\lambda)(\partial\mathbf{u}(\lambda - \Delta\lambda)/\partial x_j]$$
$$\left.+ \Delta\lambda\left(\sum_{j=1}^{N} u_j(\lambda - \Delta\lambda)(\partial\mathbf{u}(\lambda - \Delta\lambda)/\partial x_j) - \mathbf{f}\right)\right\}.$$

Setting $\delta u(\lambda) = (\delta\mathbf{u}(\lambda), (\lambda - \Delta\lambda)\delta p(\lambda))$, this problem also reads:

$$\begin{cases} \textit{Find } (\delta\mathbf{u}(\lambda), \delta p(\lambda)) \in H_0^1(\Omega)^N \times L_0^2(\Omega) \textit{ such that} \\ -(1/(\lambda - \Delta\lambda))\Delta\delta\mathbf{u}(\lambda) + \sum_{j=1}^{N}[u_j(\lambda - \Delta\lambda)(\partial\delta\mathbf{u}(\lambda)/\partial x_j) + \delta u_j(\lambda)(\partial\mathbf{u}(\lambda - \Delta\lambda)/\partial x_j)] \\ \quad + \mathbf{grad}\,\delta p(\lambda) = (\Delta\lambda/(\lambda - \Delta\lambda))\left[\mathbf{f} - \sum_{j=1}^{N} u_j(\lambda - \Delta\lambda)(\partial\mathbf{u}(\lambda - \Delta\lambda)/\partial x_j)\right] \quad \textit{in } \Omega, \\ \operatorname{div}\delta\mathbf{u}(\lambda) = 0 \quad \textit{in } \Omega, \\ \mathbf{u}(\lambda) = \mathbf{0} \quad \textit{on } \Gamma. \end{cases}$$

Note that this problem is analogous to one iteration of Newton's algorithm.

Remark 6.1. By using a suitable discrete derivative, a Newton-type algorithm can also be derived to solve non differentiable schemes like the ones analyzed in Sections 3.4 and 5.1. Under adequate hypotheses a nearly quadratic convergence can be achieved (cf. Girault & Raviart [34]).

Remark 6.2. We can also solve (6.55) with a Runge-Kutta method or with an explicit multistep method. The proof of the corresponding error estimate is pretty much like above.

References

[1] R. Adams. *Sobolev Spaces*. Academic Press. (1975).

[2] D. Arnold, F. Brezzi, M. Fortin. A stable finite element for the Stokes equations. (to appear in Calcolo).

[3] J.-P. Aubin. Behavior of the error of the approximate solutions of boundary value problems for linear elliptic operators by Galerkin's and finite difference methods. Ann. Scuola Sup. Pisa 21, pp. 599–637. (1967).

[4] I. Babuška. The finite element method with Lagrangian multipliers. Numer. Math. 20, pp. 179–192. (1973).

[5] I. Babuška, J. Osborn, J. Pitkäranta. Analysis of mixed methods using mesh-dependent norms. Math. of Comp. 35, pp. 1039–1079. (1980).

[6] A. Bendali, J. Dominguez, S. Gallic. A variational approach for the vector potential formulation of the Stokes and Navier-Stokes problems in three dimensional domains. Ecole Polytechnique report n° 95. (1984).

[7] M. Bercovier, O. Pironneau. Error estimates for finite element method solution of the Stokes problem in the primitive variables. Numer. Math. 33, pp. 211–224. (1979).

[8] C. Bernardi. Méthode d'éléments finis mixtes pour les équations de Navier-Stokes. Thèse. Univ. Paris. (1979).

[9] C. Bernardi. Optimal finite element interpolation on curved domains. (To appear).

[10] C. Bernardi, G. Raugel. Analysis of some finite elements for the Stokes problem. Math. of Comp. 44. (1985).

[11] J. Boland, R. Nicolaides. Stability of finite elements under divergence constraints. SIAM J. Numer. Anal. 20, n° 4, pp. 722–731. (1983).

[12] J. Boland, R. Nicolaides. Stable and semistable low order finite elements for viscous flows. Univ. of Connecticut, preprint. (1984).

[13] F. Brezzi. On the existence, uniqueness and approximation of saddle-point problems arising from Lagrange multipliers. R.A.I.R.O., Anal. Numer. R2, pp. 129–151. (1974).

[14] F. Brezzi, J. Rappaz, P.-A. Raviart. Finite-dimensional approximation of nonlinear problems. Part I: branches of nonsingular solutions. Numer. Math. 36, pp. 1–25. (1980).

[15] F. Brezzi, P.-A. Raviart. Mixed finite element methods for 4-th order elliptic equations. *Topics in Numerical Analysis III* (J. Miller, ed). Academic Press. (1978).

[16] G. Carey, R. Krishnan. Penalty approximation of Stokes Flow. Part I: Stability analysis. Part II: Error estimates and numerical results. TICOM Report. Univ. of Texas. (1982).

[17] H. Cartan. Calcul Différentiel. Herrmann. (1967).

[18] L. Cattabriga. Su un problema al contorno relativo al sistema di equazioni di Stokes. Rend. Sem. Mat. Univ. Padova 31, pp. 308–340. (1961).

[19] P. Ciarlet. *The Finite Element Method for Elliptic Problems*. North-Holland. (1977).

[20] P. Ciarlet, P.-A. Raviart. A mixed finite element method for the biharmonic equation. *Mathematical Aspects of Finite Elements in Partial Differential Equations*. (C. de Boor). Academic Press. (1974).

[21] P. Clément. Approximation by finite element functions using local regularization. R.A.I.R.O. Anal. Numer. 9 R2, pp. 77–84. (1975).
[22] M. Crouzeix. Approximation de problèmes faiblement non linéaires. Pub. Univ. Rennes. (To appear).
[23] M. Crouzeix, P.-A. Raviart. Conforming and non conforming finite element methods for solving the stationary Stokes equations. R.A.I.R.O. R3, pp. 33–76. (1973).
[24] J. Dominguez. Formulation en potentiel vecteur du système de Stokes dans un domaine de R^3. Anal. Numer. L.A. 189. Paris. (1983).
[25] J. Douglas, T. Dupont, L. Wahlbin. Stability in L^q of the L^2-projection into finite element function spaces. Numer. Math. 23, pp. 193–197. (1975).
[26] G. Duvaut, J.-L. Lions. *Les Inéquations en Mécanique et en Physique*. Dunod. (1972).
[27] G. Foïas, R. Temam. Remarques sur les équations de Navier-Stokes stationnaires et les phénomènes successifs de bifurcation. Annali Scuola Normale Superiore Pisa, Serie IV vol. V n° 1, pp. 29–63. (1978).
[28] M. Fortin. An analysis of the convergence of mixed finite element methods. R.A.I.R.O. Anal. Numer. 11 R3, pp. 341–354. (1977).
[29] M. Fortin. Calcul numérique des écoulements des fluides de Bingham et des fluides Newtoniens incompressibles par la méthode des éléments finis. Thèse Univ. Paris. (1972).
[30] M. Fortin. Old and new finite elements for incompressible flows. Int. Jour. Numer. Meth. Fluids, 1, pp. 347–364. (1981).
[31] M. Fortin. Résolution numérique des équations de Navier-Stokes par des méthodes d'éléments finis de type mixte. Proc. 2nd Int. Symp. Finite Elements in Flow Problems, S. Margherita Ligure, Italy. (1978).
[32] V. Girault, P.-A. Raviart. *Finite Element Approximation of the Navier-Stokes Equations*. Lect. Notes in Math. vol. 749. Springer-Verlag. (1979).
[33] V. Girault, P.-A. Raviart. An analysis of a mixed finite element method for the Navier-Stokes equations. Numer. Math. vol. 33, pp. 235–271. (1979).
[34] V. Girault, P.-A. Raviart. An analysis of upwind schemes for the Navier-Stokes equations. SIAM J. Numer. Anal. vol. 19 n° 2. pp. 312–333. (1982).
[35] R. Glowinski. Approximations externes par éléments finis de Lagrange d'ordre un et deux, du problème de Dirichlet pour l'opérateur biharmonique. Méthodes itératives de résolution des problèmes approchés. *Topics in Numerical Analysis*, pp. 123–171. Academic Press. (1973).
[36] R. Glowinski. *Numerical Methods for Nonlinear Variational Problems*. Springer Series in Comp. Physics. Springer-Verlag. (1984).
[37] R. Glowinski, O. Pironneau. *Energy Methods in Finite Element Analysis*. (R. Glowinski, E. Rodin, O. Zienkiewicz). Wiley, pp. 243–264. (1979).
[38] R. Glowinski, O. Pironneau. On a mixed finite element approximation of the Stokes problem I. Convergence of the approximate solution. Numer. Math. 33, pp. 397–424. (1979).
[39] R. Glowinski, O. Pironneau. Numerical methods for the first biharmonic equation and for the two-dimensional Stokes problem. SIAM Review vol. 21 n° 2, pp. 167–212. (1979).
[40] J. Gobert. Sur une inégalité de coercivité. J. of Math. Anal. and App. 36, pp. 529–555. (1971).
[41] C. Goldstein. Variational crimes and L^∞ error estimates in the finite element method. Math. of Comp. vol. 35 n° 152, pp. 1131–1157. (1980).
[42] P. Grisvard. *Boundary Value Problems in Non-Smooth Domains*. Univ. of Maryland, Dept. of Math. Lecture Notes n° 19. (1980).
[43] P. Grisvard. Singularités des solutions du problème de Stokes dans un polygone. Univ. de Nice. (1978).
[44] P. Hood, C. Taylor. A numerical solution of the Navier-Stokes equations using the finite element technique. Comp. and Fluids 1, pp. 73–100. (1973).
[45] E. Hopf. On Non-Linear Partial Differential Equations. Lecture Series of the Symp. on Partial Diff. Equations. Berkeley. (1955).
[46] C. Johnson, J. Pitkäranta. Analysis of some mixed finite element methods related to reduced integration. Math. of Comp. vol. 38, pp. 375–400. (1982).

[47] C. Johnson, J. Saranen. Steamline diffusion methods for the incompressible Euler and Navier-Stokes equations. Preprint.
[48] O. Ladyzhenskaya. *The Mathematical Theory of Viscous Incompressible Flow*. Gordon & Breach. (1969).
[49] P. Lax, N. Milgram. Parabolic equations. Contributions to the Theory of Partial Differential Equations. Princeton. (1954).
[50] M. Lenoir. Optimal isoparametric finite elements and error estimates for domains involving curved boundaries. Preprint.
[51] P. Lesaint, P.-A. Raviart. On a finite element method for solving the neutron transport equation. *Mathematical Aspects of Finite Elements in Partial Differential Equations*. (C. de Boor). Academic Press. (1974).
[52] J.-L. Lions. *Quelques Méthodes de Résolution des Problèmes aux Limites Non Linéaires*. Gauthier-Villars. (1969).
[53] J.-L. Lions, E. Magenes. *Problèmes aux Limites Non Homogènes et Applications*. Dunod. (1968).
[54] J.-L. Lions, J. Peetre. Sur une classe d'espaces d'interpolation. Pub. de I.H.E.S. vol. 19, pp. 5–68. (1964).
[55] D. Malkus, E. Olsen. Incompressible finite elements which fail the discrete LBB condition. ASME Symp. on Penality Finite Element Methods in Mechanics, WAM, Phoenix. (1982).
[56] L. Mansfield. Finite element subspaces with optimal rates of convergence for the stationary Stokes problem. R.A.I.R.O. Numer. Anal. 16 n° 1, pp. 49–66. (1982).
[57] J. Nečas. Equations aux Dérivées Partielles. Presses de l'Université de Montréal. (1965).
[58] J. Nečas. *Les Méthodes Directes en Théorie des Equations Elliptiques*. Masson. (1967).
[59] J.-C. Nedelec. Mixed finite elements in R^3. Numer. Math. vol. 35, pp. 315–341. (1980).
[60] J.-C. Nedelec. Elements finis mixtes incompressibles pour l'équation de Stokes dans R^3. Numer. Math. 39, pp. 97–112. (1982).
[61] J. Nitsche. Ein kriterium fur die quasi-optimalitat des Ritzchen Verfahrens. Numer. Math. 11, pp. 346–348. (1968).
[62] J. Nitsche. L^∞ convergence of finite element approximations. *Mathematical Aspects of Finite Element Methods*. (I. Galligani, E. Magenes). Lecture Notes in Math. vol. 606, pp. 261–274. Springer-Verlag. (1977).
[63] J. Peetre. Espaces d'interpolation et théorème de Soboleff. Ann. Inst. Fourier vol. 16, pp. 279–317. (1966).
[64] R. Peyret, T. Taylor. *Computational Methods for Fluid Flow*. Springer Series in Comp. Physics. (1983).
[65] J. Pitkäranta, R. Stenberg. Error bounds for the approximation of the Stokes problem using bilinear/constant elements on irregular quadrilateral meshes .Preprint.
[66] R. Rannacher, R. Scott. Some optimal error estimates for piecewise linear finite element approximations. Math. of Comp. vol. 38, pp. 437–445. (1982).
[67] J. Rappaz. Approximation of a nondifferentiable nonlinear problem related to MHD equilibria. Numer. Math. 45, pp. 117–133. (1984).
[68] G. de Rham. *Variétés Différentiables*. Hermann. (1960).
[69] F. Roux. Utilisation d'éléments finis conformes H(**rot**) pour la résolution de l'équation de Navier-Stokes tridimensionnelle. Sém. Anal. Numér. Univ. Paris. (1984).
[70] R. Sani, P. Gresho, R. Lee, D. Griffiths, M. Engelman. The cause and cure (?) of the spurious pressures generated by certain FEM solutions of the incompressible Navier-Stokes equations. Int. J. Numer. Meth. Fluids vol. 1, pp. 17–43 and pp. 171–204. (1981).
[71] A. Schatz, L. Wahlbin. Maximum norm estimates in the finite element method on plane polygonal domains. Math. of Comp. vol. 32 n° 141, pp. 73–109. (1978) and vol. 33 n° 146, pp. 465–492. (1979).
[72] R. Scholz. A mixed method for fourth order problems using linear finite elements. R.A.I.R.O. Anal. Numer. 12, pp. 85–90. (1978).
[73] R. Scott. Optimal L^∞ estimates for the finite element method on irregular meshes. Math. of Comp. vol. 30 n° 136, pp. 681–697. (1976).

[74] K. Smith. Inequalities for formally positive integro-differential forms. Bull. Amer. Math. Soc. 67, pp. 368–370. (1961).
[75] R. Stenberg. Analysis of mixed finite element methods for the Stokes problem: a unified approach. Math. of Comp. vol. 42, pp. 9–23. (1984).
[76] A. Stephens, J. Bell, J. Solomon, L. Hackerman. A finite difference Galerkin formulation for the incompressible Navier-Stokes equations. J. of Comp. physics 53, pp. 152–172. (1984).
[77] G. Strang, G. Fix. *An Analysis of the Finite Element Method.* Prentice-Hall. (1973).
[78] L. Tartar. *Topics in Nonlinear Analysis.* Pub. Math. d'Orsay. Univ. Paris-Sud. (1978).
[79] L. Tartar. Nonlinear Partial Differential Equations Using Compactness Method. Report Univ. of Madison n° 1584.
[80] A. Taylor. *Introduction to Functional Analysis.* Wiley. (1958).
[81] R. Temam. *Theory and Numerical Analysis of the Navier-Stokes Equations.* North-Holland. (1977).
[82] F. Thomasset. *Implementation of Finite Element Methods for Navier-Stokes Equations.* Springer Series in Comp. Physics. (1981).
[83] R. Verfürth. Error estimates for a mixed finite element approximation of the Stokes equations. R.A.I.R.O. Numer. Anal. 18 n° 2, pp. 175–182. (1984).
[84] K. Yosida. *Functional Analysis.* Springer-Verlag. (1965).
[85] A. Zine. Méthodes hybrides pour les éléments finis non conformes. "Résolution d'un système de Stokes". Thèse Univ. Paris. (1984).

Index of Mathematical Symbols

Index

Springer Series in Computational Mathematics

Computational Mathematics is a series of outstanding books and monographs which study the applications of computing in numerical analysis, optimization, control theory, combinatorics, applied function theory, and applied functional analysis. The connecting link among these various disciplines will be the use of high-speed computers as a powerful tool.

The following list of topics best describes the aims of *Computational Mathematics:* finite element methods, numerical methods of optimal control, multigrid methods, numerical solutions of time-dependent equations, partial differential equations, multivariate splines and applications, software packages for quadrature, elliptic P.D.E. solvers and nonlinear programming.

Computational Mathematics is directed towards mathematicians and appliers of mathematical techniques in disciplines such as engineering, computer science, economics, operations research and physics.

Springer-Verlag
Berlin Heidelberg
New York Tokyo

Springer

Springer Series in Computational Mathematics

Volume 1

QUADPACK

A Subroutine Package for Automatic Integration

By **R. Piessens, E. de Doncker-Kapenga, C. W. Überhuber, D. K. Kahaner**

1983. 26 figures. VIII, 301 pages. ISBN 3-540-12553-1

Volume 2

J. R. Rice, R. F. Boisvert

Solving Elliptic Problems Using ELLPACK

1985. 53 figures. X, 497 pages. ISBN 3-540-90910-9

Volume 3

N. Z. Shor

Minimization Methods for Non-Differentiable Functions

1985. VIII, 162 pages. ISBN 3-540-12763-1

Volume 4

W. Hackbusch

Multi-Grid Methods and Applications

1985. 43 figures, 48 tables. XIV, 377 pages. ISBN 3-540-12761-5

GPSR Compliance

The European Union's (EU) General Product Safety Regulation (GPSR) is a set of rules that requires consumer products to be safe and our obligations to ensure this.

If you have any concerns about our products, you can contact us on ProductSafety@springernature.com

In case Publisher is established outside the EU, the EU authorized representative is:

Springer Nature Customer Service Center GmbH
Europaplatz 3
69115 Heidelberg, Germany

Batch number: 10620143

Printed by Printforce, the Netherlands